GEOGRAPHY AND REVOLUTION

Geography
and Revolution

Edited by David N. Livingstone and Charles W. J. Withers

The University of Chicago Press
Chicago and London

David N. Livingstone is professor of geography and intellectual history at the Queen's University of Belfast. **Charles W. J. Withers** is professor of geography at the University of Edinburgh. The editors have collaborated previously on *Geography and Enlightenment,* also published by the University of Chicago Press.

The University of Chicago Press, Chicago 60637
The University of Chicago Press, Ltd., London
© 2005 by The University of Chicago
All rights reserved. Published 2005
Printed in the United States of America
14 13 12 11 10 09 08 07 06 05   5 4 3 2 1

ISBN (cloth): 0-226-48733-4

Library of Congress Cataloging-in-Publication Data

Geography and revolution / edited by David N. Livingstone and
    Charles W. J. Withers.
        p.   cm.
    Includes bibliographical references and index.
    ISBN 0-226-48733-4 (cloth : alk. paper)
        1. Geography—Philosophy.   2. Science—Philosophy.   3. Discoveries in science.
    4. Revolutions—Philosophy.   I. Livingstone, David N., 1953– .   II. Withers,
    Charles W. J.
    G70.G4419   2005
    910'.01—dc22

                                                            2005003443

♾ The paper used in this publication meets the minimum requirements of the American National Standard for Information Sciences—Permanence of Paper for Printed Library Materials, ANSI Z39.48-1992.

CONTENTS

Preface and Acknowledgments   *vii*

1. ON GEOGRAPHY AND REVOLUTION   *1*
    David N. Livingstone and Charles W. J. Withers

PART I.   GEOGRAPHY AND SCIENTIFIC REVOLUTION:
           SPACE, PLACE, AND NATURAL KNOWLEDGE   *23*

2. SPACE, REVOLUTION, AND SCIENCE   *27*
    Peter Dear

3. NATIONAL STYLES IN SCIENCE
    A Possible Factor in the Scientific Revolution?   *43*
    John Henry

4. GEOGRAPHY, SCIENCE, AND THE SCIENTIFIC REVOLUTION   *75*
    Charles W. J. Withers

5. REVOLUTION OF THE SPACE INVADERS
    Darwin and Wallace on the Geography of Life   *106*
    James Moore

PART II.   GEOGRAPHY AND TECHNICAL REVOLUTION: TIME,
           SPACE, AND THE INSTRUMENTS OF TRANSMISSION   *133*

6. PRINTING THE MAP, MAKING A DIFFERENCE
    Mapping the Cape of Good Hope, 1488–1652   *137*
    Jerry Brotton

7. **REVOLUTIONS IN THE TIMES**
    Clocks and the Temporal Structures of Everyday Life   *160*
    Paul Glennie and Nigel Thrift

8. **PHOTOGRAPHY, VISUAL REVOLUTIONS, AND
    VICTORIAN GEOGRAPHY**   *199*
    James R. Ryan

PART III. GEOGRAPHY AND POLITICAL REVOLUTION:
          GEOGRAPHY AND STATE GOVERNANCE   **239**

9. **GEOGRAPHY'S ENGLISH REVOLUTIONS**
    Oxford Geography and the War of Ideas, 1600–1660   *243*
    Robert J. Mayhew

10. **EDME MENTELLE'S GEOGRAPHIES AND
    THE FRENCH REVOLUTION**
    Michael Heffernan   *273*

11. **"RISEN INTO EMPIRE"**
    Moral Geographies of the American Republic   *304*
    David N. Livingstone

12. **ALEXANDER VON HUMBOLDT AND REVOLUTION**
    A Geography of Reception of the Varnhagen von
    Ense Correspondence   *336*
    Nicolaas Rupke

    AFTERWORD: REVOLUTIONS AND THEIR GEOGRAPHIES
    Peter Burke   *351*

Contributors   *363*
Bibliography   *367*
Index   *417*

PREFACE AND ACKNOWLEDGMENTS

Questions to do with the placed nature of intellectual endeavor, with what we may call the "geography of knowledge," have assumed considerable significance in a number of disciplines in recent years. Geographers, historians, historians of science, and others have been attentive in a variety of ways to the importance of place, to the transmission of ideas over space and to their reception in different places and social spaces. There has been renewed critical interest in the spatial nature of historical periods such as "the Renaissance," "the Enlightenment," and "the Scientific Revolution." Matters of geography and of geographical thinking have been employed in understanding terms like "knowledge" and "science," with how they varied over space, with how their meanings moved and with what social and epistemological consequences. For reasons discussed in what follows, it may be too much to claim that there has been a "spatial" or a "geographical revolution" in these subjects, but the fact of shared interests around questions to do with the power of geographical thinking is undeniable.

This book has its origins, then, in a range of recent and general conceptual and theoretical interests and in a particular concern to explore the ways in which ideas of geography and of revolution, and the relationships between them, might be understood. More particularly still, it has its beginnings in a conference held in the Department of Geography at the University of Edinburgh in July 2001, in which earlier versions of the chapters here presented were given. The meeting, held over four days, allowed for fruitful exchange between scholars from different disciplines and intellectual traditions. As it turned out, the meeting allowed shared interests to emerge that centered less upon geography in any strictly disciplinary sense, and much more upon geography's practices and upon the language and concepts of geography as means to the explanation of revolutions.

As might be expected of any collaborative enterprise, the editors and contributors have incurred many debts. Our contributors have generally acknowledged the help of others in their respective chapters. As editors, we would like to thank the two anonymous readers for the University of Chicago Press for their perceptive, stimulating, and encouraging comments on earlier drafts. We would like to thank Caitlin DeSilvey for her energies and enthusiasm as the conference assistant. For permission to reproduce illustrations in their care, we would like to thank the British Library for the figures in chapter 6 and in chapter 10; the Royal Geographical Society (with the Institute of British Geographers) for figures 8.1, 8.2, 8.5, and 8.6; Mrs. Daphne Foskett for figure 8.3; the University of Chicago Press for figure 3.1; and Princeton University Press for figure 3.2. Several academic bodies provided generous support in one way or another for the conference on which this book is based. We gratefully acknowledge the support of the British Academy, the University of Edinburgh through its Moray Endowment Fund and Research Projects Grant, the Royal Geographical Society (with the Institute of British Geographers), and finally the Historical Geography Research Group and the History and Philosophy of Geography Research Group of the RGS-IBG, whose funds allowed graduate students to attend. Charles Withers would like to thank the Institute for Advanced Studies in the Humanities of the University of Edinburgh for the sabbatical fellowship, which allowed time and space for the redrafting of his chapter, and for much of the joint editorial work, and the British Academy for the award of a British Academy Research Readership.

Most important, we owe a great deal to the support and encouragement—and patience—we have at all times had from Christie Henry of the University of Chicago Press and her colleagues Jennifer Howard and Stephanie Hlywak. We are pleased too that Peter Burke accepted our invitation to write the final chapter, and we thank him for "rounding off" the book and for offering further insights in the ways he has. Our final thanks must go to our authors, in part for their forbearance, but chiefly for their scholarship.

— 1 —

ON GEOGRAPHY AND REVOLUTION

...

David N. Livingstone and Charles W. J. Withers

A state of warfare generally produces the first improvements in [a country's] geography.

William Roy, 1785

To a revolutionary degree man changes geography as he goes along.

Isaiah Bowman, 1946

This book is an exploration of the ways in which ideas of geography and of revolution and the relationships between them may be understood. It is an attempt to bring together insights from geographers, historians, and historians of science concerning the importance of geographical thinking and recognition of the difference that space makes to an understanding of the nature of revolutions, however that term has been used. In developing these ideas, an important initial distinction may be made between the geography *of* revolution and geography *in* revolution.

In the first sense, questions to do with the relationships between the celestial and terrestrial worlds and with the establishment of "modern" methodological procedures for the study of nature in what has traditionally been considered the Scientific Revolution have been shown to have taken shape in particular locations, to have traveled unevenly and to have been received differently across Europe. So, too, the "Technical Revolution" that comprised the printing press and the printed book had varying locational and distributional expression across the globe. Print—not least printed maps—helped revolutionize conceptions of the known world. This is true not just in the sense of the new technical forms that books or maps assumed. Historians of the book have considered the attendant notion of the "Reading Revolution"—a revolution not just in how but also in where the printed word was read—silently in private, aloud and to others in public spaces, and so on. Similarly, what economic and social historians term the "Industrial Revolution" was not just a matter of shifts in the technologies of production and in the social consequences for the workforces involved in new systems of organization and management. The Industrial Revolution was also, profoundly, a matter of geography: of systems of industrial production that relocated people and machines as never before, of delivery mechanisms that acted to diminish the costs of space—even to "collapse" geography—and of independent innovations by others elsewhere and at other times. There are, then, recoverable geographies of revolutions in science, the printed word, reading, industrial production, and technical change. Put in general terms, these geographies *of* revolution concern the sites of production, whether of ideas, books, or factory systems, the movement over space of thoughts and things, and the sites and social spaces in which these developments were differently received in different places. In one way or another and in a variety of geographical and historical contexts, these matters of production and reception *in* space and of movement *over* space are the central concerns of this book.

In the second sense, geography as a form of knowledge has been deeply implicated in revolutions of various sorts. In the Scientific Revolution, for example, the subject had close associations with Newtonianism. Concerns to bring mathematical precision to the mapping of the globe, the correction of nautical charts, and the empirical harvesting of the world's natural phenomena were all part and parcel of the "Newtonian Revolution." In the "Darwinian Revolution," questions of biogeographical distribution, the relationships between organism and habitat, and explanations rooted in the determining agency of geographical difference are central. In the political upheaval that was the American Revolution, or in mid-seventeenth-century England, geography books were used as vehicles of debate concerning the nature of political constitutions, the right of the individual, and matters of national identity. To explain such concerns is not to see a centrality to the "discipline" of geography, not least because most historians of the subject do not now subscribe to the view of a single essential subject unchanged over time and space. It is, rather, to identify the role of what was taken to be geography, at different times and in different places, in respect to different revolutions and to consider how geographical knowledge *in* such contexts had a bearing upon the forms taken by the revolutions themselves.

The chapters that follow explore the connections between the geography of and in revolution in a variety of ways. Underlying all the studies are questions to do with geography understood as a set of related practices by which the world has been encountered and represented, and with the language and concepts of geography as an aid to the explanation of revolutionary phenomena. The chapters have been grouped together in three parts addressing scientific, technical, and political revolutions. The introduction to each part elaborates in more detail upon the individual concerns and intentions of the authors and discusses how their particular studies relate to our larger concerns. In this introductory chapter, however, in order to establish further the connections between geography and revolution, we begin by reviewing the diverse literatures regarding the term "revolution."

DEFINING REVOLUTION

Agricultural, chemical, Copernican, green, industrial, information, military, Neolithic, political, reading, scientific, urban: the term "revolution" is associated with a variety of intellectual and practical circumstances. Although the term was already in use during the late fourteenth century in reference to celestial bodies, the label "revolution" established its own distinctive identity with the publication in 1543 of Copernicus's account of the motion of the heavenly spheres, *De Revolutionibus Orbium Caelestium.* Within half a century, the word was being applied to affairs of state in a way that departed materially from its earlier associations.[1] In its Italian form, *rivoluzione,* the term had been in use during the late Middle Ages as a neutral description of change in sovereignty. But from the late sixteenth century, revolution began to acquire its modern political resonance as the overthrowing of one regime and its replacement by a successor. Its deployment in the mid-seventeenth century to describe those events in England that are routinely gathered under the rubric of the "English Revolution" was crucial in this regard. So, too, was its association with the overthrow in 1688 of the Stuart dynasty—the "Glorious Revolution"—and, perhaps even more important, with the French Revolution of the late eighteenth century.

Two further developments in the eighteenth century acted to cement the political connotations of the term. The first was the work of French Enlightenment thinkers like Denis Diderot, whose entry "Révolution" in the *Encyclopédie* associated the word with "le gouvernment d'un état" and the Baron Montesquieu who deployed the term in his influential *L'Esprit des Lois* (1748) to mark fundamental political change. Such writings helped voice a conviction that the uncovering of the fundamental laws of nature that had been secured by the new natural philosophy of the previous century—the "Scientific

Revolution" as it later came to be known—necessarily presaged the unlocking of the laws of the social order. The revolution in understanding nature prefaced revolutions in understanding humans—in their social and political organizations as well as in their place in nature.

The second development was the retrospective application of the term "revolution" to capture the events surrounding the American War of Independence, a rhetorical affirmation that did much to fix the political coordinates of the label in the minds of contemporaries. The appearance of such works as Richard Price's *Observations on the American Revolution* (1784) and of David Ramsay's *History of the American Revolution* (1789) confirmed just how successful commentators such as Thomas Paine had been in reflecting upon the ideas of both revolution and republic. Revolution had hitherto denoted, in one form or another, the conception of a completed historical cycle. Even in the political realm, it conveyed the sense that the transition from one regime to another—even if it involved violence—resolved itself in the restoration of an original state. The English Civil War, for example, found its culmination with the restoration of the monarchy. But now, with the French Revolution and with Paine's post facto apologia, the idea took hold that revolution necessarily involved innovation and replacement, not a return to a previous state or condition. In *The Rights of Man* (1791), Paine observed:

> What we formerly called Revolutions were little more than a change of persons, or an alternation of local circumstances. They rose and fell like things of course, and had nothing in their existence or fate that could influence them beyond the spot that produced them. But what we now see in the world . . . are a renovation of the natural order of things, a system of principles as universal as truth and the existence of man, and combining moral with political happiness and natural prosperity.[2]

For Stan Taylor, drawing out the wider implications of Paine's exegesis, revolution would henceforth be associated with "social restructuring of a universally-applicable kind." In the hands of Hegel and Marx, of course, revolution's ties with social progress through political overturn would be forever secured.[3]

Underlying the transfer of the language of revolution from celestial mechanics to the world of politics was a widespread belief in the intrinsic connections between the microcosm and the macrocosm, and the astrological and social conviction that the motions of the stars had a correspondence in human affairs. In his contemporary account of the history of the English rebellion, for example, the Earl of Clarendon insisted in the famous ninth book of his *History of the Rebellion and Civil Wars in England begun in the Year*

1641 that the "motions of these last twenty years . . . have proceeded from the evil influence of a malignant star."[4] When conjoined to the accepted and widespread imperatives of Christian eschatology, which laid out in chiliastic fashion a succession of world empires moving irresistibly toward Armageddon, the idea of revolution bound together into intellectual coherence the astronomical, the theological, the social, and the political.

The concept of revolution in these senses has not remained restricted to the celestial and political spheres. It has found favor among historians as an explanatory device in relation to technological and intellectual affairs. In large measure, this owes much to Marxist theory, which, in a variety of ways, has insisted on the intimate links between political, economic, and cognitive matters.[5] It is just this combination of technical developments and their related social transformations, of course, that inaugurated the concept of an "Industrial Revolution."[6] Among the technological innovations that in much of western Europe transformed production techniques and the social relations of manufacture were the spinning jenny, the flying shuttle, the water wheel, the power loom, the steam engine, and latterly, electricity. The revolutionary character of each of these historical-technological moments has been challenged by various scholars, concerned as they have been with the precise definition of *Technik* and with the fact that the machine and transport changes central to such industrial transformation after around 1750 had earlier proto-industrial roots.[7] Nevertheless, "Industrial Revolution" has become an established term within the historical lexicon, notably in relation to devices designed to dominate nature—"working-machine technology" as Paulinyi calls it—and in related shifts in technical and productive capacity. More than has been the case for cognate upheavals—such as the "Commercial Revolution," the "Price Revolution," and the "Trading Revolution"—the Industrial Revolution (and, to some degree, perhaps, the "Agrarian" or "Agricultural Revolution"), has been considered as a geographical matter.[8] In the case of Britain at least, the Industrial Revolution has been mapped as a set of processes—an "industrializing" rather than an industrial revolution in any fixed sense—with differing locational expression and underlying geographical causes.[9] The technical bases to the Industrial Revolution have, of course, contemporary counterparts in the new time regimes associated with the shift from moral to political economies of labor—human adjustments to machine speed, imposition of "standard time," regulated hours of work, and so on. Such a revolution in industrial time is a social as well as a technical thing.[10] Modern parallels exist, too, in the ways in which the "Information Revolution" has been realized as a shift from making and moving goods to making and moving information—as stocks and shares in exchanges, as data sets via the Internet, and as media commodities in a networked society.[11] Significant as this work on the Information Revolution has been, the utility of information lies less in its production than in the use to which it is put. This realization advertises the significance of recent work on the geography of meaning. Historians of the book have contributed to this process in a variety of ways. They have, of course, studied book production as a matter of technological innovation in information. They have also complemented such work by studies of the movement of books and of other forms of print in the public sphere in a "communications circuit." Crucially, they have also attended to changes in the cultural and social practices of reading. In this last respect, as suggested above, some have argued for a "Reading Revolution," in late-eighteenth-century western Europe at least, characterized by new forms of print, an emphasis upon reading as "useful," and by a widening of the reading publics.[12]

In the domain of intellectual endeavor, the idea that there has been a "Scientific Revolution"—with associated Newtonian, Copernican, and, later, Darwinian variants—has had a powerful grip on historical enquiry.[13] Those conditions that have been drawn together at one time or another under the label "Scientific Revolution" have been construed as the consequence of a profound epistemological reorientation and the generation of new metaphysical categories. In seeking an explanation for such transformations, scholars have made use of the language of paradigm shift, or Gestalt switch, or sweeping metaphor replacement, all of which owe much to Thomas Kuhn's classic analysis, *The Structure of Scientific Revolutions* (1962). Others have been more inclined to look toward religious changes at both the continental and national scales in efforts to uncover the origins of modern science. In some cases, the Reformation has been taken as critical. In others, a Puritan *mentalité* has been isolated as the key factor. In yet others, a shift in scriptural hermeneutics has been held to presage a revolution in the reading of nature. The list of explanatory factors could be expanded ad libitum: the changing role of arts and crafts, the legacy of the voyages of reconnaissance, the development of a print culture, the emergence of capitalism, the lingering repercussions of hermeticism, the breakup of feudal Europe, dialogues with Islam, and many others.[14] The interrogation and variable interpretation of different historical sources has even resulted in a challenge to the very existence of *the* Scientific Revolution in any simple canonical sense. Steven Shapin, for example, has signaled its dissolution in beginning his book *The Scientific Revolution* with the provocative assertion: "There was no such thing as the Scientific Revolution, and this is a book about it."[15] Challenges of this sort have, in their turn, called forth either outright rebuttals or modulations of this claim by other commentators.[16]

Given these associations with different domains, it is hardly surprising that establishing the necessary and sufficient conditions for revolutionary states of affairs has proven to be elusive. Students of political revolution have been beset with such definitional anxieties over how best to draw the boundary line between revolution and such close conceptual neighbors as "rebellion," Antonio Gramsci's "passive revolution," "regime change," and "social transition." Is revolution always marked by violence? Does revolution necessitate transfers of power? Is uncontrollability a cardinal feature?[17] How coherent is the idea of a "long revolution"? Is success a necessary condition for particular circumstances to be labeled a revolution? Just what is it that changes during a revolution, and, closely related, what is the appropriate unit of analysis at which to conduct inquiries into revolutionary circumstances? Can we, indeed, theorize "revolution"?[18] Given this irresolution attending the nature of revolution whether in technology, politics, or intellectual affairs, it is perhaps to be expected that theories purporting to explain the revolutionary condition have proliferated.[19] Since a wide range of forces—social, economic, psychological, political, religious, intellectual—are ordinarily implicated in narratives of radical transformation, it is not surprising that different subject areas have offered their own distinctly disciplinary causal explanations.

Sociological accounts, for example, have been advanced since at least the work of such classical social theorists as Marx, Weber, and Durkheim. Then, and since, a variety of interpretive sociological stances has been adopted, the most prominent of which were built either on structural-functionalism or on conflict-coercion theory. The reconstitution of social systems, frequently understood in terms of the changing structural dynamics of class and state, has been central to such diagnoses.[20] By contrast, psychological explanations have routinely revolved around matters of cognitive dissonance or frustration-aggression.[21] In both cases, the motive force behind revolutionary change has been sought in mental states and processes.

Advocates of economic explanations have found such arguments unconvincing and have tended to focus on the significance of economic behavior, drawing on rational choice theory to illuminate what they considered to be a basic urge toward the maximization of utilities. Econometricians and economic theorists have applied their models either to the explanation of electoral behavior or to understanding collective actions in terms of aspirations toward private gain.[22]

The analyses of political theorists are different yet again. For them, the mainsprings of revolution are to be found either in conspiratorial high politics or in mass disaffection arising either from deprivation or a sense of structural inequality. In cultivating such conceptual terrain, scholars have resorted either to political functionalism of various sorts or to ideas of political conflict. In the former case, the failure of a given social system to provide the functional necessities for social life and the attendant collapse of consensus are taken as crucial.[23] In the latter, explanations revolve around the ways in which different groups compete for social power.[24]

Particular subject areas have had their own revolutions. The late eighteenth-century "Chemical Revolution," for example, prompted by Antoine Lavoisier's "overthrow" of the phlogiston theory and its replacement by the oxygen theory of combustion has been regarded as a classic conceptual change in science. Others have seen the industrialization of chemical manufactures in the later nineteenth century as marking a second "Chemical Revolution."[25] In the earth sciences, some would look to James Hutton and to field sites in Scotland for the "revolution in time" that inaugurated modern geology. Others would look to England, to William Smith, canal building and coal mining for the stratigraphic basis for that revolution, if such, indeed, it was.[26] In biology, Charles Darwin's theory of evolution by natural selection has been widely considered a revolutionary "moment" in shifting explanations of natural history from a creationist to a naturalistic episteme.[27] Geography experienced a "Quantitative Revolution" in the decade after the mid-1960s. The term has an enduring if limited currency. Its events never registered across all quarters of the discipline and the take-up of those new statistical methods justifying the use of the term (to the practitioners anyway) varied greatly by individuals and departments and added to rather than replaced geography's ways of knowing.[28]

It is clear from even this summary review of the term that while "revolution" has a distinctive intellectual genealogy, it also has many diverse connotations and no single shared meaning. Indeed, there has been a conflation of ideas as to what the term signifies: political or intellectual upheaval, sometimes violent and rooted in "popular" dissent (with or without a return to the prior condition), an innovative "moment" in scientific enquiry, something (relatively) brief in duration. Even so, the fact that the term is such a distinctive feature in a variety of intellectual contexts ensures its continuing significance. It is noteworthy, too, that scholars exploring these different contexts have increasingly discerned diversity within and between their respective objects of study: "revolutions" rather more than the "Revolution." What has been less evident is any sustained attention to the connections between geography and revolution, to the place of geography in understanding revolutionary phenomena. It is to a general consideration of some of these issues that we now turn.

GEOGRAPHY AND REVOLUTION

Whatever their utility, and however unstable the entity they are called upon to illuminate, it is arguable that what unites the disciplinary and theoretical standpoints on revolution reviewed above is their relative insensitivity to questions of space, place, and geography. The comments of Roy Porter and Mikuláš Teich on the historiography of the Scientific Revolution are illustrative of this point:

> Cultural topography and geo-politics are, not surprisingly, almost totally neglected in the great synoptic histories of the Scientific Revolution . . . works that accented the inner adventures of ideas and their transformation. For intellectual historians of this kind, the Scientific Revolution was a revolution in and of the mind; if there was a geography of thought, it was to be traced on maps that were metaphysical, *metaphorical.*[29]

At least in part, such a diagnosis constitutes a call to attend to revolutionary features on an individual basis. Just what the appropriate scale of spatial analysis is in the investigation of any particular state of intellectual or political or technological affairs remains an open question, of course. Arguably, scholars of history and of political science have been most attentive to questions of geography, to the idea of the nation, and to differential access to resources over space in explaining revolutionary phenomena. Such questions have certainly informed analysis of the so-called Great Revolutions—the English Revolution, the American Revolution, the French Revolution, as well as the revolutions of Mexico, of China, Cuba, Iran, and of Eastern Europe and the former Soviet Bloc.[30]

In the study of scientific knowledge, geographical issues of location, of the traveling nature of knowledge and the situated production and reception of meaning in science, have been the subject of considerable attention.[31] Indeed, what may be thought of as a "spatial turn" in science studies provides an important context in which we may consider the several ways in which phenomena such as political and scientific revolutions occur in space and through which the role played by geographical knowledge in moments of revolutionary transformation may be better understood. In one sense, this is a matter of the utility of geography's language. The language of geographical imperative has often been deployed in support of revolutionary regimes. As an illustration, let us return briefly to Thomas Paine. In his 1776 *Common Sense,* for example, Paine frequently resorted to naturalistic arguments to underwrite the legitimacy of republican independence. His "idea of the form of government," he insisted, was drawn "from a principle in nature." He was certain that "the folly of hereditary right in kings" was exhibited in the fact "that nature disapproves it, otherwise she would not so frequently turn it into ridicule by giving mankind an ass for a lion." In irreducibly naturalistic terms, Paine declared: "[I]n no instance hath nature made the satellite larger than its primary planet, and as England and America, with respect to each Other, reverses the common order of nature, it is evident they belong to different systems: England to Europe—America to itself." That geographical distance would presently enter the fray in the cause of rebellion was, for Paine, entirely predictable:

> The blood of the slain, the weeping voice of nature cries, 'tis time to part. Even the distance at which the Almighty hath placed England and America, is a strong and natural proof, that the authority of the one, over the other, was never the design of Heaven. The time likewise at which the continent was discovered, adds weight to the argument, and the manner in which it was peopled increases the force of it. The reformation was preceded by the discovery of America, as if the Almighty graciously meant to open a sanctuary to the persecuted in future years, when home should afford neither friendship nor safety.[32]

In a further sense, there is considerable evidence that geographers and others have, at one time or another, sought to mobilize geographical knowledge for more or less radical political purposes. The language of Enlightenment geography in Britain, for example, is suffused with concerns about the nature of political sovereignty, the rights of kings and peoples, and the utility of geography to the political management of states.[33] In the nineteenth century, the French geographer, socialist republican, and Paris communard Elisée Reclus, although professing to have been involved personally with revolution only "in an indirect way," nonetheless considered himself "a revolutionary by principles, tradition and solidarity." Dunbar observes that for Reclus, "geography and anarchism are closely related, because the more one understands the world and its inhabitants, the more . . . prejudices and antagonisms decline."[34] Reclus's anarchist enthusiasms predisposed him to reject Malthusian explanations and to attribute social evils to the maldistribution of resources rather than to overpopulation. His friendship with Peter Kropotkin helped to sustain such radical inclinations. A Russian aristocrat whose firsthand observation of poverty in Finland in 1871 helped reverse his ambition to become secretary to the Imperial Geographical Society in St. Petersburg, Kropotkin marshaled his own ideas about cooperative (as opposed to competitive) evolution in the cause of revolutionary politics. For Kropotkin, the facts of biogeography in extreme climates confirmed the value of sociability as a survival mechanism and impressed upon him the value of mutual aid in the political sphere. It was in Siberia that he lost "whatever faith in state discipline" he had hitherto cherished: the geography of the harsh northern reaches swept him into anarchism.[35] In the 1885 article "What Geography Ought to Be," he declared that the subject

> must teach us, from our earliest childhood, that we are all brethren, whatever our nationality. In our time of wars, of national self-conceit, of national jealousies and hatreds ably nourished by people who pursue their own egoistic, personal or class interests, geography must be . . . a means of dissipating those prejudices and of creating other feelings more worthy of humanity. . . . It is the task of the geographer to bring this truth, in its full light, into the midst of the lies accumulated by ignorance, presumption, and egotism.[36]

The dissident geographical traditions that figures such as Kropotkin represent have continued to thrive in certain branches of geography over the past century.[37]

Building upon these concerns, let us return to the two themes noted at the beginning of this introduction, namely, the geography *of* revolution and geography *in* revolution. The interweaving of geography and scientific revolutions affords one arena in which issues to do with the revolutionary aspects of scientific knowledge may be understood as a matter of geography. The mapping of certain items in physical space—people and practices, agents and instruments—can throw light on the social spaces in which scientific revolutions occurred. Such operations can help disclose, for example, how theories spread from place to place and can thereby demonstrate how locally made knowledge acquired seeming universality. By the same token, querying the definite article in *the* Scientific Revolution—as Shapin has so notably done—opens up questions around just what enterprises and exercises can be subsumed under the rubric of science and how they were performed in different locations. There are, then, questions to be explored to do with the geographies of scientific revolution and the related place of geography in the Scientific Revolution, insofar as either term may be taken as a single essential phenomenon.

This localist emphasis is a common feature of work in the geographies of science. As David Turnbull has recently argued:

> What we now count as a specifically authoritative form of knowledge—Western or modern science—is a tradition which has devised social strategies, narrative forms and instrumental practices that enable local knowledge to travel, to

be assembled at a centre of calculation and then to be put into use or transmitted as a unified body to other centres.[38]

The different ways that knowledge of nature was put together in museums and laboratories, in the field and in the garden, further contributes to the querying of a single and rather placeless Scientific Revolution. The conditions of knowledge making in different locales attests to what might be called the geography of rationalizing practices.[39] The salience of location has thus done much to refocus attention on the meaning of natural philosophy in particular places. It has also moderated the inclination to consider the dissemination and diffusion of ideas and instruments across space an unproblematic process. Indeed, the circulation of warranted knowledge was always a troublesome thing, and various strategies to overcome the tyranny of geographical space had to be put in place. No less than its making in different sites, the spread of science is always an achievement.

Precisely, however, because of the relationships between local claims—what Turnbull refers to as the way "different accounts vie with one another for dominance as their narrators and practitioners struggle for authority"[40]—it is important not to lose sight of the relationships between different geographical scales. What may be of interest is to know the connections between local and national geographies of science. Across Europe as a whole, practices of natural knowledge were far from seamless. Galileo's endeavors at the Italian court were rather different from the nautical pursuits that characterized the Iberian peninsula. The different religious contexts of natural philosophy in England and France were such that when Voltaire crossed the English Channel in the early eighteenth century, he sensed that he had entered a different intellectual world. National "styles" of science, national schools, and national traditions may thus have a place in considering the advancement of scientific learning. In contrast, however, it is quite possible that, through the networks of influences that sustained local sites of science's making and reception over long distances—think, for example, of the role of scientific correspondence in the eighteenth century in the making of the international "Republic of Letters"—notions of national science and even of the nation itself will be lost.[41]

At the same time, geographical knowledge had a vital role to play within what have been held as canonical scientific revolutions. Information about distant and domestic lands—their description, their measurement, their faunal and floral inhabitants—constituted a vital strain of data that fed into the making of modern science. The rubric governing the Savilian Chair of Astronomy at Oxford, established in 1619, stipulated that the subject should encompass geography as well as optics and the rules of navigation. Newton devoted his energies to an updating of Varenius's *Geographia Generalis* (1650) in order to advance his own version of the mechanical philosophy and to provide further refutation of Aristotelianism. The teaching of geography was central to the curriculum at Gresham College.[42] And as geography was differently placed within what was held to be Newtonian science, so we may be able to discern geographies of Newtonianism in its variant forms.[43]

During the so-called Darwinian Revolution,[44] biogeographical data from across the globe provided fundamental geographical data in support of evolutionary theories of various sorts. The very basis to such a revolution lay in the role of geographical difference. Alfred Russel Wallace's vocabulary—he spoke, for example, of the "geography of life"—and the phrasing and essential conceptualization of Darwin's theory of natural selection were replete with the geopolitical language of territorial invasion, militant colonization, and imperial conquest. Darwinism and evolutionary theory generally had its adherents and opponents, and because this was so, it has been possible to consider not just the variable reception over space that Darwinism had but also the effect of such thinking in other discourses and in everyday life.[45]

It is for just such reasons of place, movement, and geographical difference that the various revolutions in printing are an obvious subject for geographical interrogation. The replacement of hand copying by printing, though accomplished in a relatively short space of time, had a particular locational expression. Beginning in a few urban centers, printing diffused across much of western Europe during the final decades of the fifteenth century.[46] By the same token, the technical shifts that made possible the art of pamphleteering played a key role in political revolution in particular places—not least during the American Revolution. Lord Grenville, for instance, gave contemporary voice to this sentiment when he observed in 1817 that "the press was the most powerful of agencies which produced the Revolution in France."[47] Printing also had other unanticipated consequences on the diffusion of texts. Placing a work on the Catholic Church's Index of Prohibited Books did wonders for sales in Protestant locations and among Catholics with a taste for what Elizabeth Eisenstein calls "forbidden fruit."[48] Book history or, more widely, print history, is also, then, a book or print geography.

What we might, for convenience' sake, term the "Print Revolution" as shorthand for a whole suite of changes was not, however, simply about technical breakthrough. Printers in different sites had to work hard to invest their products with an authority that rendered them worthy of credit. For others elsewhere, believing a text was no straightforward thing. The geographical import of such questions becomes clear when we recall that local cultures constructed their own meanings with, and for, textual objects. Not the least of the geographical factors bound up with print cultures is the geographies of reading. In different geographical and social spaces—national, regional, local, domestic— the printed word was encountered and read in different ways.[49] At the same time, new dedicated reading sites began to make their appearance as various printing innovations became available: reading clubs, lending libraries, salon-based *conversazione,* and the like. In these diverse domains, different conventions of reading were pursued in different social spaces. Even as print generated in urban workshops enabled local knowledge to achieve global reach, through the capacity of the written word to traverse space and time, it could still help consolidate cultural boundaries. As Adrian Johns observes, "by appreciating the different practices by which readers in various times and places attribute meanings to the objects of their reading," the possibility of explaining "the global by rigorous attention to the local" is rendered altogether plausible.[50]

If matters of this sort testify to the significance of the geography of print, they also highlight the related place of geography in print. Whether in the form of texts delivering knowledge of global geography, in cartographic portrayals of the world or sections of it, or in pictorial representations of faraway places and peoples,[51] the circulation of geographical knowledge in printed form has reinforced the transformations that the "Print Revolution" occasioned. Maps and, more generally, what David Buisseret has considered an emergent "map consciousness" in early modern Europe—allowed control of space and did so for different nations and rulers at different times: "[I]t is possible quite accurately to discern the stages by which map-consciousness reached the elites of early modern Europe. Like so much else, this development began in Italy, and spread first to Germany, and then . . . to the northern 'new monarchs' such as Francis I and Henry VIII. It came late but powerfully to Sweden, and even later to Russia."[52] Here, if you will, is one "map" of the map's influence as a form of printed and state authority. Maps had different value depending on context. Considerable attention has been paid, for example, to the role of the map (and of spatial thinking generally) in the so-called Military Revolution in warfare that has been used to explain the "rise of the West" in early modern Europe.[53] Maps, in combination with written texts, were directly implicated in redefinitions of European conceptions of cultural and ethnic difference—not as "reflections" of the real world, but as powerful constituents of it. And having access to the printed word and map was a different route to what was held to be "reliable" knowledge than, say, trust placed in the spoken words of others.

One of the most notable technical revolutions, certainly in the nineteenth century, was the invention of photography. If, initially, photography was seen as a means to extend established artistic notions of visual awareness in landscape perception and in portraiture, the camera was soon recognized as a form of scientific instrument that could both measure and categorize. In certain fields of science, photography prompted major conceptual changes: in meteorology, for instance, the ability to capture flashes of lightning in photographic form provided a new stimulus to the understanding of atmospheric processes, while in bacteriology, photography provided a means to capture the microscopic world pictorially.[54] In anthropology and in ethnography, photography helped establish pictorial conventions for the portrayal of the colonized and the exotic.[55] Photographic images and their textual accompaniments were potent means—if also ambivalent ones—to the making of imagined and real geographies. Because this is so, technical change was also a matter of measuring and accommodating new ways of being in the world. How one measured time, for example, was crucial to the pace of the machines and, thus, to one's own working pace as a "hand" in the factories of the Industrial Revolution. Lengthy historical transformations in the means of timekeeping were intimately associated with the moral economy of particular communities of practice. Knowing the time meant knowing one's place, not least in terms of longitude and oceanic navigation.[56] Bookkeeping in the form of ship's logs and captain's diaries was a familiar form of writing by which one knew one's place (and time) in the world.

Books of geography, on the other hand, could not afford to be all at sea in their depiction of other countries. The utility of geography's texts depended upon the accuracy of their textual description. But, at the same time, geography's books could be mobilized in the cause of political revolution. Changes in the method of compilation of geography books in the early modern period have been documented, changes that, if they did not themselves amount to a "revolution" in method, were certainly novel departures from previous textual forms.[57] In the later eighteenth century, too, distinctive forms of geographical text—the "grammar" and the "gazetteer" for instance—helped codify the languages of geography.[58] Other examples may be cited, some forming the subject of more detailed examination in this volume. During the English Revolution, books of geography were routinely employed as a vehicle through which ecclesiastical quarrels, with all their political ramifications, were rehearsed. The political languages of church-state relations in the period are readily detectable in geography's books. Writers of geography books throughout the "long eighteenth century" were embroiled almost as a matter of routine in political controversy. Because geography books commonly surveyed governmental histories, church constitutions, and religious disputations, they offered ample scope for their authors to intervene in the ecclesiastical politics of Enlightenment England.[59] As later chapters show of other places—in Revolutionary France, in the early years of the American Republic, and in Germany in the "Year of Revolutions," 1848—geography has been called upon at various times to service the needs of republican regimes, to generate and consolidate senses of national identity, and to further imperial apologetics.[60]

Geography and Revolution makes no claims to comprehensiveness in addressing the connections between the two terms in the ways outlined in this introductory chapter or in its choice of subject in each of the more detailed chapters that follow. Our intention is much more to open out paths rather than to presume them now, or to leave them in the future, well trodden. Geography is not taken here in any strictly disciplinary sense, nor held to be just a matter of things having a location or distribution in space. We want to raise questions about the difference that space makes in understanding revolutionary phenomena and to consider, in different geographical and historical contexts, how people looked to geography and to geographical insights for an explanation of what they took to be "revolution." It is our hope that these initial forays into this terrain will encourage others to follow.

...

Notes

The first epigraph to this chapter is from William Roy, "An Account of the Measurement of a Base Line on Hounslow Heath," *Philosophical Transactions of the Royal Society of London* 75 (1785): 385. (Roy, involved in the Triangulation Survey that led to the Ordnance Survey, was referring to the threat posed to Britain by Napoleon's forces.) The second aphorism, the words of the leading American political geographer Isaiah Bowman, is quoted in Neil Smith, *American Empire: Roosevelt's Geographer and the Prelude to Globalization* (Berkeley: University of California Press, 2003), 27.

1. The following sources survey the history of the idea of revolution: Arthur Hatto, "'Revolution': An Enquiry into the Usefulness of an Historical Term," *Mind* 58 (1949): 495–516; Melvin J. Lasky, "The Birth of a Metaphor: On the Origins of Utopia and Revolution," *Encounter* 48 (February and March 1970): 35–45 and 30–42. For one accessible guide, see the essays in Roy Porter and Mikuláš Teich, eds., *Revolution in History* (Cambridge: Cambridge University Press, 1986).

2. Thomas Paine, *The Rights of Man* (London: Pelican, 1971), 166.

3. Stan Taylor, *Social Science and Revolutions* (London: Macmillan, 1984), 2.

4. Quoted in Felix Gilbert, "Revolution," in *Dictionary of the History of Ideas,* ed. Philip P. Wiener (New York: Scribner's, 1973), 4:153.

5. See Robert C. Tucker, *The Marxian Revolutionary Idea* (New York: Norton, 1969).

6. Especially significant here was Arnold Toynbee's *Lectures on the Industrial Revolution in England: Popular Addresses, Notes and other Fragments* (London: Rivingtons, 1884).

7. See Maxine Berg and Kristine Bruland, eds., *Technological Revolutions in Europe: Historical Perspectives* (Cheltenham: Edward Elgar, 1998).

8. Akos Paulinyi, "Revolution and Technology," in Porter and Teich, *Revolution in History*, 261–89. For various perspectives, see Samuel Lilley, "Technological Progress and the Industrial Revolution," in *The Fontana Economic History of Europe*, ed. Carlo M. Cipolla (London: Fontana, 1973), 3:187–254; Albert E. Musson and Eric Robinson, *Science and Technology in the Industrial Revolution* (Manchester: Manchester University Press, 1969); Brooke Hindle and Steven Lubar, *Engines of Change: The American Industrial Revolution, 1790–1860* (Washington, DC: Smithsonian Institution Press, 1986); Hartmut Kaelble, *Industrialization and Social Inequality in Nineteenth-Century Europe* (New York: St. Martin's Press, 1986); Anthony Wrigley, *Continuity, Chance and Change: The Character of the Industrial Revolution in England* (Cambridge: Cambridge University Press, 1988). On the nature of the agricultural revolution in one national context, see Mark Overton, *Agricultural Revolution in England: The Transformation of the Agrarian Economy, 1500–1850* (Cambridge: Cambridge University Press, 1996).

9. See, for example, John Langton and R. J. Morris, eds., *Atlas of Industrializing Britain, 1780–1914* (London: Methuen, 1986). For more recent and wider considerations of the geography of the Industrial Revolution, see Patricia Hudson, *Regions and Industries: A Perspective on the Industrial Revolution in Britain* (Cambridge: Cambridge University Press, 1989); Patrick O'Brien, ed., *The Industrial Revolution in Europe*, 2 vols. (Oxford: Blackwell, 1994).

10. See, for example, David S. Landes, a leading historian of industrial transformation, *Revolution in Time: Clocks and the Making of the Modern World* Cambridge: Harvard University Press, 1983).

11. On these points, see for example, Chris Freeman, *As Time Goes By: From the Industrial Revolutions to the Information Revolution* (Oxford: Oxford University Press, 2002); Manuel Castells, *The Rise of the Network Society* (Oxford: Blackwell, 1996); Manuel Castells, *The Internet Galaxy: Reflections on the Internet, Business and Society* (Oxford: Oxford University Press, 2001); Carlotta Perez, *Technological Revolutions and Financial Capital* (Cheltenham: Edward Elgar, 2003).

12. For a recent review of trends in book history, see David Finkelstein and Alistair McCleery, eds., *The Book History Reader* (London: Routledge, 2002). On the question of there being a "Reading Revolution" at the end of the eighteenth century, see Reinhard Wittman, "Was There a Reading Revolution at the End of the Eighteenth Century?" in *A History of Reading in the West*, ed. Guglielmo Cavallo and Roger Chartier, trans. Lydia G. Cochrane (Cambridge: Polity, 1999), 284–312. For a full discussion of what has become known (after Robert Darnton, the leading historian of the book) as the "Darnton debate" over the relationships between books, reading, and revolutionary change in late-eighteenth-century Europe, see Haydn T. Mason, ed., *The Darnton Debate: Books and Revolution in the Eighteenth Century*, Studies on Voltaire and the Eighteenth Century (Oxford: Voltaire Foundation, 1988).

13. Classic works include Herbert Butterfield, *The Origins of Modern Science, 1300–1800* (London: G. Bell, 1949); Hugh Kearney, *Science and Change* (London: Weidenfeld and Nicolson, 1971); A. Rupert Hall, *The Revolution in Science, 1500–1750* (London: Longman, 1983); I. Bernard Cohen, *Revolution in Science* (Cambridge: Harvard University Press, 1985); I. Bernard Cohen, *The Newtonian Revolution* (Cambridge: Cambridge University Press, 1980); Gertrude Himmelfarb, *Darwin and the Darwinian Revolution* (London: Chatto and Windus, 1959); Michael Ruse, *The Darwinian Revolution* (Chicago: University of Chicago Press, 1979).

14. For these various perspectives, see Butterfield, *Origins of Modern Science;* A. Rupert Hall, *The Scientific Revolution, 1500–1800* (London: Longman, 1954); Thomas Kuhn, *The Structure of Scientific Revolutions* (Chicago: University of Chicago Press, 1962); Mary B. Hesse, *Models and Analogies in Science* (Notre Dame: University of Notre Dame Press, 1966); Richard Westfall, *The Construction of Modern Science: Mechanisms and Mechanics* (Cambridge: Cambridge University Press, 1971); Reijer Hooykaas, *Religion and the Rise of Modern Science* (Edinburgh: Scottish Academic Press, 1972); Carolyn Merchant, *The Death of Nature: Women, Ecol-*

ogy and the Scientific Revolution (New York, Harper and Row, 1980); Mary B. Hesse, Revolutions and Reconstructions in the Philosophy of Science (Brighton: Harvester, 1980); Elizabeth L. Eisenstein, The Printing Revolution in Early Modern Europe (Cambridge: Cambridge University Press, 1983); William R. Shea, ed., Revolutions in Science: Their Meaning and Relevance (Canton, MA: Science History Publications, 1988); David C. Lindberg and Robert S. Westman, eds., Reappraisals of the Scientific Revolution (Cambridge: Cambridge University Press, 1990); Floris Cohen, The Scientific Revolution: A Historiographical Inquiry (Chicago: University of Chicago Press, 1994); Ernst Mayr, "The Advance of Science and Scientific Revolutions," Journal for the History of the Behavioral Sciences 30 (1994): 328–47; Steven Shapin, The Scientific Revolution (Chicago: University of Chicago Press, 1996); Peter Harrison, The Bible, Protestantism and the Rise of Natural Science (Cambridge: Cambridge University Press, 1998).

15. Shapin, Scientific Revolution, 1.

16. For example, Peter Dear entitles his recent account Revolutionizing the Sciences: European Knowledge and Its Ambitions, 1500–1700 (Princeton: Princeton University Press, 2001). See also the critical commentaries in Margaret J. Osler, ed., Rethinking the Scientific Revolution (Cambridge: Cambridge University Press, 2000).

17. Eric J. Hobsbawm, "Revolution," in Porter and Teich, Revolution in History, 5–46.

18. See the essays in John Foran, ed., Theorizing Revolutions (London: Routledge, 1997).

19. See the reviews in Laurence Stone, "Theories of Revolution," World Politics 18 (1966): 159–76; Barbara Salert, Revolutions and Revolutionaries: Four Theories (New York: Elsevier, 1976); Taylor, Social Science; Michael S. Kimmerl, Revolution: A Sociological Interpretation (Oxford: Polity, 1990).

20. See Seymour Martin Lipset, Revolution and Counter-Revolution: Change and Persistence in Social Structure (London: Heinemann, 1969); Chalmers Johnson, Revolutionary Change (London: Longman, 1983); Theda Skocpol, States and Social Revolutions: A Comparative Analysis of France, Russia and China (Cambridge: Cambridge University Press, 1979); Theda Skocpol, Social Revolutions in the Modern World (Cambridge: Cambridge University Press, 1994).

21. Key early statements here include Leon Festinger, The Theory of Cognitive Dissonance (Stanford: Stanford University Press, 1967); James Davies, "Towards a Theory of Revolution," American Sociological Review 27 (1962): 5–18; Ted Robert Gurr, Why Men Rebel (Princeton: Princeton University Press, 1971).

22. So, for example, Anthony Downs, An Economic Theory of Democracy (New York: Harper and Row, 1957); Gordon Tullock, The Social Dilemma: The Economics of War and Revolution (Blacksburg, VA: University Publications, 1974); Morris Silver, "Political Revolutions and Repression: An Economic Approach," Public Choice 14 (1974): 63–71.

23. Samuel P. Huntington, Political Order in Changing Societies (New Haven: Yale University Press, 1968).

24. See Charles Tilly, From Mobilization to Revolution (Reading, MA: Addison-Wesley, 1978). For a wide-ranging discussion of the term "revolution" in political science, see Rosemary H. T. O'Kane, ed., Revolution: Critical Concepts in Political Science, 4 vols. (London: Routledge, 2000).

25. Carleton E. Perrin, "The Chemical Revolution," in Companion to the History of Modern Science, ed. Roger C. Olby et al. (London: Routledge, 1996), 264–77; Archibald Clow, The Chemical Revolution: A Contribution to Social Technology (London: Batchworth, 1952).

26. On Hutton, see Stephen Baxter, Revolutions in the Earth: James Hutton and the True Age of the World (London: Phoenix, 2004); on William Smith, see Simon Winchester, The Map that Changed the World: The Tale of William Smith and the Birth of a Modern Science (London: Viking, 2001). For a more measured, geographically wider, and "evolutionary" discussion of the nature of the earth sciences in this period, see Rhoda Rappaport, "The Earth Sciences," in The Cambridge History of Science, vol. 4, Eighteenth-Century Science, ed. Roy Porter (Cambridge: Cambridge University Press, 2003), 417–35.

27. On such matters, see Neal C. Gillespie, *Charles Darwin and the Problem of Creation* (Chicago: University of Chicago Press, 1979).

28. On the geographical and social variability of geography's "Quantitative Revolution," see Stan Gregory, "Quantitative Geography: The British Experience and the Role of the Institute," *Transactions of the Institute of British Geographers* 8 (1983): 80–89; Mark D. Billinge, Derek J. Gregory, and Ronald J. Martin, eds., *Recollections of a Revolution: Geography as Spatial Science* (London: Macmillan, 1984).

29. Roy Porter and Mikuláš Teich, introduction to *The Scientific Revolution in National Context*, ed. Roy Porter and Mikuláš Teich (Cambridge: Cambridge University Press, 1992), 2–3.

30. For a summary of recent work on these revolutions, see the essays that comprise part 7 and part 8 (chapters 46–57) in O'Kane, *Revolution*, 3:5–326. On Eastern Europe, see Padraic Kenney, *A Carnival of Revolution: Central Europe, 1989* (Princeton: Princeton University Press, 2002).

31. See, for example, Marie-Noëlle Bourguet, Christian Licoppe, and H. Otto Sibum, eds., *Instruments, Travel and Science: Itineraries of Precision from the Seventeenth to the Twentieth Century* (London: Routledge, 2002); William Clark, Jan Golinski, and Simon Schaffer, eds., *The Sciences in Enlightened Europe* (Chicago: University of Chicago Press, 1999); David N. Livingstone, *Putting Science in Its Place: Geographies of Scientific Knowledge* (Chicago: University of Chicago Press, 2003); Jan Golinski, *Making Natural Knowledge: Constructivism and the History of Science* (Cambridge: Cambridge University Press, 1998); Steven Shapin, "Placing the View from Nowhere: Historical and Sociological Problems in the Location of Science," *Transactions of the Institute of British Geographers* 23 (1998): 5–12; Crosbie Smith and Jon Agar, eds., *Making Space for Science: Territorial Themes in the Shaping of Knowledge* (London: Macmillan, 1998); Charles W. J. Withers, *Geography, Science and National Identity: Scotland since 1520* (Cambridge: Cambridge University Press, 2001), esp. 1–29.

32. Thomas Paine, *Common Sense* (1776; London: Penguin, 1986), 87.

33. Robert J. Mayhew, *Enlightenment Geography: The Political Languages of British Geography, 1650–1850* (London: Macmillan, 2000).

34. Gary S. Dunbar, *Elisée Reclus: Historian of Nature* (Hamden, CT: Archon, 1978), 91. The quotation from Reclus is taken from this source.

35. Peter Kropotkin, *Memoirs of a Revolutionist* (1899; London: Cresset Library, 1962), 148.

36. Peter Kropotkin, "What Geography Ought To Be," *Nineteenth Century* 18 (1885): 942–43.

37. An introduction to this tradition is provided in Alison Blunt and Jane Wills, *Dissident Geographies: An Introduction to Radical ideas and Practice* (Harlow: Prentice Hall, 2000).

38. David Turnbull, "Travelling Knowledge: Narratives, Assemblage and Encounters," in Bourguet, Licoppe, and Sibum, *Instruments*, 288.

39. Livingstone, *Putting Science in Its Place*.

40. Turnbull, "Travelling Knowledge," 288.

41. Lewis Pyenson, "An End to National Science: The Meaning and Extension of Local Knowledge," *History of Science* 40 (2002): 251–90.

42. See Lesley B. Cormack, *Charting an Empire: Geography at the English Universities, 1580–1620* (Chicago: University of Chicago Press, 1997); William Warntz, "Newton, the Newtonians, and the *Geographia Generalis Varenii*," *Annals of the Association of American Geographers* 79 (1989): 165–91; Charles W. J. Withers, "Reporting, Mapping, Trusting: Practices of Geographical Knowledge in the Late Seventeenth Century," *Isis* 90 (1999): 497–521.

43. On this, see for example, Mary Fissell and Roger Cooter, "Exploring Natural Knowledge: Science and the Popular," in Porter, *Cambridge History of Science*, 4:129–58, esp. 134–39.

44. The idea of a "Darwinian Revolution" is challenged in Peter J. Bowler, *The Non-Darwinian Revolution: Reinterpreting a Historical Myth* (Baltimore: John Hopkins University Press, 1988).

45. For example, Ronald L. Numbers and John Stenhouse, eds., *Disseminating Darwinism: The Role of Place, Race, Religion, and Gender* (Cambridge: Cambridge University Press, 1999); John Laurence and John Nightingale, eds., *Darwinism and Evolutionary Economics* (Cheltenham: Edward Elgar, 2001).

46. See the maps in Eisenstein, *Printing*, 15–17. Similar issues are raised by the maps and discussion of the location of printers and the spread of printing by Lucien Febvre and Henri-Jean Martin, eds., *The Coming of the Book: The Impact of Printing, 1450–1800,* trans. David Gerard (London: New Left Books, 1976), esp. 167–215.

47. Quoted in Arthur Aspinall, *Politics and the Press c. 1780–1850* (London: Home and Van Thal, 1949), 1.

48. See the comments in Elizabeth L. Eisenstein, "On Revolution and the Printed Word," in Porter and Teich, *Revolution in History,* 186–205; the phrase is from page 195.

49. See James A. Secord, *Victorian Sensation: The Extraordinary Publication, Reception, and Secret Authorship of "Vestiges of the Natural History of Creation"* (Chicago: University of Chicago Press, 2000); Fernando F. Segovia and Mary Ann Tolbert, eds., *Reading from This Place,* vol. 1, *Social Location and Biblical Interpretation in the United States;* vol. 2, *Social Location and Biblical Interpretation in Global Perspective* (Augsburg: Fortress Press, 1995).

50. Adrian Johns, *The Nature of the Book: Print and Knowledge in the Making* (Chicago: University of Chicago Press, 1998), 385.

51. See Jerry Brotton, "Printing the World," in *Books and the Sciences in History,* ed. Marina Frasca-Spada and Nicholas Jardine (Cambridge: Cambridge University Press, 2000), 35–48; Jerry Brotton, *Trading Territories: Mapping the Early Modern World* (London: Reaktion, 1997).

52. David Buisseret, *The Mapmakers' Quest: Depicting New Worlds in Renaissance Europe* (Oxford: Oxford University Press, 2003), 69.

53. Ibid., 113–51; Geoffrey Parker, *The Military Revolution: Military Innovation and the Rise of the West, 1500–1800* (Cambridge: Cambridge University Press, 1988); Brian M. Downing, *The Military Revolution and Political Change* (Princeton: Princeton University Press, 1992).

54. Jennifer Tucker, "Photography as Witness, Detective, and Imposter: Visual Representation in Victorian Science," in *Victorian Science in Context,* ed. Bernard Lightman (Chicago: University of Chicago Press, 1997), 378–408.

55. See, for example, James R. Ryan, *Picturing Empire: Photography and the Visualization of the British Empire* (London: Reaktion; Chicago: University of Chicago Press, 1997); Joan M. Schwartz and James R. Ryan, *Picturing Place: Photography and the Geographical Imagination* (London: I. B. Taurus, 2003).

56. On the longitude question in the eighteenth century, see Dava Sobel, *Longitude: The True Story of a Lone Genius Who Solved the Greatest Scientific Problem of His Time* (London: Penguin, 1995). For one corrective to Sobel's linear narrative of the longitude problem being simply a matter of competition between Nevil Maskeleyne and John Harrison, and for attention to the geography of places involved in the time trials designed to verify Harrison's claims over his chronometer's accuracy, see Jim Bennett, "The Travels and Trials of Mr Harrison's Timekeeper," in Bourguet, Licoppe, and Sibum, *Instruments,* 75–96.

57. Robert J. Mayhew, "Geography, Print Culture and the Renaissance: 'The Road Less Travelled By,'" *History of European Ideas* 27 (2001): 349–69.

58. Robert J. Mayhew, "Geography Books and the Character of Georgian Politics," in *Georgian Geographies: Essays on Space, Place and Landscape in the Eighteenth Century,* ed. Miles Ogborn and Charles W. J. Withers (Manchester: Manchester University Press, 2004), 192–211.

59. Ibid.; Robert J. Mayhew, "The Character of English Geography, c. 1660–1800: A Textual Approach," *Journal of Historical Geography* 24 (1988): 385–412.

60. For example, see David M. Hooson, ed., *Geography and National Identity* (Oxford: Blackwell, 1994); Avril M. C. Maddrell, "Empire, Emigration and School Geography: Changing Discourses of Imperial Citizenship, 1880–1925," *Journal of Historical* Geography 22 (1996): 373–87; Anne Buttimer, Stanley D. Brunn, and Ute Wardenga, eds., *Text and Image: Social Construction of Regional Knowledges,* special issue of *Beiträge zur Regionalen Geographie* 49 (1999): 179–91; Felix Driver, *Geography Militant: Cultures of Exploration and Empire* (Oxford: Blackwell, 2001); Withers, *Geography, Science and National Identity.*

PART I

GEOGRAPHY AND SCIENTIFIC REVOLUTION

Space, Place, and Natural Knowledge

Science is supposed to be placeless. Whether experiments are carried out in Boston or Beijing does not matter to the results that are derived; whether specimens are collected in Birmingham, England, or Birmingham, Alabama, does not matter to where they fit into the taxonomy of life. Scientific recipes work the same everywhere. The fact that it was in the German Electron Synchrotron in Hamburg that the first experimental traces of gluons that bind quarks together to form protons and neutrons were found in 1979 is hardly relevant to the proof for their existence. This intuition is, in a fundamental way, the triumph that the Scientific Revolution of the seventeenth century secured. And until relatively recently this conviction remained an unquestioned tenet of modern scientific culture. The locations where science is carried out was thought to have nothing other than the most trivial bearing on scientific knowledge, and those who thought otherwise were either foolish or mistaken. For these reasons, the connections between geographical analysis and scientific knowledge remained ruptured. To be sure, some environmental determinists sought to explain scientific developments by reference to geographical conditions like climate or topography, but these ventures simply lacked all conviction.

The chapters that comprise this first part of *Geography and Revolution* seek to revisit the interactions between space, place, and natural knowledge in a variety of different keys in order to determine how geographical factors may indeed cast light on a range of canonical scientific moments. The strategy is twofold: these chapters in various ways examine both the geography *of* scientific revolutions and the role of geography *in* scientific revolutions.

Peter Dear inaugurates the discussion in chapter 2 by probing the idea of revolution itself, noting the significance of Thomas Kuhn's celebrated account *The Structure of Scientific Revolutions.* His insistence on the need to ground so-called revolutions in the particularities of place leads him to an analysis of some of the different ways in which scientific enterprises bear the impress of the geographical. Consider the cartographic impulse, for example, and the ways in which mapping may open up a variety of key issues in the history of scientific transformations. As Dear points out, the map of the diffusion of Robert Boyle's air-pump across Europe turns out to be less a visual representation of the trail of a scientific instrument than a cartographic depiction of what he calls "the exporting of local conditions." Again, the mapping of the London lodgings of the in-set of geologists who built up knowledge of the Devonian system in the early nineteenth century unveils systemic connections between physical site, social location, and intellectual positioning. Geography has been crucial to science in other ways too. As Dear shows, the development of natural history has been shaped by spatial factors. Dutch success in the cultivation of tropical botany during the eighteenth century, compared with the French, for example, had much to do with Dutch trade patterns and the lack of geographical sensitivity on the part of the leading French botanist at the Jardin du Roi. In these ways grand narratives of cognitive scientific revolutions are disrupted by the particularities of space and place.

The role of national styles of inquiry is the subject taken up by John Henry in chapter 3. Reviewing the variety of ways in which scholars have approached the issue of national scientific styles—focusing on regional modes of reasoning, educational systems, forms of patronage, social organization, and so on—he turns to a comparison between England and France during the seventeenth century and to how culture shaped the different modes of scientific inquiry in these national settings. Different religious and political traditions had a crucial role to play in how experimental philosophy developed on either side of the English Channel, not least because they had a direct bearing on how Aristotelianism was handled in France and England. The French emphasis on causal explanations demonstrated the continuing influence of Aristotle mediated through the Catholic Church, while the English rhetoric of theory-free induction reflected an Anglican penchant for what Henry calls "doctrinal minimalism." Thus the Cartesian impulse toward a rationalistic experimentalism contrasted sharply with Newton's famous disinclination to elaborate hypotheses and his willingness to allow for the action of unexplained properties of matter. Given the significance of French and English culture on approaches to experimental inquiry, Henry goes on to speculate that rival national styles might well have been a prerequisite for the cognitive transformations that attended the Scientific Revolution. Europe's unique political geography thus emerges as a decisive factor in the rise of science in the West.

Given Henry's attention to questions at the national scale, it is appropriate that in chapter 4 Charles Withers focuses his attention on Scotland in order to elucidate something of the dynamic interplay of geography *in* and the geography *of* the Scientific Revolution. His argument is that in fundamental ways the Scientific Revolution of the seventeenth century was as much a geographical as a historical or philosophical phenomenon. To tease out these relationships he concentrates attention on three key components of Scottish science in the decades around 1700. First, he investigates the parliamentary promotion of the geographical work of Robert Sibbald and John Adair and the character of their endeavors in chorography and practical mathematics—projects that in crucial ways helped constitute the idea of Scotland itself. Second, he turns to what might be called textual space and to the production of geographical books in Scotland. Here his endeavors connect with recent inquiries into the history and geography of book production in order to bring within the frame of analysis not simply texts and authors, but publishers, printers, patrons, and audiences. Finally, the place of geographical instruction in the Scottish universities—Edinburgh in particular—falls under scrutiny, and his elucidation of the character of geography teaching at the time discloses the subject's intimate interest in debates over the nature of matter and celestial mechanics, as well as navigation and chorography. His chapter thus serves to demonstrate not only the place of geographical scholarship in the new natural philosophy, but also the critical role of such local spaces as the field, the lecture hall, and the library, in the constitution of the Scientific Revolution itself.

Chapter 5 rolls the calendar forward to perhaps the greatest "scientific revolution" of the nineteenth century—the theory of evolution by natural selection. This is James Moore's quarry in an analysis that examines the geographical underpinnings of Darwin's and Wallace's achievements. The pervasive and politically charged spatial language of invasion, aliens, foreign immigrants, and so on in the modern lexicon provides Moore with the analytical tools to deconstruct the crossovers between the imperial and the ecological vocabulary of Victorian Britain's two foremost evolutionary theorists. In each case, biogeography and geopolitics were tightly interwoven. The critical links Darwin perceived between translocation and transmutation, between migration and mutation, in the evolution of species were conceived at the height of Britain's imperial drama, and it is no accident that his thoughts on ecological geography were redolent with the imperatives of military geography. Darwin conceived of living organisms behaving much like Englishmen—invading, multiplying, colonizing. With Wallace, the outcome was different. In his vision, the imperial was less sharply focused, no doubt because he was less enthralled with Britain's overseas crusade and more likely to see corruption than civilization on the imperial frontier. But his outlook was no less molded by geography. A surveyor fascinated by the power of the line to cut through landholdings, social classes, racial groups, and animal species, his famous biogeographical line gave cartographic expression to his understanding of "the geography of life."

Taken as a set, these four chapters open up questions about the reciprocal ties between geography and scientific revolutions. They are not the last word on the subject. But as the first word, they show something of the rich potential in exploring the role of place and space in a range of knowledge enterprises.

— 2 —

SPACE, REVOLUTION, AND SCIENCE

...

Peter Dear

Revolution in History

While Jared Diamond's bestseller, *Guns, Germs, and Steel* (1997), takes geographical determinism a lot further than most historians or social scientists approve of, it nonetheless remains the case that place, locality, geographical issues in general, can be of great significance in understanding historical change. Indeed, such issues have always played a role in historical explanation. The question is how to understand them in a more systematic fashion, so as to gain a greater sense of what can and cannot be done with historical analyses that take into account the characteristics of place. Since we are concerned here with analyses of "revolution," however, we should first of all consider what that term means in standard treatments of historical processes and events.

A good indication of the state of affairs in American universities is the fact that Thomas Kuhn's *Structure of Scientific Revolutions* (1962) is used in graduate courses concerned with historical methodology—and this despite the fact that Kuhn himself, quite rightly, insisted that the originality of his book lay only in the fact that it talked specifically about *science;* his interpretive models of paradigms and revolutions were simply borrowed, he said, from existing ways of understanding things like political revolutions.[1] The use of Kuhn's book in history courses seems to be evidence, in fact, that historians are unsure of what they mean by the term "revolution" and thus have to resort to anything that can help them to conceptualize it more clearly.

In its broadest sense, historians and others use "revolution" to indicate some sort of discontinuity, or rupture, as E. J. Hobsbawm noted.[2] But as Hobsbawm also observed, it sometimes takes quite a long time to certify whether a revolution really did occur at some particular time in the past—to be sure, that is, that the new regime or state of affairs that the candidate "revolution" installed really did represent a permanent change, and was not simply swept away by a later counterrevolution that showed it to be evanescent or superficial. From this point of view, the political-social historian—or, to be more precise, the Marxist or crypto-Marxist historian—is reluctant to regard revolutions as being of central, driving importance in the ongoing processes of history. The real historical changes are macrosocial ones, such as the transition from feudal to capitalist societies, and revolutions are simply what Hobsbawm called "incidents in macro-historical change,"[3] the moments at which the rupture of an old and increasingly insupportable system occurs and a new system comes into being to take its place.

From that perspective, the most important question concerns the logical relationship between two social systems in the terms of which one of them can be seen to emerge inevitably from the problems inherent in the other. The actual revolution, the transition itself, is then not very important, because it was bound to happen in some form or another sooner or later; the details of how it actually occurred are therefore of little more than antiquarian, or commemorative, interest. But most working historians have a vested interest in regarding such details as important; theoretical models are all very well, they tend to think, but only detailed historical investigation is going to reveal the things that are really important. Inevitability, as opposed to contingency, is the enemy of the historian's project, a truth that has always given trouble to Marxist historians.

One perennial difficulty that non-Marxist historians, and nonhistoricist historians in general, have long confronted is that close historical studies of unquestionable revolutions—the French Revolution being the most chewed over of them all—always demonstrate how many institutions, practices, and assumptions failed to change significantly as a result of the revolution in question; that many underlying social institutions that were central to the old regime kept on going, below the radar, in the new regime.[4] If there is no absolute rupture between two fundamentally different ways of organizing society, should the supposed revolution be seen instead as just a rhetorical imposition by the historical actors, and by historians studying them, upon events that were not in themselves perhaps very revolutionary at all?

Kuhn then emerges as a very useful resource, because he had a preeminent point of reference by which to judge whether an event, or complex of events, was truly revolutionary. His "scientific revolutions" marked the break between two paradigms, or "disciplinary matrices," that represented the way knowledge was made in some particular area of science in some particular period. And because it was scientific *knowledge* that was the central feature in Kuhn's picture, rather than some other focus, such as the social organization of scientific disciplines, then to identify a scientific revolution all that needed to be done was to look at what people accepted as valid scientific ideas at different times and analyze the relationships of those sets of ideas to one other.

But then, of course, Kuhn left himself with plenty of additional problems that compromised his special kind of "revolutionariness." The things he called revolutions were, unfortunately, of very different orders.[5] The shift from classical to relativistic mechanics was for Kuhn a large-scale revolutionary paradigm shift, but the discovery of X-rays by Röntgen also represented a scientific revolution. The broad applicability of the category necessarily followed from Kuhn's characterization of the essential nature of a revolution in science: the interrelation of concepts in a paradigm meant that it could not accommodate the introduction of genuinely new concepts (including supposed new things in nature) without disrupting the integration of all the others—that is what a scientific revolution *was* for Kuhn.[6] In effect, Kuhn avoided any charge of imprecision in his idea of scientific revolutions—an imprecision that would stem from the idea's too-liberal use, whereby "revolutions" could take place on practically any scale whatsoever—by being able to identify and recognize them with analytical precision. This meant that, in the end, revolutions would turn up practically everywhere.

But such difficulties arise only for people who worry about solid definitions and abstract theoretical models, whether of concepts or of social structures. Fortunately, we now live in a (somewhat aging) postmodern era where structural coherence is gone, the inevitability of social revolutions is a thing of the past, and contingency and groundlessness remain the order of the day. This contingency appears most clearly in cultural history, with its uneasy relationship to the ever-shifting monster known as "cultural studies."

Cultural history, particularly of the American variety, lives in a region of meanings that exist largely above and apart from social history. Cultural historians are interested in investigating shifting meanings, which means, especially, looking at signs and their uses in texts of all sorts. Part of the independence of cultural history from social history rests in the working idea that social categories are, at least in part, constituted by the makings of meaning concerning them.[7] In this way, cultural history cannot be regarded either as simply an epiphenomenon of social history or as something that disregards social history. Instead, it becomes a kind of interacting coequal of social history. Finally, we should note that the adoption of cultural historical approaches for the study of canonical or otherwise "important" texts today defines a dominant style of intellectual history.[8]

All of this is as much as to say that, as an analytical category, "revolution" fares none too well these days among historians, whether of science or of anything else. Nonetheless, "revolution" still has a life besides its role as analytical category. Apart from the French Revolution, one can point to the Russian Revolution, the Glorious Revolution, the American Revolution (for Americans), or Lavoisier's Chemical Revolution. In other words, there are revolutions as historical actors' categories, where the use of the word, as well as its contemporary meanings, form an object of study in themselves. Additionally, there are revolutions that receive that label some time after the fact, but in a way that becomes historically significant in later periods. The term "Copernican Revolution" appeared in the eighteenth century, two hundred years after the events being designated as such, and then played a major role in self-conceptions of science in the nineteenth century.[9] The twentieth-century Marxist historian Christopher Hill significantly called the English Civil War the "English Revolution,"[10] a term that, as late as the earlier decades of the twentieth century, used itself to be applied to the Glorious Revolution. There is a lot to be said about "revolutions" as cultural-historical "facts on the ground."

Mapping History

One way of approaching the theme of geography and revolution is in terms of *mapping* and revolution. Mapping is a practice that gives some historical reality to issues of place, locality, and geography. Many historians make use of maps, and revolutionary changes, changes that are seen as in some sense fundamental (as well as changes seen as more subtle and superficial), can always, and profitably, be mapped. Ideas become established or new modes of social organization develop in ways that can be investigated in relation to the ways in which they take hold over greater or smaller spatial regions. Such geographical development can either parallel or be independent from a spread through social spaces, for example. So particular political views might spread from a metropolitan center through the surrounding hinterland, or they might spread through a particular social or professional class within a single city. There are interesting ways in which microlevel phenomena can be mapped, too.

Figure 2.1 offers a classic example from Martin Rudwick's 1985 book *The Great Devonian Controversy*. Rudwick provides a map of central London that displays the lodgings of Darwin and various other leading geologists around 1840, as well as the locations of the scientific meeting places that were central

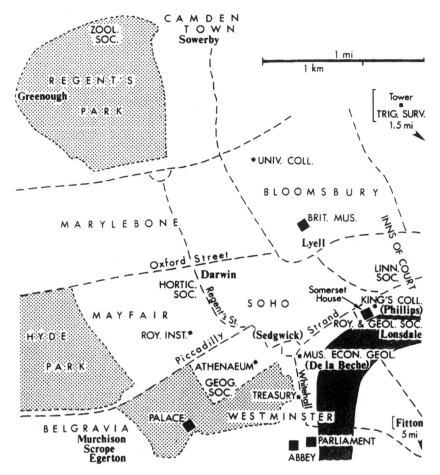

Figure 2.1. Part of London, ca. 1840, from Martin J. S. Rudwick, *The Great Devonian Contro-versy: The Shaping of Scientific Knowledge among Gentlemanly Specialists* (Chicago: University of Chicago Press, 1985), 35.

to their lives. Rudwick uses this approach as a way of displaying the quotidian interrelations of this group of individuals, a group that we know had routine and substantive dealings with one another; in a sense, Rudwick's mapping adds a dimension to what can be known from the correspondence of these geologists, including the extensive Darwin correspondence. Knowing not only whom Darwin knew and interacted with, but also the physical proximity of these various people and the places where they gathered, serves to emphasize the physical reality—one might say "intimacy"—of their intellectual life. The tenor of that life, and the concrete meaning of the interrelations among these individuals—what it would have meant to have been on bad terms with one or other of them, for example—is vividly displayed in Rudwick's cartographic representation. This is not metaphorical space, but geographical space, and Rudwick here demonstrates its importance.

Notice, too, that an example of this kind adds another dimension to the ways in which historians, and certainly historians of science, think. Historians of science like looking at scientific ideas, and the immediate way of doing that is through intellectual history, the history of ideas. In the past thirty years this approach has been supplemented, or overtaken, by social history of ideas and sociology of knowledge, where the social locations of people are used as ways of understanding their intellectual positions. But Rudwick's use of geographical locations is in some ways distinct from that of social and intellectual locations, which usually make little explicit reference to places in a concrete, spatial sense. (There are, of course, many qualifications that could be added to that remark, primarily for social-historical and sociological approaches, but the specific point still holds.) A number of scholars in the history and sociology of science since Rudwick have been drawing maps of various kinds. Some of these have been regional, national, continental, and even global in scope.[11] But the scale itself makes a significant difference to what these maps can actually do for us.

Take another map well known to historians of science, from Shapin and Schaffer's 1985 *Leviathan and the Air-Pump.* Figure 2.2 reproduces the authors' depiction of the locations of known air pumps in Europe in the 1660s. Between each of these places are traced out lines that represent journeys by particular individuals. Someone who had successfully made an air pump in one place literally carried the hands-on knowledge of how to do it to another place, where the locals could be taught to do it too. Shapin and Schaffer's central point, which concerns tacit skills and their transmission through social interactions, did not, in fact, require the drafting of a map with lines drawn on it, although the result makes a striking and concrete representation. A table, listing the places and the people who performed the knowledge transference, would have done as well. The actual relative distances between the various cities on the map are irrelevant; all that matters is that people traveled between them. This map, then, is rather less specifically "maplike" in its use than is Rudwick's map of central London. Distance is made real in Rudwick's map by virtue of the means used to traverse those distances, something that plays little significant role in the air-pump map; for Darwin and his colleagues, the importance lies in the fact that they could all walk to each others' houses.

Rudwick's Darwin map also exemplifies another point concerning the relationship between knowledge and space: maps that show things relevant to

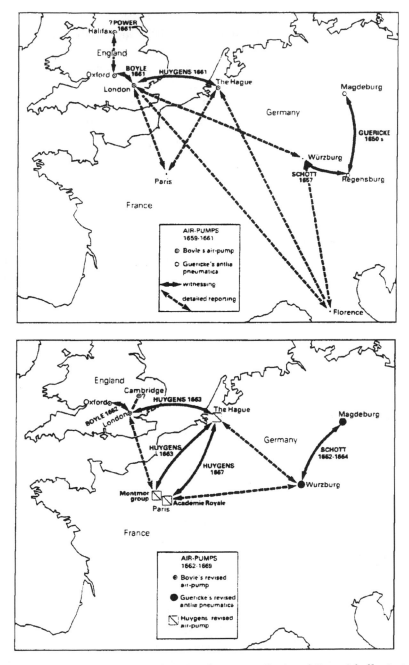

Figure 2.2. Air pumps in Europe in the 1660s, from Steven Shapin and Simon Schaffer, *Leviathan and the Air-Pump: Hobbes, Boyle, and the Experimental Life* (Princeton: Princeton University Press, 1985; reprint ed., 1989), 228. By permission of Princeton University Press.

the establishment of ideas do not necessarily need to detail a "spread" of ideas from some central point of origin to a multiplicity of new, increasingly distant places (a "diffusion" model, whether active or passive).[12] Instead, the ideas can become constituted over an entire distributed region of space, or at least represented so: Rudwick's map locating his various actors is intended to help us understand how the interactions between them helped to constitute particular ideas and modes of expression that these people wrote down in various forms. The ideas might seem to have appeared from, say, Darwin's mind, according to an idealist, intellectualist reading, but to make sense of them historically, one has to see how they could become constituted among the group of people indicated on the map, a constitution defined by the physical constraints and facilitations that the map displays.

Furthermore, to take a different tack, if one's object of investigation were the establishment of a new form of social organization, that process need not be understood as originating in one place and spreading by emulation to other sites. Instead, a kind of Marxist structuralist account of the Industrial Revolution, for example, might represent widespread change in modes of production and the setting up of factories all over northern England: these developments might then be seen to have given rise to a new class consciousness—the making of the English working class, à la E. P. Thompson.[13] But note that, although this may not have been the kind of process that springs up in one place and spreads elsewhere in serial fashion, it can still be mapped; however, the time dimension works differently here than in the "diffusion" model, and the significance of space is different, too— specific or relative spatial distances are less important (although not entirely unimportant).

One of the central puzzles in science studies is the question of how scientific ideas can come to seem universal.[14] Science studies has spent a lot of time over the past quarter century showing how scientific belief, scientific knowledge, gets created in particular local settings, settings that enable us to understand how and why scientific ideas originate or how they become stabilized within a particular community. That stress on localism, which mirrors the longstanding vogue for "microhistory" among cultural historians, leaves such work open to the charge that it ignores one of the most characteristic and remarkable features of modern science—its apparent universality: scientific knowledge appears to be equally valid everywhere, not just in its place of origin. This claim is at the heart of standard arguments for scientific realism, among other things, and so the sociologist of scientific knowledge had better have an answer. The basic solution to the apparent difficulty concerns the exporting of local conditions—we might remember Shapin and Schaffer's map of air-pump distribution in the 1660s; the skills had to be moved around before an effective apparatus could become widely distributed, and only then could the scientific facts about air pumps and what they might show about nature become widely distributed themselves. Locally created knowledge must be exported to many local sites, being reproduced in each, and only in that way is the illusion of universality created.

As the air-pump example itself indicates, this is a perspective that lends itself to graphical representation, including geographical mapping—the mapping out of networks connecting local sites, seen as nodes in the network. The most influential instantiation of this approach is the Actor Network Theory (ANT) of Michel Callon and Bruno Latour, another invention of the 1980s that still leaves its mark. In its general form, ANT is rather vague as to what its networks are, or can be, composed of; the only real ontological category in terms of which the nodes of the network can be described is that of "actant," a kind of instantiated agency. What the lines that connect the nodes represent is even less clear, and seems to involve co-optations and resistances of various kinds (in practice, social and individual interests and purposes, although Latour would reject such a naive sociological interpretation).[15] But the clearest and simplest examples of such networks are, unsurprisingly, the literal geographical ones, where the nodes correspond to places and the lines to spatial distances. Latour, drawing on slightly earlier work by John Law, talks in his 1987 book *Science in Action* about "centers of calculation," and one setpiece illustration of this notion involves a late-eighteenth-century voyage of discovery.[16]

This voyage was carried out by the French explorer Lapérouse in the western Pacific, on behalf of the French crown. Leaving aside most of the quite significant details, Latour's basic point is that by sending quantitative geographical positions back to the court at Versailles, Lapérouse enabled geographers at that center of calculation to produce maps of the region that gave them greater control over those distant places when they sent out subsequent expeditions. The new expeditions would send back yet more information to create even greater detailed control, in an endless cycle. The image is of a spider's web, with the lines extending on a map out to a multitude of distant locales, each of which has come under the control of the center by virtue of what the map is used to represent by that center and its proxies (the people who actually sail out to the faraway places). This is a particularly elegant illustration of the making of a kind of local knowledge that is also, at the same time and necessarily, universal.

Taking this model of networked interconnectedness literally implies, of course, a corresponding understanding of what a true "revolution" would involve regarding the things that it represents: a revolution would have to be a catastrophic, wholesale restructuring and reconfiguration of the entire network. In fact, in uses of the ANT model such issues are seldom examined, it being much more plausible to see change in terms of piecemeal shifts in the elements of the network, or in what Latour calls "displacements."[17] Perhaps typical uses of maps themselves could be expressed in similar terms, when a particular type of map begins to be used for purposes somewhat different from those for which it was originally made. Such a process might involve changes in conventions of representation, as well as changes in the things being represented, as one predominant use gives way to another, or as two uses begin to split apart to yield distinct genres of map. Then the practices of mapping would themselves represent shifts in social practices related to countless other kinds of shift.

Consider the following as an example of the continuity underlying classic political revolutions: in the later eighteenth century, prior to the French Revolution, a massive mapping project succeeded in producing a map of the whole of France in such a way as to include, among other things, economically significant information. Nearly the entire map, all except Brittany, was published in 1783; when Brittany was eventually printed, the map ran to 182 large sheets (scale 1:86,400). This map of the territory and polity of France was a centralized project promoted and, on the rare occasions that the crown could afford it, partly financed by the French crown. Nonetheless, following the Revolution, the royalist map proved to be of central value to the National Assembly in 1790. It was used for the political redivision of France into the departments that structured the new French state.[18] The map, in other words, was an instrument of change even though it was entirely a product of the prerevolutionary dispensation. The reason for this being an unsurprising outcome is, of course, that many of the purposes for which the map was created were continuous with similar purposes on the parts of political leaders and political interests following the Revolution; these purposes bear witness to the ways in which the Revolution was more of a realignment within a complex of ongoing processes than a fundamentally fresh start.

SPATIAL HISTORY AND NATURAL HISTORY

Spatial considerations, mapping, and historical change, whether or not that change is labeled "revolutionary," play a central role in some recent work in the history of science that concerns natural history. A number of scholars have studied ways in which botanical work, particularly collection and classification, relates to imperial themes in European national histories, including especially studies of Kew Gardens in the nineteenth century.[19] Some studies on the eighteenth century, however, show in starker, because simpler, detail, the sorts of enterprises that continued on a larger scale later on.

This work includes studies of Joseph Banks and his botanical voyaging (in the case of British imperialism), while Lisbet Koerner has completely rewritten our picture of Linnaeus and his natural-historical project in terms of a curious inverted colonial endeavor aimed at bringing the rest of the world inside the boundaries of Sweden.[20] Also, Kapil Raj has recently investigated early-eighteenth-century French botanical projects related to mercantile endeavors, chiefly in the Indian Ocean.[21]

In an article in the French popular-science magazine *La Recherche*, Raj presents a map that amounts to a literal, spatial rendering of an abstract idea employed by Latour: the "obligatory passage point."[22] The map shows crucial ports of call for trading ships in the Indian Ocean region in the eighteenth century, places that were essential for the routine conduct of European trade in the region. Among other things, these ports, from the Cape of Good Hope eastward, needed to be able to supply the ships that passed their way with food and, especially, with medicines. The Dutch in this region had already moved quickly in the seventeenth century to identify Asian plants that possessed medicinal properties. By the beginning of the eighteenth century, they had already begun to introduce such plants into places lacking them, such as the Cape of Good Hope and Batavia (Jakarta), locations that were climatically similar to the plants' native habitats. These botanically colonized ports were important "obligatory passage points" for merchant ships out from Europe— all the more obligatory now that they had been turned into depots supplying needed drugs and medicines. Systematic Dutch botanical treatises dealing with these tropical plants were published as part of this enterprise.[23]

By contrast, the French were rather slower off the mark in getting to grips with tropical botany, in large part owing to the leading French botanist of the early eighteenth century, Antoine de Jussieu, who was in charge of the Jardin du Roi in Paris (the forebear of the modern Museum of Natural History). Jussieu had the opportunity to help promote this kind of botanical imperialism, as Raj explains, but he ignored it because of his conception of the nature of botanical variety and its geographical distribution. He reckoned that every region of the world contained essentially the *same* variety of plant species, even when their appearances differed. Consequently, communications from botanists abroad were not of fundamental botanical interest to him, since the most they could do was to facilitate the identification of foreign plants with their French equivalents—including pharmaceutical equivalents. So, for example, Jussieu said that Ipecacuanha, the handy but exotic purgative, was nothing but a version of the simple European violet. Jussieu cared little for foreign botany, because France already had it all.[24]

Jussieu's attitudes and beliefs were not idiosyncratic, as Koerner's studies of Linnaeus have shown clearly. One thing that can easily be overlooked when looking at taxonomic charts in natural history from the seventeenth century onward is how they are entirely separated from notions of place—notions of geographical distribution around the globe.[25] Linnaeus's versions of botanical and zoological classification in the middle of the eighteenth century is entirely in keeping with this approach, and his explicit views on the geographical ranges of plants and animals is quite analogous to Jussieu's. Although Linnaeus did not go as far as to hold that all the plants in the world could be found in some form or another inside the confines of his native Sweden, he did think that any plant in the world could be naturalized into growing happily in his own country. Like Jussieu, Linnaeus did not regard plants as geographically specific; they were kinds of beings in the world, in a sort of idealist, Platonic sense. Koerner tells the tale of how Linnaeus made several attempts over the years to get hold of tea plants from China, so that he could attempt to raise them at home and gradually acclimatize them to Sweden, thereby making Sweden self-sufficient in what was otherwise an expensive import. Linnaeus had plans to do this with other plants too, and his biggest project, happily encouraged by the Swedish government, was to try to get crops like wheat to grow not just in fertile southern Sweden, but also in its barren northern region of Lapland.[26]

As Koerner presents it, Linnaeus's project was a kind of internal imperialism, a counterpart to the external imperialism of Spain followed by France, Britain, and the Netherlands. In Linnaeus's version, the exploitable resources of other lands would be brought home; rather than having trading posts on the coast of India, or plantations in Virginia, the Swedes, he hoped, could do just as well growing tea or tobacco near Stockholm. As for the exploitation of Lapland, this was a matter of making the maximum use of domestic resources for the same purposes, together with an element of internal colonization. Linnaeus's plans were a complete failure, but the motivations were the same as those of successful imperial enterprises.

The places of plants, therefore, or their *non*-places, were significant issues in eighteenth-century imperialist projects; if imperialism can be regarded as a form of enforced revolutionary change wrought upon colonized lands and peoples, here is a direct and substantive correlation between revolution and issues of place. From a scientific standpoint too, as Kapil Raj's map of the Indian Ocean indicates, geographical issues, the employment of maps and of conceptions of place, space, and distance, made geography itself, as both a discipline and as a practice, directly relevant to the making of natural knowledge in botany—this quite apart from the expeditions in the 1730s to Peru and Lapland to collect measurements relating to the shape of the earth and to Newtonianism.[27] Other, more elaborate cases of the intersection of geography and natural knowledge occur in the nineteenth century, with the Humboldtian project of grasping the diversity of the globe in botany, zoology, and geology, including such topics as geomagnetism.[28] Astronomy, of course, had always been intimately linked with geography, and expeditions to remote parts of the world by European astronomers first became common in the eighteenth century, continuing with ever-greater vigor in the nineteenth, for observing such things as transits and solar eclipses.[29]

GEOGRAPHY AND ASTRONOMY

The relationship between astronomy and geography is worth remembering, because it establishes an important issue regarding the kind of knowledge that geography, as an academic discipline, represented all the way up to the nineteenth century. Geography, like astronomy, counted from classical antiquity onward as a mathematical science. For a long time, at least through the seventeenth century, that had real practical consequences: mathematical sciences were commonly understood in early modern European universities, following a particular reading of Aristotle's philosophy, as failing to provide any knowledge of the inner natures, or essences, of the things they talked about. They discussed measurable, quantitative features of things, but could not address questions of what those things really *were*—that was the job of "natural philosophy."[30]

Geography on the Greek model was a kind of offshoot of spherical astronomy (celestial globes rather than terrestrial ones are therefore, in a sense, more prototypical). Because of this, geography incorporated strict mathematical restrictions on what its proper cognitive competences should be (such as specifying the precise locations of places, as is done in Ptolemy's *Geography*).[31] Qualitative accounts of parts of the world, in contrast to this quantitative enterprise, were therefore hived off into their own special discipline: chorography was a field that fitted in many ways the "travelers' tales" model of place description and provided little in the way of causal explanation.[32]

There is a sense in which the classic "revolution" of the Scientific Revolution in the seventeenth century was all about establishing descriptive knowledge, especially mathematically descriptive knowledge, as real natural philosophy. In other words, descriptive disciplines such as astronomy and geography acquired elevated status in the eighteenth century as potentially explanatory enterprises. Place did not merely describe, as previously, *where* things were; place could now, perhaps, help to explain *why* things were.

The image of the terrestrial globe is an attractive model in terms of which to understand the ways whereby knowledge enterprises involve place and locality. Change in human knowledge systems, as well as in social systems, and whether catastrophic or gradual, needs to be traced in its development by way of localities. Those localities acquire their uniqueness from the intersecting array of contingencies that happen to make them different from other localities. Such differences, understood in terms of historical contingencies, are powerful resources for explaining grander issues that might at first appear to be too big, too abstract, or too disembodied to come to grips with. Like a globe, any putative revolution is a finite topical subject of investigation; but traveling around in it and mapping out its local particularities is a field of unlimited possibilities.

...

Notes

1. Thomas S. Kuhn, *The Structure of Scientific Revolutions,* 2nd ed. (Chicago: University of Chicago Press, 1970), 208.

2. Eric J. Hobsbawm, "Revolution," in *Revolution in History,* ed. Roy Porter and Mikuláš Teich (Cambridge: Cambridge University Press, 1986), 5–46.

3. Ibid., 7.

4. Alexis de Tocqueville's *L'ancien régime et la révolution* is the classic example from the nineteenth century of this argument.

5. Compare the classic discussions of this point in Imre Lakatos and Alan Musgrave, eds., *Criticism and the Growth of Knowledge* (Cambridge: Cambridge University Press, 1970).

6. Mary B. Hesse, *The Structure of Scientific Inference* (London: Macmillan, 1974), presents a "network model" for discussing more formally the interconnectedness of concepts.

7. See Lynn Hunt's introduction to *The New Cultural History,* ed. Lynn Hunt (Berkeley and Los Angeles: University of California Press, 1989); also, as a classic exemplar, Carlo Ginzburg, *The Cheese and the Worms: The Cosmos of a Sixteenth-Century Miller,* trans. John and Anne Tedeschi (1976; Harmondsworth: Penguin, 1982).

8. See, for a good example of this by now well-established scholarly genre, Dominick LaCapra, *Soundings in Critical Theory* (Ithaca: Cornell University Press, 1989).

9. Kant described the recognition of the transcendental categories' necessity as his "Copernican Revolution" in the *Critique of Pure Reason* (1781). On the history of the theme in science, see I. Bernard Cohen, *Revolution in Science* (Cambridge: Harvard University Press, 1985).

10. The term was used in the titles of a number of Hill's books from 1949 onward, most famously in Christopher Hill, *Intellectual Origins of the English Revolution* (Oxford: Clarendon, 1965).

11. Most similar to Rudwick's is Susan C. Lawrence, "Entrepreneurs and Private Enterprise: The Development of Medical Lecturing in London, 1775–1820," *Bulletin of the History of Medicine* 62 (1988): 171–92; see also Steven Shapin's use of a map of central London in Shapin, "The House of Experiment in Seventeenth-Century England," *Isis* 79 (1988): 403. On these themes, see especially David N. Livingstone, *Putting Science in Its Place: Geographies of Scientific Knowledge* (Chicago: University of Chicago Press, 2003). Recent work by historians of science on themes relating to scientific "spaces" both literal and metaphorical appears in *Making Space for Science: Territorial Themes in the Shaping of Knowledge*, ed. Crosbie Smith and Jon Agar (Basingstoke: Macmillan, 1998).

12. On the diffusion model and its methodological shortcomings, see Bruno Latour, *Science in Action: How to Follow Scientists and Engineers through Society* (Cambridge: Harvard University Press, 1987), 141–45, 164–65.

13. E. P. Thompson, *The Making of the English Working Class* (London: Gollancz, 1963).

14. Jan Golinski calls this "the problem of construction," in *Making Natural Knowledge: Constructivism and the History of Science* (Cambridge: Cambridge University Press, 1998), 33 and passim.

15. Steven Shapin presents such a reading in his review of Latour's *Science in Action*, "Following Scientists Around," *Social Studies of Science* 18 (1988): 533–50.

16. John Law, "On the Methods of Long Distance Control: Vessels, Navigation and the Portuguese Route to India," in *Power, Action and Belief: A New Sociology of Knowledge?* ed. John Law (London: Routledge and Kegan Paul, 1986), 234–63; Latour, *Science in Action*, 215–23.

17. For example in Bruno Latour, *The Pasteurization of France*, trans. Alan Sheridan and John Law (Cambridge: Harvard University Press, 1988), 68.

18. Charles Coulston Gillispie, *Science and Polity in France at the end of the Old Régime* (Princeton: Princeton University Press, 1980), 115–17; on related issues, see also Ken Alder, "A Revolution to Measure: The Political Economy of the Metric System in France," in *The Values of Precision*, ed. M. Norton Wise (Princeton: Princeton University Press, 1995), 39–71.

19. Most prominently, Lucile H. Brockway, *Science and Colonial Expansion: The Role of the British Royal Botanic Gardens* (New York: Academic Press, 1979); Richard Drayton, *Nature's Government: Science, Imperial Britain, and the "Improvement" of the World* (New Haven: Yale University Press, 2000).

20. See Lisbet Koerner, *Linnaeus: Nature and Nation* (Cambridge: Harvard University Press, 1999). On botanical voyaging, see especially work by John Gascoigne, in particular his *Science in the Service of Empire: Joseph Banks, the British State and the Uses of Science in the Age of Revolution* (Cambridge: Cambridge University Press, 1998). An important collection of related articles is David Philip Miller and Peter Hanns Reill, eds., *Visions of Empire: Voyages, Botany, and Representations of Nature* (Cambridge: Cambridge University Press, 1996).

21. Kapil Raj, "Histoire d'un inventaire oublié," *La Recherche*, no. 333 (July/August 2000): 78–83, map on page 81. A longer treatment of the same subject is Raj's "Surgeons, Fakirs, Merchants, and Craftspeople: Making L'Empereur's Jardin in Early Modern South Asia," in *Colonial Botany: Science, Commerce, and Politics in the Early Modern World*, ed. Claudia Swan and Londa Schiebinger (Philadelphia: University of Pennsylvania Press, 2005), 252–69. A recent article that deals with maps and trade routes in the Indian Ocean around 1700 is Benjamin Schmidt, "Inventing Exoticism: The Project of Dutch Geography and the Marketing of the World, circa 1700," in *Merchants and Marvels: Commerce, Science, and Art in Early Modern Europe*, ed. Pamela H. Smith and Paula Findlen (New York: Routledge, 2002), 347–69.

22. For a particularly clear exposition of this concept, see Michel Callon, "Some Elements of a Sociology of Translation: Domestication of the Scallops and the Fishermen of St. Brieux Bay," in *Power, Action and Belief: A New Sociology of Knowledge?* ed. John Law

(London: Routledge and Kegan Paul, 1986), 196–29; also Latour, *The Pasteurization of France,* 43–49.

23. Raj, "Histoire d'un inventaire oublié"; also Schmidt, "Inventing Exoticism."

24. See Raj, "Histoire d'un inventaire oublié," and especially "Surgeons, Fakirs."

25. Noted in Peter Dear, *Revolutionizing the Sciences: European Knowledge and Its Ambitions, 1500–1700* (Princeton: Princeton University Press, 2001), 128–29.

26. Koerner, *Linnaeus;* see also Lisbet Koerner, "Purposes of Linnaean Travel: A Preliminary Research Report," in *Visions of Empire: Voyages, Botany, and Representations of Nature,* ed. David Philip Miller and Peter Hanns Reill (New York: Cambridge University Press, 1996), 117–52.

27. Various discussions of this matter may be found in Rob Iliffe, "'Aplattisseur du Monde et de Cassini': Maupertuis, Precision Measurement, and the Shape of the Earth in the 1730s," *History of Science* 31 (1993): 335–75; John L. Greenberg, *The Problem of the Earth's Shape from Newton to Clairaut: The Rise of Mathematical Science in Eighteenth-Century Paris and the Fall of "Normal" Science* (Cambridge: Cambridge University Press, 1995); Mary Terrall, *The Man Who Flattened the Earth: Maupertuis and the Sciences in the Enlightenment* (Chicago: University of Chicago Press, 2002).

28. Susan Faye Cannon, "Humboldtian Science," in *Science in Culture: The Early Victorian Period* (New York: Science History Publications, 1978), 73–110; Michael Dettelbach, "Humboldtian Science," in *Cultures of Natural History,* ed. Nicholas Jardine, James A. Secord, and Emma C. Spary (New York: Cambridge University Press, 1996), 287–304.

29. Harry Woolf, *The Transits of Venus: A Study of Eighteenth-Century Science* (Princeton: Princeton University Press, 1959); Rob Iliffe, "Science and Voyages of Discovery," in *The Cambridge History of Science,* vol. 4, *Eighteenth-Century Science,* ed. Roy Porter (Cambridge: Cambridge University Press, 2003), 618–45; Alex Soojung-Kim Pang, *Empire and the Sun: Victorian Solar Eclipse Expeditions* (Stanford: Stanford University Press, 2002).

30. See Dear, *Revolutionizing the Sciences,* 65–66; Peter Dear, *Discipline and Experience: The Mathematical Way in the Scientific Revolution* (Chicago: University of Chicago Press, 1995). Lesley B. Cormack, *Charting an Empire: Geography at the English Universities* (Chicago: University of Chicago Press, 1997), provides a study of geography as part of the academic mathematical sciences.

31. A convenient recent resource is *Ptolemy's Geography: An Annotated Translation of the Theoretical Chapters,* ed. and trans. J. Lennart Berggren and Alexander Jones (Princeton: Princeton University Press, 2000). During the renaissance, Ptolemy's *Geographia* was much more popular, judged by the number of editions as well as by numerous surviving copies, than his nowadays more celebrated astronomical work, the *Almagest.*

32. The subject of travelers' tales is discussed, with further references, in Steven Shapin, *A Social History of Truth: Civility and Science in Seventeenth-Century England* (Chicago: University of Chicago Press, 1994), 243–58. See, on geography, astronomy, and mapping as mathematical pursuits in this period, Jim Bennett, "Projection and the Ubiquitous Virtue of Geometry in the Renaissance," in Smith and Agar, *Making Space for Science,* 27–38; also Bennett's "Practical Geometry and Operative Knowledge," *Configurations* 6 (1998): 195–222, and "The Challenge of Practical Mathematics," in *Science, Culture and Popular Belief in Renaissance Europe,* ed. Stephen Pumfrey, Paolo L. Rossi, and Maurice Slawinski (Manchester: Manchester University Press, 1991), 176–90.

NATIONAL STYLES IN SCIENCE

A Possible Factor in the Scientific Revolution?

...

John Henry

NATIONAL STYLES AND THE HISTORIOGRAPHY OF SCIENCE

It would be wrong, as well as politically incorrect, to assert that there are natural differences between peoples of different nationalities. Certainly there are no significant biological differences between the English and the French, say. Nevertheless, just as each of us is shaped by our own individual life histories, so the people of a nation are shaped by the history of their country. There can be no doubt that the vicissitudes of historical contingency over the centuries have ensured that the collective experience of the English has been very different from that of the French, with the result that, generally speaking, the English and the French are very different from one another. Try as we might to resist making glib assumptions about what the Norwegians, or the Italians, or whoever, are like, we usually recognize fairly consistently (though not necessarily truthfully) what somebody means if they refer to the Italian personality or the Nordic type. It may well be that what the French regard as the Greek type does not conform to the way the Germans typify the Greeks, but this does not so much invalidate any suggestion that there is a Greek type as it confirms the typical differences between the French and the Germans.

Such historically shaped national personality types have even inspired some notable historical theories to explain them. After detailing the historical development of the great Italian city-states in the Renaissance, Jacob Burckhardt saw this history as the principle factor in "the development of the individual" and "the awakening of personality":

In the character of these States, whether republics or despotisms, lies, not the only, but the chief reason for the early development of the Italian. To this it is due that he was the first-born among the sons of modern Europe.[1]

Similarly, but much more thoroughly, the historical sociologist Norbert Elias showed how European national traits, among other things, emerged as part of the "civilizing process" that he, like Burckhardt, saw as linked to the precise political development of the nation-states from the Renaissance onward. More recently, John Hale, in his magisterial survey *The Civilization of Europe in the Renaissance,* used the historical appearance of beliefs about different national personality traits as further proof of the emergence of nationalism in the Renaissance.[2]

Given the generally acknowledged importance of social and cultural context in the formation and development not only of scientific institutions and practices but also in the formation and development of scientific knowledge, it seems reasonable to expect there to be such a thing as national styles in science. Although never a prominent aspect of the historiography of science, this assumption has indeed attracted a number of studies concerned with national differences in styles of scientific thinking. The recently published *Reader's Guide to the History of Science* has a survey of literature entitled "National Styles of Reasoning." The author of this short piece, Michael Donnelly, expresses some frustration at the ragbag of different approaches to "national styles" in the available literature, confirming my own feeling that this is not yet a fully matured aspect of the historiography of science. For one thing, it is not clear to me how *nationality* asserts itself in these circumstances: perhaps we should simply talk in terms of locality.[3] But Donnelly is in no doubt of the validity of the notion, and of its value in understanding the development of modern science. He also quotes Bertrand Russell to marvelous effect, regretting the intrusion of national attitudes even into experimental work:

Animals studied by Americans rush about frantically, with an incredible display of hustle and pep, and at last achieve the desired result by chance. Animals observed by Germans sit still and think, and at last evolve the solution out of their inner consciousness.[4]

Lorraine Daston and Michael Otte in their introduction to *Style in Science,* the special 1991 issue of *Science in Context,* say that "[n]ational styles in science present a rare example of a phenomenon that eludes precise description, but nonetheless lends itself to detailed causal analysis." Furthermore, they argue,

[i]n the case of national styles, affinity and geography are brought into alignment by the shared experience of education, career trajectories, and professional organizations that teach, articulate and reward a certain kind of science. . . . The institutions that create and sustain a style may or may not be national, but once the nation takes charge of this aspect of culture, its prevailing values often color scientific style.[5]

The earliest studies of national styles in science were usually concerned with differences arising from local variations in the organization of science, but later studies began to consider the impact of national styles on the sociology of scientific knowledge.[6]

Jonathan Harwood has seen one major source of differences in national styles in differing ways of demarcating disciplinary boundaries. Differences between genetics as it was pursued in America and in Germany, he suggests, can be reduced to differences about the theoretical scope of genetics arising from different ways of demarcating genetics from embryology, evolutionary biology, and so on.[7] Harwood suggests further that stylistic differences will most likely appear in weakly institutionalized and therefore younger disciplines. Well-established fields, he believes, are more likely to have been homogenized by the undeniably internationalist ethos of modern science. Certainly, those historical studies of science conducted so far, even if they are not obviously concerned with disciplinary demarcation, seem to suggest that new approaches to natural knowledge are most likely to bring out stylistic differences between the nations involved. An obvious example here is provided by the response of British chemists to the innovations of the French chemists, as usually summed up by the reaction of Joseph Priestley to Antoine Lavoisier.[8] It seems hardly coincidental that the theory that ill-health was the result of internal conflicts within the newly discovered cellular structure of the body and the rival belief that it was the result of invasion of the body from outside by newly discovered microbes called up champions in late-nineteenth-century Europe who were German (Rudolf Virchow) and French (Louis Pasteur) respectively.[9] Much research has shown why, not only Charles Darwin and Alfred Russel Wallace had to be Englishmen, but so did Charles Wells and Patrick Matthew, who also arrived independently at the principle of natural selection.[10] Similarly, Paul Forman's famous paper on Weimar physics shows how major aspects of the new quantum theory seemed much more plausible and natural to post–World War I Germans than to nationals of other European countries. Furthermore, Andrew Warwick has shown how British mathematical physicists, trained in

Maxwellian mathematics, had the utmost difficulty in recognizing any value in Einstein's relativity theory.[11]

If national styles are most evident in the case of new ideas, and reactions to them, or in newly formed subdisciplines breaking away from an older tradition, it seems highly likely that they should have manifested themselves on numerous occasions during the period known to historians (though to some only through the intellectual equivalent of gritted teeth) as the Scientific Revolution. Here was a period of time when almost everything in science, or natural philosophy, was new. This was the period when Scholasticism, the traditional natural philosophy of Aristotelianism, was giving way to new methods of doing science, to new theoretical visions of how the world picture should look, and to new discoveries that went hand in hand with those new visions but were incompatible with the old world picture. This was a period of massive redrawing of the boundaries between different parts of knowledge, changing old relationships and creating many new subjects. In principle, therefore, it should be easy to point to different national styles as an aspect of the Scientific Revolution. Indeed, in the recently published *Encyclopedia of the Scientific Revolution,* there is an article entitled "Styles of Science: National, Regional and Local" by Maurice Crosland, editor of one of the earliest collections of comparative essays organized along nationalistic lines. Unsurprisingly, Crosland declares that "there is a strong argument for the existence of different 'national styles' in science."[12] Similarly, Roy Porter and Mikuláš Teich in their introduction to *The Scientific Revolution in National Context* (1992), remark that history of science "needs to evaluate the role of particular and disparate national and cultural traditions of thinking and mental work . . . that operated within discrete language groups and under distinctive political jurisdictions." They point out that

> the relations between the staggering scientific changes wrought between Copernicus and Newton, and, in the widest sense, the political diversity and change, the chaos and "search for stability" that characterized Europe in the century after the Reformation remain neglected. And this is so even though every historian of European politics emphasises the magnitude of the transformations in the nature of the state and the bases of princely power that Europe underwent during the early modern centuries.[13]

The Scientific Revolution, after all, began in the late Renaissance period and continued to run its course until the end of the seventeenth century. It covered, therefore, the very period that the considerable historical intuitions of Burckhardt and Elias perceived as the crucial "moment" in modern European history—*world* history—when people began to see themselves as personalities, whose individual (and individualistic) contributions to the running of society and the state constituted and consolidated the emerging nation-states. Clearly, national styles only become possible with the rise of nations, but they are also most important when the national identity needs to be forged. Here again, therefore, it seems likely that national styles in science should be discernible in the period of the Scientific Revolution.

I would like to suggest, however, that national styles were not simply concomitants of the Scientific Revolution, but were also causative factors in the development of the Scientific Revolution. I take inspiration here from a short but highly suggestive paper delivered at the Fifteenth International Congress of the History of Science, in Edinburgh in 1977. In this paper, the author, M. G. Yaroshevsky, argues that "schools in science are an essential constant factor of its progress."[14] He was talking, of course, about schools of thought: "a scientific trend which has emerged in one country and differs in its approach to various problems, concepts, and methods, from the practices of scientists in other countries." Such schools, he pointed out, "are called national."[15] Taking as an example the interactions between the German, French, and Russian schools in physiology in the late nineteenth century, Yaroshevsky shows how Ivan Mikailovich Sechenov, founder and leader of the Russian school, learned from the other national schools of thought in physiology, and thereby "polished his own programme and his own approach, which determined the originality of the Russian school."[16] "In the real historical process", Yaroshevsky writes,

> the splitting up of the community into schools and their constant opposition meant the continuous building up of science through the interaction (confrontation, intersection, synthesis) of the different trends of scientific thought rather than the disintegration of the subject-logical basis of science.[17]

In Yaroshevsky's assessment, the different schools of physiology each had their own preoccupations and practices, but he denies the allegedly Feyerabendian adage "as many schools so many truths." On the contrary, he suggests,

> [t]he multitude of schools, provided they are truly scientific reveal certain mechanisms which make it possible to consolidate knowledge . . . irrespective of the assigned objectives of these schools. The different approaches exhibited by the national schools did not result in the dissolution of physiology as a uniform discipline, but determined its outstanding progress.[18]

While Yaroshevsky's vision is avowedly not Feyerabendian, it seems to be decidedly Lakatosian. Like Imre Lakatos, Yaroshevsky seems to believe that criticism (rather than full-blown Popperian falsificationism) leads to the growth of knowledge.[19] It is precisely the interplay between different schools of thought, with significant differences in their perspectives and approaches, that leads to the progress of scientific knowledge. This chapter is not an exercise in the philosophy of science, and so this is not the place to again go over the pros and cons of these (and other) philosophical attempts to codify how scientific discovery takes place. Suffice it to say that for our purposes, as historians, Yaroshevsky's proposal seems entirely plausible and workable and is, therefore, potentially highly fruitful for understanding developments in the history of science, and particularly, as I have indicated, during the period of the Scientific Revolution.

Similar suggestions can be found in Kostas Gavroglu's recent study of the origins of physical chemistry, where he emphasizes the role of scientific controversy and says he is "not averse to the suggestion" of a "'national level' where particular philosophical, cultural, and aesthetic trends have been 'condensed' into the practice of the community." He concludes that "the many controversies that came into being during the long developmental period of physical chemistry are but indications of the cultural pluralism that determined such a development."[20] Even more forcefully, Gideon Freudenthal has seen a seventeenth-century controversy on the compounding of forces as a single element in "a much more comprehensive controversy involving many scholars belonging to different 'schools' and ranging over many topics. At stake were the concept of force in mechanics, the notion of a new kind of magnitude ('vectors') that seemed incompatible with the previously accepted notions, the role of conservation principles in science, among others." For Freudenthal, this wider controversy partly explains the richness of innovation in these areas:

> [T]he resolution of a controversy rarely simply proves one side right, the other wrong. Rather, a genuine resolution of a controversy is achieved within a reformed conceptual system that supersedes the system in which the controversy arose and yet [shows] a relatively greater influence of one of the positions over the other. This is the reason why controversies may prove productive for the theory or discipline involved.[21]

In the following section, I want to try to bring out differences in national styles during the early modern period by looking at some characteristic differences between English and Continental, particularly French, science in the seventeenth century. I hope also to indicate how these differences could sometimes be brought together in a fruitful way to lead on to new developments. What follows, however, should not be seen as a systematic attempt to explain the Scientific Revolution in terms of national styles. The aim—complementary to Withers's concerns in chapter 4 to think geographically about the Scientific Revolution—is simply to suggest that differences in national styles may be another factor worth exploring and, I hope, to stimulate others to explore it further. I do not claim, therefore, to have discovered the answer to the problem as to why the Scientific Revolution occurred only in post-Renaissance western Europe, but merely wish to suggest that differences in national styles in early modern European natural philosophy may repay further historical research.

NATIONAL STYLES IN THE SCIENTIFIC REVOLUTION

Differences in national styles seem to be particularly evident in the seventeenth century, the period when a new way of doing science was being developed in western Europe. As the traditional Scholasticism of the premodern period was being rejected and replaced by a "new philosophy," different national styles of thinking gave rise to markedly different conceptions of the correct way to approach an understanding of the natural world. If we focus upon the situations in England and France, we can see that, to a large extent, these differences had their origins in the very different religious and political histories of the two nations. Arguably the most distinctive feature of the new approach to natural philosophy in the seventeenth century was the rejection of ancient authority, along with a new emphasis upon the importance of observation and other means of determining natural phenomena for oneself. Experimentalism has long been recognized, therefore, as a hallmark of the new science of the seventeenth century. The experimental philosophy in England, however, was markedly different from that professed in the rest of Europe. Before we look at the peculiar way in which the English developed their experimental philosophy, and the reasons for it, let us consider the development of experimentalism in France.

From about the 1620s, the experimental approach to understanding nature began to take off. To begin with, however, it emerged in the universities among professors who were still committed to the fundamentals of Aristotelianism. The Scholasticism of the universities as it had been developed since the thirteenth century had recently suffered numerous blows. A number of new physical discoveries ran counter to Aristotle's teachings, and the Renaissance recovery of the writings of other ancient philosophers suggested numerous alternative accounts of natural phenomena. The dominance of Aristotle seemed to be over. Nevertheless, the tendency in the universities, particularly in a Roman Catholic country like France where Aristotelianism was bound up with religious and political orthodoxy, was to make adjustments and refinements to Aristotelian theory to accommodate the new changes. Since Aristotle himself always emphasized the importance of the senses for establishing the truth, it was easy for university professors of natural philosophy to lay claim to being experimentalists and Aristotelians, and the theory was always flexible enough to accommodate the new discoveries.[22]

Adherence to Aristotle was almost certainly connected to religious concerns. Anti-Aristotelianism, particularly in the early part of the century, was associated with Protestantism. After the assassination of Henry IV in 1610, there was a noticeable increase in opposition to non-Aristotelian positions. Although the situation eased after 1630 when the crown and its ministers insisted upon the relative independence of the state from the church, it was still not easy for French intellectuals to embrace unorthodox positions. The leading proponents of new systems of philosophy capable of replacing Aristotelianism tout court, Pierre Gassendi (1592–1655) and René Descartes (1596–1650), both encountered severe opposition. For all Gassendi's attempts to rehabilitate the ancient atomist and reputed atheist Epicurus (ca. 341–270 BC), his atomistic system was denounced by the religious authorities as atheistic.[23] Similarly, Descartes' corpuscular philosophy presented problems for the doctrine of transubstantiation. Consequently, his writings were condemned at Rome in 1663 and banned from teaching in France in 1671.[24]

Perhaps the most important aspect of the continuing predominance of Aristotelianism in France was the undiminished emphasis on causal explanations in natural philosophy. Virtually all intellectuals agreed with the Aristotelian principle that a confident knowledge of something is only possible when we know the cause on which that thing depends. Demonstrative knowledge of a fact could only be established by showing how the fact followed from the operation of a specific cause, and how, as a result of the operation of that cause, the fact could not be other than it is. With the exception of skeptical philosophers like Marin Mersenne (1588–1648), all exponents of the new experimentalism in France, whether Jesuits, Cartesians, or autodidacts like Blaise Pascal (1623–62), subscribed to this principle of epistemology. In the case of Descartes and his followers, this was seen as an important aspect of their attempt to create a complete alternative to the Aristotelian system: if Cartesianism provided no demonstrable knowledge, it could hardly hope to win the support of the Roman Catholic Church, which was always concerned with certainties.[25]

The result of this attitude in practice was that experiments were always presented in the writings of French natural philosophers as demonstrations of lawlike behavior. The experimental setup was not regarded as a unique individual trial that took place at a certain time and place, but as a representation of general principles, embodying a universal claim about how things happen. French experiments, therefore, always went hand in hand with rational arguments that both dictated the setup of the experiment and, if successful, explained the outcome of the experiment. The general aim of a French experimental report, as Peter Dear has pointed out, was to provide a confirmatory illustration of a theoretically based claim about the behavior of natural objects. As the leading English scientist Robert Boyle (1627–91) complained about Pascal, he might not have actually tried his experiments, but merely "set them down as things which *must* happen, upon a just confidence that he was not mistaken in his Ratiocinations." For the Englishman Boyle, however, it was not possible to be justly confident that one was not mistaken.[26]

In England generally, things were very different. In the disruption of the Civil War period and the subsequent Interregnum, the new natural philosophies had come to be seen as either atheistic, associated with radical sectarianism, or perhaps worse, affiliated with Roman Catholicism (on these points, see also the issues raised by Mayhew in chapter 9). All of these associations were regarded as highly subversive to sound religion and the state. The radical sects often embraced the antiestablishment philosophical and religious theories of the Swiss alchemist and reformer Paracelsus (1493–1541). This was bad enough, but orthodox thinkers were even more worried by the fear that legions of atheists were promoting the new mechanical philosophies being developed on the Continent by thinkers like Gassendi and Descartes. Indeed, many English thinkers at this time believed that the new materialist French philosophies were deliberately being promoted in England by Roman Catholics to divert the best minds to natural philosophy, making it easier for Jesuits to enter the country secretly and reconvert the people to Catholicism. As Thomas Barlow, bishop of Lincoln (1607–91), wrote: "It is certain this New Philosophy (as they call it) was set on foot and has been carried on by the Arts of Rome."[27] Shortly after the Royal Society of London was founded in 1660, one of its fellows reported that it was widely regarded as "a Company of Atheists, Papists, Dunces, and utter enemies to all learning."[28] Clearly, English natural philosophers, after the Restoration and a return to comparative stability, had to replace this negative image of science with one that linked their new philosophy to the best interests of church and state.

One of the ways they did this was by developing a unique kind of experimentalism. The perfect English experiment was simply a detailed historical

account of exactly what happened on the particular occasion or occasions being described by the experimenter. The alleged concern was not with any particular theory or hypothesis about how the world worked, but merely with the
so-called matters of fact. The experiment was intended only to establish what
could clearly and undeniably be seen, not to confirm a particular interpretation of what must "therefore" be the underlying reality. As is well known, English thinkers were able to draw upon the earlier attempts of Francis Bacon
to draw up a new method of natural philosophy that would avoid the Scholastic pitfall of interpreting everything in Aristotelian terms. Accordingly, Bacon
conceived of experiment, not as a means of testing a hypothesis, but simply
as a means of gathering data. "I contrive that the office of the sense shall be
only to judge of the experiment," Bacon wrote, "and that the experiment itself
shall judge of the thing." While a Galileo, or later, a Pascal, might have considered an experiment to have failed if it did not confirm the theory in question, for Baconians experiments never "miss or fail," because whichever way
they turn out, they furnish the required data.[29] Now, in practice this notion of
what an experiment should be is hardly tenable. More often than not, English
experimenters smuggled theoretical interpretations into their accounts of the
"matters of fact," and their insistence that they eschewed theoretical presuppositions was largely rhetorical. Nevertheless, the way the English presented
their experiments and the way they professed to conceive of them, and
in many cases the way they did actually practice them, conformed to this
"theory-free" method of establishing matters of natural fact.[30]

So, how did this methodology help to promote the new science among English contemporaries? In order to understand this, we need to consider England's history as a Protestant country. England was unique in being the only
Protestant country whose religious reformation was not based upon doctrinal
grounds. When Henry VIII declared himself head of the Church of England in
1534, to legitimate his divorce from Catherine of Aragon and marriage to Anne
Boleyn, he severed the English church from Rome while still upholding the
fundamental doctrines of Catholicism. Subsequent tensions between English
Calvinists who desired a genuine reform of the English church and church
leaders who continued to favor Romanism resulted in the famous compromise
position, initiated in Edward VI's reign and finally worked out in Elizabeth I's
reign. From then on the Anglican Church continually tried to present itself as
the true via media between two extreme and mistaken positions: Roman
Catholicism on the one hand and Calvinism on the other. Accordingly, the rhetorical defense of the English church and its "middle way" became highly important in efforts to maintain the peace, particularly in the troubled times of
the seventeenth century, before the Civil War and after the Restoration.

Two important aspects of that rhetoric were to influence the English way of doing science. First, the founders of the Anglican compromise developed a notion of doctrinal minimalism. Dispute over theological niceties was never-ending and seemingly irresolvable. The compromise solution was simply to declare that only a few basic beliefs are essential for salvation and must, therefore, be accepted by all believers. All other doctrines, including all those that led to dissension, were declared to be *adiaphora,* things indifferent to salvation. Efforts to determine the few "common notions" to which all English believers could subscribe were not themselves without contention, of course, but English theologians endlessly repeated that such fundamental beliefs were obvious and undeniable to everyone.[31]

A second feature of the rhetoric of theological compromise was the insistence on the belief that some things are immediately obvious to "common sense" and do not need to be established by elaborate rational arguments. The important thing to bear in mind when trying to understand this attitude is that both the Catholics and the Calvinists claimed that their theological principles were securely founded upon "reason." English theologians were deeply suspicious of elaborate arguments based upon long series of ratiocinations. The suspicion was that such arguments could always be made to support any case and, more important, never succeeded in resolving dispute, but merely served to heighten it. Subtlety of reasoning was regarded as beguiling and treacherous. Significantly, the prime example of such arguments was the Scholastic "disputation," that is to say, the kind of arguments used in the universities to defend Aristotelian principles. The result was a no-doubt naive insistence that all important truths can be, indeed should be, immediately obvious to "common sense."[32]

The group of English natural philosophers who became the founding fellows of the Royal Society in 1660 sought to improve the image of English science by adapting the Anglican Church's method of resolving religious dispute and thereby establishing truth. By no means all English philosophers carried out their work in accordance with the experimental method as it was expounded by the leading public spokesmen of the society, but if they joined the society they were effectively assenting to its professed methodology. Thus, as Thomas Sprat (1635–1713) wrote in his *History of the Royal Society* (1667), the society was so "backward from settling of Principles, or fixing upon Doctrines" that it could even be said, "they have wholly omitted Doctrines." The Royal Society therefore embraced the Anglican stratagem of doctrinal minimalism. Sprat went on to say that the experiments performed at the society were concerned only to establish the "matters of fact." The assembled witnesses of the experiment do not get involved in tendentious "rational" interpretations, but restrict themselves to "the plain objects of their eyes."[33]

Elsewhere, we can see other leading fellows of the Royal Society distancing themselves from the use of rational argumentation. Robert Hooke (1635–1703), one of the greatest of the society's experimenters, wrote that "[a]rguing, concluding, defining, judging and all other degrees of Reason are liable to the same imperfection, being, at best, either vain or uncertain."[34] Robert Boyle dismissed French disputes between Cartesians and Gassendists as to whether atoms were indivisible as irresolvable by experiment and only likely to perpetuate dispute. He even refused to be drawn into arguments as to whether there really was a vacuum inside the air pump that he and Hooke used to such advantage in their experiments. Similarly, his experiments with the air pump established "that the air hath a spring," but he refused to commit himself to an explanation of what caused the spring. All such explanations, such as the claim that the particles of air were shaped like coiled springs or that the particles were continually vibrating back and forth, were merely hypothetical interpretations, for as he kept insisting, Boyle was concerned only with matters of fact.[35]

In their efforts to establish the intellectual authority of their new philosophy, then, the spokesmen for the Royal Society drew inspiration from the earlier efforts of their church to end theological dispute and establish what they thought was the true religion. It was the pursuit of this enterprise that made experimentalism in England so different from that in France. If English experimentalists had presented their experiments as confirmations of rational demonstrations in the French manner, their fellow countrymen, used to the Anglican way of establishing truth, would have been suspicious that they were being led astray by beguiling ratiocinations, supported by elaborately conceived experimental trickery. By insisting that their experiments simply revealed obvious matters of fact, with no tendentious theoretical presuppositions, orthodox English suspicions were allayed, and the experimental philosophy came to be accepted as an unbiased "objective" way of establishing truth.

The fact that this nationally idiosyncratic version of the experimental method was self-consciously modeled on the efforts of English theologians to establish the authority of the national church over all believers is perfectly evident from Thomas Sprat's *History of the Royal Society,* where he wrote that the Royal Society and the Church of England "arose on the same Method," and that the one had achieved a reformation in religion and the other in philosophy. It is important to note that the idiosyncrasy of this was by no means Sprat's alone. He wrote under the strict supervision of Thomas Wilkins, bishop of Chester and leading founder of the society. But Wilkins was entrusted with this duty by the other leading fellows because they saw him as representative of their collective ideals.[36] Sprat spoke for them all, therefore, when he wrote that the seeds of the Royal Society had been sown in Edward VI's and Elizabeth I's reigns and that "[t]he Church of England therefore may justly be styled the Mother of this sort of Knowledge."[37] It is perfectly clear that what Sprat meant by "this sort of Knowledge" was knowledge that could be freely accepted by all dissenting parties because it did not go beyond obvious and undeniable claims. In the case of natural philosophy, such claims were either statements of simple fact or minimalist interpretations of sense data, which (supposedly) could not be interpreted any other way.

The most famous example of a fellow of the Royal Society arguing in this way is, of course, Isaac Newton. Newton was taken to task by Huyghens and Leibniz, two Continental philosophers who subscribed to a generally Cartesian way of understanding the natural world, for introducing an unexplained occult force back into natural philosophy, Newton refused to comply to Continental demands to discuss causes. Leibniz was neither French nor Catholic, but he had his own reasons for upholding Aristotelian intellectual values, and like Descartes, Pascal, and others, he insisted that natural philosophy should offer causal explanations of natural phenomena. Newton's account of gravitational attraction, for all its mathematical success, was nonetheless incomplete, according to Leibniz, because Newton had not explained the cause of gravity. Newton's answer is famous: *hypotheses non fingo*—"I do not dream up hypotheses." Relying upon the Royal Society tradition—effectively the English tradition—in natural philosophy, Newton was able to insist that he only dealt in facts: "[F]or us it is enough that gravity does really exist." That matter attracts other matter comes under the category of a minimalist interpretation that could not be interpreted any other way. To offer accounts involving the continual pressure applied by streams of invisible particles (as in the Cartesian explanation) would no longer be an approved, English, way of proceeding.[38]

The details of Newton's response to Leibniz and the background to it have been seen in "stylistic" terms by the leading Newton scholar I. Bernard Cohen. For Cohen, however, it is simply the "Newtonian style." As I have pointed out elsewhere, however, Newton can be seen to be simply adopting the ready-made English style developed by the leading members of the Royal Society.[39] Certainly, the considerable differences in approach between Huyghens, Leibniz, and other Continental Cartesians on the one hand, and Newton and his fellow countrymen on the other, testify to the fact that we are dealing here with something fit to be attributed to differing national styles.

DIFFERENCES IN NATIONAL STYLES
AND THE CROSS-FERTILIZATION OF IDEAS?

There can be little doubt that there were national stylistic differences at work throughout the Scientific Revolution. What we now need to consider is whether there is any evidence that developments in natural philosophy came about as a result of something like a cross-fertilization of national styles. An obvious place to begin is with matter theory, since this can be readily acknowledged to be a major aspect of the changes in natural philosophy during the period. It is well known that the Cartesian system of philosophy depended upon completely passive and inert particles of matter. Descartes was committed to explaining physical changes in terms of changes in the arrangement in space of the particles he envisaged as constituting all bodies, that is, changes in their motions. Those changes of movement, however, could never be considered to be spontaneous or self-initiated in any way by the particles themselves. Clearly, such movements could not occur if the particles were inert. It followed that changes of motion must be caused by something external to the particles in motion.

Descartes dismissed the idea that there were external forces operating on particles because he would have had to locate these forces somewhere, and the only conceivable place available (according to Descartes) was in body or matter. This would have undermined his prior assumption that matter was inert, however, and so he did not adopt it. The only thing external to particles of body and capable of moving them, therefore, were other particles of body. So only force of impact was allowed in the Cartesian system to account for changes of motion. It also followed from this conception of the world system that new motion could never be generated; it could only be transferred from one particle of matter to another. What seemed like new motion in one part of the system had to be "bought" at the expense of motion elsewhere, and there could be no deficits and no unexplained inputs.

Undeterred by the obvious difficulties of this conception, Descartes was able to develop a system that he believed to be capable of accounting for all known physical phenomena.[40] According to the majority of English thinkers, however, this was clearly unworkable. It seemed obvious that ignited gunpowder did not throw a heavy cannon ball a distance of half a mile as a result of the sudden impact of particles converging on the rear end of the cannon, nor could the flame of the match be said to bring the necessary amount of motion required to balance the Cartesian books. Suggestions that motion was somehow trapped in the gunpowder during its manufacture, ready to be released at the stimulation of a match, again seemed tendentious and unrealistic. The Cartesian account simply seemed too fanciful and completely unjustified. Within the English tradition, the emphasis was upon the matter of fact that gunpowder simply seems to be capable of creating new motion. In short, what we see in England is a willingness to allow unexplained properties of bodies, or so-called occult phenomena, providing there was sufficient evidence of a daily, routine, kind to enable one to confirm the reality (and, in many cases, the precise behavior) of these things.

A clear example of the difference between the Cartesian and the English way can be seen in the case of magnetism. Magnetism was always regarded as an occult quality, but Descartes explained the actions of magnets in terms of particles swirling around the magnet, from pole to pole in an invisible vortex. What is more, he supposed the particles must be of two sorts: they must all be screw-shaped, but some with a left-hand thread and some with a right-hand thread. He used differences of thread in particles and in invisible pores or channels in iron or in other magnets to explain why magnets sometimes attracted and other times repelled. Nowhere, to my knowledge, does Descartes explain what keeps the vortex around magnets going. He seems to be so confident of the workability of his system that he can safely assume there is some input of motion delivered by other particles somehow, which enables the magnet to keep throwing out particles that then incessantly return to the other pole of the magnet.[41]

In England, one of the leading members of the Royal Society and an ingenious natural philosopher in his own right, Sir William Petty, sketched out his own system of mechanical philosophy in the 1670s. In his system, he assumed that all bodies were made up of invisibly small particles that could combine and recombine with different arrangements to give rise to different sensory phenomena. What made his system different from Descartes' and, indeed, from all earlier atomist or corpuscularist systems, was the fact that the invisible particles were all assumed to be tiny spherical magnets. So instead of seeking to give a mechanistic account of how magnets work, Petty took the observational and experimental data about magnets and their behavior for granted (they were the matters of fact) and used them to build a system of natural philosophy. Condensation and rarefaction, or expansion and contraction, could be explained in terms of altered alignments of the magnetic particles. When opposite poles were lined up, contraction or condensation took place as the magnets drew toward one another. If like poles were aligned, the particles repelled one another and rarefaction or expansion took place. Petty went on to try to account in similar ways for elasticity, chemical affinities, planetary movements, and much more besides. In the end, Petty's scheme failed to convince, but not because it relied upon the taken-for-granted occult nature of magnets. It was, simply, too speculative and too far removed from what might be construed as undeniable matters of fact. It is one thing to explain rarefaction in terms of self-repelling magnets, it is quite another to explain how such magnets can unfailingly be lined up the right way every time water turns into steam, for example. Yet Petty's scheme nicely shows how English thinkers were perfectly happy to accept, on phenomenalist grounds, the existence of unexplained active principles in matter. Furthermore, we can see the fruitfulness of this when combined with Cartesian ways of thinking by considering the development of Newton's thought.

It is well known that Robert Hooke conceived of a way to explain planetary movements fully consistent with Kepler's laws of planetary motion by assuming an attractive force operating between the sun and planets that varied inversely as the square of the distance between them. No Continental thinker would have considered such an occult account of cosmology. The influential French style demanded an explanation in terms of a balance of forces in toward the sun and out from the sun, keeping the planets in their orbits. But these balanced forces were forces of impact—there was an outward pressure and an inward, caused by the crowding and jostling of moving particles in the plenist Cartesian universe. Hooke simply assumed an attractive force operating across a notionally empty space whose effect was modified by the tangential inertial motion of the planet, so giving rise to an elliptical orbit. In a brief exchange of letters with Isaac Newton in 1679, Hooke told Newton and eventually convinced him (not without difficulty) of the fruitfulness of this approach (and Newton subsequently wrote the *Principia Mathematica,* but that is another story). Before this correspondence, Newton was effectively outside the English tradition. Being largely an autodidact in the new philosophy, Newton had taught himself Cartesianism, and while he was highly critical of it, he nonetheless accepted that planetary cosmology was to be explained by a Cartesian balance of forces of impact. Once he had been provided by Hooke with an entry into the English way of doing things, Newton changed overnight. As Richard S. Westfall pointed out in his magisterial and authoritative biography of Newton, after this correspondence with Hooke, Newton began to use occult actions-at-a-distance in all his speculations.[42]

If we allow that Newton was outside the English tradition that was developed earlier by the leading lights of the Royal Society (including Hooke) until his unwitting initiation by Hooke, we can see Newton's great achievement as the outcome of cross-fertilization between two rival national styles of thinking. But it also enables us to understand previously puzzling inconsistencies in Newton's approach. These inconsistencies, I suggest, derive from the fact that he was a thinker who set out with one style of thinking and switched to another. In these circumstances, we might well expect there to be inconsistencies arising from irreconcilable features of the two styles. In some cases, such inconsistencies might prove fatal to philosophical coherence. In other cases, however, they might simply prove puzzling to the onlooker, but do little harm to the philosophical system as a whole. In still other cases, of course, as Yaroshevsky suggested, they might lead to brilliant innovations.

The innovatory character of Newton's thought can hardly be doubted, but does it derive from a cross-fertilization between scientific styles of thinking? This is something we must consider. But first, let me take notice of a seeming inconsistency that suggests we are indeed dealing with a thinker who was thinking in two styles at once. In the last of the famous "queries" that complete the *Opticks* (Query 31 of the 1717 edition), Newton insists that the universe cannot continue indefinitely without winding down. In some places in his work, Newton uses this way of thinking in order to combat atheistic suggestions that the universe can run without God's intervention.[43] But this is not at issue in the particular passage I have in mind, where he invokes active principles as the natural phenomena required to keep the universe from winding down:

> Seeing therefore the variety of Motion which we find in the World is always decreasing, there is a necessity of conserving and recruiting it by active Principles, such as are the cause of Gravity, by which Planets and Comets keep their Motions in their Orbs, and Bodies acquire great Motion in falling; and the cause of Fermentation, by which the Heart and Blood of Animals are kept in a perpetual Motion and Heat; the inward parts of the Earth are constantly warm'd, and in some places grow very hot; Bodies burn and shine, Mountains take fire, the Caverns of the Earth are blown up, and the Sun continues violently hot and lucid, and warms all things by his Light. For we meet with very little Motion in the World, besides what is owing to these active Principles. And if it were not for these Principles, the Bodies of the Earth, Planets, Comets, Sun, and all things in them, would grow cold and freeze, and become inactive Masses; and all Putrefaction, Generation, Vegetation and Life would cease, and the Planets and Comets would not remain in their Orbs.[44]

I hope readers will agree that we can see in this passage the alchemical, organicist, even vitalist, Newton. Here is a Newton who is willing to move away from the mechanical philosophy and to accept unexplained occult qualities at work in the world.

The invisibly small particles that constitute the universe in mechanical philosophies are there in Newton's philosophy too, but qualified thus:

These Particles have not only a *Vis inertiae,* accompanied with such passive Laws of Motion as naturally result from that Force, but also they are moved by certain active Principles such as is that of Gravity, and that which causes Fermentation, and the Cohesion of Bodies.

Ever mindful of the requirements of the Cartesian style, however, Newton immediately goes on to insist that these active principles are *not* occult qualities:

These Principles I consider, not as occult Qualities, . . . but as general Laws of Nature, by which the Things themselves are form'd; their Truth appearing to us by Phaenomena, though their Causes be not yet discover'd.[45]

What we have here, then, is Newton saying these "active Principles" are not occult as, in the same breath, he says their causes are unknown. For a Continental writer, this would be tantamount to saying something like "These are not occult qualities but they are occult." It is interesting to note, furthermore, that Newton attributes the workings of these active principles to "Laws of Nature," but he is here using the term in an entirely traditional way. The notion of laws of nature had been used in this loose way since ancient times. In this usage, to say something is a law of nature is simply to say that things always happen that way. Descartes, however, had radically changed the notion to make laws of nature entirely specific and the foundation for his entire system of physics.[46] Newton's great achievement was to rewrite the Cartesian laws in a way that proved entirely workable until the advent of relativity theory. And yet here we have seen him effectively denying that a fully comprehensive physics can be built on the principle of inertia and his three laws of motion, and slipping into talking of laws of nature simply as general principles of uniformity.

I believe we can understand what is going on here by seeing it as the result of Newton's adopting a second style of thinking after his initial intellectual commitment to the style of Descartes. The principle of inertia and the laws of nature and rules of impact developed by Descartes are crucially important in the physics of inert particles he created.[47] Newton must have absorbed its importance when he familiarized himself with Descartes' system and during the period when he was working effectively as a Cartesian. The active principles, however, almost certainly derive from his alchemical interests (as Westfall and Betty Jo Dobbs have both shown).[48] Although Newton's alchemical studies predate the correspondence with Hooke, there is no evidence that Newton saw how to combine his alchemical work with his essentially Cartesian natural philosophy until, as Westfall said, he saw a place for actions-at-a-distance in natural philosophy.[49] From then on, Newton became a thinker in the English style, but one who was able to introduce into it features from the French style.

This returns us to the more innovatory aspects of Newton's work. Can they be seen in the same way, as the outcome of interaction between two styles of thinking? Surely they can. It is hard to imagine how a thinker committed to the anti-Cartesian belief that new motions in the world system could be initiated in an occult way could ever imagine that the world system might be explained in terms of three laws of motion. A thinker who accepted that the flight of a cannonball could be initiated by the entirely occult behavior of a substance like gunpowder, or that some bodies could spontaneously move themselves by an occult power, the way magnets evidently did, or that bodies like light could disseminate themselves spontaneously and incessantly throughout space, or through other bodies, performing various physical acts as they go (including initiating new motions), seems unlikely to have believed it possible to build an entire system of physics on three laws of motion. And yet in Newton we find someone who did believe these things, and who nevertheless built a system of physics on three laws of motion (and, we should recognize, Einstein notwithstanding, a pretty successful system too).

It is generally accepted that the reason for Newton's great success was the fact that he recognized the fundamental unworkability of the entirely inertial and kinetic mechanical philosophy of Descartes and transformed it by introducing principles of activity, including actions-at-a-distance into the mechanical philosophy.[50] There has been a tendency in the past to see Newton's innovation merely as the result of his genius, but if we hope to find a more intellectually satisfying understanding of his achievement, perhaps it resides in the fact that he was uniquely placed to combine elements from two rival national styles of scientific thinking.

If the foregoing account has any value, it should be possible to find other examples of scientific advance stimulated by the interaction between different styles of thinking. Perhaps this should be considered for future research. In the meantime, it is worth pointing to one other example that might be seen in these same terms.

It is already generally acknowledged that the controversy in the early decades of the eighteenth century between Newtonians and Leibnizians over the correct analysis of force, the so-called vis viva controversy, was not simply a difference over technical niceties. It has been seen as a fundamental clash of worldviews, involving differences not only over the nature of matter and force, and therefore over the nature of the mechanical philosophy itself, but also over the nature of God and other metaphysical principles, such as the nature of space, causality, and the conservation of motion. In a classic statement of the background to the vis viva controversy, Carolyn Iltis argues that "intellectual and metaphysical positions taken by scientists can provide emotional psychological commitments." It is clear to her that such psychological commitments were not the result of the idiosyncrasy of the individual but were, rather, social phenomena. "The metaphysics behind these concepts of force," she writes, "had developed from widely differing intellectual traditions."[51] The story Iltis tells is one of rival factions in Britain and Europe, which certainly indicates that national styles of thinking were at work. That this rivalry eventually led to cross-fertilization and an advance in scientific thinking is indicated by her concluding remark: "It was not until the 1740s that integrations between the two systems of nature began to occur, as a few natural philosophers recognized the validity of both interpretations of 'force.'"[52]

David Papineau in his summary of the vis viva controversy notes that "[b]y and large, affiliations in the dispute went by nationalities, with English Newtonians and French Cartesians following the 'old opinion' (that force is proportional to mass times velocity), while Dutch, German and Italian scientists favoured the 'new opinion' put forward by Leibniz." He concludes:

> The long persistence of the dispute . . . was due to no confusion, or lack of objectivity. It was simply that two alternative modifications of the Cartesian theory of impact were proposed when the latter was seen to be inadequate. Both these alternatives merited consideration, and time was needed for their implications and possible refinements to be explored and evaluated.
>
> Whether there would eventually have been an agreed conclusion to the debate must remain an unanswered question. For, as we have seen, the *vis viva* controversy closed, not with the victory of one side, but with a fundamental revision of physical thought resulting in the repudiation of both.[53]

Similarly, although Steven Shapin in examining the rivalry between Leibniz and Newton tends to emphasize the rhetorical dimension of the dispute rather than the underlying intellectual differences, he has no trouble showing the political and nationalistic dimension to the rivalry.[54]

For a study that attempts to show the links between the differing political backgrounds of the Newtonians and the Leibnizians and their metaphysical and physical (not merely rhetorical) differences, we need look no further than the unfairly neglected book by Gideon Freudenthal, *Atom and Individual in the Age of Newton*. In what is a highly ambitious thesis, Freudenthal links what he sees as Newton's fundamental scientific presupposition to the developing bourgeois politics of contemporary England. Newton's conception of

"essential properties," as distinct from "universal properties," defines properties that belong to every atom, even in the hypothetical case of a single unique
atom in otherwise empty space. For Newton, inertia is an essential property,
but gravity, which could not be manifested in a single, lone atom, is declared
to be merely "universal." Freudenthal then seeks

> to demonstrate the genetic dependency of Newton's assumption that essential
> properties would belong to a single particle upon the assumption of social phi-
> losophy that essential properties would be attributable to a single individual;
> this will conclude the reconstruction of the mediated dependency of Newton's
> theory of space on social relations.[55]

In short, Freudenthal sees a clear connection between the concepts of Newtonian physics and the contemporary development of the political theory of
bourgeois individualism.[56]

Given the very different political circumstances surrounding Leibniz, it is
hardly surprising that he should develop a significantly different worldview.

> [N]o detailed investigation of the social history of Germany after the Thirty
> Years War is needed to maintain that in Germany with its division into small
> states no open struggle between the bourgeoisie and the feudal nobility can be
> ascertained and by no means an ascendancy of the bourgeoisie.

The decisive result, according to Freudenthal, is that the appropriate terminology for Leibniz is never "atom and individual" but "element and system."
"In all central questions Leibniz did not subscribe to the classical bourgeois
position," Freudenthal writes. "It is decisive for Leibniz that he takes the state
of nature [in the Hobbesian sense] to be already a social condition and that he
does not see the contemporary society as composed of equal and independent individuals."[57] Likewise, Leibnizian physics is concerned with the interplay of elements in a complex system.

In view of the analyses of Iltis, Papineau, Shapin, and Freudenthal, it seems
impossible to deny that the scientific and philosophical differences between
the Newtonians and the Leibnizians can be seen in terms of differences in national styles of thinking. These differing national styles resulted in divergence
to begin with, but eventually, as Iltis and Papineau independently pointed
out, resulted in a highly fruitful reconciliation, giving rise to modern dynamic
theory. Here, then, we seem to have another clear example of the kind of fruitful interaction between different national schools of thought that Yorashev-
sky saw as crucially important in scientific advance.

NATIONAL STYLES AND THE "WHERE"
AND "WHEN" OF THE SCIENTIFIC REVOLUTION

One way of testing the validity of claims about the importance of differing national styles in science would be to consider cultures where such proximate,
and rival, national differences were not found. The expectation should be, in
such cases, that science would not have progressed so fruitfully as it had
in the case of western Europe. This, of course, immediately brings us to the
ongoing efforts of historians to explain why the Scientific Revolution had to
await the rise of post-Renaissance western Europe, and why it did not occur
previously in any of the other likely centers of advanced civilization, such as
China, Islam, Byzantium, or even the western Europe of the Middle Ages.

Before going any further, it should be said that the answer to the question
as to why the Scientific Revolution occurred when and where it did is undoubtedly to be found in a countless number of historical contingencies. Just
as the late Stephen Jay Gould believed that if the history of the world were to
run again, evolution might take a completely different course, and that the
successful flora and fauna of that replay might bear little relation to the successful species of our own timeline, so must it be with human history. If history were to run again, things might be very different.[58] Accordingly, what
follows is not presented as a single key to understanding the nature of the
Scientific Revolution. It is merely suggested that differences in national
styles, or the lack of them, might be worth considering as a so far unconsid-
ered factor among the many that contributed to the geographical and historical origins of the Scientific Revolution. It should also be obvious that, in the
absence of the necessary comparative histories of other civilizations, all that
can be offered here are a few very tentative (possibly ill-founded) indications
in support of this hypothesis.

It is easy to call upon the authority of Burckhardt, Elias, and Hale to support the claim that the countries of western Europe in the Middle Ages did
not have the requisite sense of national identity to be able to develop national
styles in natural philosophy. The idea of a nation-state and a geographically
distinct country are not the same thing, of course, and identifying oneself
with one's country requires the kind of awakening of personality that Burckhardt saw as first happening in the city-states of Renaissance Italy, and that
Elias saw as occurring in a symbiotic process with the development of nation-
states.[59] If their analysis is correct, and the historical research of John Hale
and numerous other historians vigorously supports it, then there could be no
national styles in Western science before the Renaissance.[60] If, therefore, we

accept that rival national styles were an important factor in the Scientific Revolution, then it is less surprising that it did not occur earlier in western Europe, say, at the point in the Middle Ages when the Latin Europeans came into contact with the ancient Greek writings they had recovered from the retreating Arab civilization in Spain and in Sicily.

The imperial unity of Byzantium might also be sufficient to account for its lack of national styles and so, perhaps, its lack of a scientific revolution. Certainly, there seems to have been no scope for any kind of public, much less locally representative science. Donald Nicol, a prominent Byzantinist, has said that "[b]eyond the discipline of the master-pupil relationship there was little observable cooperation among Byzantine scholars. Each worked alone as an individualist . . . only rarely did they collaborate in their researches."[61] More to the point, there are no indications of local rivalries between these Byzantine individualists, which might have led to innovation in natural philosophy.

Consider also China, which is always regarded, with hindsight, as one of the most likely sites for a scientific revolution, but where such an epoch-making change never came to pass. Again, what we are dealing with here is a vast unified empire, not a collection of geographically near but politically separate nation-states. As Joseph Needham wrote in his *Grand Titration*, "in that society, the conception of the city-state was totally absent; the towns were purposefully created as nodes in the administrative network, though very often no doubt they tended to grow up at spontaneous market centres." Similarly, he wrote, "the spatial range of public works . . . in Chinese history transcended time after time the barriers between the territories of individual feudal or proto-feudal lords. It thus invariably tended to concentrate power at the centre, i.e., in the bureaucratic apparatus arched above the granular mass of "tribal" clan villages." Needham denies suggestions that the village communities in China were "autonomic," except in very restricted ways. For the most part, the imperial state apparatus dominated everything at a local level. China has always been, he wrote, "a one-Party State."[62] Again, this does not sound like the kind of setting where different national styles, or their equivalent (regional styles?), might develop and creatively interact with one another.

This is reinforced in the conclusions of Richard E. Nisbett in his *Geography of Thought* (2003), a study of different styles of thinking between Westerners and Asians. Nisbett writes:

> even today 95 per cent of the Chinese population belongs to the same Han ethnic group. Nearly all of the country's more than fifty minority ethnic groups are in the western part of the country. A Chinese person living in the rest of the

country would rarely have encountered anyone having significantly different beliefs or practices. The ethnic homogeneity of China seems at least partly explicable in terms of the centralized political control.[63]

Similarly, Jared Diamond, the evolutionary biologist turned cultural historian, has pointed to the "astonishing" cultural unity of China, which he attributes largely to the fact that it was politically unified in 221 BC and has effectively remained so ever since. If China was ever a melting pot, it was so long ago that it has long since undergone what Diamond calls a "drastic homogenization."[64]

Such cultural unity is bound to be reflected in styles of thought. Needham even points to the vastly different effect of Chinese scientific innovations on Chinese society and in Europe:

> These many diverse discoveries and inventions had earth-shaking effects in Europe, but in China the social order of bureaucratic feudalism was very little disturbed by them. The built-in instability of European society must therefore be contrasted with a homoeostatic equilibrium in China.[65]

As Roy Porter and Mikuláš Teich have suggested,

> It may well have been crucial to the development of modern science in Europe—contrast China—that distinct intellectual traditions were able to flourish in a multiplicity of polities.[66]

Perhaps the built-in instability of European society was what allowed national styles to develop, or perhaps contrasting national styles contributed to that instability. Either way, it does *not* seem as though national styles in science, and the interactions between them that lead to new innovations and new ways of seeing the world, were ever a feature, or ever could have been, in imperial China.

This brings us, finally, to the civilization of Islam. There is no denying the incredible advances in knowledge of the natural world made by the Arabs under the stimulus of the growing new religion of Islam, from the ninth century onward. Islamic philosophy and science was so advanced that it has to be seen as one of the most important potential historical and geographical sites where a scientific revolution might have occurred. But why didn't it? None of the suggested answers to this question has ever achieved consensus. It is evident that a great deal of historical research needs to be done before we can speak confidently about this important issue. Nevertheless, there are at least some indications that the kind of national rivalries between intellectuals that have been suggested here as an important element in stimulating continued scientific innovation may have been largely absent from this civilization.

If Islam defined a unified empire, then perhaps we ought not to expect to find any variation in local styles within Islam. According to the economic historian Robert Reynolds, the civilization of Islam was a single Caliphate that "extended its boundaries to the Atlantic, north into Central France, across the whole south and over both the east and west ends of the Mediterranean, up to the Gobi desert, and out to northern India. All of this was one area of government." Small wonder, then, that "across the world over which the caliph ruled there came to be a common art, a common literature and theology." This state of cultural affairs continued, according to Reynolds, even after geopolitical fragmentation began to take place: "[E]ven when political break-up came, the Saracenic culture remained intact."[67] Indeed, Howard Turner, in his study of science in medieval Islam, describes the development of independent states within Islam as we see them today as the result of the disintegration of the Islamic Empire and the colonization of its territory by Western powers. Modern Islamic states are the legacy of recent European political maneuvering, not something that emerged within Islam itself.[68] This accords with the more general observation of the economist E. L. Jones that the nation-state is a "purely European form which has been exported to [other] parts of the world."[69] Similarly, the distinguished historian of science, A. I. Sabra, in a recent paper on the role of locality in the development of Arabic science, declared:

> As far as science is concerned, it seems to me that important considerations lead us to say that we have to do with a single unitary tradition. There are considerations of language, which—for science and philosophy—was for the most part one language (Arabic), and of Islamic religion as an ever present point of reference . . . in addition to considerations of the dominance of dynastic rules over large regions for extended periods of time.[70]

It is important to note, however, that an alternative view is possible. Majid Fakhry, a leading historian of Islamic philosophy, has insisted that "Islamic philosophy is the product of a complex intellectual process in which Syrians, Arabs, Persians, Turks, Berbers, and others took active part," and he points to the importance of noting "the role of each racial group in the development of Islamic philosophy."[71] So although the notion of *national* differences may be inappropriate for Islam, there may well have been similar kinds of intellectual rivalry between different ethnic groups to those we have seen in western Europe between different nationals. Unfortunately, a rapid survey of the readily available summaries of the history of Islamic thought reveals that putative rivalries between thinkers from different racial groups, or between representative individuals from different groups—a Persian Newton and a Turkish Leibniz, perhaps?—has not so far attracted the attention of Arabist historians of science.[72] Clearly, research along these lines, to see whether there were such rivalries, and whether they might have contributed to the advance of Arabic science, is a desideratum.

In the meantime, however, it is worth pointing out that Fakhry's comment about the contribution of different ethnic groups to the development of Islamic science does not necessarily invoke the kind of active rivalry we have seen taking place in seventeenth-century Europe. A superficial reading of the available literature suggests a picture of Islamic science that flourishes for a time in different geographical locations, at different periods of Islamic history, depending upon the patronage of enlightened rulers. Consider, for example, this comment by Sabra:

> In Islam, whether in ninth- and tenth-century Baghdad, eleventh-century Egypt and central Asia, twelfth-century Spain, thirteenth-century Maragha in northwestern Iran, or fifteenth-century Samarkand, the major scientific work associated with the names of those who were active at those times and places was carried out under the patronage of rulers whose primary interest lay in the practical benefits promised by the practitioners of medicine and astronomy and astrology and applied mathematics.[73]

This is perhaps what Edward Grant, leading historian of medieval European natural philosophy, had in mind when he wrote that "over the centuries, the number of identifiable Islamic philosophers, is relatively small," being scattered over distances of time and space.[74] Furthermore, as in Byzantium, natural philosophy was rarely taught publicly: "Many of the foremost Muslim scientists and natural philosophers, including al-Biruni, Avicenna (Ibn Sina), and Al-hazen (Ibn al-Haytham)," Grant says, agreeing with Sabra, "were supported by royal patronage and did not teach in schools." Accordingly, natural philosophy was "taught privately and quietly, rather than in public, and it was taught most safely under royal patronage."[75] Similarly, Howard Turner writes that "Muslim students in what today we call the sciences received instruction almost entirely outside of the regular educational system, usually at an installation sponsored by a princely court, or from individual, often court supported, scholars."[76] This does not look like a setting where national styles might develop, much less where disputes between rival approaches might be immediately thrashed out and lead to innovative compromise views. Clearly, only thorough research can settle this issue one way or the other. For the time being, however, it seems that there are sufficient indications to allow us to reiterate that the Islamic world can be seen as a unified civilization with "a common art, a common literature and theology," and perhaps also, therefore, a common science.[77]

But these comments are simply impressions based upon very limited reading about the geopolitical setup and its links with intellectual life in civilizations beyond Renaissance western Europe. Clearly, if there is any value in this approach, it will need to stand up to much closer scrutiny of the history of the political and cultural geography of China, Islam, and Byzantium. It seems to me, however, that the very limited survey here is sufficient to indicate that such a scrutiny might well be worthwhile. It is interesting to note that Joseph Needham dismissed out of hand any suggestions that the reasons for the rise of the West could be attributed simply to contingencies of geography. It is clear, however, that what he had in mind were physical geography and the associated climate, since he insisted that the variation in seasons and in climactic and agricultural conditions in China was comparable to that in Europe. He might, however, be willing to consider a different line on the effect of political geography on the development of the sciences. The world today is divided into nation-states, but if we were to look at a changing political map of the world from the early Middle Ages to the present, we would see the first nation-states developing in late Renaissance and early modern western Europe. What is more, they would for a long time be the only nation-states. Given the plausibility of the effectiveness of interaction between national styles for the creative development of science, it seems hard to deny that we can find an important factor in the rise of science in the West in the unique political geography of western Europe.

...

Notes

This paper owes a lot to the responses of colleagues who heard an earlier version at the colloquium, The Scientific Revolution in Multicultural Perspective, held at the University of Oklahoma in April 2003. In particular, I wish to thank Peter Barker, Floris Cohen, Jack Goldstone, Florence C. Hsia, and Robert S. Westman for encouragement and support. And for such patient advice about the history of Islamic science to an ignoramus, I must thank Sonja Brentjes, S. Nomanul Haq, Ahmet Karamustafa and, especially, Jamil Ragep. I must also apologize to them, however, for continuing to include Islamic science in the argument of the final section.

1. Jacob Burckhardt, *The Civilization of the Renaissance in Italy* (London: Phaidon, 1945), 87.

2. See Norbert Elias, *The Civilizing Process: State Formation and Civilization* (Oxford: Blackwell, 1976); and John R. Hale, *The Civilization of Europe in the Renaissance* (London: Harper Collins, 1993), 51–68. See also the discussion of the emergence of nation-states in Eric L. Jones, *The European Miracle: Environments, Economies and Geopolitics in the History of Europe and Asia* (Cambridge: Cambridge University Press, 1981), 127–49; and Guy Hermet, *Histoire des nations et du nationalisme en Europe* (Paris: Seuil, 1996).

3. A. I. Sabra reveals similar frustrations in "Situating Arabic Science: Locality versus Essence," *Isis* 87 (1996): 654–70. Perhaps the emphasis on nationality in the literature is a reflection of discussions of the supposedly international character of contemporary science. On which, see, for example, Elisabeth Crawford, Terry Shinn, and Sverker Sörlin, eds., *Denationalizing Science: The Contexts of International Scientific Practice* (Dordrecht: Kluwer, 1993).

4. Arne Hessenbruch, ed., *Reader's Guide to the History of Science* (London: Fitzroy Dearborn, 2000), 501; Bertrand Russell, *An Outline of Philosophy* (London: Allen and Unwin, 1927), 13.

5. Lorraine Daston and Michael Otte, introduction to *Science in Context* 4 (1991): 228, 230.

6. For early studies on the organization of science, see, for example, J. T. Mertz, *A History of European Thought in the Nineteenth Century,* 4 vols. (Edinburgh: Blackwood, 1896–1914); Pierre Duhem, *The Aim and Structure of Physical Theory* (Princeton: Princeton University Press, 1954); Nathan Reingold, "National Styles in the Sciences: The United States Case," in *Human Implications of Scientific Advance,* ed. Eric G. Forbes (Edinburgh: Edinburgh University Press, 1978), 163–73.

One of the earliest studies on the sociology of scientific knowledge was M. G. Yaroshevsky, "National and International Factors in the Development of Scientific Schools of Thought," in Forbes, *Human Implications,* 174–81, but see also Paul Forman, "Weimar Culture, Causality and Quantum Theory, 1918–1927: Adaptation of German Physicists and Mathematicians to a Hostile Environment," *Historical Studies in the Physical Sciences* 3 (1971): 1–115; Jonathan Harwood, "National Styles in Science: Genetics in Germany and the United States between the Wars," *Isis* 78 (1987): 390–414; Jonathan Harwood, *Styles of Scientific Thought: The German Genetics Community, 1900–1933* (Chicago: University of Chicago Press, 1993); Jane Maienschein, "Epistemic Styles in German and American Embryology," *Science in Context* 4 (1991): 407–27; Nathan Reingold, "The Peculiarities of the Americans; or, Are There National Styles in the Sciences?" *Science in Context* 4 (1991): 347–66; and Andrew Warwick, "Cambridge Mathematics and Cavendish Physics: Cunningham, Campbell and Einstein's Relativity," part 2, "Comparing Traditions in Cambridge Physics," *Studies in History and Philosophy of Science* 24 (1993): 1–25. For more general considerations of the importance of locality in understanding the nature and development of scientific knowledge, see Adir Ophir and Steven Shapin, "The Place of Knowledge: A Methodological Survey," *Science in Context* 4 (1991): 3–21; and Sabra, "Situating Arabic Science."

7. Harwood, "National Styles in Science."

8. Stephen Toulmin, "Crucial Experiments: Priestley and Lavoisier," *Journal of the History of Ideas* 18 (1975): 205–20; Pierluigi Barrotta, "Scientific Dialectics in Action: The Case of Joseph Priestley," in *Scientific Controversies: Philosophical and Historical Perspectives,* ed. Peter Machamer, Marcello Pera, and Aristedes Baltas (New York: Oxford University Press, 2000), 153–76.

9. For the political background to the very different pathologies of these leading medical thinkers, see Paul Weindling, "Theories of the Cell State in Imperial Germany," in *Biology, Medicine and Society, 1840–1940,* ed. Charles Webster (Cambridge: Cambridge University Press, 1981), 99–155; and Gerald Geison, *The Private Science of Louis Pasteur* (Princeton: Princeton University Press, 1995).

10. David Kohn, ed., *The Darwinian Heritage* (Princeton: Princeton University Press, 1985); Dov Ospovat, *The Development of Darwin's Theory: Natural History, Natural Theology and Natural Selection, 1838–1859* (Cambridge: Cambridge University Press, 1981); Robert M. Young, "Malthus and the Evolutionists: The Common Context of Biological and Social Theory," in *Darwin's Metaphor: Nature's Place in Victorian Culture*, ed. Robert M. Young (Cambridge: Cambridge University Press, 1985), 23–55.

11. Forman, "Weimar Culture"; Warwick, "Cambridge Mathematics."

12. Maurice Crosland, "Styles of Science: National, Regional and Local," in *Encyclopaedia of the Scientific Revolution, from Copernicus to Newton*, ed. Wilbur Applebaum (New York: Garland, 2000), 622. See also Maurice Crosland, ed., *The Emergence of Science in Western Europe* (Basingstoke: Macmillan, 1975).

13. Roy Porter and Mikuláš Teich, eds., *The Scientific Revolution in National Context* (Cambridge: Cambridge University Press, 1992), 2, 4.

14. Yaroshevsky, "National and International Factors," 177.

15. Ibid., 175.

16. Ibid., 176.

17. Ibid., 177–78.

18. Ibid., 178.

19. For details of these differing positions in the philosophy of science, see Paul Feyerabend, *Against Method: Outline of an Anarchistic Theory of Knowledge* (Atlantic Highlands: Humanities Press, 1975); Karl Popper, *The Logic of Scientific Discovery* (London: Hutchinson, 1959); and Imre Lakatos and Alan Musgrave, eds., *Criticism and the Growth of Knowledge* (Cambridge: Cambridge University Press, 1970).

20. Kostas Gavroglu, "Controversies and the Becoming of Physical Chemistry," in Peter Machamer et al., *Scientific Controversies*, 180, 197.

21. Gideon Freudenthal, "A Rational Controversy over Compounding Forces," in Machamer et al., *Scientific Controversies*, 126, 127–28.

22. Peter Dear, "The Church and the New Philosophy," in *Science, Culture and Popular Belief in Renaissance Europe*, ed. Stephen Pumfrey, Paolo Rossi, and Maurice Slawinski (Manchester: Manchester University Press, 1991), 119–39; Dennis Des Chene, *Physiologia: Natural Philosophy in Late Aristotelian and Cartesian Thought* (Ithaca: Cornell University Press, 1996).

23. Lynn Sumida Joy, *Gassendi the Atomist: Advocate of History in an Age of Science* (Cambridge: Cambridge University Press, 1987).

24. Nicholas Jolley, "The Reception of Descartes' Philosophy," in *Cambridge Companion to Descartes*, ed. John Cottingham (Cambridge: Cambridge University Press, 1987), 393–423.

25. The classic statement of Descartes's lingering Aristotelianism is Étienne Gilson's *Études sur le role de la pensée médiévale dans la formation du système cartésien* (Paris: J. Vrin, 1930); but see also Des Chene, *Physiologia*.

26. Robert Boyle, *Hydrostatical Paradoxes* (Oxford, 1666), 4–5; quoted in Peter Dear, "Miracles, Experiments and the Ordinary Course of Nature," *Isis* 81 (1990): 675. Cf. Steven Shapin, "Pump and Circumstance: Robert Boyle's Literary Technology," *Social Studies of Science* 14 (1984): 481–520; John Henry, "England," in Porter and Teich, *Scientific Revolution*, 178–210.

27. Thomas Barlow, *The Genuine Remains . . .* (London, 1693), 157; quoted in Michael R. G. Spiller, *"Concerning Natural Experimental Philosophie": Meric Casaubon and the Royal Society* (The Hague: Martinus Nijhoff, 1980), 30.

28. Joseph Glanvill to Henry Oldenburg, 31 January 1670, in Henry Oldenburg, *Correspondence*, ed. A. R. and M. B. Hall (Madison: University of Wisconsin Press, 1969), 6:456; quoted in Michael Hunter, *Science and Society in Restoration England* (Cambridge: Cambridge University Press, 1981), 138.

29. Francis Bacon, *Works,* ed. J. Spedding, R. L. Ellis, and D. D. Heath (London, 1857–61), 4:26 (quotation from the *Novum Organum*).

30. See, for example, Thomas Sprat, *History of the Royal Society of London* (London, 1667), 99–100; and Henry, "England," 197–98. For further exposition of the emphasis upon so-called facts, see Steven Shapin and Simon Schaffer, *Leviathan and the Air-Pump: Hobbes, Boyle and the Experimental Life* (Princeton: Princeton University Press, 1985); Shapin, "Pump and Circumstance"; Peter Dear, "Totius in Verba: Rhetoric and Authority in the Early Royal Society," *Isis* 76 (1985): 145–61; Paul B. Wood, "Methodology and Apologetics: Thomas Sprat's *History of the Royal Society*," *British Journal for the History of Science* 13 (1980): 1–26.

31. Henry, "England"; Henry R. McAdoo, *The Spirit of Anglicanism: A Survey of Anglican Theological Method in the Seventeenth Century* (London: A. and C. Black, 1965).

32. Henry, "England"; Lotte Mulligan, "'Reason,' 'Right Reason,' and 'Revelation' in Mid-Seventeenth-Century England," in *Occult and Scientific Mentalities in the Renaissance,* ed. Brian Vickers (Cambridge: Cambridge University Press, 1984), 375–401.

33. Sprat, *History,* 107, 99; see also 31–32, 62.

34. Robert Hooke, *Micrographia* (London, 1665), sig. av.

35. John Henry, "Occult Qualities and the Experimental Philosophy: Active Principles in Pre-Newtonian Matter Theory," *History of Science* 24 (1986): 335–81, esp. 360–61.

36. Michael Hunter, "Latitudinarianism and the 'Ideology' of the Early Royal Society: Thomas Sprat's *History of the Royal Society* (1667) Reconsidered," in *Establishing the New Science: The Experience of the Early Royal Society,* ed. Michael Hunter (Woodbridge: Boydell, 1989), 45–71.

37. Sprat, *History,* 372. See also Peter Anstey, "The Christian Virtuoso and the Reformers: Are there Reformation Roots to Boyle's Natural Philosophy?" *Lucas: An Evangelical History Review* 27/28 (2000): 5–40.

38. Henry, "Occult Qualities," 362–63. For another aspect of Newton's conformity with the English style of philosophizing, see Dear, "Totius in Verba," 154–55.

39. This is first discussed in I. Bernard Cohen, *The Newtonian Revolution, with Illustrations of the Transformation of Scientific Ideas* (Cambridge: Cambridge University Press, 1980), but is perhaps more forcefully stated in I. Bernhard Cohen, "The *Principia,* Universal Gravitation, and the 'Newtonian Style,' in Relation to the Newtonian Revolution in Science: Notes on the Occasion of the 250th Anniversary of Newton's Death," in *Contemporary Newtonian Research,* ed. Zev Bechler (Dordrecht: Reidel, 1982), 21–108; cf. Henry, "Occult Qualities," 358–59.

40. As he claims in proposition 199 of the fourth part of *Principia Philosophiae* (1644): "That no phenomena of nature have been omitted in this treatise" (René Descartes, *Principles of Philosophy,* trans. V. R. and R. P. Miller [Dordrecht: Reidel, 1983], 282–83).

41. See the discussion of magnetism in proposition 133 of the fourth part of Descartes' *Principia* (Descartes, *Principles,* 242–43).

42. Richard S. Westfall, *Never at Rest: A Biography of Isaac Newton* (Cambridge: Cambridge University Press, 1980), 388. In spite of this comment, it is my opinion that Westfall's overall assessment of Hooke's influence on Newton plays down the former's importance (see ibid., 382–90). For a full assessment of Hooke's influence on Newton, see Ofer Gal, *Meanest Foundations and Nobler Superstructures: Hooke, Newton and the "Compounding of the Celestial Motions of the Planetts"* (Dordrecht: Kluwer, 2002).

43. Isaac Newton, *Opticks; or, a Treatise of the Reflections, Refractions, Inflections and Colours of Light* (1730; New York: Dover, 1952), 397.

44. Ibid., 399–400. It should be noted that the term "fermentation" in Newton's day was simply taken to refer to an exothermic chemical reaction. Nobody then had any inkling that it was associated with the activity of living organisms.

45. Ibid., 401.

46. Edgar Zilsel, "The Genesis of the Concept of Physical Law," *Philosophical Review* 51 (1942): 245–79; John R. Milton, "The Origin and Development of the Concept of the Laws of Nature," *Archives Européene de Sociologie* 22 (1981): 173–77; John R. Milton, "Laws of Nature," in *The Cambridge History of Seventeenth-Century Philosophy,* ed. Daniel Garber and Michael Ayers (Cambridge: Cambridge University Press, 1998), 680–701; and John Henry, "Metaphysics and the Origins of Modern Science: Descartes and the Importance of Laws of Nature," *Early Science and Medicine* 9 (2004): 73–114.

47. I am aware that in attributing the principle of inertia to Descartes I am speaking rather loosely. For a full discussion of this issue, see Alan Gabbey, "Force and Inertia in the Seventeenth Century: Descartes and Newton," in *Descartes: Philosophy, Mathematics and Physics,* ed. Stephen Gaukroger (Hassocks: Harvester, 1980), 230–320.

48. Richard S. Westfall, "Newton and the Hermetic Tradition," in *Science, Medicine and Society in the Renaissance,* ed. Allen G. Debus (New York: Science History Publications, 1972), 2:183–98; Richard S. Westfall "Newton and Alchemy," in Vickers, *Occult,* 315–35; Westfall, *Never at Rest;* Betty Jo Teeter Dobbs, *The Janus Faces of Genius: The Role of Alchemy in Newton's Thought* (Cambridge: Cambridge University Press, 1991).

49. Dobbs spoke of the general assumption among historians that, to some extent, Newton compartmentalized his work, so that alchemy did not interfere with his mathematical physics, and so on (Dobbs, *Janus Faces,* 9).

50. Richard S. Westfall, *Force in Newton's Physics: The Science of Dynamics in the Seventeenth Century* (London: Macdonald, 1971).

51. Carolyn Iltis, "The Leibnizian-Newtonian Debates: Natural Philosophy and Social Psychology," *British Journal for the History of Science* 6 (1973): 376, 347.

52. Ibid., 376–77. For a full examination of this later and reconciliatory aspect of the story, see Thomas L. Hankins, "Eighteenth-Century Attempts to Solve the *Vis Viva* Controversy," *Isis* 56 (1965): 281–97.

53. David Papineau, "The *Vis Viva* Controversy," in *Leibniz: Metaphysics and Philosophy of Science,* ed. Roger S. Woolhouse (Oxford: Oxford University Press, 1981), 139.

54. Steven Shapin, "Of Gods and Kings: Natural Philosophy and Politics in the Leibniz-Clarke Disputes," *Isis* 72 (1981): 187–215.

55. Gideon Freudenthal, *Atom and Individual in the Age of Newton: On the Genesis of the Mechanistic World View* (Dordrecht: Reidel, 1986), 172.

56. Freudenthal's summing up can be found in his afterword (ibid., 205, but also passim).

57. Ibid., 190, 196. Freudenthal shows Leibniz explicitly denying the validity of Hobbes's individualistic and, therefore, presocial conception of the "state of nature" (191–92).

58. See, for example, Stephen Jay Gould, *Wonderful Life* (London: Hutchinson, 1990), and *Full House* (New York: Harmony Books, 1996). For the opposite view, which seems to want to suggest that there is something about humankind which will guarantee our reemergence and our renewed dominance in any new world, see Daniel Dennett, *Darwin's Dangerous Idea* (New York: Simon and Schuster, 1995).

59. Burckhardt, *Civilization;* and Elias, *Civilizing Process.* For a sociopolitical discussion of the notion of nationhood and what it involves, see Benedict Anderson, *Imagined Communities: Reflections on the Origin and Spread of Nationalism* (London: Verso, 1983).

60. Hale, *Civilization;* Hermet, *Histoire.* See also, for example, Robert L. Reynolds, *Europe Emerges: Transition towards an Industrial World-Wide Society, 600–1750* (Madison: University of Wisconsin Press, 1961); John H. Elliott, *Europe Divided, 1559–1598* (London: Fontana, 1968).

61. Donald M. Nicol, *Church and Society in the Last Centuries of Byzantium* (Cambridge: Cambridge University Press, 1979), 47–48.

62. Joseph Needham, *The Grand Titration* (London: Allen and Unwin, 1979), 196, 204, 205–6. See also, for example, Reynolds, who states, "The Chinese have had a marvellous degree of cultural unity, with a single kind of writing, a single kind of family organization, and usually allegiance to a single state" (*Europe Emerges,* 344).

63. Richard E. Nisbett, *The Geography of Thought: How Asians and Westerners Think Differently . . . and Why* (New York: Free Press, 2003), 31. Significantly, Nisbett goes on to suggest that Chinese thought style stems from the lack of significant disagreement or difference of opinion in everyday life.

64. Jared Diamond, *Guns, Germs, and Steel: The Fates of Human Societies* (New York: Norton, 1997), 323, 324.

65. Ibid., 214.

66. Roy Porter and Mikuláš Teich, introduction to *Scientific Revolution,* 5.

67. Reynolds, *Europe Emerges,* 298, 299.

68. Howard R. Turner, *Science in Medieval Islam* (Austin: University of Texas Press, 1995), 206.

69. Jones, *European Miracle,* 127.

70. Sabra, "Situating Arabic Science," 669.

71. Majid Fakhry, *History of Islamic Philosophy* (New York: Columbia University Press, 1970), 1.

72. I have consulted the following: Henry Corbin, *A History of Islamic Philosophy* (London: Kegan Paul International, 1993); Seyyed Hossein Nasr and Oliver Leaman eds., *History of Islamic Philosophy* (London: Routledge, 1996); Roshdi Rashed, ed., *Encyclopaedia of the History of Arabic Science,* 3 vols. (London: Routledge, 1996); and M. M. Sharif, ed., *A History of Muslim Philosophy,* 2 vols. (Wiesbaden: Otto Harrassowitz, 1963).

73. Sabra, "Situating Arabic Science," 662.

74. Edward Grant, *The Foundations of Modern Science in the Middle Ages: Their Religious, Institutional and Intellectual Contexts* (Cambridge: Cambridge University Press, 1996), 177.

75. Ibid., 178, 182.

76. Turner, *Science in Medieval Islam,* 203.

77. Reynolds, *Europe Emerges,* 299; Sabra, "Situating Arabic Science," 669.

— 4 —

GEOGRAPHY, SCIENCE, AND THE SCIENTIFIC REVOLUTION

...

Charles W. J. Withers

In the autumn of 1690, David Gregory, professor of mathematics at the University of Edinburgh, was tidying his rooms and papers preparatory to taking up the Savilian Chair of Mathematics at Oxford. Among the material that Gregory was preparing to pass into the care of the librarian at Edinburgh were several maps. These included "a Plane" (plan) of the "Physick Garden" (Edinburgh's first botanical garden), and a plan of the University of St. Andrews. Three other maps were duly handed over: one of the grounds of the University of Edinburgh, a plan of "Heriots Hospitall & its Inclosures," and a plan of nearby Lady Yester's Church and the high school.[1] These maps of sites within Scotland's capital were undertaken by Gregory's mathematics students. For Gregory and his students, geography, geometry, and practical mathematics were closely affiliated concerns. At much the same time, "John Adair, Geographer," and Sir Robert Sibbald, Geographer Royal, were donating books to the university's library—books later used by James Gregory, brother to David, in his own mathematics classes in Edinburgh the early 1700s. Since the 1680s, both Adair and Sibbald had been dealing at altogether larger scales with the power of geography, having been encharged with the description and measurement of the nation itself. Twenty years later, in October 1711, the by then gout-ridden Adair donated a further six maps of eastern Scotland, the result of survey work begun by him under Sibbald's direction in 1681, but never completed or published in full.[2] Elsewhere in the town, an anonymous Edinburgh bookseller was selling an Adair map to a member of the public for one shilling. Among geography books also sold by that unknown bookseller was Patrick Gordon's *Geography Anatomiz'd; or, A Compleat Geographical Grammar*. This work, first published in 1693, was written for the children of the gentry and would, by 1754, reach its twentieth edition. Gordon's book aimed to present "a short and exact analysis of the whole body of modern geography, . . . whereby any person may in a short time attain to the knowledge of that most noble and useful science."[3] For the four copies of Gordon's book sold at the Edinburgh bookshop between July 1715 and April 1717, one purchaser is known: Mr. David Freebairn, later to become bishop of Edinburgh, was in April 1717 minister of the parish of Gask, Auchertarder, and Dunning in southwest Perthshire.[4]

Nothing, at least directly, connects these instances in the social history and historical geography of early modern geography. For David Gregory, the first professor publicly to lecture on Newtonian philosophy and an important advocate of Newtonianism in England as well as in Scotland, the geographical practice of mapmaking was a means to instill mathematical principles. Adair's mapping—commissioned as it was by the Parliament of Scotland and the Privy Council—was a national expression of a similar need to know one's bounds. Sibbald and Adair worked closely together, at least initially. For these men, and others like the cleric Freebairn, geography books were part of a virtuoso's library, part of contemporary interest in the mathematical bases to natural knowledge, and one expression of what constituted proper philosophical enquiry. What does connect these men, then, is a shared interest in and engagement with geography at a time of wider intellectual change. This chapter suggests that, albeit in different ways and places and for different men—the Newtonian Gregory, the Baconian Sibbald, and Adair, the underachieving underfunded civil servant—geography was part of what we have come to term the "Scientific Revolution."

The Scientific Revolution is generally recognized and widely discussed as a *historical* phenomenon. Even admitting of the view of one leading scholar that whatever it was, it may not have existed,[5] the Scientific Revolution has been treated in an extensive range of work over the past twenty years or so as both a historical problem and a historiographical moment.[6] As Margaret Osler puts it:

> The Scientific Revolution is probably the single most important unifying concept in the history of science. . . . Not itself an explanatory concept, the Scientific Revolution has become the reference point for questions that guide historians of science, questions about what it was, what exactly happened, why it happened, and why it happened when and where it did.[7]

Questions to do with the Scientific Revolution as a *geographical* phenomenon have received less attention. Although it is not true that questions of a geographical nature have been entirely neglected, "where" questions of the Scientific Revolution stand a poor third behind the "what" and the "why." This is, I suggest, unfortunate, since the Scientific Revolution occurred in space as much as it did in time and ought to be located as well as dated.

There are several reasons why it may be useful to think of the Scientific Revolution as a geographical matter. One has to do with the central role of space and of changing conceptions of it in the Scientific Revolution. What scholars have seen as the essential feature of the Scientific Revolution, namely, the "mathematicization" or "geometricization" of space in the overthrow of Aristotelian natural philosophy and Ptolemaic astronomy, and their replacement both by Copernican astronomy and, latterly, by Newtonian mechanistic natural philosophy, may be considered geographical given the consideration of the celestial place of the Earth and other heavenly spheres.[8] It was, of course, also cosmographical in Ptolemaic terms and was understood as such by those late Renaissance mapmakers and mathematicians who drew upon and departed from the different editions of Ptolemy's *Geographia* in advancing practical mathematics.[9] Mapmaking and navigation were closely related geographical concerns in this respect, being both an expression of the relevance of practical geometry and essential to those voyages of discovery through which, from about 1500, conceptions of terrestrial space and of human diversity were so dramatically enlarged. A second reason has to do with new work on the nature of geography in the early modern period. This work has extended our understanding of what contemporaries then took to be the principal geographical practices—descriptive geography, mathematical geography, and chorography, the art of regional description.[10] A further reason concerns the recent attention of some historians of science to the situated nature of science's making and to questions of how knowledge travels. In this new historiography, concerns of a geographical nature are central to explaining the placed nature of knowledge making in general and of science in particular.[11] One scholar has even gone so far as to state that "by the end of the seventeenth century the language of one science, geography, routinely pervaded the language of all the sciences."[12] Although this may be establishing an unwarranted primacy for geographical matters, it is indicative of current interest in thinking geographically about the nature, practice, and reception of science and of the revitalized history of geography.

It may be helpful, then, in thinking further about these issues in relation to the Scientific Revolution to distinguish between several things. There were different conceptions of celestial space in the shift from Aristotelianism to Newtonian natural philosophy. Geography—however understood—was bound up with the enlargement of conceptions of terrestrial space. New ideas about the size and diversity of the world depended upon a role for the practical geographical and mathematical sciences. We should note, too, that scientific endeavor—including geography—was itself a located phenomenon: in certain universities and towns, in laboratories, in "the field." Scientific work has always had a geographical expression in the sites of its making and reception. At the same time, the subject "geography" has received hardly any formal attention within the traditional historiographies of the Scientific Revolution. Two of its constituent practices, mapmaking and navigation, are usually relegated to the role of secondary discourses well below physics and astronomy. Yet new work on geography's early modern history has highlighted a complexity for the subject not before revealed.

Precisely because it is possible to note such distinctions, I want to argue that we ought not to continue to leave geographical considerations off the map of knowledge concerning the Scientific Revolution. Given work alluded to above and given the "geographical" turn in the history of science, I want to establish a case for considering the Scientific Revolution as a geographical phenomenon. In so doing, I want, initially, to distinguish between geography *in* the Scientific Revolution and the geography *of* the Scientific Revolution. By the first, I mean that evidence for an understanding of what geography and its constituent practices was held to be between, broadly, 1500 and 1700, in which period the conceptual and institutional foundations for modern science were laid in and through the Scientific Revolution. In these terms, "What was geography?" and "Where was geography?" By the second, I mean the geographical expression and constitution of the Scientific Revolution. This is not to presume it to be alone a matter of national difference, at which geographical scale it has usually been considered. Roy Porter and Mikuláš Teich have rightly cautioned that "[i]t is at our peril that historians of science neglect the geo-cultural element" and that "[w]e will not gain a full grasp of the special filiation of that much-maligned but still useful beast of historical burden, the Scientific Revolution . . . until we take into account its *where,* as well as its when and how."[13] But their preferred scale of analysis, the "national context," is not the only way of framing the Scientific Revolution. I want here to suggest that the geography of the Scientific Revolution also encompasses the local sites and social spaces involved in the promotion and the reception of new ways of thinking. In this second sense, it is legitimate to think of geography within the Scientific Revolution as itself having a discernible historical geography: of people doing geography in particular places and certain ways, of the public buying geography's books and instruments, of students learning it and using it, of mapmakers at work in the field and so on.

In order to illustrate these two concerns, this chapter focuses on one particular geographical context, namely, Scotland between the 1681 appointment of John Adair by the Privy Council to undertake detailed maps of the country, along with the 1682 appointment of Sir Robert Sibbald as Geographer Royal, and the Union of the Parliaments in 1707. As Paul Wood has shown, the period is a significant one for the Scientific Revolution in Scotland.[14] What is also clear is that different sorts of geography and conceptions of science were at work in this period. In the second part of the chapter, I consider geography's texts, the mathematical and mapping practices of Adair and others and the teaching of natural philosophy in university curricula.[15] Let me begin, however, by considering the geography *of* the Scientific Revolution in a little more detail.

The Scientific Revolution as a Matter of Geography

"Where" questions to do with the Scientific Revolution are apparent but not explicitly stated throughout H. Floris Cohen's monumental *The Scientific Revolution: A Historiographical Inquiry* (1994). In general terms, his "where" questions consider either the causal agencies informing the variant expressions of the Scientific Revolution in different national contexts—Puritanism, technology, the advance of capitalism, the printing press, and the voyages of discovery are cited in this respect—or they address the supranational issue of why the Scientific Revolution "eluded" the non-Western world (issues also discussed above by John Henry in chapter 3). Cohen's attention to the work of leading scholars on the Scientific Revolution also documents what we might call the causes of the Scientific Revolution as a matter of geographical difference:

> Problems of place and problems of explanatory scope are another matter, however. Even if we were to accept every explanation *qua* explanation, it can still be seen that each and every one of our explanations is incomplete in that it addresses only part of the problem at issue. Both the "Archimedean" explanation and its "Aristotelian" counterpart picture Galileo as the culminating point of a particular tradition in Renaissance thought. Both are devoted exclusively to Italy; neither has anything to say about other centers of the Scientific Revolution. By contrast, the "sceptical" explanation, while capable of being extended to the whole of Europe, is chiefly about France . . . Again, the "Hermetic" explanation . . . is conspicuous in leaving Galileo out of the picture altogether. Only the "Copernican" explanation covers the Scientific Revolution in its full

geographical scope, in that the effort to meet objections to the Copernican hypothesis led to a reconceptualization of standing scientific ideas by pioneers all over Europe.[16]

Cohen has added to this sense of geographical variation by considering different national interpretations of the Scientific Revolution (and again, see Henry here, chapter 3). British historians' approaches have been distinguished by their being so overwhelmed with empirical material as to be incapable of establishing a historical thesis to explain it: "It is my impression," Cohen notes, "that this not uniquely yet altogether rather peculiarly British unease with historical theses has contributed its bit to the present quandary in which, historiographically speaking, the so-called Scientific Revolution finds itself."[17]

We might find ourselves in less of a quandary if the distinction was made more clear between geographical difference as causal agency and geographical difference as consequence of the Scientific Revolution's outcomes and between geographical information within the revolution in the sciences overall. For Reijer Hooykaas, for example, new geographical facts consequent upon the Portuguese voyages of discovery were a key element in the Scientific Revolution. New facts derived through geography "triggered off a movement which, growing into the avalanche of upheaval in sixteenth-century geography, opened the way for the reform, sooner or later, of all other scientific disciplines."[18] More recent work on geography books has likewise shown how the impact of New World knowledge transformed the nature and methods of geographical writing.[19] Richard Westfall has shown in his assessment of the 630 individuals making up the "Western scientific community" between Copernicus's birth in the 1470s and 1680 that practitioners of navigation and of mapmaking represented one in seven of the total.[20] And as Jim Bennett has shown in respect to instrumentation and practical geometry within the Scientific Revolution, the shift away from Ptolemaic concerns in the mapmaking and atlas publications of the mid-sixteenth century and the role of mathematical geography as "operative knowledge" means we should now situate geometry, mapmaking, and surveying more centrally in the history and practical development of mechanistic natural philosophy.[21]

Lesley Cormack, in examining the nature of geography and the geographical "community" in Cambridge, Oxford, and in Gresham College, London, between 1580 and 1620, has identified three sorts of geography and geographers in overlapping groups. The first and smallest group focused on mathematical geography and its evident utilitarian connections. A second larger group was concerned with descriptive geography. The third group focused on chorography, or regional description. For Cormack, chorography was "the most wide-ranging of the geographical arts, in that it provided the specific detail to make concrete the other general branches of geography."[22] It was, indeed, an essentially conservative form of geographical inquiry since it depended upon knowledge of and from the leading social groups of the time and was part of earlier narrative traditions of *descriptio* and of late Renaissance "self-fashioning" rather more than of new ideas within natural philosophy.[23]

For Cormack, geography in these terms was central in inculcating a sense of English national identity that was inward looking in its attachment to local place and country and outward looking in its attention to the nascent British Empire. Noting the courtly and practical side of geography in these terms reinforces Hooykaas's claim to the importance of geographical discovery in extending conceptions of terrestrial space. Isaac Newton, Lucasian Chair of Mathematics and Natural Philosophy at Cambridge, was charged with instruction in geography as part of his teaching responsibilities. As William Warntz has shown, Newton's textual revisions and additions to Bernhard Varenius's *Geographia Generalis* of 1650, together with later amendments and editions by James Jurin, Richard Bentley, Edmond Halley, and Roger Cotes, were part of shared interests in the application of geometry to geographical phenomena. "The *Geographia generalis* commanded significant concern and respect from Newton and the Newtonians. Their attention to it was not aberrant or spasmodic but an integral part of their concerted thrust in science."[24]

On the one hand, then, mapping the heavens, establishing secure maritime trade routes and surveying one's nation each depended upon the mathematization of space in practical context at cosmological, geographical, and chorographical scales. Yet, on the other hand, as Peter Dear has reminded us, "the mixed mathematical sciences were generally held to be of a lower status [than 'physics' or natural philosophy] because they did not, as natural philosophy did, concern the essences of the things and processes of which they spoke; rather than giving true causal explanations of physical phenomena, rooted in the real natures of the things involved, they just coordinated quantities."[25] In these terms, recovering geography's place in the Scientific Revolution is also to raise questions about what are the "essential" subjects constituting that revolution. Given that geographical knowledge was also implicated in questions of natural magic and of astrology, in the rise of systematic natural history and in the advance of what the later seventeenth century understood as "political" and "natural arithmetic,"[26] there might now be grounds for revising upward the relative position of geography within the hierarchy of subjects constituting the Scientific Revolution, or at least for taking it more seriously than hitherto. Further, thinking geographically—about the subject and about the location and movement of scientific knowledge—is more common than once it was. Such a claim with respect to the Scientific Revolution is supported by recent shifts in its interpretation concerning the role of local agency and particular contexts. Such shifts, as Osler argues, might even be construed as geographical in nature:

> The new historiography is characterized by an increasing awareness of the importance of the intellectual and social context within which ideas have developed, along with a renewed respect for the presuppositions and concepts of the historical actors rather than those of historians. It takes the history of science to places where it has not usually been seen before: into the courts, into the streets, into the countryside, and into local societies.[27]

Such concerns with context and with the local and specific find ready parallel expression in the work of geographers on the social and situated nature of knowledge, and in the work of historians of science whose social constructivist interests are apparent in attention to the geography of science. Among geographers, David Livingstone has taken the lead in studying the social spaces of scientific knowledge and the regional variations in the reception of scientific theory.[28] Among historians of science, Crosbie Smith and Jon Agar have reviewed territorial themes in the making of science under two headings, "Of the Territory" and "Of Privileged Sites." Steven Shapin and Jan Golinski have paid attention to the spatial constitution of natural knowledge.[29] Others such as Bruno Latour have been interested in the mobility and accumulative nature of scientific knowledge, and with how, in Golinski's words, things travel "beyond the laboratory walls."[30]

A central feature of this work has been its localist nature. In discussing the nature of cartographic knowledge in and of early modern Europe, for example, David Turnbull has commented that "the picture of science that has emerged from empirical investigations of both contemporary and historical scientists is that all knowledge is constructed at specific sites through the engagements of particular scientists with particular skills, material tools, theories and techniques. . . . Thus a fundamental characteristic of scientific knowledge is its localness."[31] As Adir Ophir and Steven Shapin noted in 1991, "[T]his influential *localist* genre, marked by attention to national and regional features of an enterprise once regarded as paradigmatically universal," is relatively recent. It is, as they also note, an extension of the relativist agenda established by sociologists of scientific knowledge: "[R]elativism can be practically defined through the notion that all knowledge claims and judgments secure their credibility not through absolute standards but through the workings of *local* causes operating in contexts of judgment."[32] Such claims are open to empirical examination, of course, since the making of science—in this case understanding better the geography *of* the Scientific Revolution and the place of geography *in* the Scientific Revolution—may depend upon showing in which sites exactly the new natural philosophy was advanced, upon knowing the different forms taken by scientific and geographical knowledge in those sites, and upon tracing the connections between different sites and the people and ideas found there.

For Steven Shapin, a focus upon the local nature of knowledge's making is valuable:

> Suppose one regarded it as established beyond doubt that science is indelibly marked by the local and spatial circumstances of its making; . . . and that scientific knowledge is made by and through mundane—and locally varying—modes of social and cultural interaction. If one granted all this, one would be treating the "localist" or "geographical" turn in science studies as a great accomplishment—telling us a series of important things about science which previous understandings have systematically ignored or denied.

But it may not be enough:

> And yet I also want to say that it is still incomplete and that it is in danger of missing something very important about science. The problem here is not that the geographical sensibility has been taken too far but that it has not been taken far enough. We need to understand not only how knowledge is made in specific places but also how transactions occur between places.[33]

In understanding the Scientific Revolution, then, attention to the local and to the situated nature of the new science's making may well, to paraphrase Osler, take us to places where we have not been before and to questions of the movement of knowledge *over* space as well as to its making and reception *in* certain places. If the "national context" is insufficient and a problematic way of framing the making of science generally, not just of the Scientific Revolution, then we may indeed need to take the issue of localism seriously. This does not mean, however, that we must neglect wider connections.[34] In introducing in 1996 a set of papers on the Scientific Revolution as narrative, Steven Harris posed the question, "Are we to understand the history of early modern science in terms of the global framework provided by grand narratives of the Scientific Revolution, or ought we to think solely in terms of local knowledges rooted in particular places and times and bounded by particular social contexts?"[35] In thinking about the connections between the Scientific Revolution as grand narrative or its local sites and contexts, Harris turns to ideas around "the geography of knowledge" and notes a distinction I make here between the knowledge of geography and the geography of knowledge. The latter, in general terms, "entails the spatial and temporal distribution of people, graphics, and objects required in the making of all sorts of natural knowledge, not just geographical knowledge." Given, he notes, that such questions involve the movement and not just the distribution of things, "we might distinguish three related approaches to the geography of knowledge," outlined thus:

> In the first instance, it means a static geography of place: where did people "do science"? where were they when they aimed telescopes and recorded observations, logged positions and sketched in charts, performed dissections or prepared medicaments, executed experiments and calculations, or wrote and published the accounts of their activities? In the second, it means a kinematic geography of movement: whence came the constituents of scientific practice and knowledge, the measuring or cutting instruments, the authoritative texts or latest correspondence, the exotic natural curiosity or well wrought experimental apparatus, the returning botanist or navigator? In the third sense, geography of knowledge also means the dynamics of travel: why and by what means did all these movements take place? what was the *anima motrix* responsible for the multiple peregrinations of the elements of knowledge?[36]

Harris illuminates this conceptual thinking (and acknowledges his debt to the work of John Law and Bruno Latour) by reference to tangled "thread maps" of knowledge movement "knotted" in particular places as a result of the coming together of people, knowledge, and artifacts. Empirical evidence centers upon the role of "long-distance corporations"—such as the Dutch East India Company or the Society of Jesus—which, located in certain places, coordinated the ebb and flow of knowledge into and from particular institutional sites.

For Harris, attention to site and to movement demands attentiveness to the questions of geographical scale and explanatory scale. "Just as we imagine a simple spatial metric for the operation of a corporation, so too we can imagine a similar metric for a given scientific practice: that is, what was the geographical extent and temporal duration of the practices required in a particular set of astronomical observations, the construction of an instrument, or the writing of a scientific treatise?" Based upon the different geographical and temporal scales of practice, distinction may be made, he suggests, between the "big sciences"—such as stellar and planetary astronomy, cartography, mathematical and descriptive geography, natural history, and mixed mathematics—and the "small sciences" of experimental philosophy, anatomy and surgery, and most of "pure" mathematics. Small does not imply insignificance: "Rather it refers to the number of observations, . . . their geographical range, . . . the length of time needed to gather pertinent information, or the space in which the crucial observations were gathered."[37]

These claims return us to the place of thinking geographically in the historiography of the Scientific Revolution. For Harris, most work within the traditional historiography of the Scientific Revolution *and* by social constructivists attentive to the local nature of science's making has focused on the "small sciences"—that is, it has excluded geography. Further, the work of the spatially attuned social constructivists, far from challenging traditional views about the nature and subject content of the Scientific Revolution, may have reinforced them:

> The "localist thrust" . . . has not only predisposed researchers to choose research sites that are spatially and temporally circumscribed, it has also encouraged the selection of scientific practices that were themselves spatially and temporally circumscribed. Consequently, the constructivist program has made a selection of sciences that is, if anything, more restrictive than that found in the historiography of the Scientific Revolution. . . . Thus we would seem to have a grand narrative blind to big sciences and microhistories unacquainted with scientific practices that extended beyond the laboratory, court, or academy.[38]

Siting geography in the Scientific Revolution is, then, a matter of discursive range, of explanatory scale, and of historiographical significance. It is not, in any simple sense, a question of recovering precise subject boundaries or of the making and movement of science within given national contexts without reference to local sites or to connections above and beyond the nation. Thinking of the Scientific Revolution as a geographical phenomenon may be to think of it differently epistemologically. "If we wish to tell *longue durée* stories of early modern sciences, and not just the ultimate origin story of modern science or a series of unconnected microhistories trapped in their respective 'black holes of context,' then we will need to find both an epistemology and a narrative format capable of moving across scale."[39]

I do not want here to privilege local sites and social spaces, any more than I want to give undue weight to the Scientific Revolution as a "grand narrative" at a national scale in which geography, a form of "big science," is written out. Neither do I simply see this study of geography and of the Scientific Revolution in Scotland addressing all these claims, let alone providing a new narrative format for that revolution understood as a geographical phenomenon at different scales. I am not claiming that Gregory with his students, Adair with his governmental support, and others with their interest in geography during the later seventeenth and early eighteenth centuries together "fixed" the place of geography in the Scientific Revolution in one country. Yet late-seventeenth- and early-eighteenth-century Scotland does merit examination. As Wood has shown, Newtonianism was early and enthusiastically welcomed in Scotland for several reasons: its favorable reception among Calvinists, the coincidental reform of the university system between 1690 and 1715, and that highly developed network of patronage and scientific communication centered upon Newton himself. Even so, the Scientific Revolution is as much a "historiographical minefield" for Scotland as for other countries. Scotland is anyway an anomalous example of a nation having lost her independent statehood through parliamentary union with England in 1707.[40] Yet Scotland was small and "local": as a territory, in the number of its universities, and in the size of its "scientific community." It was also "global" given its connections with England and with Europe, particularly in the training of its medical men, and its own (failed) attempts at colonial expansion in 1699 and 1700 through the activities of that short-lived "long-distance corporation" the Scots Company Trading to Africa and the Indies, usually known as the Darien Company.[41] I hope to show how, in several ways and in several places, geography was part of the situated and changing nature of science.

Geography and Science in Scotland, 1681–1707

I explore here three related themes to do with geography and science at the end of the seventeenth and beginning of the eighteenth century: the parliamentary and "official" promotion of geography in the work of Robert Sibbald and John Adair; the production of geography books in Scotland; and the teaching of geography in the universities, particularly in relation to the teaching of natural philosophy at the University of Edinburgh. The dates of 1681 to 1707 are partly matters of convenience, since connections between geography, science, and national identity are apparent before and after that span.[42] The appointment by royal warrant of Sir Robert Sibbald as Geographer Royal (and King's Physician) in 1682 might itself suggest that geography was recognized by leading figures as of intellectual value by the late seventeenth century. As I have elsewhere shown, however, Sibbald's geographical work, based on chorographical inquiry with a view to an overall national understanding, was only ever partially successful, and Sibbald's sedentary empiricism was not the only sort of geographical knowledge at work.[43] Yet this period encapsulates the features central to understanding the Scientific Revolution as a geographical phenomenon in Scotland. There is evidence for the advance of Newtonian over Cartesian natural philosophy and evidence that geography was implicated in that move. The regenting system of university teaching—through which we can best see the correspondence between course content and books and instruments purchased—was only discontinued in Edinburgh in 1708 when the university undertook curriculum reforms. As authors wrote geography books of one sort or another, geographers were at work elsewhere charting the nation's bounds.

Geography in the Field:
Chorography and Practical Mathematics in the Work of Sibbald and Adair

Sibbald's commission upon appointment was to "publish the naturall history of ye Country and the geographical description of the kingdome." Emphasis was laid upon knowing through geographical inquiry the "virtues and uses" of Scotland's national products. Geography was considered crucial to the intellectual and material wealth of the nation.[44] Sibbald's warrant was, effectively, a royal proclamation of the benefits of a utilitarian and empiricist geography. The substance of Sibbald's knowledge depended almost entirely, however, upon his not being "out there" in the field, but upon his remaining in Edinburgh and receiving the responses to circulated queries on Scotland's geography addressed to credible persons and institutions. In contrast, John Adair, who worked closely with Sibbald—that "Mathematician and skilful Mechanick" as Sibbald described him[45]—depended for his geographical understanding on being mobile, on making accurate maps as the result of direct and instrumental encounters with the shape of the nation.

Sibbald's geographical work was undertaken through the circulation to leading groups in Scottish society of queries that, in theory anyway, followed a predetermined format. "Omit nothing," he cautioned his respondents, "that can any Way contribute to the Knowledge of the present Geographical face of our Country."[46] Examination of the responses and of the respondents reveals three main characteristics. Some of Sibbald's correspondents were drawing upon earlier geographical and chorographical work, notably that relating to a failed attempt by the Church of Scotland in the 1630s and 1640s to undertake a parish-by-parish geographical survey of the country. Several of Sibbald's correspondents were themselves reliant upon native knowledge and upon smaller networks of credible persons in their local areas. To a degree, Sibbald's own position in Edinburgh as a "center of calculation" (or "knot" of knowledge, to echo Harris) was replicated on a smaller scale elsewhere in Scotland.[47] A small handful of Sibbald's respondents were actively engaged in geographical work. Among these were Alexander Pennecuik, whose *A Geographical and Historical Description of the Shire of Tweedale* (1715) was to have contained (but never did) maps by John Adair; James Wallace senior and junior, the latter of whom published *A Description of the Isles of Orkney* in 1693 based on his father's work; and William Geddes, who prepared but seems not ever to have published a work entitled *Geographical and Arithmetical Memorials*.

In these terms, and at one level, Scotland's geography was being together by what David Lux and Harold Cook have termed "the strength of weak links" sustaining small and locationally disparate networks of individuals.[48] Sibbald's central accumulative function was mirrored at a smaller scale by other men, his reliable correspondents, most of whom were resident in the places they were describing. Their language of measurement was textual description and potential economic utility. At the same time, but in other ways, Scotland's geography was being constituted through practical mathematics in the survey and mapping work of John Adair. He, too, was dependent upon distant others. For engraving, Adair was reliant upon James Moxon, brother of the mapmaker and the Hydrographer Royal Joseph Moxon, and upon the London-trained James Clark, Engraver to the Scottish Mint.[49] Adair's "metric," his "kinematic geography of movement," to borrow from Harris, was wider still. He bought his mapping and surveying instruments from the Low Countries.[50] Adair's work demanded—and, in several ways, its and his success depended upon—a language of exactitude. That this was so is clear from the act of 15 June 1686 passed by the Parliament of Scotland: "In Favours of John Adair, Geographer, for Surveying the Kingdom of Scotland, and Navigating the Coasts and Isles thereof," with its emphasis upon "Exact Geographical Descriptions" and remarks about "[t]he want of such exact Maps, having occasioned great losses in times past."[51] This is not to say that Adair had no native help "on the ground." The assistance he received in mapping the Western Isles from Martin Martin, the Gaelic-speaking author of two books on the Hebrides, was the local expression on Scotland's margins of broader networks of natural knowledge coordinated by the Royal Society of London.[52]

It is possible to see in the collaborative sedentary activities of Sibbald and in Adair's field-based instrumental encounters two broadly different means to the constitution of "the field" of geography. For the first, the knowledge was made by credible locals and circulated back to the resident Sibbald as, by intention anyway, standardized responses to written questions. For the second, knowledge was made by being active in the field, by a dependence upon one traveling local, and by reliance upon instrumental accuracy using imported equipment. It is true, of course, that neither Sibbald nor Adair ever completed his geographical endeavors. Both men, effectively geographical civil servants appointed by the king and Parliament respectively, were underfunded by the very bodies who commissioned them. Yet the fact that geography was being done at all is important given the absence of geography from our current understanding of the Scientific Revolution in Scotland more generally.[53]

What is less easy to see from either man's work is any precise and formal engagement with advances in mechanistic natural philosophy. The activities of Sibbald and Adair and those others with whom they worked may be said to demonstrate those roots to modern empiricism as a matter of direct sense experience *and* of social warrant that Shapin emphasizes.[54] The emphasis on formal methodological rules to discipline the production of natural knowledge by managing the effects of human intervention, an emphasis Shapin has also seen as central to the advance of the new natural philosophy, is easier to see in Sibbald's queries than in Adair's. The challenge posed to extant authoritative claims by seeking evidence for oneself in the field is clearer in Adair and in his maps than it is in Sibbald's correspondents' work or in his books. In these terms, the picture we have of geography "in the field" is of a Baconian empiricism resolutely grounded in the potential benefits to Scotland of secure understanding of the nation's bounds. Geography was part of contemporary scientific investigation. But its making in the field by Adair in person or through Sibbald's correspondents did not afford those two geographers an opportunity either to muse upon the celestial order or to secure continuing patronage for geography's position as a formal practice in the "new" science.

Textual Traditions: Books of Geography

Recent work on geography's textual tradition has enumerated the publication of books of special geography and shown the central place of works of descriptive geography in maintaining the classical tradition in, for example, the teaching of history in late-seventeenth- and eighteenth-century England.[55] Almost no work has been done on the few books of geography published in Scotland before the later works of men such as William Guthrie.[56] My attention here to three Scottish geographical texts does not aim to document a complete book history for the works. Rather, I want to suggest how geography's books—and the social connections through which they were produced and moved—ought to be included within those wider social and intellectual networks concerned with what geography was and what it could do.

James Paterson's *A Geographical Description of Scotland* was published in 1681 with revised and expanded editions in 1685 and 1687. Paterson was an Edinburgh mathematician who advertised his work "at the Sign of the Sea Cross-Staff and Quadrant." He also wrote *The Scots Arithmetician; or, Arithmetic in All Its Parts* (1685) and, in the same year, *Edinburgh's True Almanack; or, A New Prognostication for the Years 1685–1692*. Paterson's *Geographical*

Description is, essentially, an almanac with tide tables, dates of fairs, and so on, "Exactly Calculate and formed, for the use of all Travellers, Mariners, and others, who have any Affairs, or Merchandizing in this Kingdom of Scotland."[57] Similar work was being done for Glasgow by James Corss, author of a Glasgow almanac in 1662 and, in 1666, of *Practical Geometry*. Corss also advertised a teaching capacity: "Arithmetique Geometrie Astronomie and all uther airts and Sciences belonging thereto as horometrie Planimetrie Geographie Trigionometrie," as his advertisement of April 1658 put it.[58] Corss's teaching was perhaps intended to rectify what he saw as a lamentable lack of mathematical prowess among his countrymen, John Napier apart. In 1662, he commented, "I have often lamented with myself to see so many Learned Mathematicians to rise in sundry parts of the world, and so few to appear in our Native Country. In other things we are parallel with (I shall not say in a superlative degree far above) other Nations; but in Arts and Sciences Mathematical, all exceed us."[59]

The work of Paterson and of Corss was in keeping with contemporary interest in practical mathematics and in geography's place as a basis to commerce. Such attention to almanacs reflects that wider interest in the "curious nature" of things, as Hunter has shown of Robert Boyle's engagement with natural magic and with the problem of second sight in late-seventeenth-century Scotland.[60] I am not equating the work of Paterson and Corss with that of Boyle. I am noting that in their investigations we can see shared attempts being made to define the boundaries of the natural and observable realm and, therefore, to know what was the subject of proper investigation in natural philosophy. Since the boundaries were often imprecise, it was quite natural that apparently different intellectual traditions should be found in the work of one individual. Adair's attention to mathematical exactitude in his mapping, for example, did not stop him circulating queries whose essential focus, in contrast to those queries circulated by Sibbald, was the aberrant and the curious. Even as he wrote his books and taught, Paterson ran a shop in Edinburgh that included weatherglasses among other mathematical instruments.[61] If we are to credit that other evidence cited above, however, Paterson, the local supplier, could not meet the needs of John Adair, Scotland's "official" mapmaker at the time.

Paterson's rhetoric in understanding Scotland's geography was the language of mathematical description and commercial utility. In contrast, Matthias Symson's *Geography Compendiz'd; or, The World Survey'd*, published in Edinburgh in 1702, looked to list the countries of the world overall and to a more formal language of philosophical speculation. Symson also produced the *Encheiridion Geographicum* in 1704, a digest of geographical terms, and in the same year as *Geography Compendiz'd,* a short work entitled *Caledonia's Everlasting Almanack; or, A Prognostication Which May Serve for ever the Kingdom of Scotland.* Symson's geographical work was not original, being as the 1702 book stated, "[c]ollected from the most approv'd writers on that subject." In that respect as well as in its dedication to a patron of influence—"for the Use of the Marquess of Douglass"—Symson's work was typical of its time. The fact that his 1702 book is subtitled "Volumn I" suggests it was part of a never-completed larger project. It is noteworthy, however, that like Gordon in 1693, Symson terms geography a "science." In the prefatory material, which discusses the essential character of geography as the study of the Earth, there is recognition of the questions posed by debate on Ptolemaic and Copernican astronomical systems. "But whether [the Earth] be immoveable and seated in the Center of the Universe, according to *Ptolemy* and *Tycho;* or between the Orbs of *Mars* and *Venus,* making a Diurnal Rotation about its own Axis, and an Annual Revolution about the Sun, according to *Copernicus;* belongs rather to *Astronomy* than *Geography.*"[62]

This statement appears in similar form in the opening paragraphs of Gavin Drummond's *A Short Treatise of Geography, General and Special* (1708), which was, the author notes, "[c]ollected from the best Authors upon that subject, for the use of Schools." Like many works at this time, Drummond's view of geography's utility for schooling was that it should be an aid to history, "in understanding of the Roman Authors and Modern Histories, and reading them with that Desired profit and pleasure," as he put it:

> Geography is a Science, which teacheth the Description of the exterior part of the Earthly Globe, as it is composed of *Land* and *Water,* especially the former. It differs from *Cosmography,* as a part from the whole, from *Chorography* and *Topography,* as the whole from its parts. The *Terraqueous* Globe is situated, according to *Ptolemy* and *Tycho,* in the *Centre* of the World, but according to Copernicus, between the *Orbes* of *Mars* and *Venus.*[63]

This level of textual exegesis matters. For it is possible that one of "the best Authors" was Matthias Symson, since Drummond's *Short Treatise* was printed by Andrew Symson, Matthias's father. Andrew Symson's geographical interests extended beyond printing others' works. Prior to being a printer, Symson had been a parish minister at Kirkinner in Wigtonshire. In 1684, in response to Sibbald's directives, Andrew Symson undertook a "Description of Galloway." From 1686, Symson was minister in Douglas, Lanarkshire, under the patronage of James, Marquis of Douglas. As an Episcopalian clergyman, however, Andrew Symson was outed from his parish in 1690, and he relocated to

Glenartney in southwest Perthshire, where he revised his geographical materials before forwarding them to the Geographer Royal in 1692.[64] Symson's library catalog, published at his death in 1712, shows that he owned a copy of
his son's work among other geographical texts.[65]

Several implications arise from these remarks. What, or perhaps who,
should we study in geography's book history? Given that Symson's and Douglas's books drew upon established descriptive traditions and upon others'
works, examining the contents of geography's books is important. Yet it may
also be important to know how geography was regarded in the late seventeenth and early eighteenth centuries to trace the social networks through
which such geographical knowledge was produced and through which it
moved. The fact that an anonymous bookseller in early-eighteenth-century
Edinburgh was selling four copies of Gordon's *Geography Anatomiz'd*, several
of Laurence Eachard's *The Gazetteer's; or, Newsman's Interpreter* (first published in 1692 but in its twelfth edition by 1724), and John Harris's *The Description and Uses of the Celestial and Terrestrial Globes* is evidence of public
interest in geography as a practical subject. In the first place, it was a subject
of polite learning useful in understanding one's contemporary world, in the
second a matter of mathematical inquiry.[66] The presence of these geography
books reinforces the need to revise our assessment of the place and nature of
the Scientific Revolution in Scotland and to look not simply at texts themselves and at their authors, but at the publishers, patrons, and printers, at
books' reception, and at who may have been the audiences. Such evidence demands that we look at the public engagement with geography, not just at
more prominent figures such as Adair and Sibbald. Library lists such as that
surviving for Andrew Symson or for Sibbald,[67] may not document how books
were used. But in the absence of other sources, they can show what those
people involved in the promotion and reception of new ideas had in their possession as one means of engagement with contemporary advances in natural
philosophy. This engagement is all the more important when links can be
made between particular texts and their use in teaching.

Teaching Places: Natural Philosophy and Geography
in Scottish Universities before 1707

Evidence for geography teaching within Scotland's universities in this period is uneven in geographical coverage and detail. At King's College, Aberdeen, the regent and later subprincipal, William Black, taught geography and
cosmography in 1692–93, organizing his geographical material as a series of
"propositions" to be debated and solved.[68] This teaching method mirrored the

established practice of instruction through disputation and the power of logical argument, rather than the demonstration from direct observation and measurement as practiced, for example, by David Gregory's students at Edinburgh and at St. Andrews. Lecture notes belonging to one Alexander Maclennan, a student with the Aberdeen regent George Skene, show that Skene's geography lectures, delivered between 1701 and 1704, were likewise given as practical propositions to be solved through mathematical and logical reasoning.[69] There is no certain evidence for geography teaching at Glasgow University at this time. At St. Andrews, James Gregory, the professor of mathematics, is recorded as giving "some succinct institutions in Astronomy and Geography," in 1685. Nothing is known of their content, nor is there evidence of his students producing maps of the college bounds as his brother was to do in the town in 1690.[70]

The clearest picture of geography's teaching in the universities in this period comes from Edinburgh, given that university's gradual ascendancy over others in Scotland and the extensive evidence for the teaching of natural philosophy that has survived, including lists of books and student dictates. Using such sources, it is possible to document the relationship between texts and teaching in several subject areas, particularly before the end of the regenting system at Edinburgh and other structural changes in the university in 1708, and perhaps especially for the period 1693–1707.[71]

In the teaching of logic and in metaphysics at Edinburgh, the scholastic tradition prevailed, and even when the Cartesian method was mentioned in the dictates and student theses of the 1670s and 1680s, it was, in the teaching of logic most notably, commonly included within an Aristotelian framework. Cartesian ideas in metaphysics seem to have been accepted from the early 1670s, but there is no evidence of any dramatic shift in the replacement of Scholasticism. Many of the books used were concerned more with theology than with metaphysics, a reflection, as Shepherd puts it, of "a recurring doubt, which appears in metaphysics dictates and theses throughout the century, about what metaphysics should encompass."[72]

A more certain departure from established tradition is apparent in the teaching of ethics (in the third year of the course) and, particularly, in the fourth-year natural philosophy teaching.

Viewed in relation to the wider intellectual developments of the seventeenth century, the natural philosophy dictates and theses are perhaps the most significant part of Scottish university teaching of that period. Through investigation of them we can see what impact the scientific revolution had in university circles, how soon and to what extent Cartesian gave way to Newtonian ideas, and how much notice was taken of experimental science, the practical application of science, and events such as the creation of the Royal Society.[73]

Before the 1660s, the natural philosophy dictates and theses were largely Aristotelian, with initial references to Cartesian ideas being treated circumspectly. By the later 1670s and 1680s, Cartesian ideas had been accepted, and many of the lectures and theses were wholly Cartesian in their physics, in turn giving way to Newtonianism from the late 1680s onward. The university library acquired a copy of Newton's 1687 *Principia* in 1690. There are numerous references within this overlapping shift from Aristotelianism to Cartesianism to Newtonianism to the experiments and writings of contemporary natural philosophers—Boyle's experiments with the air pump are frequently mentioned, for example—and to the activities of the Royal Society of London.

Geography can be sited within this general picture. Evidence comes from the natural philosophy teaching of John Wishart, variously regent and professor of humanity and of philosophy at Edinburgh between 1654 and 1667 and again from 1672 to July 1680. "Dictates on Geography," taken by one of Wishart's students, John Cranstoun, show Wishart's teaching to have focused upon cosmography, with regional and local descriptions of Great Britain in which "Scotia" as a kingdom figured centrally, and in his later material, upon Cartesian natural philosophy.[74] Wishart's attention to geographical concerns structured around the classical distinctions of cosmology, geography, and chorography was one element of what Shepherd has seen as the varying treatment of new ways of thinking. Wishart outlined Cartesian philosophy in his teaching in the later 1660s and early 1670s. He described, for example, Descartes' theories of local motion and extension, his views on matter, form, and the physical earth, but he did not adopt Descartes' new philosophy, chiefly on the grounds that it limited God's power. In his dictates and in his graduation theses, we can see, as Shepherd puts it, "a curious mixture of the old philosophy and the new." Cartesian physics was rejected in Wishart's 1667 dictates, yet Newton was invoked, if more for his value as an opponent to Descartes than for his own work. In 1675, Wishart mentioned five possible world systems: those of Ptolemy, Copernicus, Tycho Brahe, the Cartesian vortex theory, and a new system that placed the moon at the center of the universe. By the 1680s, Wishart was reporting to his students the results of experiments by Joseph Moxon, the Hydrographer Royal, and was citing Newton's work on the theory of light in the *Philosophical Transactions*. As Shepherd remarks, "For one who was prepared to give Newton's theory of light an airing in his lectures, the failure to give his full assent to Copernicus' heliocentricity is indeed surprising."[75]

Andrew Symson was one of Wishart's students, graduating in 1663.[76] Had Symson been a student of natural philosophy at Edinburgh in an earlier period, it is probable that Aristotelian works, such as Honter's *Rudimentorum Cosmographicum* (1531), which was used to teach "geographie" to final-year students in the 1620s, would have been used, as it was by the historian George Buchanan at Glasgow between 1574 and 1577 and by the churchman Andrew Melville at St. Andrews in the 1580s.[77] Unfortunately, book purchase records in Edinburgh do not mention works of geography until 1664, one year after Symson graduated. Maps were bought and displayed in the university library in 1664, and in 1669, payment was approved for "2 reed [red] skinns to cover the great glob[e]s."[78] In 1671, library acquisition lists record that Blome's *Geographical Description of the World* was purchased, and, in the following academic year, Joseph Moxon's *Tutor to Astronomy and Astronomy.*[79] The title of Blome's work would suggest that this was his *A Geographical Description of the Four Parts of the World,* which was published in London in 1670 and which largely derived from the work of the French geographer Nicholas Sanson. Blome later went on to produce, in 1680–82, *Cosmography and Geography in Two Parts,* the first English-language translation of Berhard Varenius's *Geographia Generalis* of 1650. The Moxon work is interesting. Moxon had published *A Tutor to Astronomie and Geographie,* subtitled *Or An Easie and Speedy Way to Know the Use of Both the Globes, Celestial and Terrestrial* in 1659. This was in a second edition by 1670, with later editions appearing until 1698. The wording of the reference in the Edinburgh acquisitions catalog suggests the book to be Moxon's 1665 *A Tutor to Astronomy and Geography.* The difference is important. Whereas the 1659 work outlined basic geographical terms and discussed Ptolemaic cosmology as modified by Tycho Brahe, the 1665 work favored Copernician cosmology and dealt carefully with the theological arguments as to why one should favor such thinking.[80]

In 1686, the university purchased two celestial globes, one Copernican, the other Ptolemaic. These were foreign-made instruments: the university authorities are recorded as paying "expences of entry & customes."[81] Different settings of celestial globes (and terrestrial globes) in such ways were not unusual. Different positionings for the elliptic and different gearings allowed for different modeled representations of the sphere of the sun. Although the purchase of two differently calibrated instruments is interesting given the other evidence presented here concerning mathematicians' engagement with geography, we must note, however, that many globe makers, even as late as the nineteenth century, "continued to produce Ptolemaic globes as if no Copernican revolution had taken place."[82] The provenance of the globes is consistent with what is known more generally about the trade in scientific instruments in Scotland at this time and with what little we know of Adair's purchasing of his instruments.[83] Later entries also point to books being purchased from Holland.[84] Overseas connections are again apparent in the purchase in 1696 of *Divers ouvrages de mathematiques et de physique et observations de l'astronomie et la geographie, par Messieurs de l'Academie Royale,* published in Paris in 1693. In 1698, the library in Edinburgh was purchasing a new copy of Philip Cluverius's *An Introduction into Geography, both Ancient and Moderne,* to replace the old one first published in Oxford in 1654. Several geography books were included in those mathematics and geometry texts purchased in London in the autumn of 1703 by James Gregory, by then professor of mathematics at Edinburgh. It is likely, too, that Gregory was behind the purchase of a pair of celestial and terrestrial globes in February 1705. This purchase was facilitated by John Vallange, an Edinburgh stationer and bookseller. In 1702, Vallange was the central coordinating figure behind the printing and sale in the city of Matthias Symson's *Geography Compendiz'd.*[85]

Records of book and instrument purchases are no secure guide to how the artifacts in question were used, or to readers' and users' comprehension. Evidence relating to one university is not straightforwardly applicable to another. It is clear, however, that Edinburgh was the first to embrace Cartesian and, later, Newtonian thinking in its teaching and that, in respect of the latter, it did so by the 1690s. Geography was included in natural philosophy teaching about the nature of celestial and terrestrial space, as is chiefly apparent in Wishart's engagement with Cartesian natural philosophy from the later 1660s. Practical use of the globes was undertaken in the teaching of mathematics before David Gregory took his mathematics class out to measure the extent of their own university's estate and that of local churches. His brother, James Gregory, wanted to take geography's utility further still. He argued for the potential benefits of such training to the Scottish economy in advocating the establishment of a "Navigation and Writing School" by which young men might be trained in mathematics and other practical skills. Scotland in 1699 had, he said, "now a fair prospect of a considerable foraigne trade to the Indies, Africa, etc."[86]

The failure of Scotland's colonial ambitions in the Darien disaster and, thus, of her own efforts to have an overseas empire does not diminish the sense in which geography was judged important to national and natural knowledge. It was seen as a means to get to know one's nation as well as to instruct one's students in questions of natural philosophy. Matthias Symson's declared uncertainty in his 1702 book over the proper place for discussion of celestial systems may reflect the fact that the distinctions within and between what we take to be modern "subjects" were less clear then than now. But the place of geography as a form of "science" and the incorporation into what was held to be geography of concerns then central to competing notions of celestial mechanics was not at issue. In university teaching spaces in particular— indeed, the university's own grounds were used in teaching space—geography and practical mathematics continued to be linked. In his 1741 lecture course in Edinburgh, for example, Colin Maclaurin, professor of mathematics and a significant proponent of Newtonianism in Scotland, incorporated geography as part of the first course of three making up his teaching program in mathematics.[87] At the same time, geography's connection with chronology and, thus, with history was being taught as part of ancient history by the professor of humanity at Edinburgh, John Ker, and by several other professors at other universities, including Thomas Reid at King's College, Aberdeen, in 1752.[88] Geography's place in the Scientific Revolution in Scotland thus underpinned geography's place in the Scottish Enlightenment.

* * *

Attending to questions to do with the Scientific Revolution as a geographical phenomenon and moving between "grand narratives" and "local context" may be possible and useful if we take note of the several distinctions relating to the term "geography" that have been proposed and discussed here.

There is, first, the sense in which scientific knowledge has a geography, a geography both of production and of reception among different audiences in different places.[89] In this first sense, questions of geography are to be understood as a matter of site and of situatedness, of knowledge making *in* space. We may, second, note the geographical movement of knowledge *over* space, in which the focus would be the traveling nature of science and the criteria and the people that allow its successful displacement—publishers, printers, booksellers, and in Edinburgh from the 1690s at least, the purchasing policy of university librarians. There is, third, the place of geography itself in the Scientific Revolution, as a subject that had specific concerns and that was debated and practiced in given sites and intellectual contexts. In this sense, there was an important role for geographical knowledge, understood both as the practical description, measurement, and explanation of the Earth and of its human diversity and the central role played by new geographical facts in challenging established belief. Finally, we should recognize the important question of the geographical *scale* of the Scientific Revolution—local, national, and inter-, or perhaps more usefully supranational—and of the relationships between such scales.

Applying these distinctions to understanding the place of geography, science, and the Scientific Revolution in late-seventeenth- and early-eighteenth-century Scotland would allow us to make sense of the several features disclosed here. In general, developments in natural philosophy teaching in Scotland between 1660 and 1707 were uneven in pace and geography. We should think more of the *overlapping* of Aristotelian and Cartesian and Newtonian natural philosophies in this period than of revolutionary "moments" or of epistemological crisis phases. Within this picture, teaching staff at Edinburgh seem to have led the way among Scotland's universities, followed by King's College and Marischal College in Aberdeen and then by St. Andrews and Glasgow.

Geography as a recognized area of intellectual and practical concern was taught in the universities in the 1680s—as it had been in the 1580s—and was associated in the late seventeenth century particularly with advances in natural philosophy. It was taught in association with astronomy, was chiefly seen as a part of practical mathematics, and to a lesser extent, was employed in history teaching. Geography in these places was mathematical and descriptive and, at Edinburgh, was used in the classroom if not also on the street by leading advocates of Newtonianism to instill principles of mathematical reasoning. At much the same time, an emergent public interest seems to have been apparent beyond the university's precincts. In Edinburgh on 16 November 1670, the Town Council gave license to Mr. George Sinclair "to profess several usefull sciences," which, in addition to exploring the "strange and wonderfull effects and causes . . . of The Pneumaticall or air pump" and the "study of hydrostaticks," included "Mathematicks, Geometrie, Astronomy, Special and Theoricall [theoretical] Geography."[90] Sinclair had taught mathematics at the University of Glasgow before moving to Edinburgh as an extramural lecturer.[91]

What is clear is that geographical knowledge as taught in the universities was both a matter of formal instruction and of direct practical encounter, for a few students anyway. Such geographical movement as generally took place, however, was of books, maps, and instruments. It was not in any formally disciplinary sense of people who called themselves "geographers." Beyond the classroom and the library, however, different sorts of geography were being undertaken. One involved the international movement of instruments and national fieldwork for its practitioners. At the same time, in the descriptions of their localities, men like William Geddes, Alexander Pennicuik, James Wallace, and Robert Edward, who produced *A Description of the County of Angus* in 1678, reflected at a smaller scale the chorographic work of the nation's Geographer Royal. Sibbald was reliant upon local informants being "out there." John Adair was reliant upon being "out there" himself.

Siting geography in these ways in relation to the Scientific Revolution—in Scotland or anywhere else—might be construed as a belated historiographical response to the "physics envy" that commentators on the Scientific Revolution as a whole have seen to mark the secondary position of the historical sciences (and the lowly profile of the "geographical sciences") in relation to the physical sciences.[92] Siting geography's place more highly in what we as modern scholars take to be the Scientific Revolution is not necessarily to accord with what contemporaries thought of it, and understanding geography's place in one national context may also depend upon comparison with other countries. Yet it is clear that the picture afforded here of Scotland is broadly comparable with France, England, and Holland at this time: an emergent public interest, geography allied with mathematics in the teaching of natural philosophy in the universities, handfuls of geographical authors and of booksellers to push their works, maps being drawn, and projects never completed for lack of funds.[93]

Thinking about the geography of the Scientific Revolution as more than geography in the Scientific Revolution and as more than a question of sites and of movement is also to think of it as a matter of different scales. Where one looks determines what one finds and, thus, what one can say about the Scientific Revolution as a geographical phenomenon. My focus here has not been to suppose a "grand narrative" at the national scale for the Scientific Revolution in Scotland. Neither has it been to highlight the local at the expense of addressing wider connections, although we have to recognize, of course, that the "patchiness" of source survival lends one's analysis an inevitable geographical bias. Simply, we should not take it for granted that geography's place in the library was the same as its use in the classroom, or that the university classroom was the same as the public teaching space. The mobile instrument in the field—either the geographer himself or his scientific equipment—was a yet different site of geographical activity. Where, then, should we look to site geography? In its books or in the library? In the field or in the study? Answering these questions of Scotland and for other countries may first demand that we do not pose them as strict "either/or" questions. Second, it may demand that we look beyond the subject itself to the social conditions of its making—to booksellers, instrument makers, audiences, and students. Such connections may even have been familial. John Adair was by marriage distantly related to the Gregory family, and one Gregory child, James, later an advocate, married in 1702 the daughter of Alexander Penicuik (sometimes

"Pennecuik" or "Pennycook"). He, as we have seen, was the author in 1715 of *The Description of Tweedale,* a work for which the by-then ill Adair never completed the maps.[94] Focusing on one text such as Matthias Symson's 1702 *Geography Compendiz'd* may not be as useful to a wider understanding of how geographical knowledge concerning the science of planetary bodies was understood as knowing the people involved in the book's production and reception: the patron (who secured a living for the author's father) and the bookseller who sold the book in town (as, at the same time, he worked with the university library to secure geography and other texts used in teaching). In considering who and what is studied within the Scientific Revolution, Rupert Hall remarked, "Is *pointillisme* the best historiography?"[95] This chapter has suggested that one way of showing how the Scientific Revolution was geographical is in terms of what the Scientific Revolution contained. Knowing how geography in one setting was connected with other sites and with other knowledge and treating science as a matter both of local context and of wider movement is to place questions to do with geography higher up the historiographical agenda than has been the case hitherto. In that sense, to recover geography's place in the Scientific Revolution is also to think of the Scientific Revolution as having a geography.

...

Notes

For comments upon earlier drafts, I am grateful to Paul Wood, Peter Dear, John Henry, Nicolaas Rupke, David Livingstone, Jim Moore, Mike Gallagher, Diarmid Finnegan, and the two anonymous reviewers for the University of Chicago Press. I am grateful to the Trustees of the National Library of Scotland for permission to quote from manuscripts and other material in their care, and to the staff of the libraries and special collections at the Universities of Aberdeen, Edinburgh, Glasgow, Oxford, and St. Andrews. I acknowledge with particular thanks the help of Richard Ovenden, formerly director of special collections at the University of Edinburgh, for drawing "Henderson's Benefactions" (EUL, MS Da.1.31) to my attention. The final preparation of this chapter and the overall editing of the volume was undertaken while I was in receipt of a British Academy Research Readership, and I am grateful to the British Academy for its support.

1. Edinburgh University Library (hereafter EUL), MS Da.1.31, fol. 46, 11 August 1690.

2. EUL, MS Da. 1. 31, fol. 50, 26 November 1694; fol. 69, 4 October 1711.

3. Patrick Gordon, *Geography Anatomiz'd; or, A Compleat Geographical Grammar* (London: Printed by Robert Morden and Thomas Cockerid, 1693), i. For a summary publication history of the work, see O. Francis G. Sitwell, *Four Centuries of Special Geography* (Vancouver: University of British Columbia Press, 1993), 261–66.

4. National Library of Scotland (hereafter NLS), MS Acc 9800, fol. 24, 5 April 1717.

5. Steven Shapin begins his book on the topic with the arresting words, "There was no such thing as the Scientific Revolution, and this is a book about it" (*The Scientific Revolution* [Chicago: University of Chicago Press, 1996], 1).

6. In an extensive literature, I think of Mario Biagioli and Steven J. Harris, eds., *The Scientific Revolution as Narrative,* special issue of *Configurations* 6 (1998); Herbert Butterfield, *The Origins of Modern Science, 1300–1800,* rev. ed. (1949; New York: Free Press, 1957); I. Bernhard Cohen, *The Revolution in Science* (Cambridge: Harvard University Press, 1985); H. Floris Cohen, *The Scientific Revolution: A Historiographical Inquiry* (Chicago: University of Chicago Press, 1994); Andrew Cunningham and Perry Williams, "De-centering the 'Big Picture': The *Origins of Modern Science* and the Modern Origins of Science," *British Journal for the History of Science* 26 (1993): 407–32; Peter Dear, ed., *The Scientific Enterprise in Early Modern Europe: Readings from "Isis"* (Chicago: University of Chicago Press, 1997); John Henry, *The Scientific Revolution and the Origins of Modern Science* (Basingstoke: Macmillan, 1997); David C. Lindberg and Robert S. Westman, eds., *Reappraisals of the Scientific Revolution* (Cambridge: Cambridge University Press, 1990); Margaret J. Ostler, ed., *Rethinking the Scientific Revolution* (Chicago: University of Chicago Press, 2000); Roy Porter, "The Scientific Revolution: A Spoke in the Wheel?" in *Revolution in History,* ed. Roy Porter and Mikuláš Teich (Cambridge: Cambridge University Press, 1986), 290–316; John A. Schuster, "The Scientific Revolution," in *Companion to the History of Modern Science,* ed. Roger C. Olby et al. (London: Routledge, 1990), 217–42; Shapin, *The Scientific Revolution.*

7. Margaret J. Ostler, "The Canonical Imperative: Rethinking the Scientific Revolution," in Ostler, *Rethinking the Scientific Revolution,* 3.

8. Cohen, *Revolution in Science,* 397–98; Peter Dear, *Discipline and Experience: The Mathematical Way in the Scientific Revolution* (Chicago: University of Chicago Press, 1995); Richard S. Westfall, *The Construction of Modern Science: Mechanisms and Mechanics* (New York: Wiley, 1971).

9. Jim Bennett, "Practical Geometry and Operative Knowledge," *Configurations* 6 (1998): 195–222.

10. Lesley Cormack, *Charting an Empire: Geography at the English Universities, 1580–1620* (Chicago: University of Chicago Press, 1997); Robert J. Mayhew, *Enlightenment Geography: The Political Languages of British Geography, 1650–1850* (Basingstoke: Macmillan, 2000).

11. On this topic, two useful texts are Crosbie Smith and Jon Agar, eds., *Making Space for Science: Territorial Themes in the Making of Science* (Basingstoke: Macmillan, 1998); Jan Golinski, *Making Natural Knowledge: Constructivism and the History of Science* (Cambridge: Cambridge University Press, 1998).

12. This claim is made by Richard Sorrenson, "The Ship as a Scientific Instrument in the Eighteenth Century," *Osiris,* 2nd ser., 11 (1996): 221.

13. See the introduction to Roy Porter and Mikuláš Teich, eds., *The Scientific Revolution in National Context* (Cambridge: Cambridge University Press, 1992), 2, 6.

14. Paul Wood, "The Scientific Revolution in Scotland," in Porter and Teich, *Scientific Revolution,* 263–87.

15. For a fuller discussion of Adair in particular, see Charles W. J. Withers, "John Adair, 1660–1718," *Geographers' Biobibliographical Studies* 20 (2000): 1–8.

16. Cohen, *Scientific Revolution,* 305.

17. H. Floris Cohen, "The Scientific Revolution: Has There Been a British View?—a Personal Assessment," *History of Science* 37 (1999): 112.

18. Reijer J. Hooykaas, "The Rise of Modern Science: When and Why," *British Journal for the History of Science* 20 (1987): 472–73.

19. Robert J. Mayhew, "Geography, Print Culture and the Renaissance: 'The Road Less Travelled By,'" *History of European Ideas* 27 (2001): 349–69.

20. Richard S. Westfall, "Science and Technology during the Scientific Revolution: An Empirical Approach," in *Renaissance and Revolution: Humanists, Scholars, Craftsmen and Natural Philosophers in Early Modern Europe,* ed. J. V. Field and Frank A. L. James (Cambridge: Cambridge University Press, 1993), 63–72.

21. Jim Bennett, "Practical Geometry and Operative Knowledge," *Configurations* 6 (1998): 195–222.

22. Cormack, *Charting an Empire,* 163.

23. Stephen Greenblatt, *Renaissance Self-Fashioning from More to Shakespeare* (Chicago: University of Chicago Press, 1980); Richard Helgerson, *Forms of Nationhood: The Elizabethan Writing of England* (Chicago: University of Chicago Press, 1992).

24. William Warntz, "Newton, the Newtonians and the *Geographia Generalis Varenii,*" *Annals of the Association of American Geographers* 79 (1989): 181.

25. Peter Dear, "The Mathematical Principles of Natural Philosophy: Toward a Heuristic Narrative for the Scientific Revolution," *Configurations* 6 (1998): 177.

26. David N. Livingstone, "Science, Magic and Religion: A Contextual Assessment of Geography in the Sixteenth and Seventeenth Centuries," *History of Science* 26 (1988): 269–94; David N. Livingstone, "Geography, Tradition and the Scientific Revolution," *Transactions of the Institute of British Geographers* 15 (1990): 359–73.

27. Margaret J. Osler, "The Canonical Imperative," 7.

28. David N. Livingstone, "Science and Religion: Foreword to the Historical Geography of an Encounter," *Journal of Historical Geography* 20 (1994): 367–83; "The Spaces of Knowledge: Contributions Towards an Historical Geography of Science," *Environment and Planning D: Society and Space* 13 (1995): 5–34; *Putting Science in its Place* (Chicago: University of Chicago Press, 2003).

29. Smith and Agar, *Making Space;* Golinski, *Making Natural Knowledge.*

30. Golinski, *Making Natural Knowledge,* 91–102.

31. David Turnbull, "Cartography and Science in Early Modern Europe: Mapping the Construction of Knowledge Spaces," *Imago Mundi* 46 (1996): 7.

32. Adir Ophir and Steven Shapin, "The Place of Knowledge: A Methodological Survey," *Science in Context* 4 (1991): 5–6.

33. Steven Shapin, "Placing the View from Nowhere: Historical and Sociological Problems in the Location of Science," *Transactions of the Institute of British Geographers* 23 (1998): 6, 7.

34. For a general review of these issues, see Lewis Pyenson, "An End to National Science: The Meaning and Extension of Local Knowledge", *History of Science* 40 (2002): 251–90.

35. Steven J. Harris, "Thinking Locally, Acting Globally," introduction to Biagioli and Harris, *The Scientific Revolution as Narrative,* 131.

36. Steven J. Harris, "Long-Distance Corporations, Big Sciences, and the Geography of Knowledge," in Biagioli and Harris, *The Scientific Revolution as Narrative,* 272–73.

37. Ibid., 294–95.

38. Ibid., 297.

39. Ibid.

40. Wood, "Scientific Revolution in Scotland."

41. Ibid., 266–70.

42. Charles W. J. Withers, *Geography, Science and National Identity: Scotland since 1520* (Cambridge: Cambridge University Press, 2001).

43. Charles W. J. Withers, "Geography, Science and National Identity in Early Modern Britain: The Case of Scotland and the Work of Sir Robert Sibbald, 1641–1722," *Annals of Science*

53 (1996): 29–73; and "Reporting, Mapping, Trusting: Making Geographical Knowledge in the Late Seventeenth Century," *Isis* 90 (1999): 497–521.

44. EUL, MS Laing III, 535.

45. Sir Robert Sibbald, *An Account of the Scotish Atlas; or, The Description of Scotland Ancient and Modern* (Edinburgh: Printed by David Lindsay, James Kniblo, Joshua van Solingen and John Colmar, 1693), 1–2.

46. NLS, Crawford Deposit, Crawford MS, MB 277, fol. 2.

47. The idea of centers of calculation is from from Bruno Latour, *Science in Action: How to Follow Scientists and Engineers through Society* (Milton Keynes: Open University Press, 1987).

48. David S. Lux and Harold J. Cook, "Closed Circles or Open Networks? Communicating at a Distance during the Scientific Revolution," *History of Science* 36 (1998): 179–211.

49. Alison Morrison-Low, "'Feasting My Eyes with the View of Fine Instruments': Scientific Instruments in Enlightenment Scotland, 1680–1820," in *Science and Medicine in the Scottish Enlightenment,* ed. Charles W. J. Withers and Paul B. Wood (East Linton: Tuckwell, 2002), 17–53.

50. Withers, "John Adair."

51. National Archives of Scotland (hereafter NAS), RH 14/203, 15 June 1686.

52. Withers, "Reporting, Mapping, Trusting."

53. Paul Wood makes the point that our understanding of the nature and timing of the revolution there depends upon what sciences we include within its remit. Such a view accords with Harris's point discussed here about the nature of the "big" and the "small" sciences. Wood mentions Sibbald as Geographer Royal in relation to the place of natural history. See Wood, "Scientific Revolution in Scotland," 274–75.

54. Shapin, *Scientific Revolution.*

55. Robert J. Mayhew, "The Character of English Geography, c. 1660–1800," *Journal of Historical Geography* 24 (1998): 385–412; Robert J. Mayhew, "Geography in Eighteenth-Century British Education," *Paedagogica Historica* 34 (1998): 731–69; Sitwell, *Four Centuries of Special Geography.*

56. Robert J. Mayhew, "William Guthrie's *Geographical Grammar,* the Scottish Enlightenment and the Politics of British Geography," *Scottish Geographical Journal* 115 (1999): 19–34.

57. James Paterson, *A Geographical Description of Scotland* (Edinburgh, 1681), preface.

58. Michael Wood, *Extracts from the Records of the Burgh of Edinburgh* (London: His Majesty's Stationery Office, 1940), 93.

59. David J. Bryden, *Scottish Scientific Instrument-Makers, 1600–1900* (Edinburgh: Royal Scottish Museum, 1972), 1.

60. Michael Hunter, ed., *The Occult Laboratory: Magic, Science and Second Sight in Late Seventeenth-Century Scotland* (Woodbridge: Boydell Press, 2001).

61. Morrison-Low, "Feasting My Eyes."

62. M.S. [Matthias Symson], *Geography Compendiz'd; or, The World Survey'd* (Edinburgh: Sold by Mr. Henry Know and John Vallange, 1702), [i].

63. M.G.D. [Gavin Drummond], *A Short Treatise of Geography, General and Special* (Edinburgh: Andrew Symson, 1708), 1.

64. The manuscript for Symson's geographical description of Galloway exists as NLS, Advocates' MS 31.7.17. It has been printed once: see Andrew Symson, *A Large Description of Galloway,* ed. Thomas Maitland (Edinburgh: W. and C. Tait, 1823). For a guide to the religious and political history of this period, see Clare Jackson, *Restoration Scotland, 1660–1690: Royalist Politics, Religion and Ideas* (Woodbridge: Boydell and Brewer, 2003).

104 | Charles W. J. Withers

65. Bibliotheca Symsoniana, *A Catalogue of the Vast Collection of Books, in the Library of the Late Reverend Learned Mr Andrew Symson* (Edinburgh, 1712).

66. NLS, MS Acc 9800, fols. 2r, 3r, 10r, 18v, 21r, 21v, 24r.

67. [Gavin Drummond?], *Bibliotheca Sibbaldiana* (Edinburgh, 1708).

68. Aberdeen University Library (hereafter AUL), MS K 153, fols. 164r–184v and fols. 188r–206v; Roger Emerson, *Professors, Patronage and Politics: The Aberdeen Universities in the Eighteenth Century* (Aberdeen: Aberdeen University Press, 1992), 24–26, 136, 143.

69. AUL, MS 2092, "Geographicus Tractatus," fols. 4–9v; Emerson, *Professors, Patronage and Politics*, 72, 73, 90, 93, 104–5, 126, 128–29, 143, 146.

70. Bodleian Library, Ashmolean MSS (1813), fol. 243, 3 February 1685.

71. John L. Russell, "Cosmological Teaching in the Seventeenth-Century Scottish Universities," *Journal of the History of Astronomy* 5 (1974): 122–32 (part 1) and 145–54 (part 2); Christine Shepherd, "The Inter-Relationship between the Library and Teaching in the Seventeenth and Eighteenth Centuries," in *Edinburgh University Library, 1580–1980*, ed. Jean Guild and Alexander Low (Edinburgh: Edinburgh University Library, 1982), 67–86; Christine Shepherd, "Newtonianism in the Scottish Universities in the Seventeenth Century," in *The Origins and Nature of the Scottish Enlightenment*, ed. Roy H. Campbell and Andrew S. Skinner (Edinburgh: John Donald, 1982), 65–85; Wood, "Scientific Revolution in Scotland," 269–71.

72. Shepherd, "Inter-Relationship between the Library and Teaching," 70.

73. Ibid., 73.

74. EUL, MS DK.5.27, fols. 57–70.

75. Shepherd, "Newtonianism in the Scottish Universities," 70–71.

76. EUL, Da.1.32, "Magistrands Graduation Receipts and Disbursements," fol. 53.

77. Withers, *Geography, Science and National Identity*, 46–47.

78. EUL, Da.1.33, "Matriculation Receipts and Disbursements, 1653–1693," fol. 39.

79. EUL, Da.1.33, fols. 48, 52.

80. Sitwell, *Four Centuries of Special Geography*, 106–8, 424–26.

81. EUL, Da.1.33, fol. 101.

82. Elly Dekker, *Globes at Greenwich* (Oxford: Oxford University Press and the National Maritime Museum, 1996), 6–12.

83. Morrison-Low, "Feasting My Eyes."

84. EUL, Da.1.33, fol. 116, for early 1691, for example, has notes of exchange rates and shipping costs from Holland.

85. EUL, Da.1.34, "General Book of Disbursements, Library Accession Book, 1693–1719," fols. 4, 11, 17, 21–22, 25.

86. EUL, MS Dc.1.60, fol. 751–54, James Gregory, "A Prospect for a Navigation and Writing School by the E. India Company of Scotland."

87. Maclaurin's mathematical work as a form of practical Newtonianism and as a means to measured enquiry about the state of Scotland is the subject of Judith V. Grabiner, "Maclaurin and Newton: The Newtonian Style and the Authority of Mathematics," in Withers and Wood, *Science and Medicine in the Scottish Enlightenment*, 143–71.

88. Charles W. J. Withers, "Toward a Historical Geography of Enlightenment in Scotland," in *The Scottish Enlightenment: Essays in Reinterpretation*, ed. Paul Wood (Rochester: University of Rochester Press, 2000), 63–97.

89. I think in this context, for example, of Nicolaas Rupke, "Translation Studies in the History of Science: The Example of *Vestiges*," *British Journal for the History of Science* 33 (2000): 209–22.

90. Quoted in Wood, *Extracts from the Records,* 92–93.

91. Roger L. Emerson and Paul Wood, "Science and Enlightenment in Glasgow, 1690–1802," in Withers and Wood, *Science and Medicine in the Scottish Enlightenment,* 131.

92. On this, see for example Dear, "Mathematical Principles"; Field and James's introduction to *Renaissance and Revolution,* 1–14.

93. I think, for example, of several of the essays in Pamela H. Smith and Paula Findlen, eds., *Merchants and Marvels: Commerce, Science, and Art in Early Modern Europe* (London: Routledge, 2002).

94. I am grateful to John Moore of the University Library in Glasgow for this information. See Joseph Morris, "Belfield, East Calder: The Country Mansion of the Lanton Oliphants," *Proceedings of the Society of Antiquaries of Scotland,* 4th ser., 43 (1909): 324–29.

95. A. Rupert Hall, "Retrospection on the Scientific Revolution," afterword to Field and James, *Renaissance and Revolution,* 244.

REVOLUTION OF THE SPACE INVADERS

Darwin and Wallace on the Geography of Life

...

James Moore

The immigration of a few new forms, or even of a single one, may well cause an entire revolution in the relations of a multitude of the old occupants. . . . Every where we see organic action & reaction. All nature is bound together by an inextricable web of relations; if some forms become changed & make progress, those which are not modified or may be said to lag behind, will sooner or later perish.

Charles Darwin, 1857

A t the dawn of the twenty-first century, Great Britain awoke to an invasion. Alien organisms were slipping ashore, putting down roots, and occupying the soil. They overran a green and pleasant land, turning it into a blighted "foreign country"—or so the tabloid press and Tory politicians fumed. "Bogus asylum seekers" enjoyed a "soft touch" here—balmy climate, genial politics, sumptuous welfare. They craved it so that they were literally dying to get in.

Sadly for the refugees, Britons have a long tradition of dying to keep aliens out. From 1066 to 1945 there was always a "home guard," and a comic "Dad's Army" still confronts panzer divisions in each TV episode, great vectors thrusting from France toward the white cliffs of Dover. Real invasions are less funny, Suez, the Falklands; for Americans, Cuba, Vietnam, Iraq—the list is long. The United States itself has only once faced a full-scale invasion. On Halloween eve 1938, a public inured to Buck Rogers comics, Flash Gordon films, and broadcast bulletins about Nazi conquests in Europe, suddenly heard news of an attack on their own country—by Martians. A trick radio-adaptation of H. G. Wells's *War of the Worlds* went down a treat. Millions, believing, panicked.

But people running scared of aliens is no joke either. Forty years on, a political cowboy from California announced his intention to tame the "high frontier" and rid space of Soviet missiles. In Ronald Reagan's scenario, the West would be won by high-tech death-rays, extraterrestrial thunderbolts as seen in *Star Wars,* the space western. Both of these sci-fi dramas, released in the 1980s, served to militarize the heavens, blurring the line between invasion *of* and invasion *from* space. And about this time, everyone got in on the act. Strange new noises were heard in public places, unearthly squeals and grunts from video consoles, where frantic players, lost in their private voids, parodied the U.S. president, zapping alien landers. Space Invaders now seems as quaint as skittles, though Star Wars the defense system may yet be born again.

Invasions, then, are political, or at least the way we talk about them is. Life threatening or death dealing, they occupy space on different scales, in different ways. In medicine, for instance, invading germs are said to cause "explosive epidemics," while history celebrates (as in one classic title) "the conquest of epidemic disease." "Invasive" surgery may be hazardous, but like a military strike, it is often needed to "take out" a foreign object. In bodies as in bodies politic, the enemy is the alien. Sometimes an alien presence is desirable. Britain's leading immunologist, Sir Peter Medawar, was of Arab descent and an émigré from Brazil. Having suffered racist abuse at Marlborough School, he went on to win a Nobel Prize for discovering "immunological tolerance," the possibility, as he put it, of "breaking down the natural barrier that prohibits the transplantation of genetically foreign tissues."[1] Medawar's whole life was shaped by the politics of invasion.

Geographies Converge

Foreign tissues and organisms also concern biogeographers, who map the history of life. Outside politics proper, they supply the richest examples of rhetorical space invasion.[2] In the subdiscipline of "invasion ecology," for instance, "invasiveness" and "invasibility" are key concepts; "aggression," "assault," "attack," "onslaught," and "raid" are life's strategies; and "colonization" and "naturalization" the results.[3] So contagious is this language that in times of political tension, ecological xenophobia has caught on. "We are . . . fighting a rearguard action against an invasion of sycamore," the *Guardian* newspaper cried in the build-up to the Gulf War. "The mass elm-death of the 1970s encouraged an attempted sycamore coup d'état," and sure enough, this old "cross-Channel invader" succeeded "where Hitler . . . failed," an "arboreal fascist" now ousting "our native ash and oak." "In the crazed mind of the conservationist," the *Guardian* went on uncharacteristically, "the chainsaw's whine echoes the Spitfire's scream," even despite efforts (according to the *Quarterly Journal of Forestry*) to secure the sycamore's "political rehabilitation."[4]

All of which makes semantically sensitive phytogeographers despair. Oh, for an end to "anthropocentric concepts"! Oh, for a "clearly defined," "universally acceptable," and "objectively applicable" language in which to describe alien species! To an extent they recognize that historic usage is the problem. The term "naturalized" has been misapplied since the mid-nineteenth century, and its meaning is now imprecise. The early ecologists equated "invasion" with "colonization," another supposed mistake. "Colonization," critics say, "has carried no implications of hostility, harm or encroachment since its appearance in modern usage in the 16th century," whereas such implications are what make the term "invasion" ecologically apt.[5] Others disagree. A recent survey proposes to use "invasive" for introduced species "without any connotation of impact," reserving the neutral phrase "transformer species" for hostile aliens. "Much of the debate on terminology," these authors insist, "is essentially semantic, and poses little threat to the development of an increasingly robust understanding of invasion." Better definitions will banish all ambiguities.[6]

I doubt it. When science speaks the language of politics, semantic debates are intractable. This can be shown for biogeography as readily as other fields. For instance, the simple question, "What is a weed?" goes a long way toward thwarting the quest for a universal, objective language of plant invasion.[7] More tellingly, the vocabulary of modern biogeography can be situated in its formative political context. With ecologists claiming that the term "'colonization' has carried no implications of hostility, harm or encroachment since its appearance in modern usage," we know where that context is to be found. The solecism points to an age when the behavior of nations and the behavior of life were seen to be all of a piece, thanks to the science of evolutionary biogeography as founded by Charles Darwin (1809–82) and Alfred Russel Wallace (1823–1913).

Biogeography and political geography are kith and kin. Both study the distribution of living organisms and inscribe their findings on maps. Both have served imperial masters, "geography militant" by converting terra incognita into territories, biogeography militant—"green imperialism"—by manipulating organisms for political and commercial advantage.[8] Both geographies have achieved their ends through exploration, but exploration is only invasion on a small scale; in post-Darwinian perspective, the two geographies converge. Evolutionary biogeography bids to explain the causes and the consequences of exploration itself, the "natural" conditions attendant on people's dispersal over the earth. Biological expansion of this sort, or "ecological imperialism," is the concern of human invasion ecology.[9]

Homo sapiens has been called the world's "most dangerous and unrelenting" predator. For thousands of years, waves of "portmanteau biota"—people, their pets, and pests—swept across seas and lands, culminating in a "Caucasian tsunami" as Europeans overran the globe.[10] The British were in the vanguard, with imperial geography and then evolutionary biogeography developing in their wake. Both disciplines were practical instruments of expansion, but historians have so far neglected to connect them theoretically. Even geographers, following their lead, find Darwin's theory of evolution by natural selection to have social more than spatial implications.[11] In this chapter, I establish the theoretical link by showing, first, how Darwin rendered life itself an agent and ally of empire, and second, how Wallace developed an alternative geopolitics.

DARWIN'S NEW WORLD ORDER

Biogeography asks where on earth things live and how they got there. Darwin entered the field aboard H.M.S. *Beagle* with an open mind. Although he believed in divine creation, he had no idea where the great events took place. All he knew was that clues lay in life's spatiotemporal distribution—obscure clues. Every species had been created preadapted to its physical surroundings—so natural theology taught—but many species inhabited strange places. Evidently organisms possessed adaptive flexibility. They could fly, swim, walk, crawl, or hitchhike to new pastures and still survive. The story of Noah's descent from Ararat contained this truth at least: migration occurred. What stumped Darwin and other naturalists was how far historic migrations had obscured the original pattern of creation. Distributional anomalies abounded.[12] Why did similar groups of species occupy entirely different environments? Why did similar environments contain entirely different groups of species? Was it all because of migration? Had living things climbed every mountain, crossed every sea? Could they have dispersed from a single creation center, or had God helped by creating in many places? Maybe, as some believed, God himself created the distributional anomalies. Perfectly formed organisms were put in places for which they were imperfectly adapted in order to display the power of designing Mind over matter.[13]

Voyages of discovery kept these questions to the fore. The more species were known, the greater the biogeographic puzzles. Naturalists such as Alexander von Humboldt and Augustin de Candolle tackled them famously in the early nineteenth century, and then Charles Lyell, standing on their shoulders, made a fresh start in his *Principles of Geology* (1830–33). This was Darwin's Bible on the *Beagle*. Poring over it, he learned to see the world through Lyell's eyes, even things Lyell had never observed.

What Lyell saw was the earth's crust in perpetual flux. Always and everywhere, seas rose and fell, land was submerged and elevated, surface features were formed at no faster rates and by no other means than those occurring today. The system was dynamic but stable, ordained by God to oscillate indefinitely, and living species were integral to it. They too existed in a state of equilibrium, their quantity of life fixed. Creation did not occur all at once but proceeded ad hoc. Where physical and climatic conditions were right, species appeared, adapted to those conditions. Where conditions exceeded life's design limits, species became extinct. Creation and extinction took place wherever, whenever, in order to maintain nature's exquisite balance. Any species might appear at any time under the right conditions, any might die out at any time under the wrong ones. There was no progression from low to high, simple to complex; and this, Lyell insisted, ruled out the foul, French notion that species (and by implication humans) might originate through transmutation. He did not know, or would not say, how species were created.[14]

Darwin took Lyell on board. His geology set geography in motion, giving life's distribution a complex past. Like all good students, Darwin became his teacher's critic, and in South America he constantly tested Lyell's creationist earth system. In time he worked out how gradual uplift had formed the continent—"geology of whole world will turn out simple," the twenty-seven-year-old crowed—but meanwhile the biogeographic puzzles piled up. Back in London, they triggered his long lonely quest to supply the missing part of Lyell's system, the laws of life's creation and distribution. "Zoonomia," he scribbled at the head of his first transmutation notebook in 1837, adding later, "The Grand Question, which every naturalist ought to have before him . . . is 'What are the laws of life[?]'"[15]

Those puzzles fascinated him. Take the fossil bones he dug up in South America, huge sloths, armadillos, llamas, and capybara, so reminiscent of today's pint-size species. What caused the giants to die out without a Lyellian environmental shift? Why were extinct and extant species of the same type found in the same area? Had they overlapped in time, as the two Patagonian ostriches now do in space, the smaller replacing the larger in the dry south of the country? Could one ostrich, or indeed sloth, have been created from another to adapt the type to new conditions? Or look at the mighty Andes, with similar climates but contrasting faunas on either side, yet both faunas, like the ostriches, changing from north to south. Surely here was proof that, pace Lyell, conditions do *not* govern the species God creates. Finally, Darwin's chief puzzle, the Galapagos archipelago: volcanic, of recent origin and arid climate, utterly unlike the mainland, yet sharing the same groups of birds. Did the birds originate as separate species on the mainland and then migrate to their appointed places? Or did God create each species directly on the spot? If so, why did different species now inhabit different islands, all with the same hellish environment? Why indeed bless this infernal region with a creative "halo" if single migrations followed by adaptive expansion among the islands could explain the pattern?[16]

Alert to a link between migration and transmutation, Darwin revised his *Beagle* diary for the press, playing up the distributional puzzles he observed during the latter legs of the voyage. In New Zealand, with no indigenous mammals, he found rats everywhere; and those "I was forced to own as countrymen," he frowned, were causing havoc. Introduced weeds had a similar effect, crowding out "native kinds," as on that "little world within itself," the island of St. Helena.[17] All over the globe, strangers were replacing the locals—could this be nature's way? In Australia, God saw fit to create only marsupial mammals, yet placentals now thrived there, some at the marsupials' expense. Likewise, "wherever the European has trod, death seems to pursue the aboriginal," Darwin reflected. "The varieties of man seem to act on each other; in the same way as different species of animals—the stronger always extirpating the weaker."[18]

Apt words indeed for a volume included in Robert FitzRoy's official narrative of the voyage. Darwin's *Journal of Researches* (1839) showed that FitzRoy's command, like Lyell's geology, had transformed his view of the world. The Admiralty, charged with securing Britain's presence in South America, had instructed FitzRoy to improve coastal charts, reconnoiter the disputed Falkland Islands, and fix exact longitudes. Better navigation meant better business. With easy access to ports, British merchants could usurp the Spanish and Portuguese and forestall the United States in its own back yard.[19] The *Beagle*'s mission was imperial, for crown and commerce; and FitzRoy even had a parson and three Christianized natives on board to set up an Anglican outpost in Tierra del Fuego. Darwin recorded it all patriotically. The "strength & power" of the Royal Navy at Rio de Janeiro (where FitzRoy fixed the longitude for charting the continent) gave him a thrill of "exultation."[20] In Sydney, "the power of the British nation" cast South America's rulers further into the shade, and Darwin backslapped himself for being "born an Englishman." Across the southern hemisphere, he saw "little embryo Englands," like Cape

Colony, "springing into life," and homeward bound, he basked in the knowledge that the "British flag" was bringing "wealth, prosperity, and civilization" to the globe.[21]

After FitzRoy's costly four-volume *Narrative* sold out, Darwin's *Journal* appeared in John Murray's Colonial and Home Library series, "cheap literature for all classes," the publisher puffed, and choice reading for an age of systematic colonization. Britain had long exported her felons and paupers, but by 1840 the "removal of society" itself was deemed in the nation's interest.[22] Men of Darwin's class believed the country to be suffering from an excess of people and an excess of capital. Some thought the problems connected, citing the famous essay on population by Parson Malthus. Within years, famine in Ireland and a railway boom cut the figures, yet poor rates remained high, returns on investment low, and the professions oversubscribed. Shipping men and money abroad seemed the ideal solution. All classes would emigrate, rich and poor in groups, and set up "a miniature representation of England, complete in every part," wherever the Union Jack was planted.[23] These enclaves would grow into centers of production and new markets; the class structure would ensure social stability and ward off that bane of failed colonies, democracy.

As Darwin filled his transmutation notebooks, pondering life's migrations, overseas expansion on this scale had cross-bench support. For all their differences, Tories, Whigs, and Radicals were gripped by the conviction that Britain had been chosen by God to populate the earth. "Let the sons of Albion carry civilization to . . . distant shores," one pundit urged, "and a future age will applaud their enterprize."[24] The rhetoric was pervasive and persuasive; rare indeed the gentleman who could resist its turns of phrase.

So what was going on in Darwin's private jottings? As Lyell's biogeography broke down, what took its place? The flip side of creation in Lyell's world was noncreation, the failure of species to appear where they were ill adapted to survive. Knowing this, Darwin let loose his Antipodean placental mammals. They survived mightily in New Zealand where *no* mammals had been created. Why? Because they had managed to get there—migration. Creation had nothing to do with it. But if this was so, then the absence of Antipodean placentals in the first place might have nothing to do with creation either. They had simply never arrived—nonmigration. The conclusion was irresistible: creation and noncreation explained nothing about life's distribution; migration and nonmigration everything.

And now Darwin saw clearly that, by migrating, organisms encountered conditions to which they could adapt, forming new species, as on the Galapagos. So forcefully did this strike him that he began to picture migration as wholesale *invasion*. Look at the guava introduced into Tahiti, which had claimed "all the moist & fertile land" within twenty years. Or the Spanish cardoon in Uruguay, a thistle that had rendered hundreds of square miles "impenetrable by man or beast." Was there ever a case "of an invasion on so grand a scale of one plant over the aborigines"?[25] "Study the wars of organic being," he reminded himself in the same note. "If South America grew very much hotter . . . Brazilian species would migrate south ward[,] being ready made" for the climate, "& so destroy [less adapted] individuals, wher[e]as in [the] Falkland Is[lan]d[s]," with a bracing environment, "they would change & make new species."[26]

New conditions, new opportunities, new life to exploit them—a new world order was in the making. Organisms invade places where new conditions induce physical and behavioral changes in them. Their offspring inherit the changes according to laws ordained by God to maintain the fit between the organic and inorganic worlds. Isolation, as on islands, prevents migrants and their offspring from mixing with other groups, so obliterating their differences. With long-continued isolation, new species emerge. To Darwin this was an improvement on Lyell's dynamic earth system, extending it to the origin of species and making it complete. Creation, rightly understood, was just that natural process by which organisms kept adapted to changing environments. Where adaptation was stymied by lack of isolation or by conditions changing too fast, extinction occurred. But still nature proceeded lawfully, harmoniously, just as Lyell said.[27]

Then in September 1838 (the year he was elected to the Royal Geographical Society), Darwin read Malthus—Malthus who explained why Britain was overcrowded, Malthus who had convinced pundits that, with the population doubling every twenty-five years, "two-thirds of the world . . . lying waste, and the other third very imperfectly cultivated," colonization was destiny.[28] Talk about "wars of organic being"—this was carnage! With a shudder, Darwin realized that at the core of his new world order, driving it remorselessly, was the crushing "force" of population growth. And there was nothing harmonious about it. Organisms lived in a constant pell-mell rush for food, "warring," the well-fed multiplying, invading new turf, and "thrusting out [the] weaker ones," which became extinct. "Recollect," he noted, "the multitudes of plants introduced into our gardens . . . & which might spread," or the "opportunities . . . for foreign birds & insects." Here too a "dreadful but quiet" war raged.[29] How much "more deadly," then, the human conflict caused by "immigration of other races," the fighting, the infecting, but above all the struggle of unequal intellects. Thus are inferior races "exterminated on *principles*," Darwin declared—principles "strictly applicable to the universe."[30]

Months later, when he had married and begun adding rapidly to England's expanding population, Darwin conceived those "principles" as a selective mechanism analogous to the weeding out of inferiors practiced by animal breeders. "My theory," as he called it, now became the theory of natural selection.[31]

LIFE'S CONSTANT VECTOR

At this time, natural selection did seem "strictly applicable" to human affairs. Britain waged imperial wars nonstop during the middle decades of the century. Campaigns were mounted against the Kaffirs of the eastern Cape (1834–35, 1846–48, 1850–53), a French rebellion was put down in Canada (1837), China was attacked three times (1840, 1857, 1860), and Persia once (1856). Skirmishes with the Maori turned into the longest sustained warfare in New Zealand's history (1846–70). In India the army was constantly on the march: it fought in Afghanistan (1838–42) and Burma (1853), conquered the Sind (1843) and the Punjab (1845–46, 1848–49), suppressed the bloody so-called Mutiny (1857–58), and from time to time crossed the Northwest Frontier to punish recalcitrant tribes. At home, the press gloated over each new conquest, boosting circulations with the latest technologies. Engravings of battlefield photographs brought the slaughter vividly to life; reports by electric telegraph kept readers on tenterhooks, passions inflamed. Never had so many felt so patriotically about so many conflicts so far away.[32]

These were the years in which Darwin refined his theory and readied it for publication. In 1844 he first worked out a rhetorical strategy in a long essay, using the "geographical distribution of organic beings" as his main evidence that species were "naturally formed races, descended from common stocks." His argument turned on the hypothesis that each organism was "created or produced on one spot" and then migrated "as widely as its means of transport and subsistence permitted." Barriers checked dispersal, as did "*preoccupation*," except where migrants proved themselves better adapted by "struggling with and overcoming the aborigines." Whatever the migrants' advantage—a "mere tendency to vary, or some peculiarity of organization, power of mind, or means of distribution"—it would generally be passed on to their offspring, and the group "whose place the new . . . ones are seizing, from partaking of a common inferiority, would *tend* to become rarer and rarer in numbers, and finally extinct." This "extermination" resulted not merely from "changes in the external conditions," as Lyell believed, but also "from the increase or immigration of more favoured species."[33]

Islands, formerly Darwin's chief puzzle, now became his prize exhibit. "Nurseries of new species," he dubbed them affectionately, isolated, secluded, where an "occasional . . . visitant" might multiply and then fresh "immigrants" invade to compete with the "tenants," and so on, until at last "every place or office in the economy of the island" was filled.[34] The challenge of establishing new beachheads appealed to the old salt. In 1855, faced with claims that seeds would not survive an ocean passage, Darwin put dozens of species to the test. At the end of that bitter winter when tens of thousands died in the Crimea, he began floating seeds in bottles of brine kept in tanks of snow. Weeks later, he planted the seeds and, sure enough, they sprouted. His triumphant report "Does Sea-Water Kill Seeds?" ended memorably, describing

> how beautifully pods, capsules &c. . . . close when wetted, as if for the very purpose of carrying the seed safe to land. When landed high up by the tides and waves, and perhaps driven a little inland by the first inshore gale, the pods, &c., will dry, and opening will shed their seeds; and these will then be ready for all the many means of dispersal by which Nature sows her broad fields. . . . But when the seed is sown in its new home then, as I believe, comes the ordeal; will the old occupants in the great struggle for life allow the new and solitary immigrant room and sustenance?[35]

These words were published in May 1855. In the Crimea, the siege of Sebastopol was eight months old. The English were closing from the port at Balaklava, the French attacking from their base at Kamiesh Bay. The tide of battle had just turned, though Sebastopol did not fall until September. Meanwhile Darwin continued his seed-floating experiments and published five more articles. As the war ended, he wrote up his final, magisterial report on the survival of seaborne seed invaders and read it himself in May 1856 before the Linnean Society, an unheard-of public performance.[36]

One week later he began writing a blockbuster to demolish the old static creationism and convince naturalists that life's diversity and distribution had been brought about by Lyellian natural forces working for countless ages. As the manuscript piled up, the last piece of his new world order fell into place.

For years Darwin had believed that migrant organisms produced new species by adapting to new physical conditions, competing with and finally overcoming the old residents. The process went on intermittently, as when a sudden drought or shortage of food caused beasts to invade fresh pastures. Now, by reflecting on how species were classified into groups, he saw that invasion, far from a sporadic event, was a constant vector in evolution. Small genera—groups of closely related species—were known to have narrow geographic ranges, large genera wide ranges. Why? Because, he reckoned (totting up the numbers), the big genera increased faster, producing a greater number of both species and individuals. This showed them to be more adaptable, better able to diversify into new environments. And why did diversity breed success? Because being different improved an individual's chance of surviving in the struggle for life. Natural selection favored specialists, organisms that could get to parts others couldn't reach, make a living where others failed, subsist at others' expense. This now became the "key-stone" of his theory, "the principle of divergence."[37]

Evolution as Darwin now recast it—relentless, divergent evolution—did not wait on migration, new conditions, or isolation. Organisms constantly altered their own environment even on a single patch of ground. They themselves produced the conditions to which they had to adapt, and thus they adapted to one another more than their physical surroundings. "Every where we see organic action & reaction. All nature is bound together by an inextricable web of relations."[38] Swelling populations and the struggle for resources made organisms specialize, spread apart—diverge—both intensively into local niches and extensively into new fields. Honed by competition, the most successful ventured farthest, multiplying, fanning out, beating off rivals, sweeping across the globe—nature's space invaders.

Darwin's revolutionary world order was complete. And in his manuscript for the first time, he adopted a full-blown political idiom. "It is not the oppressed & decreasing forms which will tend to be modified, but the triumphant, which are already very numerous in individuals, widely diffused in their own country & inhabiting many countries." Thus, "[i]n the great scheme of nature, to that which has much, much will be given." These expansive groups "include the ancestors of future conquering races," which in their turn will be "still more triumphant" as they seize "the places occupied by the less favoured forms . . . supplanting them & causing their extermination." For in "each country" it is "a race for life & death; & to win implies that others lose." The starting gun is an "intrusion of strangers"—an invasion—which transforms all relationships. The strangers succeed best when a compact group, for "if a whole nation migrated in a body, each might retain almost his usual habits & business, but if only a few settled in a foreign land each probably would have more or less to change his habits, & occupy a different position in society." In this way a new "division of labour" emerges; and "the more complete" the strangers' "association with foreigners," the greater the probability that selection will mold their offspring into well-adapted specialists. As generations pass, the offspring diversify into groups, the groups themselves diversify, and so on, until finally the descendants of the first invaders can be arranged "like families within the same tribes, tribes within the same nations, & nations within the same sections of the human race."[39]

None of this, however, went to press. Darwin's blockbuster, which he called *Natural Selection,* was about two-thirds finished—225,000 words—when an obscure bird collector in the Far East, Alfred Russel Wallace, sent a short manuscript that made him rush into print with an "abstract" of his opus, entitled *On the Origin of Species.* Written for general readers, the *Origin of Species* used political parlance to even greater effect, describing life's behavior in the lurid language of mid-Victorian imperial conquest.

IMPERIAL EVOLUTION

As Anglo-Indian forces mopped up in the Second Opium War (1857–60), cutting down Chinese defenders with the latest weapons from British factories, readers of the *Origin of Species* learned that successful species too were "manufactured," that the large groups to which they belonged were themselves each a "manufactory," and that the "manufacturing" process, though a "slow one," went on wherever many closely allied species "now flourish." The greater the "division of labour" among such species, as in British industry, the better the group's "chance of succeeding in the battle of life." For "widely-ranging species, abounding in individuals, which have already triumphed over many competitors in their own widely-extended homes will have the best chance of seizing on new places, when they spread into new countries."[40]

Darwin, with fresh frankness or temerity, now called this spreading "colonisation" and the invaders "colonists." Again and again in the *Origin of Species,* his colonists "beat," "conquer," and "exterminate" the "aborigines" and "natives."[41] For instance, why had more plants migrated from north to south than the reverse? Because northern plants, occupying that hemisphere's greater landmass, had "existed in their homes in greater numbers" and thus had been "advanced through natural selection and competition to a higher stage of . . . dominating power." With their "machinery of life" perfected in the "more efficient workshops of the north," these plants had "beaten the natives" in South America and like a tide had "freely inundated" the tropics, leaving their "drift in horizontal lines" visible on many mountains. "The various beings thus left stranded may be compared with savage races of man, driven up and surviving in the mountain-fastness of almost every land, which serve as a record, full of interest to us, of the former inhabitants of the surrounding lowlands."[42]

Colorful language, you may say, but not to be taken literally and on no account to be construed as illustrating how Darwin's world order really worked. The *Origin of Species* is full of fertile metaphors. No one supposes that its iconic "great Tree of Life" was a real tree, that all organisms literally struggled with one another, or that "Nature" selected. These are only engaging figures of speech. Similarly, plant "colonists," geographic "workshops," and nature's "division of labour" reveal the literary strategist at work, using a popular idiom—sometimes perhaps misleadingly—to impress a contentious theory on patriotic minds steeped in Malthusian political economy.[43] Certainly no one should imagine that Darwin himself saw imperial Britain with its competitive factories, booming population, overseas struggles, and proliferating colonies as a model of divergent evolution.

Except in the *Natural Selection* manuscript, he did have successful invaders diversifying generation by generation, so that their descendants could be grouped "like families within the same tribes, tribes within the same nations, & nations within the same sections of the human race." In his *Journal of Researches,* "human varieties" behaved "in the same way as different species of animals—the stronger always extirpating the weaker," and in his transmutation notebooks, Darwin declared the "principles" underlying such extermination to be "strictly applicable to the universe." Or as the *Origin of Species* stipulates, "What applies to one animal will apply throughout all time to all animals."[44] Even Darwin's comparing "beaten" southern plant species to "savage races of man," driven up and surviving on remote mountains—and that in a book where he shunned talk of human origins—contains more than analogy. Colorful language, yes, but Darwin saw divergent evolution and imperial conquest as of a piece, plants, animals, and British subjects all as consummate space invaders.[45]

This view is made explicit for the first time in Darwin's long-delayed work on racial evolution, the *Descent of Man* (1871), and in a number of late private letters. For some purposes or when it suited him, Darwin had always treated *Homo sapiens* as a "domesticated animal," highly bred through mate selection into different races, rather like show dogs or fancy pigeons.[46] The races themselves were more or less domesticated, "tame" or "wild," "high" or "low" on the Victorian scale of "civilization," and in the *Descent of Man* he moved effortlessly among these categories.

Humans, even in their "rudest state," were the "most dominant animal" ever. They had "spread more widely than any other highly organised form," and "all others" had yielded before them owing to the "immense superiority" of their "intellectual faculties . . . social habits . . . and corporeal structures." Yet although the geographic range of humans, considered as "a single species," was enormous, "some separate races" had "very wide ranges," and those races, said Darwin, recalling his principle of divergence, like "widely-ranging species," were "much more variable."[47] Whether races were classified as varieties of one species or as species of genus *Homo* mattered little at this level, for he held that "groups of species . . . follow the same general rules in their appearance and disappearance as do single species, changing more or less quickly, and in a greater or lesser degree."[48] The more widespread a race, therefore, the more adaptable it must be, and hence the more dominant, which "at the present day," Darwin assumed, meant more civilized.

"Civilised nations are everywhere supplanting barbarous nations, excepting where the climate opposes a deadly barrier; and they succeed mainly . . . through their arts, which are the products of the intellect." Among civilized nations, one "rises, becomes more powerful, and spreads more widely, than another" because of "an increase in the actual number of the population, on the number of the men endowed with high intellectual and moral faculties, as well as on their standard of excellence." For instance, take "the remarkable success of the English as colonists over other European nations, which is well illustrated by comparing the progress of the Canadians of English and [of] French extraction." Or look at the "wonderful progress" of the United States, whose people were conquering a continent.[49] When everywhere "we see . . . enormous areas of the most fertile land peopled by a few wandering savages, but which are capable of supporting numerous happy homes," who can deny that in the "distant future" all historical progress will "only appear to have purpose and value when viewed in connection with, or rather subsidiary to . . . the great stream of Anglo-Saxon emigration to the west"? And this notwithstanding that meantime, according to Darwin, "the civilised races of man will almost certainly exterminate and replace throughout the world" both the "savage races" and the "anthropomorphous apes."[50]

We grieve at such losses; Darwin it seems did not. Progress cost lives, of species and races as well as individuals. Loss of cultural and biodiversity was the price of his new world order and well worth paying he assumed. "Remember what risk the nations of Europe ran, not so many centuries ago of being overwhelmed by the Turks, and how ridiculous such an idea now is!" he cheered after Disraeli's foray into the Balkans in 1877 and acquisition of Cyprus (a police action that gave "jingoism" its modern sense). "The more civilized so-called Caucasian races have beaten the Turkish hollow in the struggle for existence."[51] Indeed, with "the white man 'improving off the face of the earth' even races nearly his equals," Darwin had been sanguine since the 1860s; for when "in 500 years" the "Anglo-saxon race" has "spread & exterminated whole nations," he then wrote, "the Human races, viewed as a unit, will have risen in rank," and men will look back on the Victorians "as mere savages." Smiling wryly, Darwin confessed that the thought gave him "infinite satisfaction."[52]

Even so, in the *Descent of Man,* this happy outcome was contingent. "We must remember that progress is no invariable rule." But a "cool climate" helped, Darwin hastened to add, as well as "inheritance of property," "accumulation of capital," "a good education during youth," and "open competition for all men" so that "the more intelligent members within the same community will succeed better in the long run than the inferior, and leave a more numerous progeny."[53] These not unfamiliar conditions would keep evolution on the up-and-up, and nothing, he insisted, should interfere with them. Trade unions, cooperative societies and the like, which "opposed . . . competition," were "a great evil for the future progress of mankind."[54] So a fortiori were "artificial checks" to population, which he despised. Suppose birth control had been practiced "during the last two or three centuries, or even for a shorter time in Britain"? he protested. "What a difference it would have made in the world, when we consider America, Australia, New Zealand, and South Africa!" Within a century "France will tell us the result" of using artificial checks; already indeed "we can . . . see that the French nation does not spread or increase much," and this must never be Britain's fate. "Our natural rate of increase, though leading to many and obvious evils, must not be greatly diminished by any means," for "no words can exaggerate the importance . . . of our colonization for the future history of the world."[55]

WALLACE'S GEOPOLITICS

Darwin's discourse was appropriate to the venture on which he had embarked in 1831, the formative event of his career. Traveling at his rich father's expense, he was the Admiralty's guest and the captain's companion aboard a ten-gun brig sent to measure the earth and make South America's waters safe for British trade. His laboratory was the poop cabin, his working surface the chart table, where for five years he watched hydrographic maps being drawn. He acquired a global vision himself, mapping coral reefs and continents, claiming—with Lyell's help—an earth empire of his own.[56] Great movements in time and space gripped his imagination, wholesale elevation and subsidence, and finally life's struggle to expand and diversify around the globe. From the start, Darwin's evolutionary biogeography was part of a geopolitical enterprise.

In a different way, so too was Wallace's.[57] The eighth child of an impoverished family, Alfred Russel Wallace left school at age thirteen to train as a land surveyor. The trade was booming when Victoria ascended the throne, and he roamed the country making maps for commons enclosure and tithe commutation. In Wales the backlash against these practices shocked him as tenant farmers turned guerilla fighters in the so-called Rebecca Riots, and he grew to despise his job. After a spell of schoolteaching, he set up as a roving specimen collector, sailing to Brazil on a merchantman in 1848 and to the Dutch East Indies in 1854 aboard a P. & O. steamer. Self-employed, shipping back rare birds and beetles for cash, he traveled cheaply by local transport and lived as an equal with natives.[58] He also traveled as an evolutionist in search of a theory (the anonymous hackwork *Vestiges of the Natural History of Creation* had converted him in 1845), and he finally devised one by noting, as was his wont, where things lived and how they obtained their food.

Maps imbued his mind. Always he thought of homes and habitats, autochthones and aliens, scarcity and abundance. His livelihood as a collector depended on it. In 1855 he got his first great insight, the "law," as he put it, that "every species has come into existence coincident both in space and time with a pre-existing closely allied species." A year later he glimpsed a further remarkable fact: the presence in the Malay archipelago of "two distinct faunas rigidly circumscribed" by an invisible "boundary," later to be famous as "Wallace's line."[59] His third and best-known aperçu came in 1858 as he extended his faunal boundary to separate the Malay and Papuan races. Along this line, as in rural Wales, he witnessed struggles for "a constant supply of wholesome food," struggles that, reminding him of Malthus's essay on population, suggested the "general principle" by which species succeeded one another. Darwin, who famously learned of it by post, called that principle natural selection.[60] Thus Wallace's evolutionary biogeography too had a geopolitical character, one, however, acquired in native boats and huts rather than aboard a naval hydrographic vessel.

Before going abroad, Wallace studied Lyell's *Principles of Geology,* and in the Far East he carried the cheap "Colonial and Home" edition of Darwin's *Journal.* His biogeography began where Darwin's had, in acceptance of Lyell's gradualism and rejection of his creationism, which made adaptation explain where and when species originated. Like Darwin, Wallace accounted for life's diversity and distribution by adaptive evolution consequent on migration or environmental change, and their theories of species formation looked identical.[61] But whatever the similarities, Wallace's rhetorical world was as remote from Darwin's as their social worlds—they wrote up their theories differently. Although a colonial infrastructure made much of Wallace's fieldwork possible, the solitary English collector, living alongside natives and dependent on their knowledge and skills, eschewed the rich imperial language in which

Darwin depicted evolving life.[62] Wallace thought spatially and described his theories in ways appropriate to the Welsh mapmaking enterprise from which he first learned about native habitats.

He wrote with artless clarity. One searches in vain for conquering colonial imagery in his major theoretical essays between 1855 and 1864. Here "organic beings" are continually "peopling" the earth and making it a "theatre of life." New species evolve under changed "physical conditions" in "an unbroken and harmonious system."[63] The faunas of "neighbouring countries" testify to their geological past, showing that new species were "gradually introduced" as the regions became isolated.[64] The arrival of "chance immigrants" is often followed by "natural extinction and renewal of species," and those organisms with "greater powers of dispersion" and "a greater plasticity of organization" have "extended themselves" over continents. The "regular and unceasing extinction of species, and their replacement by allied forms" is an "established fact," contingent in every case on the quantity and quality of available food.[65]

Food was the central theme in Wallace's 1858 manuscript. A species constantly increases up to the limits of its food supply in the "whole district it inhabits." The population is "generally stationary," nor does migration—of birds, for example—permit much growth, for migration would not continue unless food was lacking in the "countries" visited. But where varieties of a species coexist and the "physical conditions" deteriorate, those better able to obtain food will increase in numbers and "occupy the place" of the less able and the parent species. Under continued "adverse . . . conditions," the process repeats: new varieties successively appear with "diverging modifications of form" specialized for different modes of life. These "lines of divergence" emerge because the adaptive "principle" acts "exactly like . . . the centrifugal governor of the steam engine," checking and correcting "any irregularities almost before they become evident."[66]

Thus Wallace's species are honed to their environments by a process resembling a stationary self-adjusting mechanism rather than a dynamic thrusting invasion. His adaptive principle is "ecologically static" and so cannot be Darwin's natural selection.[67] Wallace saw evidence for divergent evolution but did not explain it. In his famous 1858 manuscript, life's vector was missing.

This difference ran deep. By 1862, when he returned to London, Wallace had read the *Origin of Species,* annotating many passages.[68] Its imperial language may not have troubled him, though soon enough he produced an alternative. Writing in 1863 for the *Natural History Review* (edited by Darwin's arch-advocate T. H. Huxley), he declared the *Origin's* two chapters on geographical distribution "in every respect satisfactory" even while pointing up "discrepancies" that still had to be explained, notably the large groups of animals and plants inhabiting alien environments.[69] He went on to distinguish indigenes from interlopers in six "regions" of endemic terrestrial life, and then without warning, in a technical passage about beetles, he mounted his soapbox:

> [T]here is an ancient insect-population in the Austro-Malayan Islands which accords in its distribution with the other classes of animals, but which has been overwhelmed, and in some cases perhaps exterminated by immigrants from the adjacent countries. The result is a mixture of races in which the foreign element is in excess; but naturalists need not be bound by the same rule as politicians, and may be permitted to recognise the just claims of the more ancient inhabitants, and to raise up fallen nationalities. The aborigines and not the invaders must be looked upon as the rightful owners of the soil, and should determine the position of their country in our system of Zoological geography.[70]

This was gratuitous but for the wider context. Even if Darwin's imagery was not his target, Wallace drew a bead on politicians. He made science his ally, not theirs. In biogeography, the foundational study of living space, the science that furnished him and Darwin with crucial clues about evolution, the earth was to be divided according to ancient right, not invaders' might.

Or so it would seem. For two months after these lines were published, in March 1864, Wallace beat Darwin to the punch and tackled human evolution. His audience was the new men-only Anthropological Society of London, with a white supremacist in the chair. Darwin's best friend J. D. Hooker thought the group sleazy, "a sort of Haymarket to which the *demi-monde* of science gravitated on its establishment";[71] and indeed, no subject was taboo here provided one had the stomach for debate. Wallace stuck his neck out and confronted local prejudice: he argued that the human races belonged to one "family," but that their physical differences had emerged from a "single homogenous" population *before* mankind's distinctive mental and moral characters appeared. He explained the physical diversity by "natural selection," though it still worked for him like a governor, keeping groups "in harmony with the surrounding universe," equipping them with traits adapted to new climates as they "ranged farther" from their ancestral tropical home.[72]

Meanwhile mind developed—social sympathies, the moral sense, intelligence to tame environments. Natural selection favored individuals and groups in whom these powers were ascendant, and eventually bodily evolution ceased. Thereafter races advanced as one humanity, as they do still: "[T]he better and higher specimens . . . increase and spread, the lower and more brutal

... give way and successively die out," from North to South, the direction in which "all the great invasions and displacement of races" have occurred. When Europeans come into contact with "savage man," they "conquer" and "increase at his expense . . . just as the weeds of Europe overrun North America and Australia, extinguishing native productions by the inherent vigour of their organisation, and by their greater capacity for existence and multiplication." This "extraordinary fact" Wallace credited to "Darwin's own book": "The intellectual and moral, as well as the physical qualities of the European are superior" and must transform the world. In the distant future, *Homo sapiens* will again be a "single homogeneous race," no individual of which will be "inferior to the noblest specimens of existing humanity." "Perfect freedom" with "perfect sympathy" will prevail; "compulsory government will have died away," and the earth will be "as bright a paradise as ever haunted the dreams of seer or poet."[73]

Wallace was something of a blotting pad, always picking up impressions. Back in London after a dozen years abroad, he read Herbert Spencer on the "social organism" and Henry Buckle on the progress of civilization, as well as Darwin. He used the resources of his new rhetorical universe to make a name for himself and, not incidentally, a living. In 1864, facing up to a racialist snakepit at the Anthropological Society, he must have known he would rattle nerves. While not sharing the group's worst prejudices, neither did he wish to offend, and an element of accommodation may be detected in his language—adaptation, as it were, to a new imperial environment. Weeds apart, however, Wallace referred only to human invasions; he did not anthropomorphize life.[74] The contrast with his previous work is well seen from the last essay he wrote before reaching London, on native trade with New Guinea.[75] More striking is the contrast between his Anthropological Society discourse and the work he dedicated to Darwin in 1869.

In *The Malay Archipelago,* Wallace's most popular and widely read book, the only "empire" is Austrian, "imperial" is a common species name, and only the Dutch, the Portuguese, and ants have "colonies." "Aborigines" are always human, "natives" are established residents (also marsupials in the Moluccas and flowers in the Himalayas), and people wage "war," "conquer," and "exterminate" one another (also the flying opossum). "Competition" too is a human prerogative, but no "invasion" crops up, nor any of its cognates. Districts may be "overrun" and indigenous populations "supplanted"; "inhabitants" and "enemies" of different species may "struggle" and "migrate." Yet Wallace is remarkably consistent—startlingly so compared to Darwin in the *Origin of Species*—in omitting to cast living organisms in imperial Britain's image.[76] In *Malay Archipelago,* he seems to have reverted to the discourse of his earlier theoretical essays, influenced perhaps by his conversion in 1865 to a spiritualism that set the moral world above the natural. Certainly his geopolitics in the book has little in common with Darwin's in the *Descent of Man.*

In a final paean to his Malaysian hosts, he made political capital from their native communities, where "all are nearly equal."

> There are none of those wide distinctions, of education and ignorance, wealth and poverty, master and servant, which are the product of our civilization; there is none of that wide-spread division of labour, which, while it increases wealth, produces also conflicting interests; there is not that severe competition and struggle for existence, or for wealth, which the dense population of civilized countries inevitably creates. All incitements to great crimes are thus wanting, and petty ones are repressed, partly by the influence of public opinion, but chiefly by that natural sense of justice and of his neighbour's right, which seems to be, in some degree, inherent in every race of man.[77]

But "progress" was on the way, and Wallace feared the consequences. Advancing civilization would improve the natives' physical condition and promote population growth, leading to a fierce Malthusian struggle, a "spirit of competition," and the usual "crimes and vices." A "high-class European example" might "obviate much of the evil," but where to find one? Britain itself was sunk in a state of moral "barbarism" beside which savage life looked progressive. "We (the English) try to force" a society from barbarism up to civilization and "our system has always failed. We demoralize and we extirpate, but we never really civilize." The same was true of other powers, though from experience Wallace judged "the Dutch system" of administration to be "the very best that can be adopted, when a European nation conquers or otherwise acquires possession of a country." This system attempts "to bring the people on by gradual steps to that higher civilization"; it "takes nature as a guide, and is therefore more deserving of success, and more likely to succeed, than ours."[78]

No human sacrifices here on the altar of an Anglo-Saxon world order. The "bitter satire" of which Engels wrote, the one that—Marx sneered—recognized "English society with its division of labour, competition, opening up of new markets, 'inventions,' and the Malthusian 'struggle for existence'" as the normal condition of life among "beasts and plants," was penned by Darwin. Although Wallace did not radically shun such imagery—he was a Victorian after all—there is at this level literally a world of difference between Darwin's biogeography and his. To speak of a "Darwin-Wallace tradition in biogeography" that ran in "parallel with the colonialistic spirit of the times" obscures that keen divide.[79]

Darwin's and Wallace's biogeographies arose at different sites using different maps inspired by different needs. Both drew on colonial resources, political and literary, but Darwin's living organisms behaved like Englishmen, invading everywhere; or rather, Englishmen to him were invasive organisms, multiplying, spreading across the earth, keeping evolution on the march. Had he not sailed the globe and admired the results? Wallace's world by contrast was less a sailor's, more a surveyor's; the imperial ethos in him was muted, and no wonder: he saw Britain as a corrupting rather than civilizing influence abroad. People in his world didn't behave like other organisms, or they could refuse to if they wished. Such a view may yet appeal in a day when empire strides back and the colonizer fears being colonized, the invader invaded.

...

Notes

The epigraph to this chapter is from *Charles Darwin's "Natural Selection": Being the Second Part of His Big Species Book Written from 1856 to 1858*, ed. Robert C. Stauffer (Cambridge: Cambridge University Press, 1975), 271–72.

This chapter draws inspiration from a manuscript by Jon Hodge, "Darwin's Science in Darwin's Society and Economy," which he kindly shared with me. Warmest thanks to Jon for alerting me to relevant literature and sharpening my ideas in marathon phone conversations. I am also grateful to Fred Burkhardt, Nigel Wace, Jim Endersby, and members of the Cabinet of Natural History at Cambridge University for generous constructive criticism.

1. Charles-Edward Amory Winslow, *The Conquest of Epidemic Disease: A Chapter in the History of Ideas* (Princeton: Princeton University Press, 1943), 362–63; Peter Medawar, *Memoir of a Thinking Radish: An Autobiography* (Oxford: Oxford University Press, 1986), 134.

2. Patrick Armstrong, "The Metaphors of Struggle, Conflict, Invasion and Explosion in Biogeography," *Ekológia* 11 (1992): 437–45.

3. David M. Richardson et al., "Naturalization and Invasion of Alien Plants: Concepts and Definitions," *Diversity and Distributions* 6 (2000): 93–107; Peter Alpert, Elizabeth Bone, and Claus Holzapfel, "Invasiveness, Invasibility and the Role of Environmental Stress in the Spread of Non-native Plants," *Perspectives in Plant Ecology, Evolution and Systematics* 3 (2000): 52–56.

4. Martin Argles, "The Bark's Worse than the Bite," *Guardian* (London), 14 December 1990, 31; Pierre Binggeli, "Misuse of Terminology and Anthropomorphic Concepts in the Description of Introduced Species," *Bulletin [of the British Ecological Society]* 25 (1994): 10–13. The metaphor enjoys a new lease of life in twenty-first-century Britain: see Mark Townsend, "Alien Invasion: The Plants Wrecking Rural Britain," *Observer* (London), 2 February 2003, 14, and Steve Farrar, "Academic Blacklisted over Threat of Invasion," *Times Higher Education Supplement,* 26 September 2003, 1–3. For other contexts, see Gert Groening and Joachim Wolschke-Bulmahn, "Some Notes on the Mania for Native Plants in Germany," *Landscape Journal* 11 (1992): 116–26; Gary Alan Fine and Lazaros Christoforides, "Dirty Birds, Filthy Immigrants, and the English Sparrow War: Metaphorical Linkage in Constructing Social Problems," *Symbolic Interaction* 14 (1991):375–93; Matthew K. Chew and Manfred D. Laubichler, "Natural Enemies—Metaphor or

Misconception?" *Science* 301 (4 July 2003): 52–53; and Philip J. Pauly, "The Beauty and Menace of the Japanese Cherry Trees: Conflicting Visions of American Ecological Independence," *Isis* 87 (1996): 51–73, incorporated in Pauly's *Biologists and the Promise of American Life: From Meriwether Lewis to Alfred Kinsey* (Princeton: Princeton University Press, 2000), 71–92.

5. Alpert, Bone, and Holzapfel, "Invasiveness, Invasibility and the Role of Environmental Stress," 53.

6. Richardson et al., "Naturalization and Invasion of Alien Plants," 94, 97, 102. Sarah Darwin, who is researching the tomato's invasion of the Galapagos, brought this literature to my attention. The debate continues among environmentalists: see Jonah H. Peretti, "Nativism and Nature: Rethinking Biological Invasion," *Environmental Values* 7 (1998): 183–92; William Throop, "Eradicating the Aliens: Restoration and Exotic Species," in *Environmental Restoration: Ethics, Theory, and Practice,* ed. William Throop (Amherst, NY: Humanity Books, 2000), 179–91; Mark Woods and Paul Veatch Moriarty, "Strangers in a Strange Land: The Problem of Exotic Species," *Environmental Values* 10 (2001): 163–91; and Ned Hettinger, "Exotic Species, Naturalisation, and Biological Nativism," *Environmental Values* 10 (2001): 193–224. I am grateful to Akihisa Setoguchi of Kyoto University for these references.

7. Martin J. S. Rudwick, "Transposed Concepts from the Human Sciences in the Early Work of Charles Lyell," in *Images of the Earth: Essays in the History of the Environmental Sciences,* ed. L. J. Jordanova and Roy S. Porter (Chalfont St. Giles: British Society for the History of Science, 1979), 67–83; James A. Secord, "King of Siluria: Roderick Murchison and the Imperial Theme in Nineteenth-Century British Geology," *Victorian Studies* 25 (1982): 413–42; Simon Schaffer, "The History and Geography of the Intellectual World: Whewell's Politics of Language," in *William Whewell: A Composite Portrait,* ed. Menachem Fisch and Simon Schaffer (Oxford: Clarendon Press 1991), 201–31; Suzanne Zeller, "Environment, Culture, and the Reception of Darwin in Canada, 1859–1909," in *Disseminating Darwinism: The Role of Place, Race, Religion, and Gender,* ed. Ronald L. Numbers and John Stenhouse (Cambridge: Cambridge University Press, 1999), 91–122. I owe the latter suggestion to Nigel Wace of the Australian National University, veteran biogeographer of Tristan da Cunha. He reports richly diverse answers from Europeans and their descendants but has yet to find words for a weed-concept in native Australian Aboriginal languages (personal communication). See N. M. Wace, "The Units and Uses of Biogeography," *Australian Geographical Studies* 5 (1967): 15–29.

8. Felix Driver, *Geography Militant: Cultures of Exploration and Empire* (Oxford: Blackwell, 2001); D. Graham Burnett, *Masters of All They Surveyed: Exploration, Geography, and a British El Dorado* (Chicago: University of Chicago Press, 2000); Richard H. Grove, *Green Imperialism: Colonial Expansion, Tropical Island Edens and the Origins of Environmentalism, 1600–1860* (Cambridge: Cambridge University Press, 1995); Janet Browne, "Biogeography and Empire," in *Cultures of Natural History,* ed. Nicholas Jardine, James A. Secord, and Emma C. Spary (Cambridge: Cambridge University Press, 1996), 305–21; Richard Drayton, *Nature's Government: Science, Imperial Britain, and the "Improvement" of the World* (New Haven: Yale University Press, 2000).

9. Alfred W. Crosby, *Ecological Imperialism: The Biological Expansion of Europe, 900–1900* (Cambridge: Cambridge University Press, 1986); Chris Bright, *Life Out of Bounds: Bioinvasion in a Borderless World* (New York: Norton, 1998).

10. Crosby, *Ecological Imperialism,* 270, 273.

11. David N. Livingstone, *The Geographical Tradition: Episodes in the History of a Contested Enterprise* (Oxford: Blackwell, 1992), 178–87. Cf. Charles H. Smith, "Historical Biogeography: Geography as Evolution, Evolution as Geography," *New Zealand Journal of Zoology* 16 (1989): 773–85.

12. Clarence J. Glacken, *Traces on the Rhodian Shore: Nature and Culture in Western Thought from Ancient Times to the End of the Eighteenth Century* (Berkeley and Los Angeles: University of California Press, 1967), 705.

13. Janet Browne, *The Secular Ark: Studies in the History of Biogeography* (New Haven: Yale University Press, 1983); Dov Ospovat, "Perfect Adaptation and Teleological Explanation: Approaches to the Problem of the History of Life in the Mid-Nineteenth Century," *Studies in History of Biology* 2 (1978): 33–56.

14. Dov Ospovat, "Lyell's Theory of Climate," *Journal of the History of Biology* 10 (1977): 317–39; Martin J. S. Rudwick, introduction to Charles Lyell, *Principles of Geology,* 1st ed. (1830–32; Chicago: University of Chicago Press, 1990), 1:[vii–lviii].

15. Paul H. Barrett et al., eds., *Charles Darwin's Notebooks, 1836–1844: Geology, Transmutation of Species, Metaphysical Enquiries* (London: British Museum [Natural History] / Cambridge University Press, 1987), 44 [RN 72], 222 [B 229].

16. Ibid., 61 [RN 127], 271 [C 106e]; M. J. S. Hodge, "Darwin and the Laws of the Animate Part of the Terrestrial System, 1835–1837: On the Lyellian Origins of his Zoonomical Explanatory Program," *Studies in the History of Biology* 6 (1983): 40–66.

17. Charles Darwin, *Journal of Researches into the Geology and Natural History of the Various Countries Visited by H.M.S. "Beagle" under the Command of Captain FitzRoy, R.N., from 1832 to 1836* (London: Henry Colburn, 1839), 511, 580, 583; cf. Richard Darwin Keynes, ed., *Charles Darwin's "Beagle" Diary* (Cambridge: Cambridge University Press, 1988), 392, 428.

18. Darwin, *Journal of Researches,* 520; cf. Keynes, *Darwin's "Beagle" Diary,* 398–99.

19. Keith Stewart Thomson, *HMS Beagle: The Story of Darwin's Ship* (New York: Norton, 1995), 48–56; P. J. Cain and A. G. Hopkins, *British Imperialism: Innovation and Expansion, 1688–1914* (London: Longman, 1993), 279–81; Rory Miller, *Britain and Latin America in the Nineteenth and Twentieth Centuries* (London: Longman, 1993), 70–96.

20. Keynes, *Darwin's "Beagle" Diary,* 78.

21. Darwin, *Journal of Researches,* 515–16, 575, 607; cf. Keynes, *Darwin's "Beagle" Diary,* 396, 424, 445–46.

22. Richard B. Freeman, *The Works of Charles Darwin: An Annotated Bibliographical Handlist* (London: Dawson, 1977), 35; *Westminster Review,* 1835, quoted in Klaus E. Knorr, *British Colonial Theories, 1570–1850* (Toronto: University of Toronto Press, 1944), 295.

23. *Hansard,* 1843, quoted in Knorr, *British Colonial Theories,* 311.

24. W. B. Cooke, 1835, quoted in Knorr, *British Colonial Theories,* 314; Trevor O. Lloyd, *The British Empire, 1558–1995* (Oxford: Oxford University Press, 1996), 139–47.

25. Barrett et al., *Darwin's Notebooks,* 262 [C 73], 267 [C 93]; Darwin, *Journal of Researches,* 38; cf. Keynes, *Darwin's "Beagle" Diary,* 190.

26. Barrett et al., *Darwin's Notebooks,* 291 [C 168].

27. Dov Ospovat, *The Development of Darwin's Theory: Natural History, Natural Theology, and Natural Selection, 1838–1859* (Cambridge: Cambridge University Press, 1981), 43–59.

28. Patrick Matthew, 1839, quoted in Knorr, *British Colonial Theories,* 274.

29. Barrett et al., *Darwin's Notebooks,* 376 [D 135e], 429 [E 114].

30. Ibid., 414 [E 64–65], 465 [T 81].

31. M. J. S. Hodge and David Kohn, "The Immediate Origins of Natural Selection," in *The Darwinian Heritage,* ed. David Kohn (Princeton: Princeton University Press, 1985), 185–206.

32. Lawrence James, *The Rise and Fall of the British Empire* (London: Little, Brown, 1994), 184–99.

33. Charles Darwin and Alfred Russel Wallace, *Evolution by Natural Selection,* ed. Gavin de Beer (Cambridge: Cambridge University Press, 1958), 171–72, 181, 216–17.

34. Ibid., 71, 197–98.

35. Paul H. Barrett, *The Collected Papers of Charles Darwin* (Chicago: University of Chicago Press, 1977), 1:258.

36. Ibid., 1:264–73; Frederick Burkhardt et al., eds., *The Correspondence of Charles Darwin* (Cambridge: Cambridge University Press, 1985–), 6:100.

37. Burkhardt et al., *Correspondence of Charles Darwin*, 7:102; Ospovat, *Development of Darwin's Theory*, 170–90; Janet Browne, "Darwin's Botanical Arithmetic and the Principle of Divergence, 1854–1858," *Journal of the History of Biology* 13 (1980): 53–89; Browne, *Secular Ark*, 195–220; Karen Hunger Parshall, "Varieties as Incipient Species: Darwin's Numerical Analysis," *Journal of the History of Biology* 15 (1982): 191–214; David Kohn, "Darwin's Principle of Divergence as Internal Dialogue," in Kohn, *Darwinian Heritage*, 245–57.

38. Stauffer, *Darwin's "Natural Selection,"* 272.

39. Ibid., 219, 227, 249, 548–49, 557, 560.

40. Charles Darwin, *On the Origin of Species by Means of Natural Selection; or, The Preservation of Favoured Races in the Struggle for Life* (London: John Murray, 1859), 56–57, 115, 128, 350, 469–70; cf. 168–69 and Barrett et al., *Darwin's Notebooks*, 209–10 [B 157–58].

41. Darwin, *Origin of Species*, 83, 110, 115, 337, 355ff., 378ff., 403ff., 477.

42. Ibid., 83, 379–80, 382. Darwin later urged New Zealand naturalists "to observe & record the rate & manner of spreading of European weeds & insects, & especially to observe *what native plants must fail:* this latter point has never been attended to. Do the introduced Hive-bees replace any other insect? &c &c.—All such points are, in my opinion, great desiderata in science" (Burkhardt et al., *Correspondence of Charles Darwin*, 11:68).

43. Margaret Schabas, "Ricardo Naturalized: Lyell and Darwin on the Economy of Nature," in *Perspectives on the History of Economic Thought*, vol. 3, *Classicals, Marxians and Neo-Classicals: Selected Papers from the History of Economics Society Conference, 1988,* ed. D. E. Moggridge (London: Edward Elgar for the History of Economics Society, 1990), 40–49, suggests how a similar excuse might be made for Lyell's *Principles of Geology*, from which Darwin learned not only how to observe, but how to describe nature memorably. Thus Lyell stated: "[I]n . . . obtaining possession of the earth by conquest, and defending our acquisitions by force, we exercise no exclusive prerogative. Every species which has spread itself from a small point over a wide area, must, in like manner, have marked its progress by the diminution, or the entire extirpation, of some other, and must maintain its ground by a successful struggle against the encroachments of other plants and animals" (*Principles of Geology*, 2:156).

44. Darwin, *Origin of Species*, 113, 382; Stauffer, *Darwin's "Natural Selection,"* 249; Darwin, *Journal of Researches*, 520; Barrett et al., *Darwin's Notebooks*, 414 [E 64–65].

45. On the constitutive, not merely colorful, role of analogy and metaphor in the *Origin of Species*, see Gillian Beer, *Darwin's Plots: Evolutionary Narrative in Darwin, George Eliot and Nineteenth-Century Fiction* (London: Routledge and Kegan Paul, 1983); and Robert M. Young, *Darwin's Metaphor: Nature's Place in Victorian Culture* (Cambridge: Cambridge University Press, 1985).

46. James Marchant, *Alfred Russel Wallace: Letters and Reminiscences* (London: Cassell, 1916), 1:181.

47. Charles Darwin, *The Descent of Man, and Selection in Relation to Sex* (London: John Murray, 1871), 1:112, 136–37.

48. Darwin, *Origin of Species*, 316; cf. Darwin and Wallace, *Evolution*, 207.

49. In the 1860s, the U.S. Civil War had tested Darwin's politics and morals. He hated slavery and admitted that a struggle might be needed to "purge a government" (Burkhardt et al., *Correspondence of Charles Darwin*, 10:625), but he resented the North's hostility to England and the distress caused in the cotton towns by its blockade of Southern ports. Eventually patriotism got the better of him, and he sought to rationalize the survival of the Confederacy. "The whole affair," he admitted, "is a great misfortune in the progress of the world; but I sh^d not regret it so much if I could persuade myself that Slavery would be annihilated" (9:215). What Darwin now hated as much as slavery was a tendency in Washington's policies that might lead

to the North "declaring war against us" (10:471), or England being "bullied & forced into a war" by a triumphant Union (11:166). England, with "fine children all over the world" (11: 582), had less to lose from a slaveocracy than from a parricidal *United* States; thus safety lay in partition. After the July 1863 antidraft riots in New York City left over a hundred dead, Darwin cast the sectional conflict in his own racial terms: "What devils the low Irish have proved themselves in New York. If you conquer the South you will have an Ireland fastened to your tail," he warned his Harvard interlocutor Asa Gray (11:582; cf. 615). Better if the slaveowning Celtic Confederacy ("rednecks" to a later generation) joined with other states to eject the Anglo-North, which then could "marry Canada, & divorce England & make a grand country, counterbalancing the devilish South" (12:319). Competition afterward would see the stronger nation prevail, and Darwin had no doubt which it would be. In 1865, with the Union secure, he backpedaled, confident that England would not be attacked after all and that reason had always told him the North was in the right. Cf. Ralph Colp, Jr., "Charles Darwin: Slavery and the American Civil War," *Harvard Library Bulletin* 26 (1978): 471–89.

50. Darwin, *Descent of Man,* 1:160, 177, 179–80, 201. Lyell had written: "A faint image of the certain doom of a species less fitted to struggle with some new condition in a region which it previously inhabited, and where it has to contend with a more vigorous species, is presented by the extirpation of savage tribes of men by the advancing colony of some civilized nation. In this case the contest is merely between two different races, each gifted with equal capacities of improvement—between two varieties, moreover, of a species which exceeds all others in its aptitude to accommodate its habits to the most extraordinary variations of circumstances. Yet few future events are more certain than the speedy extermination of the Indians of North America and the savages of New Holland in the course of a few centuries, when these tribes will be remembered only in poetry and tradition" (*Principles of Geology,* 2:175). Thus in 1859 Darwin could taunt Lyell with the evolutionary consequences: "I can see no difficulty in the most intellectual individuals of a species being continually selected & the intellect of the new species thus improved . . . I look at this process as now going on with the races of man; the less intellectual races being exterminated. . . . [T]he species of most genera are adapted at least to rather hotter & rather less hot—to rather damper & dryer climates—& when the several species of a group are beaten & exterminated by the several species of another group it will not, I think *generally* be from *each* new species being adapted to the climate, but from all the new species having some common advantage in obtaining sustenance or escaping enemies. As groups are concerned, a fairer illustration than negro & white in Liberia would be, the almost certain future extinction of genus Ourang by genus Man, not owing to man being better fitted for climate, but owing to the inherited intellectual inferiority of the Ourang-genus,—man-genus by his intellect inventing fire-arms & cutting down forests" (Burkhardt et al., *Correspondence of Charles Darwin,* 7:345–46).

51. Francis Darwin, ed., *The Life and Letters of Charles Darwin, including an Autobiographical Chapter* (London: John Murray, 1887), 1:316.

52. Burkhardt et al., *Correspondence of Charles Darwin,* 8:171, 189, 379; 10:72.

53. Charles Darwin, *The Descent of Man, and Selection in Relation to Sex,* 2nd ed. (London: John Murray, 1874), 133, 135, 140, 143, 618.

54. Quoted in Richard Weikart, "A Recently Discovered Darwin Letter on Social Darwinism," *Isis* 86 (1995): 611.

55. Quoted in Jane Hume Clapperton, *Scientific Meliorism and the Evolution of Happiness* (London: Kegan Paul, Trench and Co., 1885), 340–41; Darwin, *Descent of Man* (1874 ed.), 618.

56. Cf. Secord, "King of Siluria."

57. Peter Raby, *Alfred Russel Wallace: A Life* (London: Chatto and Windus, 2001); Martin Fichman, *An Elusive Victorian: The Evolution of Alfred Russel Wallace* (Chicago: University of

Chicago Press, 2004); Andrew Berry, *Infinite Tropics: An Alfred Russel Wallace Anthology* (London: Verso, 2002); Jane Camerini, ed., *The Alfred Russel Wallace Reader: A Selection of Writings from the Field* (Baltimore: Johns Hopkins University Press, 2002).

58. Jane Camerini, "Wallace in the Field," *Osiris*, 2nd ser., 11 (1996): 44–65.

59. Alfred Russel Wallace, *My Life* (London: Chapman and Hall, 1905), 1:358–59; Jane Camerini, "Evolution, Biogeography, and Maps: An Early History of Wallace's Line," in *Darwin's Laboratory: Evolutionary Theory and Natural History in the Pacific,* ed. Roy MacLeod and Philip F. Rehbock (Honolulu: University of Hawai'i Press, 1994), 90–92.

60. Alfred Russel Wallace, "On the Tendency of Species to Depart Indefinitely from the Original Type," *Journal of the Proceedings of the Linnean Society: Zoology* 3 (1858): 54, 55–56; James Moore, "Wallace's Malthusian Moment: The Common Context Revisited," in *Victorian Science in Context,* ed. Bernard Lightman (Chicago: University of Chicago Press, 1997), 290–311.

61. M. J. S. Hodge, *Origins and Species: A Study of the Historical Sources of Darwinism and the Context of Some Other Accounts of Organic Diversity from Plato and Aristotle On* (New York: Garland, 1991), xix–xxx, 9–18, 86–110.

62. Jane Camerini, "Remains of the Day: Early Victorians in the Field," in Lightman, *Victorian Science,* 370–71.

63. Alfred Russel Wallace, "On the Law which has Regulated the Introduction of New Species," *Annals and Magazine of Natural History,* 2nd ser., 16 (1855): 188, 192, 194, 196.

64. Alfred Russel Wallace, "On the Natural History of the Aru Islands," *Annals and Magazine of Natural History, Supplement,* 2nd ser., 20 (1857): 482, 483.

65. Alfred Russel Wallace, "On the Zoological Geography of the Malay Archipelago," *Journal of the Proceedings of the Linnean Society: Zoology* 4 (1860): 181, 182, 183.

66. Wallace, "On the Tendency of Species," 55, 56–59, 62.

67. David Kohn, "On the Origin of the Principle of Diversity," *Science* 213 (1981): 1106. Thus Darwin could not have appropriated the principle of divergence from Wallace's manuscript and then worked it into his own theory, as some have alleged, following Arnold C. Brackman, *A Delicate Arrangement: The Strange Case of Charles Darwin and Alfred Russel Wallace* (New York: Times Books, 1980). See Kohn, "Darwin's Principle of Divergence," and for the full complexity of the case, Barbara G. Beddall, "Darwin and Divergence: The Wallace Connection," *Journal of the History of Biology* 21 (1988): 1–68.

68. Barbara G. Beddall, "Wallace's Annotated Copy of Darwin's 'Origin of Species,'" *Journal of the History of Biology* 21 (1988): 265–89.

69. Martin Fichman, *Alfred Russel Wallace* (Boston: Twayne, 1981), 68.

70. Alfred Russel Wallace, "On Some Anomalies in Zoological and Botanical Geography," *Natural History Review* 4 (1864): 118–19.

71. Quoted in Driver, *Geography Militant,* 97.

72. Alfred Russel Wallace, "The Origin of Human Races and the Antiquity of Man Deduced from the Theory of 'Natural Selection,'" *Journal of the Anthropological Society of London* 2 (1864): clxv.

73. Ibid., clxiv, clxv, clxxxvi, clxix–clxx.

74. Efforts to use Wallace to make or break an ideological link between natural selection and laissez-faire political economy overlook just such considerations: cf. William Coleman, "The Strange 'Laissez Faire' of Alfred Russel Wallace: The Connection between Natural Selection and Political Economy Reconsidered," in *Darwinism and Evolutionary Economics,* ed. John Laurent and John Nightingale (Cheltenham: Edward Elgar, 2001), 36–48; and Young, *Darwin's Metaphor,* 23–55, 179–202. For the background to Wallace's views on land and human migra-

tion, see Greta Jones, "Alfred Russel Wallace, Robert Owen and the Theory of Natural Selection," *British Journal for the History of Science* 35 (2002): 73–96.

75. Alfred Russel Wallace, "On the Trade of the Eastern Archipelago with New Guinea and Its Islands," *Journal of the Royal Geographical Society* 32 (1862): 127–37.

76. The analysis is based on a key-word search of the Project Gutenberg digital text.

77. Alfred Russel Wallace, *The Malay Archipelago: The Land of the Orang-utan and the Bird of Paradise: A Narrative of Travel with Studies of Man and Nature,* 6th ed. (London: Macmillan, 1877), 595.

78. Ibid., 92, 94–95, 257, 596.

79. Frederick Engels, *Dialectics of Nature* (London: Lawrence and Wishart, 1941), 19; Karl Marx and Frederick Engels, *Selected Correspondence, 1846–1895,* vol. 9 (London: Lawrence and Wishart, 1943), 125–26; Gareth Nelson, "From Candolle to Croizat: Comments on the History of Biogeography," *Journal of the History of Biology* 11 (1978): 299.

PART II

GEOGRAPHY AND TECHNICAL REVOLUTION

Time, Space, and the Instruments of Transmission

The three chapters in part 2 consider some of the connections between "revolution," as that term is understood to result from enhanced technical capacity, and matters of geography. In several contexts, in the "Industrial Revolution" for example, or in certain areas of the "Agricultural Revolution," what has been held to count as "revolution" has readily been associated with technical change and the rapid and wholesale adoption of innovation. In this respect, "spinning jennies," power looms, seed drills, new plough types, and the like have almost mythic status as the technological or instrumental instigators of revolutions in productive capacity and in social systems of production. Even when research has pointed to varying take-up rates, to parallel innovations and to cultures of resistance to new ways of making and doing, these and other technical devices are seen as agents of technical revolution.

The chapters here are concerned less with the nature of given technical processes, however, and rather more with the revolutionary and the geographical implications of what was produced by them. Of central interest to each are the ways in which different instruments of transmission helped to direct and to promulgate different conceptions of revolution and to do so in relation to particular geographies. The instruments in question are the printed map, notably in its association with published texts, the clock and changing conceptions of clock time, and the camera as a device affording, at least initially, new conceptions of pictorial "realism." Each of the chapters offers a detailed examination of these instruments and technologies in relation to prevalent "grand theory" about, respectively, the nature of the "Information" or "Print

Revolution," the "Horological Revolution," and the "innocence" of visual representation and of systems of classification.

Brotton's attention in chapter 6 to the mapping of the Cape of Good Hope
illuminates the commercial, political, and geographical consequences of that
"revolutionary moment" in the West's contemporary geographical consciousness when, in 1488, the Portuguese navigator Bartholomeu Diaz rounded what
is known now as the Cape of Good Hope. At once dispelling ancients' claims
about the size and shape of the world and opening out new commercial pros-
pects, Diaz's encounter helped set in train new geographies of mercantilism,
but it did not immediately lead to a clear and full understanding of the Cape
region or its inhabitants. Only as demanded by the exigencies of commerce
and the contingencies of mapping did the area and its peoples "appear." With
the advent of printed maps, European geographers and merchants acted in
several ways to "position" the Cape: as a site in a system of global trade managed by long-distance corporations and as the home of peoples who, initially
anyway, appeared strangers to the benefits of trade and exchange. Maps made
men marginal. Maps like the one produced by Willem Blaeu, Brotton suggests,
helped equate geographical location with cultural difference. The marginality
of the Khoisan people was secured because of the ways in which the transformation in print culture evident in printed map and published text acted
both to "fix" them—as dirty, dangerous, on the edges of civility's geography—
and to make those fixed images move through European society. Brotton's
work is cautionary. In order to understand whether the production and circulation of print was revolutionary, we need to recognize the commercial imperatives behind the different forms of printed material and to appreciate the
ways in which geographical information was incorporated in such material.

In chapter 7, Glennie and Thrift likewise eschew simplistic notions of revolution, in their case in respect of time and in timekeeping, as a direct consequence of new technologies. They focus instead on "new senses of time,"
on the everyday practices of timekeeping, and on the emergence of those
"kinds of new common sense" that, they argue, were "the real revolutions in
history." Thus, they are less intent on discussing the so-called Horological
Revolution of the late seventeenth century, when accuracy in timekeeping improved dramatically in consequence of changes in the technical capacity of
time's measurement, than they are in demonstrating the much longer-run
changes in how time was understood. Three "revolutions in the times" are
discerned: in clock time and in the practices and sites of its constitution;
in the significance of time's measurement for different communities of practice; and in the ways in which everyday practices of timekeeping were differently learned and embodied within social behavior. Clock times were, in turn,

constitutive of different sorts of social practice, among specialized "temporal communities," in particular, communities that, in turn, had certain geographical expression and permeable social membership. The focus of their attention is early modern England, their claim that, in these three senses, there was indeed a "revolution in the times" but that it was much longer, differently embodied, and socially more variable than notions of "rapid-time" revolution might suppose.

For some nineteenth-century commentators—at just that moment in which notions of "standard time" were both more prevalent and notions of accuracy in reading time one basis to increased exactitude in science—the camera was without doubt a revolutionary device. For in photography, time was stopped. Places and peoples could be "captured," shown as they really were. Instant visions of distant geographies could travel, a powerful and portable form of "geographicacy" written not in words but with light. But while some thought the new "art-science" of photography provided the "first truly revolutionary means of reproduction," others were not so sure, seeing in it a continuation by different chemical means of longer-run practices of visual representation. What Ryan shows here in chapter 8 is how—to borrow and adapt a term from Glennie and Thrift—different communities of scientific practice used photography for different ends. Ryan's concern is with photography's application in geography, what he calls photography's "range of roles, both formal and informal" within geography. The camera was seen as an instrument of geographical exploration, but not in any simple sense. For some, photography's value lay in collaboration of the field sketch and written account, for others because it was a taxonomic device, whose very precision—whether the subject was human or botanical—could simultaneously reveal the truth and "disarm the captious critic." Just as new practices of timekeeping became socialized by virtue of repeated social exposure, so photography's revolutionary impact within geography can be traced by considering its absorption into the subject's everyday practices and by seeing photography as a technical facility that could be taught, learned, and disseminated.

What connects these chapters, then, is their insistence in addressing the social and geographical implications of what revolution was held to do and to mean in relation to certain instruments of transmission—the map, the clock, the camera and photograph—through which the world has been put to order. Simple models of technical revolution fall in the face of geographical difference. So, too, understandings of geography as a specialist intellectual pursuit and as a popular concern and of geographers as communities of practice will remain overly simplistic unless we pay attention to how and where "revolutionary" technologies were made to work and with what result.

— 6 —

PRINTING THE MAP, MAKING A DIFFERENCE

Mapping the Cape of Good Hope, 1488–1652

...

Jerry Brotton

Any approach to the question of geography and revolution has to assess the impact of print on cartography from the mid-fifteenth century onward. The early pioneering studies of Marshall McLuhan and Elizabeth Eisenstein defined what they saw as an "Information Revolution" created by the invention of the printing press.[1] According to this approach, print transformed how knowledge itself was understood and transmitted. Print, with its standard format and type, introduced exact mass textual reproduction. This meant that two readers separated by distance could discuss and compare identical books, right down to a specific word on a particular page. With the introduction of consistent pagination, indexes, alphabetic ordering, and bibliographies (all unthinkable in manuscript cultures), knowledge itself was gradually repackaged. Textual scholarship became a cumulative science, as scholars could now take a manuscript and print a standard authoritative edition based on a comparison of all available copies. Reference books and encyclopedias on subjects like language and law could now claim to reclassify knowledge according to new methodologies of alphabetical and chronological order.

In recent years, however, this approach to the printing revolution has come in for serious revision and criticism.[2] More skeptical historians of print have questioned the grand, revolutionary claims made for the ways in which print established a level of linguistic, scientific, and visual standardization, fixity and global dissemination that paved the way for the birth of the "Scientific Revolution" and the rise of the modern nation-state. These critiques have focused on more specific local contexts for the creation of knowledge in a variety of written, oral, printed, manuscript, and visual forms, while also examining in greater detail the contexts of printers, publishers, booksellers, scribes, and readers. In this shift from revolution to revision, I will in this chapter consider another specific context that has been relatively neglected within these discussions. Little attention has been paid to the ways in which the complex and gradual shift from manuscript to print in mapmaking redefines the ways in which cultural difference is represented and understood within and across different reading communities.

Here I want to examine how early printed geography responded to cultural contact and encounter, and to analyze the ways in which printed maps constructed an anthropology of subject peoples through close interaction with written texts dealing with issues of travel, diplomacy, and trade. Benedict Anderson's highly influential *Imagined Communities* (1983) developed Eisenstein's argument a step further by examining how "print capitalism" created the linguistic and philosophical conditions for the emergence of what he calls the "national imagined community" across the long sweep of European modernity.[3] Building on Anderson's ideas about the development of European mercantilism in relation to print, I will suggest that early printed maps interacted with travel, diplomacy, and trade to construct a discourse of cultural difference and "otherness" that is definably different from the ways in which manuscript culture constructed cultural difference. What follows is a specific case study of how European contact with one particular space on the map of the early modern world shows how the development of print defines a shift in European understandings of cultural difference. The place is the Cape of Good Hope, at the southernmost tip of southern Africa. The scope of my concern reaches from the first European sighting of the Cape in 1488 to just before the Dutch settlement of the region under Jan van Riebeeck in 1652. It is a period that encompasses the decline of the manuscript *mappae mundi* and the rise of the global printed atlases of Ortelius, Hondius, and Blaeu.

Encountering the Cape

The first known European encounter with the Cape came in 1488, when the Portuguese admiral Bartholomeu Diaz rounded the Cape with three ships before returning to Lisbon to report his discovery to the Portuguese court. The circumnavigation of the Cape, the southernmost tip of Africa, has often been overlooked in terms of its geographical and political importance in the light of Columbus's subsequent voyage to the "New World" just four years later in 1492. Columbus was in Lisbon when Diaz returned at the end of 1488, and noted what took place in the margins of his copy of Pierre d'Ailly's *Imago Mundi:*

[I]n December of this year 1488, Bartholomaeus [Diaz], commandant of three caravels which the King of Portugal had sent out to Guinea to seek the land, landed in Lisbon. He reported that he had reached a promontory which he called Cabo de Boa Esperança. . . . He had described this voyage and plotted it league by league on a marine chart in order to place it under the eyes of the said king.[4]

Columbus's remarks provided an early glimpse of the momentousness of Diaz's voyage to Cabo de Boa Esperança—the Cape of Good Hope, so-called because its discovery promised "good hope" of subsequently reaching the fabled spice markets of India. Contemporary commentators suggested that the discovery anticipated access to as yet unknown new worlds to the east on a par with those discovered by Columbus in the western Atlantic. Referring to Diaz's discovery, the early-sixteenth-century Portuguese chronicler João de Barros claimed that "when it was seen [it] made known not only itself but also another new world of countries."[5] Like his contemporaries, Barros realized that the geographical significance of Diaz's voyage lay in the fact that his discovery shattered the late-fifteenth-century geographical world picture established by the Greek geographer Ptolemy and his influential text the *Geographia.*

Ptolemy's text suggested that the Indian Ocean was in fact one enormous enclosed lake, which connected southern Africa with the furthest limits of Southeast Asia deep in the southern hemisphere. Late-fifteenth-century printed editions of Ptolemy's text provide a vivid image of this perception of Africa and Asia (fig. 6.1). In keeping with Ptolemy's speculative geography of the southern hemisphere, the Indian Ocean is depicted as landlocked, with Africa running along the southernmost latitude of the map, conjoined with Southeast Asia. The vague and speculative nature of the geography of the southern regions of Africa is emphasized in the nomenclature of Ptolemy's map. The territories to the south of Ethiopia on the map are simply labeled "Terra incognita." More generally, these early printed editions of Ptolemy acted as a template upon which printers, scholars, and merchants began to define a "grammar" of geographical representation through the initially inflexible medium of print.[6] While manuscript still allowed for the proliferation of styles of geographical representation, print demanded greater standardization of the representation of relief and different types of urban, rural, and hydrographic space. As I will argue, it also participated in the construction of a field of cultural difference around its portrayal of "subject" peoples.

In breaching the geographical limits of Ptolemy's vision of Africa and Asia, Diaz's voyage had enormous implications for both the geography and economy of the early modern world. The discovery allowed Portugal unrivaled seaborne

Figure 6.1. Ptolemy, "World Map," *Geographia,* woodcut, 1482. By permission of the British Library, Maps G8175.

access to the markets of the Indian Ocean, and subsequent monopolization of the trade in spices, as they circumvented the need for laborious and expensive overland transportation of goods from markets in Southeast Asia to the marts of northern Europe.[7] The Cape thus became a strategically crucial point in the establishment of the Carreira da India, a complex network of ports and trading stations controlled by the Portuguese, which stretched from Lisbon via the west coast of Africa, the Cape of Good Hope, the coastal ports of East Africa, the Red Sea, and the Persian Gulf all the way to Malacca on the Malaysian Peninsula. What was so unusual about the Cape, however, was that it lacked any discrete geographical identity itself. It was represented as a geographical means to a commercial end, a transitional point in the contemporary geographical imagination. Even its name located its importance as consequent upon somewhere else—the access it allowed to the markets of the East.

If its geographical identity depended upon somewhere else, then its inhabitants were even more peripheral to early European travelers to the East. Once established as a temperate place of refreshment in the arduous journey eastward, the Cape's inhabitants, known by contemporary anthropologists as the Khoisan, were incorporated into one of the only classical models available to European travelers sailing into terrae incognita—the pastoral conventions of Virgil's *Eclogues*.[8] In the surviving account of Vasco da Gama's 1497 voyage, the anonymous chronicler recalls that upon landing at the Cape, the assembled inhabitants "began to play on four or five flutes, and some of them played high and others played low, harmonising together very well for negroes in whom music is not to be expected; and they danced like negroes."[9] It was a convenient fiction that influenced representations of the Cape's inhabitants until their subsequent responses to the depredations of the early Portuguese callers shattered such pastoral conventions.

Less than seven years later, Balthasar Sprenger, a German merchant sailing to India via the Cape, offered a different perspective from that of the anonymous chronicler of da Gama's voyage in his description of the Khoisan. Sprenger's was one of the first printed accounts of the region and its people and was also illustrated with idyllic "native" figures.[10] He noted that

[t]he men cover their genitals with a sheath of wood, the women with hairy furs. On their heads they put the skins of sheep and of other animals, like clothes. They bind the natural part of their young men against the body.... They have bulls, cows, oxen and sheep of huge size, and some other fine animals. The country is charming, irrigated by good rivers, the air healthy and smelling sweet of herbs. Their language sounds stammering and lisping. This people has

no money made from gold or silver, but is content with iron, which takes the place of currency.[11]

Sprenger's account was economically motivated, a merchant's account produced via the printing presses of northern Germany. It established certain characteristics of the Khoisan that would define subsequent European accounts of the people they encountered at the Cape: the fixation on the genitals, and the wearing of animal skins; admiration for the landscape and its animals, and bemusement at the Khoisan's language—their use of implosive consonants—and their lack of any meaningful currency. While devoid of the hostility of later reports, Sprenger's account also lacks the more idealized pastoral descriptions that defined da Gama's account. The idyllic representation of the pastoral native can quickly collapse into moral revulsion at the "lazy native" who is content to live off the land, rather than to cultivate it.[12] But even more significant was how Sprenger's printed text began to establish a standard account of the Khoisan, one that began to receive general currency in subsequent European accounts.

What significantly characterizes these early accounts of the Khoisan is the native inhabitants' refusal to engage in what European callers regarded as consistent and mutually acceptable forms of exchange. In various accounts, this repudiation is directly contrasted with the fertility and plenitude of their environment. As a result, the early-sixteenth-century Portuguese navigator Duarte Pacheco Pereira occluded the inhabitants altogether, claiming instead in his account of the Cape that "there is no trade here, but there are many cows, goats and sheep."[13] The difficulty of assimilating the Khoisan was compounded by one particularly dramatic act of resistance to Portuguese incursions at the Cape, which is also recorded in Adamastor's warning to da Gama in the fifth canto of Luis Vaz de Camões's *The Lusiads* (1572). In 1510 Francisco de Almeida, the Portuguese viceroy to India, quarreled with the Khoisan while stopping at the Cape on his return to Lisbon. Caught in the shallow waters of Table Bay, de Almeida and more than sixty of his men were slaughtered by the Khoisan.[14] Angered and humiliated, the Portuguese chose thereafter to take their refreshment on the long voyage to the East at either St. Helena or on the east coast of Africa, preferring to give the Cape and its inhabitants a wide berth.

By the middle of the sixteenth century, the Cape and its inhabitants were already on a collision course with the commercial exigencies of the expanding European seaborne trade, supported by the increasing geographical fixity of the printed maps, globes, and atlases emerging from the presses of northern Europe. The region was established as a defining feature of the "new geography" of the period, marked prominently upon printed maps of the world, despite the mystery that surrounded its topography and inhabitants. As late as 1541 the baffled geographer Euphrosynus Ulpius noted on his new terrestrial globe that, with regard to the southernmost regions of Africa, "neither have we been able to assert anything with certainty concerning it."[15] Toward the end of the sixteenth century, however, the Cape once more became the focus of European attentions. As the Spanish crown consolidated its grip on its colonies in the Americas, the developing maritime powers of England and the Netherlands sought to break into the markets of the East monopolized for so long by the Portuguese. English merchant adventurers proposed to pursue a search for a passage to the East by sailing northward. Both a Northeast and a Northwest Passage were eagerly sought between 1550 and 1580, but with no success. This led English mercantile interests to propose direct commercial encounters with the Indian Ocean and with the Indonesian archipelago in particular.[16]

PRINTING THE CAPE, MAPPING OTHERS

Throughout the 1590s, English diplomats and geographers sought cartographic and navigational information crucial to the establishment of a route to the East via the Cape. In 1598, Richard Hakluyt assisted William Philip and John Wolfe in obtaining and subsequently translating into English a copy of John Huyghen van Linschoten's *Itinerario.* Linschoten's text, the first comprehensive hydrographic, geopolitical, and commercial account of the route to the Spice Islands situated in the Indonesian archipelago, had been published in Amsterdam in 1596. In rushing out a copy of Linschoten's text translated by Philip in 1598 entitled *John Huighen van Linschoten: His Discours of Voyages into ye East and West Indies,* Hakluyt was lauded by Philip for obtaining a text that was "not only delightfull, but also very commodious for our *English nation*," emphasizing as the text did that, in the pepper-rich Indonesian archipelago, "men might very well traffique, without any impeachment."[17]

Linschoten emphasized that the Portuguese grip on the area was in no way invincible, a fact confirmed by the success of the first Dutch voyage via the Cape in 1595, led by Cornelis de Houtman. Calling at Bantam, de Houtman used Linschoten's information to obtain a substantial cargo of pepper, nutmegs, and cloves. Again, Hakluyt obtained a Dutch copy of the account of de Houtman's voyage, which was also translated by William Philip and published by John Wolfe in January 1598, within days of their publication of Linschoten's text. The illustration of such acute commercial acumen by Hakluyt in collating materials necessary for the development of a trade route to the

East led to his informal employment by the founding members of the East India Company, in March 1600. The outcome of Hakluyt's advice was the departure of the first East India Company voyage to the Indonesian archipelago in the spring of 1601, led by James Lancaster.[18]

Here were printed texts, dealing with diplomatically and commercially sensitive material, circulating between diplomats, printers, booksellers, and "intelligencers" like Hakluyt, whose activities were shaped (and in many cases financed by) the changing auspices of late sixteenth century practices of international trade and exchange. Their work was also embedded within the new practices of the Anglo-Dutch joint-stock companies. The distinctions between the effects on eastern trade of the London-based East India Company and the Dutch Vereenigde Oost-Indische Compagnie (VOC) in comparison with the earlier Portuguese trade in the region were significant. As Neils Steensgaard has pointed out in his discussion of the structure of the VOC, it integrated the functions of a sovereign power with the functions of a business partnership. As a consequence, "political decisions and business decisions were made within the same hierarchy of company managers and officials, and failure or success was always in the last instance measured in terms of profit."[19] Commercial considerations increasingly shaped cultural encounters, as well as the kind of printed maps and texts that accompanied accounts of long-distance travel and exchange. It was this mercantile mentality that created a definable shift in Anglo-Dutch accounts of the Khoisan in a series of early-seventeenth-century reports of European encounters at the Cape.

In 1595 the Dutch captain Cornelis de Houtman successfully reached the Indonesian archipelago, stopping en route at the Cape. De Houtman recorded one of the most sustained European encounters with the inhabitants of the Cape for nearly ninety years, and his account redefined the Khoisan once more in line with changing European commercial imperatives and established a model for their future perception:

[T]he inhabitants are of small stature, well joynted and boned, they goe naked, covering their members with Foxes and other beastes tayles: they seeme cruell, yet with us they used all kind of friendship, but are very beastly and stinking, in such sort, that you may smell them in the wind at the least a fadome from you: They are apparrelled with beastes skinnes made fast about their neckes: some of them, being of the better sort, had their mantles cut & raysed checkerwise, which is a great ornament with them: they eate raw flesh, as it is new killed, and the entrailes of beastes without washing or making cleane, gnawing it like dogs, under their feet they tye peeces of beastes skinnes, in steed of shoes, that they may travel in the hard wayes: We could not see their habitations, for wee saw no houses they had, neither could we understande them, for they speake very strangely, much like the children in our Countrey with their pipes, and clocking like Turkey Cockes: At the first wee saw about thirtie of them, with weapons like pikes, with broade heades of Iron, about their armes they ware rings of *Elpen* bones: There wee coulde find neyther Oringes nor Lemons, which we purposely sought for.[20]

Gone is the pastoral idealism of the earlier accounts, replaced by a more aggressive perception of the savage and "beastly" nature of the Khoisan, with the stress placed upon their perceived lack of hygiene, their dietary habits, their apparent lack of any meaningful habitation, and their incomprehensible language. De Houtman's account divests the Khoisan of any recognizable features of culture, effectively placing them outside the realm of civilization, marking them, anthropologically speaking, barbarous. This perception is compounded by de Houtman's final point: not only are the Khoisan barbarous, but the region also fails to provide the Dutch with the produce required for their arduous journey across the Indian Ocean.

This connection between the nature of the Khoisan and the required goods they failed to provide emerges once again in the account of a subsequent Dutch voyage to the East via the Cape, undertaken in 1598. The pilot of this particular voyage was the Englishman John Davis, whose story appears in the well-known travel collection edited by Samuel Purchas, *Hakluytus Posthumus; or, Purchas His Pilgrimes:*

> The eleventh [of November 1598], we anchored in the Bay of Saldania, in thirtie foure degrees of the South Pole, ten leagues short of Cape Bona Esperanza. The people came to us with Oxen and Sheep in great plentie, which they sold for pieces of Old Iron. . . . The people are not circumcised, their colour is Olive blacke, blacker than the Brasilians, their haire curled and blacke as the Negroes of Angola.[21]

Davis's account is strikingly bereft of the "barbarous" representation of the Khoisan that characterizes de Houtman's descriptions. Instead, the English pilot's narrative defines the highly satisfactory transactions carried out between the Dutch sailors and the Khoisan as apparently "spontaneous" acts on the part of the locals, who inundate the Dutch with their oxen and sheep, which are exchanged for remarkably little on the part of the Dutch, who trade nails, iron hoops, and rusty knives for the livestock. It is this establishment of some form of exchange that appears to spare the Khoisan from the vituperative echoes of de Houtman. In establishing trade with the Khoisan, Davis does not condemn them as barbarous, but attempts to place them within the commercial contours of the trade route to the East, comparatively situating the Khoisan in relation to the "Brasilians" and "the Negroes of Angola." Drawing on his own commercial maritime experience, Davis situates the Khoisan in the middle of a trade route that took in Brazil (a frequent port of call en route to the East) and Angola (on the east coast of Africa, another stopover prior to the long voyage across the Indian Ocean).

Davis was not alone in attempting to incorporate the Khoisan within this commercial circuit. Calling at the Cape in 1609, another Dutch merchant, Cornelisz Claesz van Purmerendt, described the Khoisan as "yellowish, like the Javanese"; and as late as 1634, the English traveler Peter Mundy claimed of the Cape inhabitants that they were "in Couleur swart like those in India."[22] Traveling to the markets of the East, these voyagers attempted to emplot a geographical model of the position of the Cape in relation to a trade route that encompassed Europe, Brazil, East Africa, and Indonesia, as a way of trying to come to terms with the alterity of the people encountered at the Cape. The problem with the Khoisan was that they failed to correspond to the forms of trade and exchange that characterized first Portuguese and subsequently the Dutch and the English merchant communities as they traveled eastward. In discussing the specific social and commercial nature of early Portuguese trade, Malynn Newitt has argued:

> In West and East Africa . . . the Portuguese encountered societies with well-developed internal and external trading networks . . . it was the profits of trade, not the profits of tribute, that immediately appealed to the Portuguese. When the Portuguese reached the Far East their command of the sea and ability to levy tribute on seaborne commerce led to their creation of more formal imperial structures but they were confronted by land-based states far too powerful for them to attempt large scale conquest and once again it was the profits of trade which were to constitute their major source of wealth.[23]

The communities encountered along the Cape route to the East—in West Africa, East Africa, and Southeast Asia—practiced a complex, ritualized type of trade with European merchants, which often involved highly elaborate forms of mutually acceptable routines of gift exchange. At the defining geographical point in this complex commercial system, the Cape, European callers found, however, that the Khoisan could not—or even worse would not—be assimilated into this reciprocal system of trade and exchange. As a result, they were labeled "barbaric," outsiders to a version of civilization that drew on early modern print capitalism to assimilate or reject other cultures based on commercial conventions.

Even at the point at which the Khoisan appeared to accede to the commercial logic required by European callers at the Cape, as in the case of Davis's account, problems emerged. In 1601 James Lancaster led the first official East India Company voyage to Bantam via the Cape. In September of that year Lancaster landed at Table Bay. The records of the company note that Lancaster "[W]ent presently aland to seeke some refreshing for our sicke and weake men: wher hee met with certaine of the countrey people, and gave them divers trifles, as knives and peeces of old iron and such like, and made signes to them to bring downe sheepe and oxen." The account goes on: "[T]he people brought down beefes and muttons, which we bought of them for pieces of old iron hoopes, as two pieces (of eight inches apiece) for an oxe and one piece (of eight inches) for a sheepe; with which they seemed to be well contented."[24]

This account is devoid of de Houtman's demonizing language. Lancaster is clearly delighted with the deal struck with the Khoisan. There remains, nevertheless, a flicker of anxiety at the puzzling and apparently irrational "contentment" of the Khoisan with the trifles they receive in return for the enormous number of livestock they offer in exchange. The situation is compounded by the inequality of the exchange in terms of consumption; the English obtain food for basic survival, while the Khoisan appear to desire the hoops of iron purely for ornamentation. Within anthropological accounts of exchange and consumption, the Khoisan's attitude to this form of exchange appears even more problematic. Any exchange of gifts or goods between two communities invariably involves an attempt by one community to assert its wealth and power over another. Wealth and power is thus not necessarily indexed to what one possesses, but more significantly the ability to expend or give away conspicuous wealth. As Georges Bataille has pointed out in his discussion of gift exchange, "wealth appears as an acquisition to the extent that power is acquired by a rich man, but it is entirely directed toward loss in the sense that this power is characterized as power to lose. It is only through loss that glory and honor are linked to wealth." For Bataille, this display of "ostentatious loss" is constituted "with the goal of humiliating, defying and *obligating* a rival."[25] So, for a figure like Lancaster, the delight at the apparently irrational expenditure of the Khoisan is also tinged with anxiety at the apparent equanimity of their ability to "lose" such obviously valuable commodities as livestock.

If this troubling scenario caused anxieties to early callers at the Cape such as Lancaster, then the subsequent behavior of the Khoisan only further compounded the sense of their irrationality and barbarity in the eyes of European callers. Following Lancaster's voyage, later English callers at the Cape discovered that the Khoisan had begun to establish an obscure economy of their own in terms of which trifles they were prepared to accept in return for disproportionate numbers of livestock. Increasingly, the Khoisan refused to trade in iron and would only deal in copper. By the second decade of the seventeenth century, travelers reported that only brass was acceptable to the Khoisan.[26] The frustration of both Dutch and English mercantile travelers quickly became evident in their recourse to old pejoratives. In 1604, Jacob Pieterszoon van Enkhuisen called at the Cape, where he recorded meeting

a poor miserable folk who went quite naked, except that they had a cloak of sheep or other skin bound about their neck, and the tail of such hanging before their privies. Some had copper or ivory rings on their arms. They clucked like turkeys and smeared their bodies so that they stank disgustingly.[27]

Banishing his initially ambivalent responses to the Khoisan on his previous visit to the Cape, John Davis noted upon his second voyage in 1605 that

[i]n the time of our being there, they lived upon the guts and filth of the meate, which we did cast away, feeding in most beastly manner: for they would neither wash nor make cleane the guts, but take them and cover them over with hote ashes, and before they were through hote, they pulled them out, shaking them a little in their hands, and so eate the guts, the excrements, the ashes.[28]

This obsession with the dietary habits of the Khoisan became an increasingly prominent feature of traveler's reports. In the same year that Davis called at the Cape for a second time, Edward Michelbourne also landed at Table Bay, claiming that its inhabitants "lived upon the guts and filth of the meat that we did cast away."[29]

What these accounts establish is a picture of European responses to the Khoisan distinguished not by condescending superiority, but by violent revulsion. Throughout those travel narratives published with reference to the Cape between 1610 and 1652, virtually no account describing encounters with the Khoisan is free of appalled responses to what was perceived to be their disgusting and shocking behavior, and more particularly their dietary habits. Nicholas Downton's encounter with the Khoisan in 1610 is typical of such accounts and provides some explanation of this intense demonization:

I found that all the Devises we could use by bribes or otherwise to them . . . would pcure nothing from them for our sicke mens releife, and then 4 Cowes which we did buy were so old and so leane that there was but little goodnes in the flesh, for which they would take no Iron, but thin peeces of copper. . . . These people are the filthiest for the usage of there bodyes that ever I have heard of, for besides the naturall uncleanes (as by Sweat or otherwise) whereto all people are subject, which the most by washing cleare them selfes of, contraryewise these people doe augment by annointing there bodyes with a filthy substance which I suppose to be the Juice of hearbes, which one there bodyes sheweth like Cowe doung.[30]

Downton's account explicitly connects the refusal of the Khoisan to trade with the English, even after his humiliating attempt at bribery, with their apparently "filthy" appearance. Like Davis and Michelbourne, Downton's account enacts symbolic revenge upon the Khoisan by condemning their consumption of whatever the Europeans find useless, for example, the intestines of livestock, which, as far as the Europeans are concerned, have already been consumed. Both Davis and Michelbourne imply that, in line with their frustrating interest in trifles such as iron hoops, the Khoisan desire what the Europeans cast off. As a result, the behavior of the Khoisan appears to have retained its troubling ability to invert the conception of commercial and cultural value accepted by the Dutch and the English, not only expending apparently valuable objects like cattle, but also seeking in exchange what the Europeans perceive to be valueless. The European callers as a result adopt a compensatory mechanism of demonization, which focuses on the dietary habits of the Khoisan as symbolic of their failure to grasp culturally defined notions of value.

In defining the Khoisan as "filthy," these accounts attempted to divest them of their identity as troublingly inassimilable within the commercial logic of the Dutch and English travelers who called at the Cape. As their behaviour was classified as "dirty," so the Khoisan themselves were condemned as "dirt," wasteful matter restricting and compromising the commercial development of European initiatives. However, as Mary Douglas has argued in her classic study *Purity and Danger* (1991):

Where there is dirt there is system. Dirt is the by-product of a systematic ordering and classification of matter, in so far as ordering involves rejecting inappropriate elements. . . . [O]ur pollution behaviour is the reaction which condemns any object or idea likely to confuse or contradict cherished classifications.[31]

If, as Douglas argues, dirt can be perceived as "matter out of place,"[32] then the response of these callers at the Cape to the Khoisan was to define them as filthy figures out of place on the basis of their annoying tendency to upset prescribed norms of transaction and exchange. The constitution of marginality is, as Douglas, among others, has suggested, central to the constitution of the notion of society, and its subsequent establishment and maintenance of its own cohesiveness:

> The idea of society is a powerful image. It is potent in its own right to control or to stir men to action. This image has form; it has external boundaries, margins, internal structure. Its outlines contain power to reward conformity and repulse attack. There is energy in its margins and unstructured areas. For symbols of society, any human experience of structures, margins or boundaries is ready to hand.[33]

Printing and Marginalizing Cultural Difference

What is particularly striking in the development of printed sixteenth- and early-seventeenth-century maps is the vivid manifestation of their culture's internal structure in the definition of margins and unstructured areas.[34] The margins of seventeenth-century printed maps of Africa and Asia often single the Khoisan out with a visual treatment that relates directly to the perceptions of the Europeans who encountered them at the Cape. What has not been addressed, however, is the way in which this fixed, printed figure was disseminated to create a particularly savage and static image of the Khoisan. The machinery of the printing press singled them out for particularly savage condemnation, primarily because of their lack of use value.

By the early seventeenth century, the transformation in print culture had created a particularly close connection between written travel accounts of voyages via the Cape and cartographic production. In 1607, Willem Janszoon Blaeu, the official cartographer to the Dutch VOC, published his "Wall Map of the World" in Amsterdam. All copies of the map were subsequently destroyed, and it only exists in photographic reproductions held in the Rijksmuseum in Amsterdam, but it still remains an important example of early-seventeenth-century commercial mapmaking.[35] More than a century after the preeminence of Ptolemy's *Geographia* and its depiction of an encircled Indian Ocean, Blaeu's map enshrined the geographical and commercial significance of the Cape of Good Hope within a map sponsored by the VOC, whose success was based on the incursion into the trade route via the Cape. Drawing on accounts logged by the VOC of the voyages of its employees to the East via the Cape, Blaeu's map contained in its borders inset views of specific towns and regions, complemented by illustrations of the natives of the various places. This was a relatively new stylistic development, inherited from Georg Braun and Frans Hogenberg's *Civitates Orbis Terrarum,* an atlas of over five hundred European cities published in six volumes between 1572 and 1617. As a classic geographical statement of the development of a confident European civil society, Braun and Hogenberg's *Civitates* provided the perfect model to define other spaces as the antithesis to such civility.

Of particular significance among Blaeu's insets of subject peoples are the figures labeled "Promontorii Bonae Spei et Congo Populi." Drawing on the spate of Dutch travel narratives defining the Khoisan as filthy and abject, Blaeu portrays the Khoisan as miserable intestine-chewing savages, marginal figures represented in direct contrast to the other peoples represented on the map. Marginalized even within the border of the map, the troublesome Khoisan became the victims of what Douglas has already referred to as the systematic ordering and classification of matter. This strategy initially appears to have affinities with an earlier medieval tradition of placing the "monstrous races" within the margins of *mappae mundi.* Seen thus, Blaeu's map is in many ways the final point in a long historical development of printed cartography stretching back to printed editions of Ptolemy via Ortelius and Mercator.

This printed representation of the Khoisan offers, however, a more systematic classificatory ethnographic procedure than that of the *mappae mundi.* Blaeu's vision seeks to equate geographical location with cultural identity. While the medieval *mappae mundi* projects its fears of what lies over the horizon anywhere into its own margins, Blaeu's map is more careful to cultivate a systematic typology of cultural difference. Unlike the imaginative geography of the medieval *mappae mundi,* that constructed by Blaeu emerged under the commercial requirements of a joint-stock company that literally could not afford to endorse a belief in monsters at the edge of the world. Territories such as Africa and Southeast Asia were no longer speculative places replete with monstrous races. If medieval *mappae mundi* envisaged its world centrifugally, moving outward from sacred centers of civilization, the commercial exigencies of the seventeenth-century joint-stock companies demanded a centripetally defined world, within which no corners were defined as off-limits to the potentialities of trade and exchange.

Blaeu's map represents both a cultural and a spatial demarcation that places the figure of the Khoisan on the boundaries of what comes to be defined as white, European civilization. This geographical strategy is clearly drawn from travel narratives such as that submitted to the offices of the

English East India Company by John Jourdain upon his return from a stop-over at the Cape in 1608. Recording the collection of seals for food and fuel, Jourdain recounted:

> [W]e cutt the fatt from them for oyle, and the rest was throwne a good distance from the tents because of noysomnes; upon which fish the Saldanians fed very hartilie on, after it had lyen in a heap 15 daies, that noe Christian could abide to come within a mylle of itt. . . . my opinion is, that if without danger they could come to eat mans flesh, they would not make any scruple of it, for I thinke the worlde doth not yield a more heathenish people and more beastlie.[36]

Literally and metaphorically, Jourdain's account places the Khoisan beyond the pale of his constituted civility, feeding on the "noysome" remains of the seals and again consuming what the English discard. They are beyond the "tents" that constitute Jourdain's spatial demarcation between his civility and the Khoisan's barbarity, which is defined as so appalling that Jourdain believes that it could easily lead to cannibalism.

First Dutch and subsequently English cartography would take up this approach to the Khoisan in the maps and atlases produced in following decades. The organization of English and Dutch hydrographic offices was based on a close relationship between merchant, traveler, printer, and mapmaker. Cartographers like Blaeu were required to read a range of printed (as well as manuscript) material to ensure the accuracy of their work.[37] In 1617, Blaeu produced a series of new maps, including his highly influential *Africa Nova Descriptio*. The cultural generalizations along the margins of Blaeu's map have been intensified from the earlier world map, with the intestine-chewing Khoisan once again reproduced in the bottom right-hand corner of the map. Clearly borrowing from Blaeu's convention, Jodocus Hondius published his map *Africae Nova Tabula* in 1623, which also depicts in its bottom right-hand corner a grotesque image of a Khoisan couple dressed in rags and chewing hungrily on loops of intestine. Having clearly inherited the illustrative technique from Blaeu's influential 1607 wall map, Hondius's representation was subsequently reworked by the English cartographer John Speed in his map entitled *Africae*, printed in London in 1626 (fig. 6.2). This engraving was deeply indebted to both Blaeu and Hondius, and also drew on the printed English accounts of travelers like Jourdain.[38] Reduced to one caricatured stereotype, the singular Khoisan male on Speed's map is again depicted chewing inanely on a piece of intestine, a crudely reductive but visually arresting depiction of what the Khoisan had come to signify to both Dutch and English callers at the Cape by the 1620s. This image was distributed among the elite

Figure 6.2. John Speed, *Africae,* engraving, 1626. By permission of the British Library, Maps C7c6(1).

literate map-buying public through a series of maps and atlases issued in print runs of several thousand copies published throughout the seventeenth century.

The classificatory procedures of these widely distributed maps put the Khoisan in the place where the frustrated and revolted Europeans believed they should remain, that is, on the margins of a geography of civilization whose parameters were defined by the prejudices of the European cartographers that shaped them. With their self-appointed power to classify whole cultures and territories, such maps even placed the Khoisan apart from their putative neighbors along the coastline of Africa, who were never depicted in such barbaric fashion. As Douglas reminds us, however, in her account of the image of society, there is energy in its margins and unstructured areas. The Khoisan remained a persistent and dangerous presence on the map, a reminder of just one of the uncomfortable effects that the process of filling in the spaces on the European world map produced for the travelers who sailed via the Cape. As dirt placed on the margins, the presence of the Khoisan remained, leaving the map and its producers, in the words of Julia Kristeva, "edged by the abject."[39] In this way, these maps revealed that their notion of order and civilization was, as Douglas has pointed out, shaped by the dangerous margins that they sought to occlude.

This situation clearly became intolerable in the face of sustained European settlement at the Cape. As South African historiography has pointed out, by the middle of the eighteenth century, European and predominantly Dutch settlement at the Cape had seen the virtual eradication of any recognizable Khoisan culture. Elphick has argued that "by 1720 the transformation of the Western Cape Khoisan into 'colonial Hottentots' was almost complete. The Khoisan had been reduced to a small fraction of their former population, their ancient economic and political institutions had virtually disappeared. . . . All this had happened in the seventy years since Van Riebeeck had landed at Table Bay."[40] By 1798, the English diplomat John Barrow claimed that "[t]he name of Hottentot [Khoisan] will be forgotten or remembered only as that of a deceased person of little note."[41]

The Khoisan were among the earliest victims of the emergence of European print capitalism manifested most obviously through the discipline of cartography. Once their abject status was established in the margins of late-sixteenth and early-seventeenth-century travel accounts and printed maps, it proved impossible to dislodge. The consequences of the revolution in print for the Khoisan were particularly disastrous. Douglas's comments on the fate of filthy objects as a result of the zealous reassertion of purity stands as a sobering example of the ways in which the metaphors of revulsion, demonization, and exclusion, elaborated with such vivid force in the maps of Blaeu, Hondius, and Speed, created the conditions for the destruction of a community that was unwilling or simply not interested in acceding to the cultural and commercial demands of European travelers:

> First they are recognisably out of place, a threat to good order, and so are regarded as objectionable and vigorously brushed away. At this stage they have some identity; they can be seen to be unwanted bits of whatever it was they came from. . . . This is the stage at which they are dangerous; their half-identity still clings to them and the clarity of the scene in which they obtrude is impaired by their presence. But a long process of pulverising, dissolving and rotting awaits any physical things that have been recognised as dirt. In the end, all identity is gone.[42]

...

Notes

1. Marshall McLuhan, *The Gutenberg Galaxy* (London: Routledge, 1962); Elizabeth Eisenstein, *The Printing Press as an Agent of Change: Communications and Cultural Transformations in Early Modern Europe,* 2 vols. (Cambridge: Cambridge University Press, 1979).

2. Adrian Johns, *The Nature of the Book: Print and Knowledge in the Making* (Chicago: University of Chicago Press, 1998).

3. Benedict Anderson, *Imagined Communities: Reflections on the Origins and Spread of Nationalism* (1983; London: Verso, 1991), 5–7.

4. Henry Hart, *Sea-Road to the Indies* (London: Hodge, 1952), 40.

5. Eric Axelson, *Dias and His Successors* (Cape Town: Saayman and Weber, 1988), 3.

6. R. A. Skelton, "The Early Map Printer and His Problems," *Penrose Annual* 57 (1964): 171–84; Tony Campbell, *The Earliest Printed Maps, 1472–1500* (London: British Library, 1987).

7. Charles R. Boxer, *The Portuguese Seaborne Empire* (New York: Knopf, 1969); Bailey Diffie and George Winius, *Foundations of the Portuguese Empire, 1415–1580* (Minneapolis: University of Minnesota Press, 1977).

8. Malvern van Wyk Smith, "'Waters Flowing from Darkness': The Two Ethiopias in the Early European Image of Africa," *Theoria* 68 (1986): 67–77.

9. Axelson, *Dias and His Successors,* 8.

10. On the early representations of the Khoisan, see Andrew B. Smith, "Different Facets of the Crystal: Early European Images of the Khoisan at the Cape, South Africa," *South African Archaeological Society Goodwin Series* 7 (1993): 8–20; and on the iconography of the Khoisan, see also Van Wyk Smith, "'Waters Flowing from Darkness.'"

11. Maria E. Kronenberg, ed., *De Novo Mondo: Antwerp, Jan van Doesborch; A Facsimile of an Unique Broadsheet Containing an Early Account of the Inhabitants of South America, together with a Short Version of Heinrich Sprenger's Voyage to the Indies* (The Hague: Martinus Nijhoff, 1927), 31.

12. J. M. Coetzee, *White Writing: On the Culture of Letters in South Africa* (New Haven: Yale University Press, 1988); Kenneth Parker, "Fertile Land, Romantic Spaces, Uncivilized Peoples: English Travel-Writing about the Cape of Good Hope, 1800–1850," in *The Expansion of England: Race, Ethnicity and Cultural History,* ed. Bill Schwarz (London: Routledge, 1996), 198–231.

13. Duarte Pacheco Pereira, *Esmeraldo de Situ Orbis,* trans. G. H. Kimble (London: Hakluyt Society, 1937), 154.

14. David Quint, *Epic and Empire: Politics and Generic Form from Virgil to Milton* (Princeton: Princeton University Press, 1993).

15. Edward Luther Stevenson, *Celestial and Terrestrial Globes,* 2 vols. (New Haven: Hispanic Society of America, 1921), 1:120.

16. Rodney A. Skelton, *Explorers' Maps* (London: Routledge, 1958), 143.

17. Jan Huygen van Linschoten, *John Huighen van Linschoten: His Discours of Voyages into ye East and West Indies,* trans. William Philip (London, 1598), sig. A1; A. C. Burnell and P. A. Thiele, eds., *The Voyage of John Huyghen van Linschoten to the East Indies* (London: Hakluyt Society, 1885), 1:112.

18. Kenneth Andrews, *Trade, Plunder and Settlement: Maritime Enterprise and the Genesis of the British Empire, 1480–1630* (Cambridge: Cambridge University Press, 1984).

19. Michael N. Pearson, *The Portuguese in India* (Cambridge: Cambridge University Press, 1987), 86; Neils Steensgard, *The Asian Trade Revolution of the Seventeenth Century: The East India Companies and the Decline of the Caravan Trade* (Chicago: University of Chicago Press, 1975).

20. Cornelis de Houtman, *The Description of a Voyage Made by Certaine Ships of Holland into the East Indies,* trans. William Philip (London: John Wolfe, 1598), 4.

21. Samuel Purchas, *Hakluytus Posthumus; or, Purchas his Pilgrimes: Contayning a History of the World in Sea Voyages and Lande Travells* (Glasgow: James MacLehose, 1905–8), 2:308.

22. R. Raven-Hart, *Before Van Riebeeck: Callers at South Africa from 1488–1652* (Cape Town: Struick, 1967), 45, 140.

23. Malynn Newitt, "Mixed Race Groups in the Early History of Portuguese Expansion," in *Studies in the Portuguese Discoveries,* ed. Thomas F. Earle and Stephen Parkinson (Warminster: Aris, 1992), 1:36.

24. William Foster, ed., *The Voyages of Sir James Lancaster, 1591–1603* (London: Hakluyt Society, 1940), 80, 81.

25. Georges Bataille, "The Notion of Expenditure," in *Visions of Excess: Selected Writings, 1927–1939,* trans. Allan Stoekl (Minneapolis: University of Minnesota Press, 1985), 122, 121.

26. Richard Elphick, *Kraal and Castle: Khoisan and the Founding of White South Africa* (New Haven: Yale University Press, 1977), 76–82.

27. Raven-Hart, *Before Van Riebeeck,* 29.

28. Purchas, *Hakluytus Posthumus,* 2:347.

29. Raven-Hart, *Before Van Riebeeck,* 33.

30. Ibid., 48

31. Mary Douglas, *Purity and Danger: An Analysis of Pollution and Taboo* (London: Routledge, 1991), 36–37.

32. Ibid., 36.

33. Ibid., 115.

34. John Gillies, *Shakespeare and the Geography of Difference* (Cambridge: Cambridge University Press, 1993).

35. For a reproduction and discussion of the map, see Günter Schilder, "Willem Jansz: Blaeu's Wall Map of the World, on Mercator's Projection, 1606–07, and Its Influence," *Imago Mundi* 31 (1979): 36–50.

36. Raven-Hart, *Before Van Riebeeck,* 42.

37. Schilder, *"Willem Jansz";* Kees Zandvliet, *Mapping for Money: Maps, Plans and Topographic Paintings and Their Role in Dutch Overseas Expansion during the Sixteenth and Seventeenth Centuries* (Amsterdam: Batavia Lion International, 1988).

38. Oscar Norwich, *Maps of Africa: An Illustrated and Annotated Cartography* (Johannesburg: Donker, 1983), 74, 75.

39. Julia Kristeva, *Powers of Horror: An Essay on Abjection,* trans. Leon Roudiez (New York: Columbia University Press, 1982), 6.

40. Elphick, *Kraal and Castle,* 235. This is interesting not least given the sympathetic treatment of the "Hottentots" (as he put it) by Peter Kolb in his *The Present State of the Cape of Good Hope,* which was first published in English in 1731 (originally published in German in 1719). For a discussion of Kolb's representation of the Khoisan, or Khoikhoi, see Mary-Louise Pratt, *Imperial Eyes: Travel Writing and Transculturation* (London: Routledge, 1992), 41–49.

41. Van Wyk Smith, "'Waters Flowing from Darkness,'" 327.

42. Douglas, *Purity and Danger,* 161.

— 7 —

REVOLUTIONS IN THE TIMES

Clocks and the Temporal Structures of Everyday Life

...

Paul Glennie and Nigel Thrift

I n this chapter we aim to provide an account of the "revolutions" that took place in the practices of clock time in England between 1300 and 1800. However, in order to account for why and how these practices changed and why these changes constituted a revolution requires a certain amount of preliminary ground clearing.

To begin with, there is the notion of revolution itself. Revolutions are one of the staple tropes of historical writing, centering on the notion of dramatic change: the political turmoil of the French Revolution, the world turned upside down in the English Revolution, the bloody pall of the Russian Revolution, the far-reaching transformations and social upheavals of the Industrial Revolution or the Agricultural Revolution or the Consumer Revolution, as well as the intellectual transformations of the Renaissance, the Enlightenment, and the Scientific Revolution. These iconic ways of looking at revolution convey the idea of a once-and-for-all change, taking place relatively swiftly and usually fixed in the cultural imagination by particular representations that confirm their existence in the historical record.

This kind of thinking has certainly characterized the history of time, which continually argues for dramatic changes in the practice and perception of time by fixing on a few revolutions in temporal perception that furnish history with new forms of time, usually driven by the invention of new devices.[1] Key elements of "horological" and "disciplinary" chronologies of clock times are summarized in table 7.1.

Such a mode of describing the history of time points to a second problem:

Table 7.1. Chronologies of Timekeeping "Revolutions"

	Key changes in horological technologies	Key changes in social technologies
1200		Monastic/"Church time"
		"Merchants' time"
	First mechanical clocks	
1300		
1400		
1500		Puritan valuing of time
	First watches	
1600		Establishing global positions
		Land surveying
	"Horological Revolution": Pendulum clocks, balance-spring anchor escapement	
	More precise indication on many temporal devices	Concept of mean time
1700		Oceanic navigation
		Expanding market for domestic time pieces / Consumer revolution
	Marine chronometer	
		Factory work discipline
1800		
	Working-class clocks and watches	
		National Standard Times
		Standard International Time Zones
1900		
	Electric clocks	
	Quartz watches	
	Atomic timekeeping	

the prevalence of technological determinism. Much of the history of clock time takes its cue from horological history in a way that assumes an all-but-unproblematic relationship between the history of clockmaking technology and economic, social, and cultural change. Even the more subtle exponents usually assume, at best, a lagged relationship in which the impact of changes in timekeeping technology takes a while to take hold. But, of course, the growth of work on the sociology of science has shown just how problematic "technological" advance is and how dangerous it is to take a simple transmission model for granted.

This relationship is even more fraught in the case of the history of time because of two complicating factors. One is the difficulty of deciding what
counts as a "clock" and as clock time. Clocks and clock times can encompass
a wide range of different technologies and practices distinguished only by the
onset of a certain regularity promoted by the use of little intervals. The other
is the difficulty of deciding what counts as evidence. Too often, the literature
on the history of time has pursued a model that simply searches out confirmatory evidence (usually, it might be added, of a generally elite, textual kind,
such as literary sources) for changes in perception of time without considering counterfactuals.

In this chapter, we want to consider revolutions in time as they played out
between 1300 and circa 1800. But, as we have pointed out elsewhere, our
work does not follow the conventional account of the history of time, which
usually works to a climax in the nineteenth century somewhere around the
full fruiting of the Industrial Revolution and the spread of standard time.[2] We
have tried to show that most timekeeping practices that were regarded as coincident with the Industrial Revolution in classic accounts such as that of E. P.
Thompson had an earlier vintage, and this has led us to attempt a wholesale
rewriting of the historical geography of timekeeping practices of which this
paper is one of the fruits.[3]

In this chapter, we therefore want to retain the notion of revolutions in
time but, as is now hopefully clear, the revolutions that we describe are not
only out of synch with conventional accounts but may often seem glacial, at
least when compared with some of those other revolutions we have already
mentioned. This latter point deserves expansion.

In general, the transmission processes that established new forms of temporal doing in people's lives have remained remarkably undertheorized and,
as a result underexamined in the historical record. But models like those emanating from the history of science, from various forms of cultural history,
from the history of material culture, and from other forms of history influenced by work that takes these objects, cultures, and practices seriously suggest that this state of affairs is now changing and that the emphasis on discovery, iconic moments, and remorseless processes of transmission so typical
of historical endeavor in the field can be foresworn for something more productive of genuine historical understanding. In other words, our interest is in
longer, slower but no less effective revolutions in everyday timekeeping practice that naturalized new kinds of temporal phenomenality and promoted new
kinds of awareness of temporal objects as temporal, as timekeeping objects
were bound into everyday practice.

As recent work in "cognitive phenomenology" shows, this is no simple process since it requires the generation of new bodily dispositions as well as the kinds of cognitive understandings to be found in books and manuals.[4] If we had to look for an example, it might come from the world of music: people have to learn to play musical instruments, and this is about not just musical notation, but also the correct accommodation of lungs, lips, hands, device(s), and so on. In addition, playing musical instruments requires a certain spontaneity. In other words:

> The world is comprehensible, immediately endowed with meaning, because the body . . . has the capacity to be present to what is outside itself, in the world, and to be impressed and durably modified by it, has been protractedly (from the beginning) exposed to its regularities. Having acquired from this exposure a system of dispositions attached to those regularities, it is inclined and able to anticipate them practically in behaviours which engage a corporeal knowledge that provides a practical comprehension of the world quite different from the intentional act of conscious decoding that is normally designated by the idea of comprehension.[5]

It is the building up of these new anticipations of, or attunements to, time—these new "as-if naturalizations" if you like—that we aim here to show as present in the historical record. It is these kinds of new common senses that are, we would urge, the real revolutions in history, revolutions that historians are now beginning to study in some detail.[6]

This chapter examines a set of revolutions in everyday clock-timekeeping practices, and argues that these sum to a long-run revolution in human experience and capabilities. In the first section, we set out our general approach in a schematic fashion so as to provide a background for the three "revolutions in the times" subsequently identified in the following section. In the final section, we conclude by grounding those revolutions within broader long-run changes in everyday temporal environments, and the rewoven textures of everyday life that they engendered.

THE GENERAL APPROACH

In the following four subsections, we set out the main conceptual means by which we will be able to identify and understand the three "revolutions in the times" that are the subject of the succeeding section of the paper. Our intention is to be schematic, so as to allow us to cover a lot of ground efficiently.

Some Axioms concerning Clock Time

How should we think of "clock time"? The first thing to say is that it is not a centered object. Rather it is a plurality. There is no one clock time but rather a series of clock times that arise from different aims and imperatives. The literature, however, resolutely continues to try to tell stories of a singular clock time whose manifestations in the world can be read off unproblematically. This point leads to a second. Clock times are constituted *in practice*. That is, they consist of all kinds of practices that usually use clock time incidentally as a part of the carrying out of that practice, not as central to it. Whether the use of clock time is faint or strong, we are trying to get away from the view of clock time as just a clock-bound metric. Instead, we are focusing on a range of practices that differentially involve the measurement of time as part of how they are achieved. But a third point is that the measurement of time is— to an extent to be investigated—held in common between practices. There is some degree of standardization, and this degree has changed over time, although certainly not in a linear way.[7] As clock-time-related frames like diaries, timetables, and the like appear, so they reciprocally produce a need for more clock-time measurement. The density of clock-time-related practices increases, and they increasingly appear as reciprocally confirming. And this leads to a fourth point, the use of clock time is very often site related. Clock times are regnant in certain locations; in others they may have very little pull. There are certain locations, like the centers of large cities, where access to clock time is easy and unproblematic. But in other locations, access to clock time may require considerable effort. In other words, clock times have a geography that is not just incidental but constitutive.

Communities of Practice

It is important to note that considerable criticism has been made of practice-based approaches to history like that outlined above, usually predicated on the idea that practice approaches only offer pseudo-explanations. In particular, the criticisms are commonly made that the inference from common behavior to a supposedly underlying source in shared presuppositions cannot be sustained; that the causal powers of practices often seem mysterious; and that the transmission or reproduction of practices over time, and from one practitioner to another, cannot be accounted for.[8]

We take hold of these criticisms in two ways. First, they are an empirical challenge. The historical record can meet at least some of these objections by producing specific evidence (at least, where it is available: clearly processes like everyday talk are rarely able to be recovered). Second, they provide a theoretical challenge, which, in this paper, we meet through the concept of "communities of practice." Communities of practice are shared enterprises formed over time, contexts of significance "in which we can work out common sense through mutual engagement."[9] These shared engagements involve the negotiation of meaning (involving talk, gesture, and other forms of communication and specific dilemmas), certain levels of participation (which is not the same as collaboration), and what Wenger calls "reification," the production and circulation of objects that confirm a community's existence and occupy much of its collective imaginary.

> From entries in a journal to historical records, from poems to encyclopaedias, from names to classification systems, from dolmens to space probes, from the Constitution to a signature on a credit card slip, from gourmet recipes to medical procedures, from flashy advertisements to census data, from simple concepts to entire theories, from the evening news to national archives, from lesson plans to the compilation of textbooks, from private address lists to sophisticated credit reporting databases, from tortuous political speeches to the yellow pages. In all these cases, aspects of human experience and practices are congealed into fixed forms and given the status of object.[10]

Obviously, there are very many communities of practice and each of them, in their own ways, act as a timekeeping environment, within which particular heuristics can thrive.

Over time, reification has become a more important element of communities of practice, as these communities have generally come to involve circulations over greater and greater spatial extents and have become populated with intermediaries that allow such circulations to occur.[11] Objects like clocks both allow such circulations to occur and, at the same time, shape the character of experience. They are both process and product.

It is important to note that communities of practice are very rarely separate and distinct entities. They merge into each other, both because most individuals are part of, and are indeed made up by, several communities of practice, and because so many communities of practice came into being in relation to other communities of practice. It is also important to note that the nature of judgment prevalent in each community of practice can be very different in character: some communities are abstract, formal, "distanciated," highly textual in character, while others are chiefly based on informal circuits of talk. But, whatever the case, each community tends to have its own political order and moral disciplines, which are continually being revised through the labor of controversy.[12]

Everyday Calculations

Communities of practice are not static entities. These time- and space-producing collectives constantly evolve, learning new tricks. How can we frame this learning in such a way that we can begin to find the right historical questions to ask? We would argue that such learning (and learning what constitutes learning) takes place at three different levels and involves three different intelligibilities.

The first, which we have already foreshadowed, is corporeal learning. Many timekeeping practices are so deeply grooved into the body that they are intuitive. They emerge without conscious understanding through a kind of osmosis. Claxton calls this kind of learning "know-how without knowledge."[13] This kind of understanding is obviously difficult to describe since it lacks clarity and articulation and is often not verbalized. But it is nonetheless crucial. Practical mastery of environments (so-called 'intelligence without reason') often emerges through immersion and experimentation in ways that are not open to conscious understanding or able to be easily turned into expert knowledge.[14]

However, it is obviously absurd to suggest that all learning is of this kind. So we come to the second kind of learning about time, which we might call "cognitive." However, our sense of cognition is not a rational choice process. Rather, it consists of sets of ad hoc calculations attached to the moment, "tool-

Table 7.2. Reading Flexibly

For each entry in the left-hand column, see if you can identify the best type of reading in the right-hand column.

Type of reading material	Type of reading
Sports page of the newspaper	Plotting a route
Instructions on a packet	Looking up facts
Science fiction novel	Quick scan to find a result
A–Z map of a city	Slow step-by-step reading
Complete national rail timetable	Repeated reading, thinking, rereading
Table of library opening hours	Repeated reading to recite from memory
Technical photography manual	Fast reading without effort for hours
Crossword clues	Scan quickly to get going, referring to sections as you go along
Rules of the game Monopoly	Quick glance, pin to wall for future use
Poem in a school poetry book	Careful slow reading of selected sections, and noting things down

Source: Guy Claxton, *Wise Up: The Challenge of Lifelong Learning* (London: Bloomsbury, 1999).

Table 7.3. A Taxonomy of Work-Related Reading

Reading to identify: Glancing at a document only in order to identify what a document is or what type of document it is.

Skimming: Reading rapidly in order to establish a rough idea of what is written, to decide whether any of its contents might be useful, or whether anything needs to be read in more detail later.

Reading to remind: Reading specifically in order to remind oneself of what to do next, for example, a to-do list, shopping list, Post-it note.

Reading to search for answers to questions: Reading to search for particular information: to answer a question, for reference, or to obtain information necessary to make a decision. This kind of reading is goal-directed, ranging from very simple goals to complex decision-making or problem-solving tasks.

Reading to self-inform: Reading for the purpose of furthering general knowledge without any particular goal to which the information will be applied.

Reading to learn: Reading with the goal of being able to relate or apply information at a later date, including reading to review the basic concepts for discussion, or reading that is much more reflective in nature.

Reading for cross-referencing: Reading across more than one document or more than one page in order to integrate information. This is often done for the purpose of writing and may well include some editing activities.

Reading to edit or critically review text: Reading in order to monitor what has been written in terms of content, style, grammar, syntax, and/or overall presentation. Includes editing one's own text, seeing how one's text fits into a collaborative document, or the review of the text of others.

Reading to support listening: Reading in order to support listening to someone else talk (e.g., following a presentation by looking at a series of slides).

Reading to support discussion: Referring to a document during a discussion in order to establish a mutual frame of reference and focus for discussion. Usually takes place in a face-to-face meeting.

Source: Abigail J. Sellen and Richard R. Harper, *The Myth of the Paperless Office* (Cambridge: MIT Press, 2002).

boxes" of simple heuristics that are used to assess situations, rather than general-purpose decision-making algorithms.[15] These heuristics are "fast and frugal," often involving the gathering of very little information. They are computationally cheap rather than consistent, coherent, general—and expensive of time. They constitute a kind of "quick fix" that will work most of the time to the degree necessary to tackle a particular situation, "tools" selected to be able to influence this environment, but with minimal effort. Claxton takes the case of the practice of reading to illustrate the use of these simple heuristics (table 7.2). What is often considered to be a "one size fits all" activity actually consists of a series of heuristics adapted to particular circumstances (table 7.3).

Such heuristics are bound to be adaptive. They are adjusted to circumstance and can be swapped around. And they are "constitutively leaky" in that they depend upon constant interaction.[16] In particular, they evolve through what Gigerenzer calls "foraging" behavior.[17] Effort is adjusted to circumstance according to what information is available in the environment. Thus, for example, in the case of timekeeping, in certain circumstances, considerable effort may have to go into exact time telling, if a situation calls for it. But most of the time circumstances only call for a close approximation, which will involve the expenditure of much less effort in order to extract limited information from the environment.

But, in turn, this brings us to the third kind of learning, which we will call ecological rationality. The environment itself speaks—it is a purveyor of large amounts of information, and practices result from the interaction between corporeal logics, adaptive heuristics, and the information that the environment provides. Generally speaking, if the environment is information rich, then heuristics will be quick and simple, for example, a quick glance at a clock face. If the environment is information poor—or so complex that it is difficult to negotiate—then more complex heuristics may be needed, for example, asking a friend who knows, or knows how to access, the information.[18] In acknowledging the importance of ecological rationality, we move closer to Hutchins's and others' notion of distributed intelligence.[19] "Thinking" is distributed across environments via a range of different divisions of labor and tools. In classical "posthumanist" fashion, thinking exists as a set of spatially and temporally "distanciated" practices that do not start or end with the individual human being, and is extended through the use of various tools (like clocks) that allow the corporealities of practices to be extended and new kinds of thought and phenomenality to come into existence.

To summarize this section, we can draw on Schatzki:

[The] prioritisation of practices over mind brings with it a transformed conception of knowledge. . . . Knowledge (and truth) are no longer automatically self-transparent possessions of minds. Rather, knowledge and truth, including the scientific versions, are mediated both by interactions between people and by arrangements in the world. Often, consequently, knowledge is no longer even the property of individuals, but instead a feature of groups, together with their material set-ups. Scientific and other knowledges also no longer amount to stockpiled representations. Not only do practical undertakings, ways of proceeding, and even set ups of the material environment represent forms of knowledge—propositional knowledge presupposes and depends upon them.[20]

It follows that when we look for revolutions in timekeeping, we need to look to all three kinds of learning—corporeal, heuristic, and ecological—if we are to understand the history of telling the time, and how clocks (and related timekeeping instruments) were used by groups to construct the new forms of synchronizations and eventfulness (what Flaherty calls "routine complexity") that we now take to be normal.[21]

Clock Times as Practices

The preceding thoughts lead us to formulate the four historical questions in which we are therefore most interested, in trying to index revolutions in everyday clock-timekeeping practices. First, since we are more interested in what things do, than in what they mean, we need to ask what statements clocks and other timekeeping devices made through the ways and contexts in which they were used.[22] Second, since we are primarily interested in everyday life, and in how new things become incorporated into "normal" practice, we need to ask now how clocks and other timekeeping devices populated everyday life. Third, since we are interested in embodiment, we need to ask how clock time was taken into the body, for example, through increasing nervousness about timekeeping or punctuality, increasing impatience about waiting, more measured movements when out walking, say, and so on. We also need to ask about prevailing forms of temporal phenomenality. Last, since we are interested in communities of practice, we need to ask what timekeeping practices circulated in which communities and what significance they had. We also need to ask how these practices both related to and differentiated communities of practice from each other. All four questions involve attempting to recover the societal and geographical dimensions of clock times as practices, as best we can, from an array of documentary sources that, without exception, have purposes other than attesting the use of clock time.

At first sight, there are at least three good reasons why the agenda is an unpromising one. We are discussing something that we've characterized as corporeal, intuitive, and unverbalized, something that was often taken for granted at the time and in which many interesting questions center on non-elite people. Each of these reasons on its own provides good reason to be pessimistic about what we might extract from the documentary and artifactual records. But perhaps not. As we now show, the records can give us more than we might suppose.

THREE REVOLUTIONS IN TIMEKEEPING PRACTICES

Toward evening on 15 June 1600, a churchwarden at North Walsham, Norfolk, started a new page in the churchwardens' account book to write this description of that day's "Great Fire," explicitly addressing his distant successors:

> [I]t began about six of the clocke in the morning and went on so fiercely that in two hours the whole body of the town being built chiefly round the market place was in one flame and so in two or three hours were burnt down to the ground.[23]

While the events described are undoubtedly dramatic, there is nothing remarkable about the statement's use of clock time to locate events within the narrative. Similar instances are absolutely commonplace, incidental elements within sixteenth-century narratives and descriptions, their frequency illustrating how familiar and routine the use of clock times had become, more than half a century before the "Horological Revolution." But how old and how distributed were such practices? With what activities were clock times particularly associated?

Our strategy in answering these questions is to situate such temporal statements with regard to activities and communities of practice, and to interpret them in terms of three dimensions (fig. 7.1).

These are, first, the *cost* of individual behavior: the intellectual analysis, or—more often—the heuristics of finding the time, planning and coordinating activities, and so forth. Second, there is the *ecological rationality* of the environment, that is, the clocks, other timepieces, time signals, and temporal cues embedded within the relevant context. Third, there is *adequation:* practical notions of the sufficiency or appropriateness of accuracy for the concern of the moment. Here we identify three revolutions in timekeeping practices between the late thirteenth and late eighteenth centuries.

The Incorporation of Clock Times into Everyday Life

The first revolution involves clock times becoming incorporated into everyday life, through widespread use of equal-hours timekeeping. This was rare in early medieval Europe, though some astronomers used equal-length hours of one twenty-fourth of the day/night cycle. General use of equal hours as a uniform temporal metric for the day/night was a new development from the late thirteenth century, initially taking a variety of forms in different centers.[24]

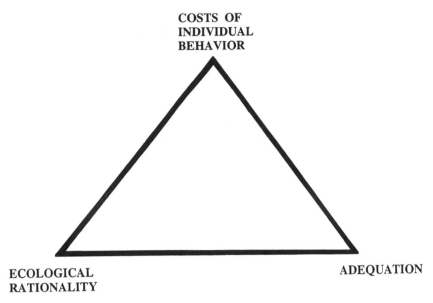

Figure 7.1. Influences on the specificity of recording of clock times.

This is emphatically *not* to suggest that people hitherto lacked potentially acute senses of time. Early medieval Europeans used three loose frameworks of daily temporality. One described the sun's light, height, and heat (e.g., first light, sunrise, "hot morning," midday, "evening cool," sunset, dusk, dark night) or the position, rising, or setting of the moon or conspicuous stars. A second temporal framework came from the canonical hours of monastic and church activity. These were unequal hours, of one-twelfth of the day or night, and so varied with day length: daytime hours at English midsummer were nearly twice as long as night hours, whereas midwinter night hours were much longer than day hours. So canonical hours varied with season and longitude, needing to be measured locally (with sundials, weather permitting). They varied in significance as well as length, because some denoted liturgical services such as prime, nones, and vespers. Monastic rules, like the sixth-century Rule of St. Benedict had both practical and moral force in scheduling everyday life.[25] Both the ideas of daily temporal frameworks and the bells that sounded monastic times spilled over into everyday life for a much wider population. Third, daily routines themselves, especially in towns, functioned as fairly predictable "alarm calls" or time cues. Cues arose both from civic signaling (e.g., market and curfew bells) and from other quasi-regular rhythms within the day, providing an approximate measure of the day, as in phrases like "at the ringing of the curfew bell" or "about the common hour of dinner." All three types of temporal indicator had long been drawn on in daily activity prior to the invention of mechanical clocks.

Thus medieval village bylaws used both natural time cues and canonical hours to regulate harvesting, grazing, livestock movement, and gleaning. At Newton Longville, Buckinghamshire, in the 1290s, for example, "anyone who wants to gather beans, peas or such like shall gather them between sunlight and prime . . . after the feast of the Blessed Virgin Mary," while at Great Horwood, no tenant "shall gather beans for his food except between sunrise and prime on pain of paying 6d." Similar stipulations survive from across medieval England, but in the fifteenth century, such bylaws commonly used clock time.[26]

Likewise, in Bristol's "Great Red Book" of market, port, and trading regulations, times like "not before the ringing of prime" were being rephrased in the late thirteenth and early fourteenth centuries as "until after ten o'clock smitten on St Nicholas church clock."[27] Evidence that clock times were superseding other expressions in plebeian speech comes from fifteenth-century York. In February 1356, Maud Katersouth deposed in a matrimonial cause dispute: "Asked at what hour she heard these words, she says *about the ringing of the curfew bells* on [the street called] Pavement, in York."[28]

A generation later, such references to time were starting to use clock times. In January 1395 John de Akom, a saddler, was accounting for his movements over an earlier weekend, deposing that

> he and others passed . . . Saturday night in the village of Sutton and in the morning of the following Sunday they crossed to the village of Crayke, where they heard mass and ate. Some time later they returned towards York. When they came to Haxby three miles from the city of York it was night so that when they came to the city of York it was *almost the tenth striking of the bell commonly called "Clokke."*[29]

Use of clock time was emphatically *not* a masculine preserve. Although relevant medieval evidence is sparse, there's sufficient to show clock times entering the speech of both country- and townsfolk, and both men and women. Diverse sources attest women's early use of clock time: indeed, Laurence reconstructs daily activity patterns for women of various social positions from depositions to courts and inquests, descriptions of work, and the like.[30] Once again, clock times often replaced phrases like "about prime" or "after the common hour of dinner" by around 1450. The changeover was gradual, not instantaneous, and on occasion people mixed, and freely translated among, clock-time metrics, other metrics, and various nonmetric environmental or embodied timings.

We can summarize use of clock-time practices around the mid-sixteenth century in six propositions and provide examples, some of which are pre- or post-midcentury.

First, *clock times were used for many different purposes by many different people*. Many uses were those familiar from the literature: timekeeping and signaling for regulatory purposes by ecclesiastical or urban authorities, with clock times replacing whatever earlier signals had been used as time cues. The availability of clock time provided a ready means of reformulating existing principles of regulation, but the regulation predated mechanical clocks.

But there were also types of use that have previously received little discussion.[31] Some urban time-signaling was explicitly connected to civic independence from ecclesiastical authorities or with the activities and self-identification of specific groups, such as lawyers.[32] But the uses of clock times went beyond such specific practices. Irrespective of the original reason for signaling, people commonly made opportunistic use of public time signals for organizational purposes of their own. For example, writers in the Stonor correspondence of the 1470s used clock times to make arrangements ("You must send a servant of yours to meet with me at Wallingford on Monday by seven of the clock"); to complain about broken arrangements ("According to the commandment of your mastership we were at Stebenhith by nine of the Clock . . . [but] your men came not").[33] Probably the most widespread of all uses were in everyday sociality: voluntarily making and coordinating prosaic everyday arrangements, as when the 1480s London merchant Thomas Betson worried about being punctual for dinner with his prospective in-laws. But such self-organization, whether utilitarian or social, is usually documented only by chance, in contrast to institutional-cum-disciplinary uses of clock times, which predictably dominate the archives of administering institutions.

Particularly revealing are indications of people's ability to improvise temporal information in constructing hypothetical or fictitious accounts of behavior. Under examination by Sir Morris Ashby, a Dorset J.P., and a parish constable on 23 September 1615, a Dorset farm laborer named Thomas Shott attempted to construct an alibi for the burglary of John Wheadon's house at Cerne some six weeks earlier. Shott's bland account of his movements that evening was probed by Richard Bartlett, the Cerne parish constable:

> Richard Bartlett . . . asked [Thomas Shott] about what time of the night he was abroad in the street when John Whedon's house was robbed, [who] answered after eleven o'clock, when he went to fetch a shirt to reap in on the morrow. . . .

[Bartlett . . . asked how he knew the hour], who answered that the folk told him it was near about midnight.

[Bartlett] . . . asked . . . what folk told him so. Shott was thereupon mute, but in the end answered again, that he heard the clock strike eleven.[34]

Shott's changing story of how he knew the time did not help his cause, but the case shows that neither the J.P. nor the constable thought it implausible for people of low status to know the times in these ways. It was the specific occurrence that was queried, not the feasibility of the claim.

Second, *the numbers and density of public clocks and associated time signals were rapidly increasing.* Though late medieval documentation is uneven, at times absent, the diffusion was clearly extensive and rapid. By the late fourteenth century, clocks had been generally taken up across European cities, as a route to simpler acoustic environments.[35] A survey covering modern-day France, Italy, Germany, England, the Low Countries, Poland, the Czech Republic, and the Ukraine establishes that more than two hundred and fifty cities had public civic clocks before 1400, and well over five hundred cities boasted them by 1450 (table 7.4).

The increase reflects increasing documentation as well as more clocks. Many large and important cities are undocumented before 1400, and although public clocks are recorded in their earliest civic or ecclesiastical finance documents, we cannot know when those clocks were constructed. Since some small towns clearly had public clocks well before 1400, early increases in

Table 7.4. Documented Public Clocks in Late Medieval Europe

	1360	1370	1380	1390	1400	1450
Europe as a whole	18		120		255	>500
Italy	10	14	22	26	33	
France	4	7	34	50	74	
Germany	1	7	18	26	32	
Low Countries	—	3	12	21	30	
England	2	8	12	15	18	
England (updated)	14	20	21	23	25	

Sources: The figures for Europe as a whole (relating to the area of modern-day Italy, France, Germany, England, Belgium, the Netherlands, Spain, Austria, Switzerland, Czech Republic, Slovakia, Poland, Ukraine, and Croatia), as well as the individual figures for Italy, France, Germany, the Low Countries, and England, are from Gerhard Dohrn-van Rossum, *History of the Hour: Clocks and Modern Temporal Orders* (Chicago: University of Chicago Press, 1996); updated figures for England are from Paul D. Glennie and Nigel J. Thrift, *The Measured Heart: Histories of Clock Times in England* (Oxford: Oxford University Press, 2005).

the numbers of public clocks were considerably steeper than suggested by table 7.4.

The density of small-town clock provision varied across Europe, for several reasons. First, in increasing order of importance, was the differential availability of technology and smithing expertise (in few parts of Europe were smithing and other metalworking skills common). Then there were differing perceptions of the worth of public time signals, and differing notions of adequation—that is, differing views on the relative sufficiency of older temporal metrics. Finally, the resources available to communities or patrons and donors varied. The expense of early clocks was considerable: indeed, the impressive finding is that so many clocks were installed and maintained.

Though increasing numbers of towns featured several public clocks, striking hours, their density was too low for clock times to be pervasive in everyday life. Late medieval clock times were identified as predominantly—not exclusively—urban. However, we depart from Gerhard Dohrn-van Rossum in emphasizing clock times as integral to growing ideas of "urban living," in contrast to his characterization of clocks as an "urban accessory."[36]

Third, *clock times were apprehended as public and natural.* Bells "broadcast" signals across urban environments. Time signaling was mainly aural rather than visual, as Maud Katersouth's, John de Akom's, and Thomas Shott's statements all illustrate, and public dispersal of temporal information by hour ringing reinforced a sense of time as a public good. Clock-time signals, unlike the crowd of specific signals that they replaced, were overseen by civic or parochial authority as a general end in their own right: they were not dictated by a particular interest or authority.

Fourth, *clock times were "natural" and taken-for-granted among English people,* because of their familiarity in everyday sociality, not just through social disciplining. The astrologer John Dee was hardly a representative Tudor Englishman, but he nicely illustrates the sheer normality of clock times in diary entries like that for 14 July 1607, which records the following instruction: "Tomorrow half an hour after 9 of the clock, give your attendance to know the Lord's pleasure."[37]

What gives this instance broader significance is that the instruction was given not by a person, but by an angel with whom Dee believed he was conversing through the skryer (medium) Bartholomew Hickman. Routinely using clock times himself, Dee saw nothing unusual in angels doing likewise, and it passes without comment. This sense of the sheer normality of clock time as self-evident for angels and men nicely demonstrates how thoroughly Dee took clock times for granted.

This latent familiarity with clock times and signals extended right across society, including such seemingly unlikely groups as shepherds.[38] It is also striking that school hours, and elements of teaching such as elementary translation exercises, presume an existing grasp of clock times among children, rather than identifying "telling the time" as requiring formal teaching.[39]

Fifth, *finding the exact time could still require considerable foraging behavior,* because clocks were unevenly distributed, and provided limited information. As timekeepers became more complex (construction of equal-hours sundials was particularly demanding), finding the time usually involved a trade-off between the accuracy of information obtained and the effort required to get it. Often it sufficed to notice something happening, like John de Akom's "striking of the bell . . . called '*Clokke*,'" or for those attending Sir Christopher Wren's mother—John Aubrey records Wren's stating that he was born near Shaftesbury on the evening of 20 October 1631—"the bell rang VIII as his mother fell in labour with him."[40] Sometimes, knowing the time required more work, as John Dee recorded:

[Prague, 3 September 1584] About 2 of the clock after noon came this letter to me, of the Emperor his sending for me. . . . The Emperor has just signified to the Spanish ambassador that he will summon your Lordship to him at 2 of the clock, when he desires to hear you. . . . Hereupon I went straight up to the Castle: and in the Ritter-Stove or guard chamber I stayed a little. *In the mean space I sent Emericus to see what was of the clock.*[41]

The autobiography of Samuel Jeake, a Nonconformist Sussex merchant provides another instance: "I was born at Rye in Sussex July 4th 1652 on the Lord's Day, $\frac{1}{4}$ of an hour past 6 a Clock in the morning, according to the aestimate time taken by my Father from an Horizontal Dial, the Sun then shining." Jeake's father rushing out to read a reliable sundial, and both father and son thinking it relevant (in 1652 and 1694 respectively) to record the type of sundial, show the importance attached to knowing an exact time for Samuel's birth.[42]

Divergent motives for knowing the time and the uneven quantity and detail of available time information created spatial variations in the trade-off between effort and accuracy in finding the time, which in turn shaped geographies of foraging behavior.

Sixth, *notions of adequation were relatively loose* most of the time. An hour or half hour usually sufficed, and greater accuracy wasn't worth the effort involved. Early instruments relied on narrowly distributed skills, and the restricted distribution had major implications for whether devices worked at all.[43] The availability (or not) of heuristics facilitating certain calculations or uses of information was also important, with a trade-off between simple and robust forms of estimation, and complex but nonrobust time-telling. We stress that this dichotomy usually was the extent of choice available.

Adequation generally entailed what Gigerenzer terms "fast and frugal algorithms"—rough-and-ready heuristics that drew on small amounts of robust temporal information. The contrast between "fast and frugal" and "precise but slow" timing is clear in Samuel Jeake's diary. He rarely used times more precise than a quarter-hour when recording business activities, social arrangements, or journeys, but carefully recorded the exact minutes of ague attacks that, as with his birth, he wished to interpret astrologically.[44] Jeake's precision in timing his ague attacks shows that his astrological analysis involved much more exacting adequation than did his other activities. The precision of recording directly reflects the strength of Jeake's motivation to obtain precise times. His recording of rounded times for other activities reflects loose adequation, not an incapability of precision.

To summarize: by the mid-sixteenth century, clock times were constitutive of all sorts of social practices; were widely though not universally signaled, becoming a general everyday resource and skill; involved varying combinations of purposeful foraging or simple heuristics, usually the latter because of a considerable trade-off between the effort involved and the adequacy of the information obtained, for different purposes.

Dividing the Hour: Quarters, Minutes, and Seconds

One way of looking at the classic, late-seventeenth-century Horological Revolution in timekeeping accuracy might be precisely as transforming the trade-off between effort and adequacy, with which we characterized the early centuries of mechanical timekeeping. Figure 7.2, an iconic graph in the horological literature, shows a decisive leap in timekeeping accuracy in after 1660, as major innovations, including the pendulum, anchor escapement, and the balance spring dramatically improved the accuracy of domestic clocks, public clocks, and watches (almost immediately, minute hands began to appear on clocks). Simultaneously, costs were lowered, and the market broadened dramatically.

However, while technical achievements during the Horological Revolution can't be gainsaid, the accuracy of pre–pendulum clocks has been widely denigrated within horology's supply-side view of times in practice. Much horological literature implies that more accurate pendulum-regulated clocks caused the minute hand to appear, in turn leading to the use of minutes in timekeeping. But this is to overlook or oversimplify several important factors.

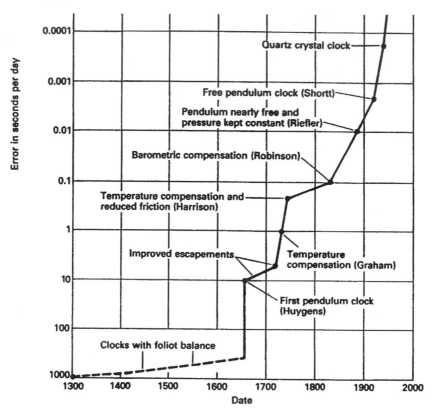

Figure 7.2. Progressive improvements in the accuracy of mechanical clocks. From Gerald J. Whitrow, *Time in History: Views of Time from Prehistory to the Present Day* (Oxford: Oxford University Press, 1988).

First, single-handed clocks could be interpreted to shorter intervals. Hour hands moved continuously, not in hourly jumps, so the hand's progress between numbered hour marks provided intermediate signals absent from aural signaling. Even for timepieces aimed at the bottom of the market, seventeenth-century clockmakers commonly engraved half- and quarter-hour marks on clocks' hour rings, and some marked half-quarters as well, making seven intermediate marks between hours. Depending on the delicacy of the hand's point, its position relative to marks and gaps enabled reading of time to within a few minutes. People reliant on one-handed timepieces were clearly not restricted to describing the time in whole hours.[45]

Second, clockmakers had been attempting to time and display minutes for more than a century. Several methods emerged, besides the familiar "two hands on one dial." The earliest known depiction of a clock mechanism displaying minutes, maybe never built, is Leonardo da Vinci's detailed drawing of around 1500.[46] Leonardo's clock mechanism drove four separate dials, respectively turning once a day (showing twenty-four hours), once an hour (showing sixty minutes), once every twenty-seven and two-thirds days (showing phases of the moon), and once a year (showing the month, date, and relevant day length). Minute-showing clocks didn't stay hypothetical for long. By 1565, one of the two masterpieces required by Nuremberg Corporation for apprentice clockmakers to gain "master" recognition was a clock that struck hours and quarter hours and showed minutes on a dial. Similar requirements for minute indication existed in other European clockmaking centers before 1600: making and use of minute watches is recorded in Venice in 1612. Nicolaus Munchen's Nuremberg masterpiece clock in 1640 struck individual minutes on a bell, as well as hours and quarter hours in any of several different hour systems, set by a control lever.[47] Of course, masterpieces were test pieces, exhibitions of virtuosity, not run-of-the-mill products. But it is significant that one of the particular skills being tested in the late sixteenth century was minute indication.

There were minute watches in Britain, too. John Aubrey relates the death of the physician William Harvey from a stroke, in 1657: "As soon as he saw he was attacked, he at once sent for his brother and nephews and gave one a watch . . . as remembrances of him. . . . *It was a minute watch with which he made his experiments.*"[48]

Third, before 1600, minutes were being used in publications directed at readers of humble status, who were sufficiently numerous to be targeted by publications such as Humfrey Lloyd's *Almanack and Kalendar Containing the Day, Hour and Minute of the Change of the Moon for Ever* (first publication 1563). This was one source for William Bourne's *Regiment for the Sea*, first published in 1574, and his astronomical and navigational almanacs, in print from 1567 to 1643. Bourne explicitly distanced his writings from the "learned sort" of men in universities or observatories, explicitly targeting non-elite readers "utterly voyde of any knowledge in the Mathematicall Science," "meaner men," and "the simplest sort of readers."[49] In other words, he sought a broad audience of practical men not possessing technical vocabulary or more than basic arithmetic. The *Regiment's* technical explanations used vernacular language, diagrams, reckoning tables, and worked examples. Key tables gave times in hours and minutes for phases of the moon—the basis for knowing tide times at different ports. Ready-reckoning tables gave the moon's age on any date, with corresponding tide times, with moonrise and hence high tides at any port, occurring forty-eight minutes later each day. Minutes also figured in the times predicted for eclipses, and the rising of key navigational constellations like the Pleiades.

Bourne envisaged that ordinary ships carried various timekeeping equipment, including an accurate two-hour watchglass, ideally reset each noon. He advocated the use of accurate shorter timings when using the English log-and-line to measure speed.[50] As he paid out a line behind the ship, attached to a float, the mariner required

> a minute of an hour glass, or else a known part of an hour by some number
> of words, or such other like, so that the line being vered out and stopped just
> with that time that the glass is out, or the number of words spoken, which done
> they . . . look how many fathoms the ship hath gone in the time: that being
> known, . . . they multiply the number of fathoms, by the portion of time or part
> of an hour.[51]

Obviously, authors could fall short of their ambitions, but Bourne's work reached a wide audience: commercial publishers issued twelve reprints of the *Regiment* between 1574 and 1631. Three were translated into Dutch as *De Const der Zee-vaerdt* (published Amsterdam 1594–1609), again targeting modestly educated readers. In 1570 a mariner from Leigh in Essex bequeathed his copy to another mariner there.[52]

Fourth, for certain specialist use, there were seconds clocks in the sixteenth century. In 1596 the Uraniborg observatory of the Danish astronomer Tycho Brahe contained "four clocks of different sizes, accurately made, which show not only the minutes but also the seconds." Their accuracy was reiterated in his *Mechanics of the New Astronomy* of 1598: "the clocks . . . show with precision, not only the minutes of the hours, but even the second parts, and they rival the uniformity of the revolutions of the heavens."[53] As Europe's leading astronomer Tycho was clearly exceptional, but there are other instances in which the limits of meaningful precision involved seconds. John Aubrey notes being told by the political arithmetician Sir William Petty that "[h]e was born . . . on Monday the 26th of May 1623, eleven hours 42′ 56″ after noon. . . . His horoscope was done, and a judgement upon it."[54]

Clearly this raises many questions. Was Petty here teasing Aubrey's credulity? What defined the moment of birth? How was seconds timing made in the 1620s? But that Aubrey recorded the information demonstrates three ideas: that events could be related to specific very short durations or instants, that distinct seconds can be measured, and that seconds could "make a difference" for astrological interpretation.

If we ask where in, say, 1750 a shopkeeper in Sussex or Lancashire encountered "seconds that mattered," the answer is most likely to have been when gambling on a footrace or a horserace. For example, a ten-mile race time of fifty-four minutes thirty seconds by Woolley Morris in 1753 was hailed by contemporary press reports as the fastest time recorded.[55] Again, the point is more the use of minutes and seconds than the accuracy of distance or time. Whether such uses connect to origins in astrology, or to a broader enthusiasm for technologies that measure remains moot.

By and large, though, the second was a distinction too far. The indication of seconds follows within a very few years of the indication of minutes on domestic clocks and watches except at the very bottom of the market. But use of seconds was very limited in the late seventeenth century, and later. Even today people who routinely pay close attention to seconds in everyday life are comparatively unusual. As a competitive runner, one of us is such a person: his coauthor thinks him mad. This example is also relevant to our next point, which is precisely that uses of precision were highly selective—both socially and for individuals.[56]

Considerable selectivity in precision is apparent throughout the documentation. For example, John Dee recorded precise minutes for the births of his children, between 1579 and 1586:[57] "Michael born, Prague, 3 hours 28 minutes after noon" (22 February 1585). Approximation was, however, acceptable for other people's children: "Mr Laward's son, Thomas, born at noon or a little after, $\frac{1}{4}$ or perhaps $\frac{1}{2}$ an hour" (18 February 1595). Michael's first birthday was marked with equal concern for minutes, but in a painfully roundabout way: "Michael Dee revolutus 9 hours 23 minutes" (22nd February 1586). The obsolete adjective "revolute" meant "that has completed a full revolution."[58] Here the earth has completed a full orbit around the sun, returning to precisely its relative position at Michael's birth. Among late-sixteenth-century astronomers, 365 days, 5 hours, and 55 minutes was the consensus figure for the precise length of the year.

But precise timing of significant births was no novelty. Its occurrence many decades earlier is shown by Dee's recording of the king of Poland's death in 1586: "King Stephen of Poland died 2 hours after midnight, in Grodno. He was born on the 13th day of January 1530, at 4 hours and 25 minutes in the morning in Transylvania, in Scholnio" (11 December 1586).[59]

Apart from the recording of minutes in 1530, the entry indicates transmission of that information over the king's lifetime, and a considerable distance, to be accessible to foreign observers half a century later. Earlier still, some early-sixteenth-century portraits record the late-fifteenth-century birth times of their sitters.[60]

The picture sketched here, of a more widespread presence of relatively precise clock times well before the late seventeenth century would not have surprised some contemporaries. In 1577, William Harrison had included as chapter 16 of his *Description of England* an essay entitled "Of Our Account of Time and Her Parts," in which he commented:

> Our common order [of time] is to begin at the minute, . . . one-sixtieth part of an hour, as at the smallest part of time known unto the people, notwithstanding that in most places they descend no lower than the half-quarter [that is, seven and a half minutes] or quarter of an hour . . . the common and natural day [being] observed continually by clocks, dials and astronomical instruments of all kinds.[61]

We see very little reason to depart from Harrison's account, contrary as this is to the impression conveyed by the horological literature, in which minute indication follows from the use of the pendulum and the anchor escapement. It is still more surprising for much historical literature, which has turned a (presumed) lack of everyday purposes for which close timing mattered, into people having no capacity to measure closely. But note in particular Harrison's important distinction between *capability for precision*—the minute as "the smallest part of time known unto the people"—and *everyday practice*—in which "they descend no lower than the half-quarter or quarter of an hour." We should be cautious, therefore, in attempting more than broad inferences about people's attitudes toward times from their nonrecording of times of day in documents such as diaries and letters.

Diversification among Communities of Practice

How and when precision in clock times mattered, and to whom, introduces our third revolution, in many ways the hardest to pin down. This defines the emergence of more specialized temporal communities, centered on practices involving small units of time, and ideas of precision and accuracy. We want to explore the dynamics of relations among specialized communities of practice, along with their interactions with everyday practices.

Early society contained growing numbers of communities of practice defined by their own priorities and practices in timing, for example those involving minutes and/or seconds. People commonly participated in several communities involving different activities and timings, witness the varied technical, political, and spiritual times of John Dee. Specialized communities of practice varied greatly in their size, composition, stability, and interaction with other temporal communities. Communities of practice formed and were transformed: they could split or merge or produce new hybrid practices; they could shrink or even disappear. They were neither mutually exclusive nor completely self-contained, and communities' geographies shaped their interactions—as where the shared interests in marine chronometry and longitude among astronomers, navigators, clockmakers, and naval bureaucrats were facilitated by their presences in and around London. Connections among communities could fluctuate dramatically over time, and could cause tensions, as well as constitute new sites for the productive spread or hybridization of practices.

Several of these considerations are illustrated by the ephemeral communities of practice created by Edmund Halley and other Astronomers Royal to time and map solar eclipses, to improve calculations of the geometry of the solar system.[62] By 1714, when Halley sought to exactly measure the moon's shadow on the earth's surface during an eclipse, he appreciated the value of accurate figures for eclipse duration. In the absence of standing observational facilities, Halley collected and mapped timings from more than thirty correspondents on 22 April (fig. 7.3), telling the Royal Society:

> I caused a . . . request to the curious . . . especially to note the time of continuance of total darkness, as requiring no other instrument than a Pendulum clock with which most Persons are furnish'd, and as being determinable with the utmost exactness, by reason of the momentaneous occultation and emersion of the luminous edge of the Sun.[63]

However, Halley regretted that "[several] observers give us no account how they measured this time, and therefore it may well be supposed . . . [some] took it in a round number, and perhaps from pocket minute watches."[64]

In 1737 the Royal Society collected eclipse durations in seconds from across northern Britain, though many observers lacked clocks showing seconds. Some observers had to judge seconds from a minute hand: "the duration was six minutes as near as could be judged by a watch that did not shew the seconds" (Hopetoun House, outside Edinburgh), and the annular appearance at Montrose continued seven minutes "as near as he could judge by an ordinary watch." Seconds could be measured without a second hand, since long-case clock pendulums were often designed to beat seconds.[65] Several observers counted pendulum swings, or escapement "ticks." At Crosby, north of Ayr, "a distinct annulus . . . continued exactly seven minutes, measured by a pendulum vibrating seconds"; at Frazerburgh, "from the time of the Ring's beginning to appear upon the lower and western part of the sun's disk, till it

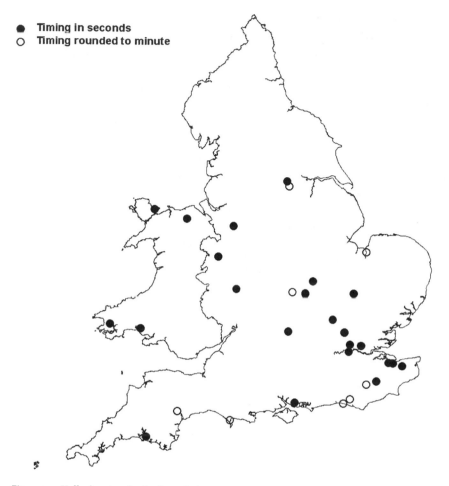

Figure 7.3. Halley's network of eclipse timings, 1714.

began to break on the east and upper part, there were 300 vibrations of a pendulum, or five minutes"; and at Longframlington, "the duration [was] 40 or 41 half-seconds, measured by a pendulum 9.81 inches long."[66]

The eclipse timings show the swelling of a dispersed network of precise timing practice, mobilized for a specific purpose, whose coordination disappeared after the eclipses. People's subsequent timing practices might or might not pay greater attention to seconds than hitherto, though the impact of precise timing on subsequent everyday practices are hard to gauge, given problems of documentation. Whether such initially specialized practices as seconds timing moved into everyday life was contingent on what such timing had to offer to the objectives of everyday activity. Useful—or apparently useful—practices were readily taken into everyday conduct.

Although specialized communities of practice dominate the literature, we have already highlighted everyday sociality as a prime site for spontaneous uses of clock times. Of course, close timing was valuable in particular tasks, from coordinating factory work to timing eclipses to maintaining public order. But we reiterate the importance of much broader communities of everyday practice in generating new timing practices (including using minutes and seconds in bets), not merely picking up timing practices that worked in everyday living. We emphasize just how socially and geographically marginal some clock-time communities were: groups such as schoolchildren, for example.[67] Such "peripheral ecological sets," as Gell might describe them, demonstrate the social and geographical "reach" of "thick" everyday temporal practices.

Everyday practices involved much more than a passive residue of "leakages" from purposeful, specialized communities into a vague "lowest common denominator" dustbin of inconsequential habits. We reject the equations of "specialized" with "purposeful" and "everyday" with "inconsequential." Rather the everyday was a site of practical negotiation and the resolution of dilemmas, where very different practices and notions of timekeeping were circulating.

Likewise, aesthetics and sensibilities could move into everyday practices, depending on whether they "worked" on an everyday basis. For example, the idea and aesthetic of seconds migrated to become widely known in everyday life, whereas practices of seconds timekeeping were less mobile, since they didn't "ground" in everyday life.

The rapid but selective take-up of minutes and seconds indicates a pent-up demand for accuracy, but accuracy was not just—often not mainly—a utilitarian issue. Several other aesthetics bore on people's stances toward clock time, and clock time mattered to different people for different reasons. Without being comprehensive, we note seven aesthetics beyond the utilitarian, often present in combination, that motivated and shaped stances toward, and practices of, close timekeeping.

First, enthusiasm for time measurement related to more general notions that measurement was a useful way of comprehending, understanding, or controlling, the temporality of things in general. Timing events or measuring durations were parts of approaching a problem, even without a specific objective. As Thompson relates, time measurement in early factories provides the classic, but not unprecedented, case. Second, as Thompson also notes, making good use of one's (God's) time was a recurrent theme among post-Reformation writers.[68] Within discourses hostile to waste and inefficiency, seconds indication was an important rhetorical point, whether or not such close timings were actually used. Third, a fascination with gadgetry and things mechanical was also endemic in early modern England. Diarists like Samuel Pepys, Claver Morris, and Anne Lister record watching watchmakers work, and their notice attests to the attraction of machines and gadgetry.

Fourth, novelty and newness were valued as interesting in themselves, especially in discourses of science and of consumption. The category "new and exciting" was produced and contingent, rather than given, but "new kinds of newness"—including precision timing—had an appeal in themselves. Fifth, and remaining with consumption, timepieces sold on facets of craft and design, satisfying their owners as sensory objects, not just on technical grounds. Cases, finishes, and dials were important elements for the experiencing of clocks as objects during the early modern "consumer revolution." Private clocks' performances as consumer goods were central to attitudes toward them. Sixth, clock times quickly became an index of people's commitment to emergent forms of politeness and civility, to which changing attitudes to punctuality were central. Conventions about promptness formalized responses to an informal quantification of impatience. Anne Lister's diary, for example, records testing social relationships by deliberately forcing issues of inclusion/ exclusion that centered on the punctuality of routines.[69]

Seventh, new precision in timekeeping was at once drawn into explorations of timing and the human body, mingling with other aesthetics— witness Pepys's enthusiastic response to his new minute watch, on 13 September 1665:

> Up and walked to Greenwich, taking pleasure to walk with my minute watch in my hand, by which I am now come to see the distances of my way from Woolwich to Greenwich. And do find myself to come within two minutes constantly to the same place at the end of each quarter of an hour.[70]

Recognizing these diverse aesthetics of timekeeping does much to explain why first public and then private timekeeping were much more widely distributed than can be explained by an exclusively technical drive to using clock times and owning clocks.

The making of clocks involved more than the makers of clocks, and it is extremely important to surmount a supply/demand-side dichotomy in analyzing clockmaking. There were potentially many voices, and various motives, in decisions about the making or purchase of particular clocks, in which the concerns and demands of various clock-using communities interacted with the clockmaking community's priorities and internal dynamics (fig. 7.4). Special-

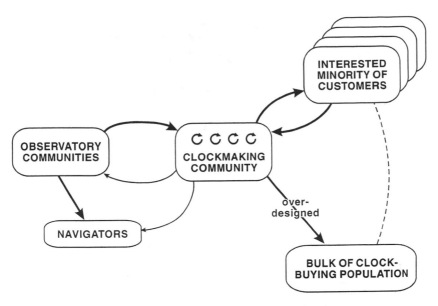

Figure 7.4. Communities of practice involved in early modern clockmaking.

ized communities of practice like astronomers worked closely with instrument makers, whose products had to be practical for less-skilled people in difficult environments (such as aboard ship).[71]

"Everyday" consumers of clocks could also bring strong interests to bear on clockmakers, artisans who usually produced clocks in conjunction with customers, not in isolation. A substantial minority of customers (like the diarists Pepys and Morris) were active and demanding consumers, wanting influence over the clocks they bought. This interest could extend as far as building their own clocks.[72]

Thus, notions of precision and the display of accuracy shaped, and were shaped by, technical and aesthetic dynamics among clockmakers and interested customers, from private individuals to specialized institutions. A capacity for precise indication swiftly became part of clockmaking aesthetics, and part of the identity of timepieces as consumer goods. For most plebeian customers, though, even cheap clocks were overspecified. By about 1770, cheap single-hand clocks cost around twenty shillings, less if they were secondhand. Such clocks had finely subdivided hour rings, providing greater precision than their owners seem to have used. As better-performing timepieces were produced in greater numbers, and as prices fell, plebeian purchasers bought off-the-shelf clocks with that "excess capacity," and haggled about prices and credit, rather than seeking simpler but (because bespoke) more expensive

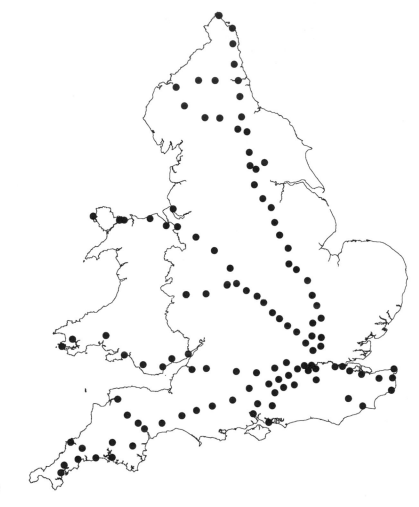

Figure 7.5. Post towns in England, ca. 1600.

clocks.[73] Even among middling-sort consumers, though, clocks commonly had a significant surplus of accuracy for most everyday uses.

   All these clock-time practices were distributed rather than aspatial, and we can begin to discuss these geographies. First, there is the presence or absence of specialized temporal communities, and how they influenced everyday temporal practices in particular places. For example, were specialized practices consciously promoted, or "leaking"? In leaking, how were they adapted or modified? We see spatially concentrated specialist communities of practice as

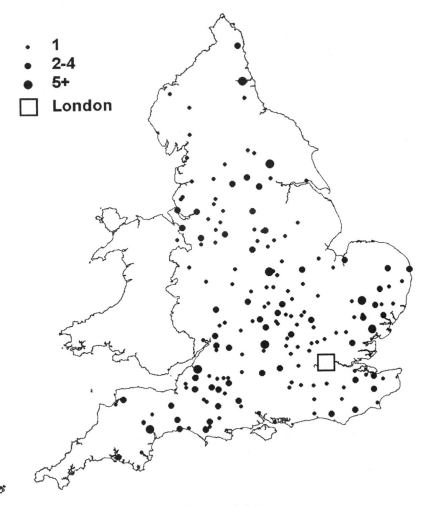

Figure 7.6. Provincial makers of surviving long-case clocks before 1700.

producing geographies of the permeability of everyday life, as practices circulated among and beyond specialized communities.

The geographies of translations among temporal communities were not merely processes of diffusion, but of translation and recasting, affected by wider sensitivities to diverse timing practices. Figures 7.5 and 7.6 indicate the national-scale geographies of two among many relevant activities: respectively, the nodes for the state recording of timings of the carriage of post around 1600 and provincial clockmakers with extant work dating from before 1700.[74]

Second, we need to examine the "thickness" of everyday timekeeping practices, or the ways in which practices mutually confirmed one another, to produce assemblages of diverse skills and embodied learning enabling people to cope with complex fittings-together of people and materials in space and time. London was the most prominent geographical focus of specialist temporal communities in seventeenth- and eighteenth-century England: the Royal Society, Gresham College, the Greenwich Observatory, the Clockmakers Company, administrators in the Royal Navy, the Exchange, the Posts, the Port, musical cultures, commercialized leisure industries, among others. Together, they created dense circulations of specialized practices within the metropolis.

Mutually confirming practices constituted part of the fabric of everyday being and acting, becoming "second nature" to people in extended distributions of various embodied skills. Those skills included first, the handling of *routines* as complicated days required actions to be carefully dovetailed in time, forming complex schedules of activities through coordination, controlled visualization, and sequencing activities and encounters. Repeated fitting-in with times created senses of right and appropriate actions through the familiarity of unanalyzed embodied experience. Second, skills of *improvisation* were the counterpart of building routines: a capacity to "decide as one goes"; being able to think of different other combinations of actions, materials, and people; juggling pressures; rapidly responding to changing circumstances and demands. Capacities to handle routines and to improvise can both be seen as primarily urban sensibilities. As urban life and conditions became more familiar, increasingly there was a *routinization of improvisation:* building a repertoire of experiences and skills that "domesticated" improvisation as familiar and not demanding of unusual effort. In turn, skills related to routine and improvisation rest on various *estimation abilities,* through which practical actions can be gauged.

To modern sensibilities these skills may appear so commonplace as to be hardly worth remark (though that is itself revealing of contemporary socialization), but in early modern England people were becoming more adept at them; they were being done better, and that made differences to lives, to social interactions, and to places. They reinforced clock times as integral elements of everyday life, but distributed according to people's situations within diverse communities of practice, and to their social skills. In short, the cumulative significance of clock times grew substantially across the population, and this growth occurred whether or not people were subject to explicit disciplinary impulses to explicit clock timekeeping.

UNDERLYING TRAJECTORIES IN EVERYDAY TEMPORAL ENVIRONMENTS

We have highlighted three revolutions: (1) the adoption of a single metric framework; (2) an increasing subdivision of that metric over time; and (3) interactions among a growing number of more specialized communities of practice. These three revolutions were grounded in long-run shifts in everyday temporal environments. Mechanical clocks provided the critical impetus to a standardization of temporal metrics, but were at first assimilated into existing sets of practices. Publicly struck hours were a routine descriptive resource, but the extensive "foraging" required to tell the time really accurately ensured that, for everyday purposes, fast and frugal algorithms based on small amounts of robust temporal information were preferred, and adequation was relatively loose most of the time. As greater numbers of better instruments circulated a richer array of information, they cumulatively produced environments that functioned more and more effectively as timekeeping devices, becoming increasingly self-referential and confirmatory.

So whereas in 1550 complicated foraging behavior was required to find the time to high levels of adequation, by 1750 the need for complicated foraging had largely disappeared into better timekeeping devices and more self-referential environments. The use of minutes in fine timing and planning no longer posed any significant difficulties. Even cheap timepieces were becoming able to do far more than was entailed in even the most sophisticated everyday practices. We can recast the "Horological Revolution" account as a relatively rapid transformation from measurable precision lagging behind generally envisaged precision in timing to a considerable surplus of measurable accuracy over the everyday uses to which it could be put. But that development was not, we contend, as important a transformation of temporal practices as those highlighted here.

By contrast, in the eighteenth century, the density of instruments was much higher; they performed much better; and they provided a much richer array of information. Many specialized communities of practice, mainly secular, formed, emergent for varied reasons, and demanding more exact timekeeping for their particular purposes. In turn, this stimulated the provision of more and better clocks, and larger volumes of more refined temporal information. Taken together, all these developments stimulated demands for better adequation, and communities were more willing to bear the costs to meet their prevailing interpretation of the appropriate level of "the right time." In the long run, what Gigerenzer might refer to as more complicated foraging behavior was offloaded into the better-performing devices.

Table 7.5. Chronology of Changes in English Clock-Time Practices

1200		
	Longstanding temporal practices using other metrics	
1300	Clock times and equal-hours reckoning pervade existing temporal practices	
		Clocks start to be used to measure out work processes
1400		
1500		"Telling the time" has become a normal procedure
	Almanacs widespread	Large number of public clocks
	Minutes generally understood	
1600		
		Diary-keeping more common
1700	Use of seconds by specialized communities of practice	Clock time in general use for work and bureaucratic processes
	Clock-time linked to rhythmic practices in general	
		Commonplace in urban areas to ask the time of people in the streets
1800		
	Allowance for local times in transport and communications	
1900		

A preliminary chronology of clock-time practices is shown in table 7.5. It clearly differs significantly from the chronologies of technology and of time disciplining summarized in table 7.1. The contrasts encompass both substantive differences in timings claimed or implied, and in interpretations of the relationships between practices and sociotechnical change.

In the long run, timekeeping and timing practices underwent a massively important revolution, but the form taken by this revolution in clock times was a "slow burn" compared with the "Horological Revolution." The revolution in clock-time practices was important partly from its "extending the importance" of the parameters of time and space. An environment much denser in material objects was rich with many sorts of temporal information.

In relating the highlighted revolutions to a complex of secular processes in the growing self-referential "confirmatory-ness" of the temporal environments of everyday life, we recognize that we are advancing a circular argument. That, though, is precisely the point: the argument is circular because the circularity is inherent in everyday timekeeping environments themselves. It is not an illegitimate imposition of circularity *from* the argument, but an intrinsic feature of everyday life itself.

Our argument is *not* that important changes in times and timing did not originate in the sphere of work. But it is that groundings of histories of practice in either technology or in social disciplining produce arguments in which any originary role for other spheres, from consumption to everyday practices, is by definition impossible. Such questions as the extent to which revolutions in times and timing were rooted in production (and work), in consumption (e.g., where times were imported into corporeality via clocks, books, and other goods), or in changing patterns of informal sociality, are questions that require sustained substantive exploration, not theoretical prejudgment.

REWEAVING THE TEXTURES OF EVERYDAY LIFE

Seen over the long run, practices of timekeeping and timing underwent a series of revolutions between 1300 and 1800. "Time" was unhinged and reconstituted as new forms of eventfulness by practices that came to rely on the clock and other allied devices as not simply representations of time (meanings) but rather changes in the form of variability brought about at the level of what Jean-Pierre Warnier calls "motricity" (doings).[75] However, the form taken by these revolutions was not explosive, simply mirroring the so-called Horological Revolution in clockmaking of the late seventeenth century. Rather, in each case, it was a long settling-in, in which technology and use coevolved, producing new practices. Intermediaries like clocks thereby became woven into the everyday practices of numerous communities of practice to the point where they (quite literally) changed the nature of what counted as time and space by weaving together interactions in many places.[76]

What seems to us as important about these revolutions is that they can be interpreted as part of a massive increase in the density of material objects that began to produce the artificial environment that we now take for granted.[77] The material objects did not just constitute an increase in material prosperity. They were also means of recording, storing and retrieving, analyzing, and communicating large amounts of information (which would result in the outpouring of "information technology" we see in the nineteenth and twentieth centuries.[78] Through these material objects, systematic practices like adding, calculation, recording, tabulating, and analyzing information started to become a part of everyday life. A metrological network was laid down.

Clocks were an essential part of the growth of this mesh of temporal information in that they provided the sense of regularity and repetition that allowed eventfulness to be modified as a new kind of measured-out ground. At first, these new ways of tagging events were still an oddity, rather like Clanchy's illustration of the gradual diffusion of dates and times on letters in the twelfth century.[79] But this state of affairs changed as temporal exactitude was linked into everyday practices in ever more prodigally suggestive ways: in the jargon, clocks became "keystone species" crucial to the swirl of the new temporal information ecology.[80]

However, as we have repeatedly pointed out, only a few communities of practice needed the kind of exactitude implied by units like the second, even toward the end of our period. The routine complexity of most practices never demanded this kind of adequation (and still generally does not, perhaps suggesting a kind of phenomenological limit).[81]

So it was that fragile techniques and anxious glances could be transformed into new phenomenalities. A temporal bracketing of what was known allowed new temporal knowledges to be produced—new bodily sensations and movements, new "cognitive" heuristics, and new kinds of temporal environment all acting in lockstep.[82] Yet this history largely remains to be written for, as Rée puts it, "The fact that the structures in which we live our lives are obvious does not mean that their significance is clear to us—for nothing is harder to understand than what is most familiar."[83]

...

Notes

1. See for example, David S. Landes, *Revolution in Time: Clocks and the Making of the Modern World* (Cambridge: Harvard University Press, 1983); David S. Landes, "Clocks and the Wealth of Nations," *Daedalus* 132, no. 2 (2003): 20–26; Stephen Kern, *The Culture of Time and Space, 1880–1913* (Cambridge: Harvard University Press, 1983); Helga Nowotny, *Time: The Modern and Postmodern Experience* (Cambridge: Polity, 1994); Hannah Gay, "Clock Synchrony, Time Distribution and Electrical Timekeeping in Britain, 1880–1925," *Past and Present* 181 (2003): 107–40.

2. Paul D. Glennie and Nigel J. Thrift, "The Spaces of Clock Times," in *The Social in Question: New Bearings in History and the Social Sciences,* ed. Patrick Joyce (London: Routledge, 2002), 151–74.

3. See E. P. Thompson, "Time, Work-Discipline and Industrial Capitalism," *Past and Present* 38 (1967): 56–97; Paul D. Glennie and Nigel J. Thrift, *The Measured Heart: Histories of Clock Times in England* (Oxford: Oxford University Press, 2005).

4. For example, Jean Petitot et al., eds., *Naturalizing Phenomenology: Issues in Contemporary Phenomenology and Cognitive Science* (Stanford: Stanford University Press, 1999).

5. Pierre Bourdieu, *Pascalian Meditations* (Cambridge: Polity, 2000), 135.

6. As, for example, in histories of shopping practices: Paul D. Glennie and Nigel J. Thrift, "Consumers, Identities, and Consumption Spaces in Early-Modern England," *Environment and Planning A* 25 (1996): 25–45.

7. Cf. Geoffrey C. Bowker and Susan L. Star, *Sorting Things Out* (Cambridge: MIT Press, 1999).

8. See, for example, Stephen Turner, *The Social Theory of Practices* (Cambridge: Polity, 1994).

9. Etienne Wenger, *Communities of Practice* (Cambridge: Cambridge University Press, 1999), 47.

10. Ibid., 59.

11. Bruno Latour, *We Have Never Been Modern* (Hassocks: Harvester Wheatsheaf, 1993).

12. See the essays in Christopher Lawrence and Steven Shapin, eds., *Science Incarnate: Historical Embodiments of Natural Science* (Chicago: University of Chicago Press, 1998).

13. See Guy Claxton, *Hare Brain, Tortoise Mind: Why Intelligence Increases When You Think Less* (London: Fourth Estate, 1997); and *Wise Up: The Challenge of Lifelong Learning* (London: Bloomsbury, 1999).

14. For example, Tim Ingold, *The Perception of the Environment* (London: Routledge, 2001).

15. Gerd Gigerenzer, *Adaptive Thinking* (New York, Oxford University Press, 2000); Gerd Gigerenzer and Peter M. Todd, *Simple Heuristics that Make Us Smart* (Oxford: Oxford University Press, 1999); Gerd Gigerenzer and Reinhard Selten, eds., *Bounded Rationality: The Adaptive Toolbox* (Cambridge: MIT Press, 2001).

16. Andy Clark, *Mindware* (New York: Oxford University Press, 2001).

17. Gigerenzer, *Adaptive Thinking*.

18. Bonnie A. Nardi and Vicki L. O'Day, *Information Ecologies: Using Technology with Heart* (Cambridge: MIT Press, 1999).

19. See Edwin Hutchins, *Cognition in the Wild* (Cambridge: MIT Press, 1995).

20. Theodore R. Schatzki, *Social Practices* (Cambridge: Cambridge University Press, 2001), 12.

21. Michael G. Flaherty, *A Watched Pot: How We Experience Time* (New York: New York University Press, 1999).

22. In the interest of succinctness, hereafter our use of the term "clocks" includes watches and other mechanical timepieces.

23. Norfolk Records Office, MF/RO 461/4.

24. See Gerhard Dohrn-van Rossum, *History of the Hour: Clocks and Modern Temporal Orders* (Chicago: University of Chicago Press, 1996), 113–17.

25. Justin McCann, *The Rule of St. Benedict* (London: Sheed and Ward, 1976).

26. Warren O. Ault, *Open-Field Farming in Medieval England: A Study of Village Bye-Laws* (London: Allen and Unwin, 1972), 82–94, 105, 171–74.

27. See Simon Andrew Christopher Penn, "Social and Economic Aspects of Fourteenth-Century Bristol" (PhD diss., Birmingham University, 1989).

28. Borthwick Institute of Historical Research, University of York, C.P.E.159, emphasis added. See also Peter J. P. Goldberg, *Women in England, 1275–1525* (Manchester: Manchester University Press, 1995), 106.

29. Emphasis added; the court record glosses the italicized phrase as meaning two hours before midnight (Borthwick Institute C.P.E.159). See also Goldberg, *Women in England,* 156.

30. See Anne Laurence, *Women in England, 1500–1760* (London: Phoenix, 1996).

31. See Dohrn-van Rossum, *History of the Hour.*

32. See Chris Humphrey, "Time and Urban Culture in Late-Medieval England," in *Time in the Medieval World,* ed. Chris Humphrey and W. M. Ormrod (York: York Medieval Press, 2001), 105–18; and Paul Brand, "Lawyers' Time in England in the Later Middle Ages," in Humphrey and Ormrod, *Time in the Medieval World,* 73–104.

33. Christine Carpenter, ed., *Kingsford's Stonor Letters and Papers, 1290–1483* (Cambridge: Cambridge University Press, 1996).

34. J. H. Bettey, *The Casebook of Sir Frances Ashley JP, Recorder of Dorchester, 1614–35,* Dorset Record Society Publications no. 7 (Dorchester, 1981), 15–16.

35. Similarly, there were several different metrics of clock time in late medieval Europe: days started at different points, and hours were counted in different ways, for example, 1×24, 2×12, 3×8, 4×6. See Dohrn-van Rossum, *History of the Hour.*

36. Ibid., 125.

37. Edward Fenton, ed., *The Diaries of John Dee* (Charlbury: Day Books, 1998), 299.

38. See Glennie and Thrift, *Measured Heart,* chaps. 6 and 7.

39. See Nicholas Orme, Education and Society in Medieval and Renaissance England (London: Hambledon, 1989).

40. Richard Barber, ed., *John Aubrey's Brief Lives* (Woodbridge: Boydell, 1982), 330.

41. Fenton, *Diaries,* 142. Emphasis added.

42. Michael Hunter and Annabel Gregory, eds., *An Astrological Diary of the Seventeenth Century: Samuel Jeake of Rye, 1652–1699* (Oxford: Clarendon, 1988), 85.

43. For example, in the 1370s, a clock-keeper traveled in the king of France's entourage. See Dohrn-van Rossum, *History of the Hour,* 120.

44. Glennie and Thrift, "Spaces of Clock Times," 170, esp. table 9.1.

45. Production of single-handed clocks continued into the nineteenth century, providing basic clocks cheaply for the lower end of the market. For example, in the 1720s, Andrew Knowles at Bolton, Lancashire, produced single-hand clocks, in which quarter-hour units on the hour ring were subdivided into markings at five-minute intervals. See Brian Loomes, *The Clockmakers of Northern England* (Ashbourne: Mayfield, 1997), 153. These small divisions were still several times wider than the hand tip, so further interpolation was readily available. Sundials also exhibit increasing attention to accurate clock-time indication, as in Loomes's example (page 143) of a 1760s sundial using a Vernier scale to show time to the nearest minute. In other words, sundials were affected not only by the transition from unequal to equal hours, but by more demanding standards of adequation: a dial might need to perform to a considerably higher standard to suffice in 1700 than was necessary two or three hundred years earlier.

46. Granville H. Baillie, *Clocks and Watches: An Historical Bibliography* (London: N.A.G. Press, 1951), 1:5.

47. Ibid., 32, 40.

48. Barber, *John Aubrey's Brief Lives,* 132–34, emphasis added.

49. E. G. R. Taylor, ed., *A Regiment for the Sea, by William Bourne, and Other Writings on Navigation,* 2nd ser., vol. 121 (Cambridge: Hakluyt Society, 1963), 242.

50. In principle, Bourne took the subdivision of time way beyond levels of precision with practical utility. In a brief general discussion of space and time, Bourne explains that "For as 60 minutes is a degree or an hour, so 60 seconds is a minute, and 60 thirds is a second, and 60 fourths is a third, etc." So an hour contains 12,960,000 fourths. See Taylor, *A Regiment,* 166. Bourne does not calculate anything with thirds or fourths, but he does think them worth explaining to his reader.

51. Taylor, *A Regiment,* 237–39.

52. Frederick G. Emmison, *Essex Wills: The Commissary Court, 1560–1574* (Chelmsford: Essex Record Office Publications, 1994).

53. Baillie, *Clocks and Watches,* 27, 28–29.

54. Barber, *John Aubrey's Brief Lives,* 341–42.

55. See Raymond Krise and Bill Squires, *Fast Tracks: The History of Distance Running* (Lexington, MA: Stephen Greene, 1982).

56. Marathon races instance a contemporary concern for timing in seconds among runners at all levels. Another widespread practice where seconds matter for point-scoring and success comes in computer games, a major contemporary site of children's exposure to close timing.

57. Dee used Julian calendar dates in England, but Gregorian calendar dates when in Bohemia and other countries that had shifted from the Julian to the Gregorian calendar in 1582.

58. Fenton, *Diaries,* 192 n. 3.

59. Ibid., 204.

60. For example, a precise birth time from the late 1490s is given in a portrait of Matthaus Schwarz of Augsburg painted by Christoph Amberger in 1542. Schwarz's horoscope was included in the picture, the time of the painting was given as "4.15pm on 22nd March 1542," and his exact age was recorded as "45 years, 30 days, and $21\frac{3}{4}$ hours." This Matthew Schwarz wrote the first treatise on double-entry bookkeeping published in Germany, and also wrote a pictorial chronicle of his life and clothes. He had already had himself painted on his twenty-ninth birthday, with watch and hourglass hanging at his neck. His wife Barbara was painted (also by Amberger) on her thirty-fifth birthday in 1542. Lorne Campbell, *Renaissance Portraits: European Painting in the Fourteenth, Fifteenth and Sixteenth Centuries* (New Haven: Yale University Press, 1990), 191.

61. Georges Edelen, ed., *William Harrison's "The Description of England," 1577* (Ithaca: Cornell University Press, 1968).

62. See Geoff Armitage, *The Shadow of the Moon: British Solar Eclipse Mapping in the Eighteenth Century* (Tring: Map Collector Publications, 1997). Durations of totality for the 1666 eclipse had been published in the first volume of the Royal Society's *Philosophical Transactions* (page 296), but seconds timing was only available from the Swedish astronomer Hevelius, and from the Paris observatory.

63. Edmund Halley, "Observations of the Late Total Eclipse of the Sun on the 2nd of April," *Philosophical Transactions of the Royal Society* 29, no. 343 (1715): 245–46.

64. Ibid., 254.

65. Clockmakers took 39.41 inches as the length of pendulum required to beat seconds, and 9.81 long inches to beat half-seconds. In practice, the length required varies with elevation and air pressure.

66. G. Graham et al. "Observations of the Late Total Eclipse of the Sun," *Philosophical Transactions of the Royal Society* 63 (1737): 175–201.

67. Orme, *Education and Society;* Donald Woodward, ed., *The Farming and Memorandum Books of Henry Best of Elmswell,* Records of Social and Economic History, n.s., 8 (London: British Academy, 1984); Glennie and Thrift, *Measured Heart,* chaps. 5 and 10.

68. See Thompson, "Time, Work-Discipline and Industrial Capitalism."

69. Helena Whitbread, *I Know My Own Heart: The Diaries of Anne Lister, 1791–1840* (London: Virago, 1988).

70. Robert Latham and William Matthews, eds., *The Diary of Samuel Pepys: A New and Complete Transcription* (London: Bell, 1970–83), 6:221–22.

71. Allan Chapman, *Dividing the Circle: A History of Critical Angular Measurement in Astronomy, 1500–1850* (London: Wiley, 1994); Gerald L'E. Turner, *London Instrument Makers: The Origins of the London Trade in Precision Instrument Making* (Oxford: Oxford University Press, 2000).

72. Loomes discusses a clock made in 1700, or slightly earlier, by Joseph Calvert, a farmer at Longthwaite, Cumbria, under the tutelage of John Sanderson of Wigtown (*Clockmakers,* 65–69).

73. For example, gear wheels cut by workshop "engines" were cheaper and quicker to produce, and more accurate, than nonstandard wheels filed by hand.

74. See Glennie and Thrift, *Measured Heart;* Mark Brayshay, "Royal Post-Horse Routes in England and Wales: the Evolution of the Network in the Late-Sixteenth and Early-Seventeenth Century," *Journal of Historical Geography* 17 (1991): 373–89; Mark Brayshay, P. Harrison, and Brian Chalkley, "Knowledge, Nationhood and Governance: The Speed of the Royal Post in Early-Modern England," *Journal of Historical Geography* 24 (1998): 265–88; Brian Loomes, *Clockmakers,* and *Brass Dial Clocks* (Woodbridge: Antiques Collectors Club, 1998).

75. Jean-Pierre Warnier, "A Praxeological Approach to Subjectivation in a Material World," *Journal of Material Culture* 6 (2001): 5–24.

76. Bruno Latour, "Trains of Thought: Piaget, Formalism and the Fifth Dimension," *Common Knowledge* 6 (1997): 170–91.

77. Nigel J. Thrift and Shaun French, "The Automatic Production of Space," *Transactions of the Institute of British Geographers,* n.s., 27 (2002): 309–35.

78. See JoAnne Yates, "Business Use of Information Technology during the Industrial Age," in *A Nation Transformed by Information,* ed. Alfred D. Chandler and James W. Cortada (Oxford: Oxford University Press, 2000), 107–36.

79. M. T. Clanchy, *From Memory to Written Record: England, 1066–1307* (Oxford: Blackwell, 1993).

80. Nardi and O'Day, *Information Ecologies.*

81. Flaherty, *A Watched Pot.*

82. Petitot et al. *Naturalizing Phenomenology.*

83. Jonathan Rée, *I See A Voice: Deafness, Language and the Senses—a Philosophical History* (London: Flamingo, 1999).

— 8 —

PHOTOGRAPHY, VISUAL REVOLUTIONS, AND VICTORIAN GEOGRAPHY

...

James R. Ryan

The invention of photography in 1839 is often regarded as a revolutionary moment in science, art, technology, and visual consciousness. Certainly, many Victorians heralded the development of the medium in such terms. Photography was a startlingly new means of collecting, classifying, and communicating geographical knowledge to both popular and elite audiences in the mid-nineteenth century, and its status as a revolutionary technology should not be underestimated. Armed with a camera, travelers or explorers could record the sights of their journey for audiences back home; distant geographical "realities" were thus transformed into detailed, portable pictures. These visual facts could then be incorporated into various collections of information, cataloged, classified, and redefined in any number of ways—as objects of beauty, as subjects for scientific scrutiny, as evidence of the achievements of individuals and institutions.

This chapter sets out to explore the revolutionary implications and applications of photography in the science of geography in Britain in the Victorian period. Geography—geo-*graphos*—has long been a discourse of making and interpreting visual representations of the world. Indeed, the centrality of this "picture-making impulse," as David N. Livingstone as pointed out, may be traced as far back as "the reappropriation during the Renaissance of Ptolemy's conception of geography as an enterprise essentially concerned with picturing (or representing) the world."[1] In recent years, geographers have shown increasing interest in how this impulse toward "visualization" both shapes and reflects geographical languages, practices, and ideas.[2] The metaphorical association between human vision and geographical knowledge has often been pointed out, and geographers have not been slow to note that the relationship between sight and knowledge is neither as direct nor as straightforward as is sometimes assumed. Indeed, sight is mediated through complex perceptual and cultural processes; "seeing" is not always "believing." As the historical geographer H. C. Darby, writing on the nature of geographical description, put it, citing George Perkins Marsh: "[T]he eye . . . sees only what it seeks."[3]

The mediated nature of observation and depiction is central to an important body of recent geographical scholarship that examines the representation of landscape in painting, cartography, literature, architecture, and design.[4] Historians of geography and cartography have similarly been concerned to explore how practices such as mapping construct geographical knowledge through different technical conventions and rhetorical strategies, invariably operating within settings of power.[5] It is largely in this context that interest in the place of photography in the production of geographical knowledge has emerged.[6]

This chapter also draws on recent studies that strive to place different aspects of Victorian science within a range of cultural, social, and political contexts. Historians of science, technology, and medicine, as well as historians of geography, have begun to address the ways in which Victorian science operated in the public gaze, notably though exhibitions, lectures, and museums, and how visual practices and technologies were often central to the making and display of scientific knowledge.[7] This field is producing studies that give fascinating insights into the complex debates over the uses of photography as a form of evidence within Victorian science.[8] Such work raises important questions for anyone attempting to assess the impact of photography on the science of geography in the Victorian period. It challenges us to consider fully the claims and concerns that contemporary commentators expressed about the possibilities and limitations of photography as a form of geographical evidence. It also shows how we need to pay particular attention to the ways in which photography was adopted by institutions such as the Royal Geographical Society (RGS) in London and to contemporary debates over the effectiveness of photography in geographical fieldwork and expeditions. For, as I hope to show in the examples that follow, the effectiveness of photography as a means of making, recording, and legitimating certain kinds of geographical knowledge was by no means taken for granted; its status as a revolutionary technology was therefore invariably contested.

THE ART-SCIENCE OF PHOTOGRAPHY

Space and time have ceased to exist. The propeller creates its vibrating spiral, the paddle-wheel beats the waves, the locomotive pants and grinds in a whirl-wind of speed; conversations take place between one shore of an ocean and the other; the electric fluid has taken to carrying the mail; the power of the thunderstorm sends letters coursing along a wire. The sun is a draughtsman who depicts landscapes, human types, monuments; the daguerreotype opens its brass-lidded eye of glass, and a view, a ruin, a group of people, is captured in an instant.[9]

Thus wrote Théophile Gautier in 1858, capturing the revolutionary spirit of his age and the place of photography within a modern world in which the power of nature was harnessed for human progress. Though he does not use the word, Gautier conjures a powerful sense of "revolution"—in both its older association with "revolving" and its modern meaning of "dramatic change."[10] Indeed, the association of the term "revolution" with new institutions and new technologies was itself a product of the modern age from which the new medium of photography was born. Gautier expresses clearly the widespread mid-nineteenth-century excitement with photography as a revolutionary new means of recording the world. Many contemporaries saw in photography a uniquely wonderful marriage of science and art—what the French commentator Francis Wey described as a "kind of hyphen between the two."[11] Indeed, photography was commonly described as an "art-science." Photography made it possible, for the first time, to fix permanently the image generated by the camera obscura, to make richly detailed images of the natural world, and to make multiple reproductions of the same image. Unlike other forms of visual representation, such as painting or engraving, photography did not depend entirely on a human image-maker. As Gautier put it, "the sun is a draughts-man," and the "brass-lidded eye" of the camera merely had to blink open to capture the scene before it "in an instant." More than this, photographs were not only interpretations of the world; they were material traces of that world, visual residues of reflected light. Photographs could therefore constitute knowledge independent of firsthand observation, offering a means of remote visualization and acquisition. In harness with other optical devices such as the telescope and microscope, the camera was quickly adapted to capture things beyond the usual capacity of human vision, from the surface of the moon to the pattern of a bird's wings in flight.

Many subsequent commentators have noted how photography inaugurated revolutionary changes in senses of space, time, memory, and human consciousness. In his well-known essay "The Work of Art in the Age of Mechanical Reproduction," Walter Benjamin described photography as "the first truly revolutionary means of reproduction" arguing that it was characterized by its limitless reproducibility.[12] Many historians of photography have claimed that the medium revolutionized visual perception, offering a new, expanded sense of perception to the nineteenth-century mind. As William Ivins noted: "The nineteenth-century began by believing that what was reasonable was true and it wound up by believing that what it saw a photograph of was true."[13]

However, it may be too easy to apply the term "revolutionary" to photography. There are clearly many ways in which photography did not overthrow existing conventions or knowledge. To begin with, photography stemmed from the marriage of long-established research, rather than dramatically new knowledge, in the optics of glass lenses and the chemistry of light-sensitive salts. Moreover, as a technology of visual representation, photography inherited aesthetic conventions such as linear perspective that were well established in other visual and graphic arts. Indeed, photography could not have emerged as a "modern point of view" without the pictorial conventions of perspective realism that it inherited via landscape painting.[14] It thus enhanced rather than challenged dominant conventions of naturalism and perspective realism that it inherited from Western painting. Finally, the term "photography" came to encompass a vast array of technologies and ephemera—from stereoscopes to *carte-de-visites*—that often had very different visual effects for the viewer. Photography might thus be thought of as a technology some of whose uses were symptoms, rather than causes, of more profound shifts in the reorganization of vision in the nineteenth century.[15] For all these reasons, it can be argued that photography extended and reinforced as much as it revolutionized existing regimes of visual representation and modes of knowledge.

Nevertheless, many nineteenth-century writers recognized in photography revolutionary potential in various branches of art and science. Commentators were quick to amass new subjects suitable for photographs, from paintings to public works, and new fields of study that might profit from the medium, from astronomy to geography. To nineteenth-century enthusiasts, photography seemed eminently suited to the recording and dissemination of geographical knowledge. William Lake Price, writing in 1868 noted how

[i]n a multiplicity of ways, Photography has already added, and will increasingly tend to contribute, to the knowledge and happiness of mankind: by its means the aspect of our globe, from the tropics to the poles,—its inhabitants,

from the dusky Nubian to the pale Esquimaux, its productions, animal and vegetable, the aspect of its cities, the outline of its mountains, are made familiar to us.[16]

Like steamships, railways, and telegraphs, photography seemed to dissolve the distance separating "there" from "here," bringing new audiences face to face with distant realities. Photography was thus central in forging revolutionary changes in the geographical imaginations of nineteenth-century Europeans and North Americans. In a close reading of early critical writing on photography in Britain, France, and North America, Joan Schwartz reveals that thinking about the new medium of photography implied new ways of visualizing the world. Through the processes of production, circulation, and consumption, photographs became what Schwartz terms "agents of sight" and "a functioning tool of the geographical imagination," informing and mediating engagement with the physical and human world.[17]

Geography and Photography

As a professional science, geography, like photography, was a child of the nineteenth century. When Talbot presented details of his photographic experiments at the Royal Society in 1839, the RGS in London had been in existence for less than a decade and had just moved to small premises at 3 Waterloo Place, where it conducted regular meetings for its fellows—then numbering fewer than eight hundred in total. The RGS grew in stature from the 1850s, notably under the various presidencies of Sir Roderick Murchison, placing its finances on a firmer footing and moving to larger premises at 15 Whitehall Place. In 1859, it acquired a royal charter and financial support from the Treasury. With the growth of public interest in African exploration and the missionary-explorer David Livingstone, the popularity of the society and the number of its fellows grew significantly, by 1872 numbering more than three thousand. The RGS remained the most significant (though not the only) institutional face of geography in Britain throughout the nineteenth century, overseeing the promotion of scientific expeditions, survey, and mapping work on a global scale, and in the 1870s, from its new premises at 1 Saville Row, promoting geography in schools and universities. At once a center of geographical intelligence, scientific society, gentlemen's club, and imperial staging post, the RGS came to occupy, particularly through powerful presidents such as Murchison, a pivotal position between the British scientific establishment, imperial government, and the wider public sphere. Indeed, the society nurtured and exploited the great public enthusiasm for geographical exploration and discovery that dominated the "culture of exploration" in mid-

Victorian Britain.[18] More than other scientific societies, the RGS courted both popular prestige and scientific distinction, and it is therefore not surprising that many of its fellows should have taken an active interest in the revolutionary new technology of photography.

The promise of photography as a new means of recording visible facts and synthesizing new knowledge had been noted since its invention. Geography was one of a range of sciences, including natural history, astronomy, geology, botany, and medicine, that commentators thought could benefit from photography. Geographers were not slow to show interest in the medium. In December 1838, for example, Alexander von Humboldt was part of the Committee of the Académie des Sciences sent to assess Daguerre's invention. Humboldt was much impressed and no doubt recognized the utility that Daguerre's apparatus might provide for the pursuit of geography. Not only was Humboldt an early purchaser of a Daguerreotype apparatus in Berlin, he likened Daguerre to Chimborazo—the great Andean peak that he had attempted to climb in 1802.[19] In 1841, George Greenough, the president of the RGS, noted:

> If one art more than another conveys to the mind a perception of the ideal . . . surely it is photography. Derived from the process of reflection, it gives permanence to images in either an increased or diminished ratio; distance, foreshortening, and perspective are to it as easy as the plainest operation of the draughtsman's pen; It acts, as it were, on the impulse of the moment, and with unerring certainty.[20]

Greenough's appreciation of photography stemmed largely from the fact that it shared scientific geography's two-dimensional worldview and, as Paul Carter has pointed out, was an important means of legitimating geography's "claim to reduce the world accurately to a uniform projection."[21] For men like Greenough, the main task of geography was the accurate surveying and mapping of the surface of the globe; photography was thus welcomed in circles of geographical science for its potential as a technique to assist cartography. Sir Roderick Murchison, during his second term in office as president of the RGS (1856–58), encouraged photography's use as an accurate and economical means of reproducing maps.[22] However, photography also exemplified the grammar of observation and depiction central to Victorian geography's mission to visualize the entire world. Sir John Herschel, the originator of the term "photography," commented in 1861 that perfect descriptive geography should "exhibit a true and faithful picture, a sort of daguerreotype, without note or comment."[23] The science of geography, with its emphasis on mapmaking, topographic survey, resource inventory, and strategic planning, was an intensely practical pursuit. However, as an enterprise focused upon the exploration, conquest, and comprehension of the surface of the earth, geography relied upon significant imaginative processes. Photography was thus significant within geography as much for its symbolic power to apprehend geographic reality as for its practical utility in visually mapping landscapes.

Photography was given a range of roles, both formal and informal, and currencies within geography in the nineteenth century: as a tool of exploration, a surveyor of landscape, a witness of military campaigns, a weapon of natural history, a methodology of racial science, a technology of propaganda and education. Photography played an important role in the exchange of geographical imagery and knowledge between different domains of science and popular culture. There was a constant traffic of photographic imagery between various scientific discourses, such as geography or anthropology, and forms of mass culture, including exhibitions, lantern shows, popular journals, and school textbooks. Institutions such as museums, geographical societies, and exhibition galleries occupied a crucial mediating position in these processes. The broader impact of photography within geographical discourse more generally involves a diverse range of practices, from commercial publishing to amateur photography, and institutions, from the military to missionary societies. Moreover, photographs had their own historical geography as objects with mutable meanings that could occupy, sometimes simultaneously, a number of different "discursive spaces," from the private intimacy of the domestic drawing room to the public scrutiny of the scientific society lecture theater.[24] In what follows, I want to focus on one aspect of this larger story by situating photography within British geographical science, focusing on some examples of its use as a tool of discovery in the field as well as visual evidence in the metropolitan spaces of scientific institutions, notably the RGS in London.

Photography and Exploration

Where truth and all that is abiding are concerned, photography is absolutely trustworthy, and the work now being done is a forecast of a future of great usefulness in every branch of science. What would one not give to have photographs of the Pharaohs or the Caesars, of the travelers, and their observations, who supplied Ptolemy with his early record of the world, of Marco Polo, and the places and people he visited on his arduous journey? We are now making history and the sun picture supplies the means of passing down a record of what we are, and what we have achieved in this nineteenth century of our progress.[25]

Thus declared the photographer and geographer John Thomson (1837–1921) in the lecture "Photography and Exploration" presented to the Geographical

Section of the British Association for the Advancement of Science in Cardiff in 1891. As a professional photographer, experienced traveler, and committed geographer, Thomson had long been an enthusiastic advocate of the revolutionary capacity of photography, here assigned the natural-magical title "the sun picture," to record accurately the people and places of a rapidly changing world. Through his work as a professional travel photographer, his several publications, and his activities as a fellow of the RGS, notably in his capacity as official instructor in photography from 1886, Thomson did much to promote the usefulness of photography to British geography in the second half of the nineteenth century.

By 1891 photography was no longer a specialist activity. Firms such as Kodak were making cameras increasingly inexpensive, portable, and reliable. However, the place of photography as a mode of geographical knowledge in the second half of the nineteenth century was not automatically assured, if only because what constituted "geography" was itself being redefined. The efforts of men like Thomson to promote photography as an accurate means of geographical description coincided with a growing emphasis within the RGS and the British Association from the 1870s on "scientific geography."[26] Such efforts show that the case for photography in the making and communicating of geographical knowledge was not established overnight. The place of photography in Victorian geography, particularly within practices of exploration, was closely implicated in sets of debates over geography's changing status as a science.

Photography was implicated in a geographical enterprise that took as its raison d'être the exploration and conquest of territory. "The noblest function of photography," one anonymous reviewer claimed in 1864, was "to remove from the paths of science . . . the impediments of space and of time, and to bring the intellects of civilised lands to bear upon the phenomena of the vast portion of the earth whose civilization has either not begun, or is passing away." To this end the reviewer added: "[F]or the purposes of science, an explorer and a photographer should be convertible terms."[27] As a technology based on the illuminating quality of light, photography took on particular symbolism as part of geographical discourse informed by a "providential theology of colonial praxis" whereby the mutual extension of Christian civilization and scientific knowledge was envisioned as a transference of "light" into the "dark" recesses of the globe.[28]

In saying this, I am arguing that photography became, in certain instances, a tool of both scientific empiricism and Western imperialism. To photograph the world could thus be a means of simultaneously comprehending it and controlling it. Such a view concurs with that of historians and critics of colonial discourse who cast photography as a technology of panoptic surveillance and an unambiguous purveyor of imperial vision.[29] Yet in working toward a more contextual account, I am also concerned here not to project a stereotype of photography that fixes and immobilizes what is a dynamic, heterogeneous set of practices. The work of anthropologists and historians of science here provides an important reminder of the need for engagement with the specific, material practices in which photography was embedded, from fieldwork techniques to the mutable contexts of reproduction and display.[30] For despite the adulation of its Victorian admirers, photography was by no means an uncontroversial practice of reporting in Britain during the nineteenth century. As Jennifer Tucker has noted in her pioneering work on the place of photography within Victorian science:

> Nineteenth century debates in Britain over claims made with photographs in a variety of settings, from field outposts to the laboratory to the spiritualist séance, suggest that Victorians did not, in fact, accept photographic evidence as unconditionally true and, indeed, that they interpreted facts based on photographs in a variety of different ways.[31]

Tucker's study of practices of photography within Victorian meteorology, bacteriology, and spiritualism shows that there were many complex processes involved in the production of agreement over the meaning and accuracy of photographs. Although Tucker's focus here is with photographic evidence of illusory phenomena such as lightning or spirits, I want to argue that the scientific utility of photography in geographical fieldwork was also framed by technical, practical, cultural, and political constraints. Moreover, the truthfulness or accuracy of its products was not fixed by optics and chemistry alone but developed upon exposure within specific contexts of production, interpretation, and display, where the currency of photographs was conditioned by a range of factors, including the social standing of scientists and the reigning ideologies of geographical institutions.

Photography was quickly heralded as a boon to overseas exploration. There were many calls for photography to be employed on overseas expeditions. For example, an 1846 issue of the journal *Art Union* anticipates that the "Talbotype" photographic apparatus

> will be henceforth an indispensable accompaniment to all exploring expeditions. By taking sun-pictures of striking natural objects the explorer will be able to define his route with such accuracy as greatly to abbreviate the toils and diminish the dangers of those who may follow in his track.[32]

However, early photographic equipment was expensive, bulky, and difficult to operate. Until the mid-1850s, photography was beyond the reach and skill of most ordinary travelers. Of those who did carry cameras, many had great difficulties using their burdensome equipment in frequently unfavorable conditions.

The Scottish explorer Colonel James Augustus Grant, for example, made some of the earliest photographs of Zanzibar at the start of his 1860–63 expedition with John Hanning Speke to the source of the Nile. Grant's stereoscopic views of Zanzibar show an interesting attempt to use stereoscopic photography as a descriptive geographical medium. His photograph "Slave Market-place, Zanzibar" for example, depicts a courtyard with a row of African slaves seated on the ground (fig. 8.1). The view is annotated with Grant's own observations. He thus describes how the image was "very difficult to take—slaves & arabs kept running away leaving only a line of women slaves whose legs and a face or two may be observed." The difficulties of securing the cooperation of human subjects for photography may in part explain why, despite his initial successes at using the camera, Grant abandoned it for the actual expedition, concentrating instead on making colored sketches. With its monotone register, photography could not capture color or highlight particular features of a scene. Thus Grant had to use words in his caption to describe the women's dress, hair, and body ornamentation as well as the houses of the market. Yet Grant also wanted his photographs to be an accurate and naturalistic representation; a snapshot of the explorer's own vision. Thus he ends his annotation by noting, with the characteristic moralistic tone of the British imperial vision of civilization, "an indistinct wily arab squats to the right eyeing the women." The numbering of the photograph shows that Grant intended it to be viewed in sequence with other images; the official stamp of the RGS claims this visual record for the institutional archive.[33]

The choice of stereoscopic photography on this expedition is worth noting. When mounted on a card and viewed by a single observer through a stereoscope—the conventional method of viewing such images—the two photographs, each taken from a slightly different angle by, in this instance, a dual-lens camera, merged into one image producing the illusion of three-dimensional space. The history of the stereoscope, as Jonathan Crary has pointed out, has been long elided and confused with that of photography in general, with the result that the specificity of the stereoscope as a technology of observation has been rarely considered.[34] To the contemporary observer, blinkered by the wooden eyepieces of the stereoscopic viewer, such technology produced the effect of total visual immersion within the scene. With the

Figure 8.1. James Augustus Grant, "No. 2: Slave Market-place, Zanzibar," 1860. By kind permission of the Royal Geographical Society, London.

stereoscope, photography could be used in an unprecedented way to reproduce the visual experience of space for an individual observer. The major developer of the technology, David Brewster, the scientist whose book *The Stereoscope* was published in 1856, was an enthusiastic proponent of photography on scientific expeditions.

The application of stereoscopic photography to geography was also promoted at the RGS. In 1865, the polymath scientist Francis Galton proposed a scheme for making "photographic maps" from three-dimensional topographic models, particularly of mountainous countries. In support of his idea, Galton exhibited ten stereoscopic photographs made by his cousin Robert Cameron Galton, who was, according to Francis, an "excellent amateur photographer." These included the "Island of St. Paul" (fig. 8.2), which showed an aerial view of the island and its striking topography and was reproduced in a pamphlet and journal article promoting Galton's idea. There is little attempt to conceal the constructed nature of the pair of photographs that made up the stereoscopic image. They are, after all, photographs of four other photographs (the joins being quite apparent) of a relief model made from a map. When viewed in the stereoscope, however, the photographs produced the illusion of three-dimensional space, as if the observer were looking down on the island's topography from directly above. These kinds of "photographic map," Galton

Figure 8.2. Robert Cameron Galton, "Island of St. Paul, in four parts united together, from a bronze Austrian model." From Francis Galton, "On Stereoscopic Maps, Taken from Models of Mountainous Countries," *Journal of the Royal Geographical Society* 35 (1865): 99–104. By kind permission of the Royal Geographical Society, London.

claimed, were considerably better than engravings: "I hey belong to quite another order or representation. The delicacy of their detail is far superior to the workmanship of any engraver, and the vividness of their relief is absolutely startling."[35]

With a library of such stereoscopic views available, a geographer in the map room or military commander in the field might simply deploy a scaled-down stereoscopic viewer to visualize, quickly and comprehensively, the topography of an area. Although Galton's plan did not achieve the large-scale use he envisaged, it is significant for the ways that it combined photography, stereoscopic technology, and cartography. It draws on a long tradition of cartographic-artistic projections of the earth from a remotely positioned observer and also anticipates the largely military development of aerial stereophotography in the twentieth century.[36] Such developments were clearly not unimagined by Victorian geographers. In 1891, John Thomson noted that "the ideal survey of the future will probably be carried out by an engineer aeronaut photographing from a balloon."[37]

As well as being claimed for science, however, the stereoscope was quickly taken up as a popular form of entertainment and education. Indeed, by 1858 the London Stereoscopic and Photographic Company could list a stock of over a hundred thousand views of landscapes, monuments, and people from around the world. The availability of such stores of visual entertainments shows that by the late 1850s photography, in its various guises, was well established as a commercial medium of virtual travel and that commercial opportunities as much as scientific enterprise lay behind early expeditionary use of the camera.[38]

Grant and Speke's expedition was not the first high-profile geographical expedition to set out with photographic equipment. The Zambesi Expedition, led by the missionary-explorer David Livingstone from 1858 to 1863 and sponsored by the RGS and Royal Society of London, included photographic equipment. David Livingstone's brother Charles was given the post of official photographer and cartographer. Charles used the "wet-plate" collodion process, which, while it had distinct advantages over the earlier Talbotype or Calotype processes, involved cumbersome apparatus and technical manipulation, coating, sensitizing, exposing, developing, and fixing glass plates on the spot.[39] Despite many setbacks, due largely to malaria, exhaustion, and his inexperience with photographic apparatus, Charles returned home with some forty stereoscopic negatives. John Kirk, the "economic botanist" and medical officer of the party, was an experienced amateur photographer and took his own equipment on the expedition, using the more portable dry-prepared plates as well as waxed-paper negatives.[40]

With its official status on the expedition, photography was embraced as a powerful means of recording permanently the landscapes, inhabitants, flora, and fauna of the area to be explored. In an early letter to his brother, David Livingstone suggested that from the early stages of the expedition Charles should get his photographic apparatus working in order to "secure characteristic specimens of the different tribes . . . specimens of remarkable trees, plants, grain or fruits and animals" as well as "scenery."[41] He set out similar instructions to Thomas Baines, the official artist of the expedition, a point to which I shall return, since the fortunes of the official artist and official photographer on this expedition were to be very different.

Livingstone's categorization here of the explorers' objects of attention into three visual domains, broadly ethnography, natural history, and scenery or topography, are significant, as they set out many of the future fields of vision of photography in geographical science, particularly on expeditions. For these categories are by no means exclusively "scientific" and draw extensively on the conventions of commercial photography and art. Both official photographer and official artist surveyed important topographical features, particularly those relating to the navigation of the river. In this respect, expeditionary photography was engaging with a long-established tradition of artistic practice within scientific, imperial exploration and survey, a tradition it emulated rather than overthrew.[42] However, Livingstone clearly wanted to exploit the evidential power of photographs within his survey work. David Living-

stone thus included photographs, along with maps and sketches of the Cab-
ora Bassa rapids, with his written expedition report of December 1858 to the
new foreign secretary, Lord Malmesbury, which was later exhibited at evening meetings of the RGS.[43] Describing the party's exploration and survey of
the rapids in the early months of the expedition, Livingstone expressed his
conviction that a steamer could pass over the rapids "without difficulty" when
the river was in flood. He backed up his assertion by noting that "a careful
sketch and photograph were made of the worst rapid we had then seen."[44]
David Livingstone explained that he had included watercolors and photographs with his dispatches because "I thought this the best way of conveying
a clear idea of my meaning," noting how the photographs exhibited the rocks
and channel in the river.[45] Livingstone thus attempted to use photographs to
support the accuracy of his observations and his assertion that a suitable
steam vessel could navigate the entire length of the river. Livingstone needed
to marshal a range of supporting evidence in order to support his request for
a new steamship and indeed to justify the entire expedition.[46] Thus the evidential value of the photographs had, in turn, to be proved by Livingstone.
Their authority, like that of the maps and sketches, came from their being
made "on the spot" by trusted members of the expeditionary team and then
authenticated by David Livingstone's own hand.

As with landscape survey, it had long been suggested that photography
would be of use in the recording of flora and fauna of distant lands. Photography was employed on the Zambesi Expedition, for example, as a means of
natural history observation. Charles Livingstone thus photographed animals
shot by the explorers, particularly those simply too large to be conveniently
taken home. David Livingstone included a stereoscopic photograph of a hip-
popotamus taken by Charles in an expedition report to the foreign secretary,
adding that it would prove interesting to Professor Owen. Charles also photographed trees, particularly the baobab. However it was John Kirk, "economic
botanist," who showed particular aptitude in applying photography as a
means of recording specimens of flora, notably different kinds of vegetation
and trees. One characteristic photograph, for example, depicts a landscape
view near Senna, showing a large baobab tree and large tamarind tree in the
distance. A smaller moevwa tree and a large anthill occupy the middle foreground, while the edge of a hut is just visible on the far right of the photograph (fig. 8.3). Kirk sent his photographs, together with sketches, written
descriptions, and plant specimens, to the Royal Botanical Gardens at Kew,
headquarters of Britain's empire of nature. Kirk also supplied detailed captions to his photographs to provide further economic and botanical information on the flora depicted.

Figure 8.3. John Kirk, "View near Senna, a Baobab on the left, to the right in front an anthill with Moevwa tree seen against a large Tamarind tree," July 1859. National Library of Scotland, Acc. 9942/40. By kind permission of Mrs. Daphne Foskett.

Another distinct focus of many geographical expeditions was the indigenous human inhabitants. By the mid-1850s, the refining of techniques of racial classification was already underway and photography was being incorporated into a method for securing accurate and comparable ethnological data.[47] On the Zambesi Expedition, for example, David Livingstone instructed his brother Charles to use photography to secure "characteristic specimens of the different tribes . . . for the purposes of Ethnology." "Do not choose the ugliest," he advised, "but (as among ourselves) the better class of natives who are believed to be characteristic of the race."[48] Such a reading of the human body as an index of internal, moral constitution was systematized within a number of related and overlapping fields, including phrenology, physiognomy, and ethnology, as well as systems and conventions of depicting character in art.[49]

One of Charles Livingstone's two surviving photographs from the Zambesi Expedition shows two African women, one carrying a baby on her back, standing grinding corn with two small boys seated on the ground holding baskets (fig. 8.4).[50] The women are posed and pictured as if caught mid-motion, though the figures in the group are shown to be aware of the camera and its operator. The photograph certainly evidences Charles's concern to capture

Figure 8.4. Charles Livingstone, stereoscopic view made during the Zambesi Expedition, 1858–63. John Murray, London.

what he described as "natives in their various occupations and amusements" in response to his brother's instructions. Indeed, David had advised Charles: "[I]f possible, get men, women and children grouped together."[51] Yet it is questionable just how useful such an image would be to ethnological enquiry. While the figures here are not individualized, neither are they fully presented as "characteristic specimens" in terms of the anthropometry of "racial types" that was beginning to structure some anthropological photography; there is little basis for measurement or isolated views of parts of the body, for instance. Indeed, the subject matter depicted, which includes material artifacts and signs of characteristic agricultural practices as well as human figures, seems chosen as much to represent David Livingstone's hope for a future African labor force in a European colony in the Shire Highlands.[52] The photograph was made as much for its picturesque appeal as for ethnography, resulting in a kind of ethnographic picturesque that appealed to commercial photographic studios. This would also explain Charles's comment, in a letter to his wife written in the early part of the expedition, that he had got some "good negative pictures which I hope to sell in England,"[53] which betrays some of the personal commercial motivations existing within his ostensibly scientific purpose. Charles seemed particularly interested in photographing African women, later noting: "I have been taking some prints—got 2 of women, one with their water pots on their heads & the other some in gala dresses."[54] Thus, in spite of his subsequent appeals to the scientific credibility of his photographs, Charles was less interested in serving the needs of metropolitan science than with capturing marketable scenes of the picturesque and unusual. Yet this distinction between the popular and the scientific was clearly a muddled one; negotiations around such boundaries lay at the heart of many debates at the RGS in the Victorian period and its attempts to promote its new branch of science.

Photography was therefore embraced in the field as a powerful new means of producing knowledge about the topography, natural history, and ethnology of different parts of the world. The practice of photography, as well as that of mapping, sketching, and collecting, was by no means neutral, since it was part of expeditions whose expressed goals often envisioned the establishment of European settlement and commerce. This was certainly the case with David Livingstone's expeditions to Africa, which sought to bring commerce, Christianity, and civilization to an imagined "Dark Continent." Photography became part of this ethos, as it was an ideal means of representing Africa as a blank, open space, ripe for a colonial future. Indeed, it was Livingstone's colonial vision that framed the perspective of photography in the field. The categories of ethnology, natural history, and "scenery," while they overlapped in places, were increasingly becoming discrete foci for the survey, measurement, and scrutiny of science as well as of commercial photographic aesthetics. Thus, while Charles Livingstone directed his camera at Africans, John Kirk's photographs of scenery and vegetation present a landscape largely empty of human presence. Even though signs of human settlement in the African landscape were apparent to Kirk, they were routinely excluded from his photographic frame, only intruding by accident, as in Kirk's view of landscape and trees near Senna (see fig. 8.3 above). Human figures are not even used, as they are by other photographers, to provide a sense of scale for trees or landforms. This is not because Kirk's technology did not allow him to take photographs of people. Rather, this selective vision of topographical features and botanical specimens represents Kirk's single-mindedness to keep his field of vision within the framework of his expeditionary duties as set out by Livingstone. Kirk's selective vision also represents the exercise of that "peculiar interest," as David Livingstone had put it, with which Europeans imagined Africa as a continent without history until colonization. Along with cartographic and literary representations, photography was selectively deployed within a geographical discourse that emptied lived environments of their human presence and in turn isolated indigenous peoples from their habitats.[55] Photographs of topography and botanical specimens in this sense became the counterpart to representations of ethnographic "specimens." In both cases, the camera became a powerful device for rendering unfamiliar environments familiar and for translating "undiscovered" space into known geography.

Exertions and Exposures

By the 1860s, photography was seen by many as a necessary adjunct to exploration, despite the difficulties posed by harsh environmental conditions and bulky apparatus. Far from being a threat to their status, many explorers found in photography ample scope for capturing and displaying their individual ordeals and achievements. For example, the traveler James Chapman took photographic apparatus on his hunting and trading expeditions in the interior of South Africa in 1859–63.[56] In 1859–60 Chapman accompanied Thomas Baines, after the latter had been dismissed from Livingstone's Zambesi Expedition, to the Victoria Falls on the Zambesi. In a letter of January 1860 to Sir George Grey, the governor of Cape of Good Hope, Chapman explained how problems of illness, drought, and poor sport had been compounded by a greater failure:

Of all our little disappointments I regret none more deeply, and I am sure your Excellency will sympathise with me when I say that I come back without one good photograph.[57]

Thus, as early as 1860, the failure of securing a photographic record was seen to amount to a failure of exploration itself. Chapman regretted his failure deeply since, as he told Sir George Grey in 1860: "[N]o exertion has been spared to render my efforts successful. Many whole days, again and again, have I devoted without any favourable results."[58] Chapman's photographic efforts bore fruit on his later explorations in 1860–63, when he finally managed make photographs of the Victoria Falls. One photograph shows the white waterfall and chasm of rock glimpsed beyond some dark vegetation in the foreground, as if conveying the experience of the weary explorer stumbling upon the view (fig. 8.5). Chapman went on to make a large collection of photographs, many of which were exhibited at the Paris International Exhibition in 1867. Chapman also sent photographs, via Sir George Grey, to the RGS in the hope of wider recognition of his travels.[59] Such examples show how photography was employed within geographical discourse as a form of witnessing as well as colonial prospecting. Like David Livingstone, James Chapman had dreams of utilizing a navigable Zambesi River to establish colonial settlements and trade in central Africa.

However, we should not imagine that photographs were always appropriate to the tasks of geographical discovery. Even if Charles Livingstone had not forgotten his chemical preparations when David Livingstone's expedition party visited the Victoria Falls in 1860, it is doubtful that the camera would

Figure 8.5. James Chapman, Victoria Falls, ca. 1863. By kind permission of the Royal Geographical Society, London.

have been effective in capturing the scope and magnitude of this landmark. A small watercolor and pencil sketch by David Livingstone, dating from this, his second, visit to the waterfall, combines both pictorial and cartographic conventions to map the size and extent of the river and gorges. While all visualizing technologies are perhaps inadequate to truly conjure up the barrage of sensual phenomena presented by this dramatic site, a simple, quickly executed sketch captured from different viewpoints something of the scale and spectacle of the falls in a way that photography, rooted to the ground and victim of its own limited frame, could not. Indeed, although James Chapman had finally succeeded in photographing the Victoria Falls in 1863, his small and misty photograph captured nothing of the drama, scale, or color of the scene, unlike the paintings and sketches made by his friend Thomas Baines. Similarly, although photography was useful in recording the larger types of flora, particularly trees, it would not usurp the collecting of plant specimens, where color and texture as well as seeds could be preserved.

Despite some of its obvious limitations, photography was, by 1890, firmly established as a tool of geographical exploration. Travelers and explorers could rely on gelatine film and handheld cameras to record their journeys. As John Thomson put it in 1891, the explorer "has only to refer to the photographs he has taken en route to banish doubt, disarm the captious critic, and afford enduring evidence of work faithfully performed."[60] Indeed, in order to "disarm the captious critic" and amplify their own achievements, many explorers and photographers emphasized just how difficult it was to secure photographs. This is certainly true of G. A. Farini's 1886 account of his journey, "with gun, camera, and note-book" through the Kalahari Desert, one of the most far-fetched accounts of photographic adventure and exploration. According to Farini, the expedition resulted "in the performance of the great gymnastic and photographic feat of taking views of the largest and most inaccessible Falls in the world—the Hundred Falls on the Orange River."[61] The camera thus features in Farini's accounts largely as a means of undertaking adventure, acquiring knowledge, and proving the manliness of the explorers. The skill, perseverance, and fearlessness of the explorers in capturing views of adventure, from wild animals to waterfalls, were emphasized throughout. The camera was used not merely as a means for the explorer heroes to conquer space, although Farini clearly delighted in using the camera to bewilder and dominate indigenous people.[62] Farini and his photographer companion "Lulu," who were both American, also used photography as a means of cultivating scientific respectability and heroic-explorer credentials in Britain, exhibiting their photographs at the Photographic Exhibition in London in 1885, and at the RGS in March 1886 to accompany the reading of Farini's narrative account.[63] However, Farini's narrative combined description of geography with tales of adventure so dramatic that they stretched the credibility of the expedition's scientific credentials. Claims for the photographs' veracity were put in doubt by the account of the explorer-heroes' numerous lucky escapes from death from hunger, wild animals, and savage people. Farini claimed that his expedition disproved "the long-prevailing notion that the Kalahari is a barren wilderness."[64] Yet the Kalahari Desert received little descriptive attention, serving only as a backdrop for a narrative of adventure, lucky escapes, and strange encounters. Should skeptical audiences have asked why no photographs existed of the explorer's more fanciful stories, Farini had carefully prepared excuses for the absence of the camera.[65] Such narratives stress the ways in which photographic evidence of travel and exploration did not simply speak for itself; it had to be backed up with a range of other supporting evidence.

Many commercial photographers also adopted the rhetoric of manly exploration in describing their photographic quests into the unknown. Samuel Bourne, the well-known commercial photographer based in India between 1862 and 1872, used the language of military campaigning and imperial conquest in describing his photographic expeditions in northern India in a series of articles for the *British Journal of Photography*. He told his readers how he set off on a nine-month expedition to Kashmir in 1864 with a staff of six servants, six "dandy" bearers, and forty-two porters, "quite a little army in themselves." Bourne was anxious to point out to his readers that "dandy" did not refer to him but to the vehicle for transporting him. In his account, he notes that his photographic equipment alone made up twenty full loads and that his dark tent (for developing the photographs) was ten feet high and had a tenfoot square base. In addition, he took extra tents, bedding, sporting requisites, books, camp furniture, and a good supply of Hennessy's brandy. It was advisable to take as many luxuries as possible, Bourne noted, "Seeing that I was sometimes for two months in some solitary and remote district without ever seeing a European, talking nothing and listening to nothing the whole time but barbarous Hindostani." By framing his travels as journeys into an unknown, uncivilized, and hostile environment, Bourne promoted his own discoveries of new scenes for the camera.[66]

The commercial photographer and geographer John Thomson also emphasized the difficulties he faced from tropical climates and unfriendly subjects during his photographic travels in China. A fellow of the RGS since 1866, Thomson strenuously promoted his photographs as works of both art and science. Thus, in 1873, he exhibited a collection of his photographs at the RGS to much praise as part of an account of his explorations on the Chinese island of Formosa. Thomson also promoted his work through pioneering photographically illustrated books. The publication, over the following years, of Thomson's broad photographic surveys of China (1873–74), London (1878), and Cyprus (1879) greatly helped establish his name and reputation at the RGS as a photographer and geographer.[67] Thomson saw photography as an essential means of making geographical knowledge. As he noted: "[T]he faithfulness of such pictures affords the nearest approach that can be made towards placing the reader actually before the scene which is represented."[68]

Thomson's preferred use of the wet-plate process produced finely detailed photographic images that secured their "scientific" status for the RGS. Years later, H. R. Mill recalled the significant impact of Thomson's photographs of

Cambodia and China at the society. Praising their "exquisite minuteness of detail," Mill noted, "The lantern slides in the Society's collection made from Mr. Thomson's negatives can stand inspection in any square millimetre by a powerful microscope."[69] Like other commercial landscape photographers, Thomson viewed his geography through a picturesque lens. At the same time, he was also motivated by the romance of exploring unknown landscapes with the camera. In early 1871, Thomson explored the upper part of the Yangtze-Kiang River, recounting his journey in a series of photographs and a written narrative. In *Illustrations of China and Its People,* Thomson arranged his large-format photographs and accompanying notes sequentially to produce a compelling travel narrative.[70] One page of the volume thus depicted a series of four photographs that are part of the visual narrative of his river voyage, showing scenes on the riverbank and views of vessels on the river (fig. 8.6). Thomson was clearly keen to capture the picturesque character of Yangtze,

Figure 8.6. John Thomson, "The Mi-Tan Gorge, Upper Yangtsze," *Illustrations of China and Its People* (London: Sampson Low, Marston, Low, and Searle, 1873–74), vol. 3, nos. 36–39. By kind permission of the Royal Geographical Society, London.

yet his photographs were also designed to map the river and its suitability for steam traffic. By placing a series of photographs on one page, Thomson presented the viewer with images located in both time and space, restaging his journey as a series of views. Linking these numbered photographs to his written description of his expedition, Thomson charted the river and possible obstructions to steam vessels, including rocks in the channel, alluvial banks, and sections of rapids. That Thomson's photographs were recognized as scientific documents as well as artistic scenes was further confirmed when *Illustrations of China and Its People,* together with four photographs of Chinese scenery, were exhibited in August 1875 as part of the RGS collection at the International Geographical Congress in Paris.

By combining colonial prospecting with picturesque views, Thomson's photographic work reminds us that many Victorian landscape photographs were regarded by their makers, as well as their audiences, not merely as "landscapes" to be enjoyed as tasteful scenes but as "views" to supply information about the geography and resources of unexplored parts of the world. Moreover, any boundary between the "view" in science and the "landscape" in art was not fixed; both operated within a range of contexts, including those of artistic genre and scientific record.[71] Such photographic narratives may be placed within a longer tradition of the making and display of picturesque "views" that was intimately associated with imperial exploration and survey.[72] For example, Commander William Allen's *Picturesque Views on the River Niger* (1840) used a sequence of numbered views and accompanying captions to guide the reader on an exploration up the river. Accompanying maps indicated the locality of the views, and its author described the work as an "endeavour to delineate the features of the country, and the manners of the people."[73]

Many travel photographers followed in this tradition, making landscape views that were picturesque but also full of empirical detail, especially through being linked to accompanying narratives. In 1891, Theodore Hoffman, for example, of the Indian-based photographic firm Johnston and Hoffman, accompanied Mr. C. White, British Resident at the Court of Sikkim, on an expedition to the Sikkim Himalayas. Hoffman's resulting photographs, sketch map, and descriptive narrative were presented to the RGS, of which Hoffman was a fellow, in 1892.[74] The society duly praised the collection, noting that

every view in the series is not only a beautiful specimen of photography, but, from a geographical point of view, gives a more accurate idea of the physical features and grand mountain scenery of Sikkim than could possibly be done be either a verbal or written description.[75]

It was precisely this ability of photography to stand in for firsthand observation that prompted the RGS from the late 1870s to build up collections of photographs made by commercial travel photographers of different parts of the world.

Visual Authority and the Geography of Display

Much of the meaning and revolutionary impact of photography within geographical circles depended not only upon practices of observation in the field but also upon contexts of display and reproduction enacted in metropolitan centers of science. During the course of their expeditions, some British explorers sent photographs back "home" to London-based institutions, such as the RGS, the Foreign Office, or Kew. However, for the majority, the printing of all negatives had to wait until the end of the expedition and their return.

It was also once explorers were back home that they could start promoting the value of their photographic work to science. Late in 1863, after his return from the Zambesi Expedition, Charles Livingstone wrote to the Foreign Office seeking remuneration for the cost of printing what he referred to as his "photographic specimens," claiming: "I have been engaged in making arrangements for printing about 40 different stereoscopic photographs of the natives in their various occupations and amusements, some remarkable trees, rocks, etc. for the use of Sir Roderick Murchison and Professor Owen."[76] Richard Owen, the well-known naturalist, was quick to support Charles's claim:

> With respect to the photographs, as these are most useful & faithful records of the physical characters of the native tribes, I suggested the desirability of their being printed, in the interest of Ethnology. I have no doubt that the photographs of rocks would thereby be made equally useful to the Geologist and of the trees to the Botanist."[77]

Following such a commendation, Charles was duly awarded an extension to his expedition salary and the cost of printing the photographs. As the official photographer, Charles Livingstone's productions were officially the property of the Foreign Office, and while he was happy to exploit this in terms of gaining recognition and remuneration for printing costs, he also, at least initially, had his own ideas about the commercial potential of his photographs.

As many of the major geographical disputes of the nineteenth century show, the production of reliable geographical knowledge depended on shared social maps of "trustworthy practice."[78] The projection of such conceptual maps was controlled in large measure by institutions such as the RGS, which not only promoted expeditions in the first place but also could bestow or withhold its powerful stamp of authority on the results. Thus, as an important venue and gatekeeper for the representation of geographical knowledge, the RGS framed the meaning of photographs through its regulations and conventions, notably those limiting membership and scientific respectability to specific groups of "gentlemen." Women remained excluded from becoming fellows throughout the Victorian period, an issue that aroused considerable debate in the 1890s.[79] Notions of gentility and manliness were thus central in determining who could be a reliable geographer and who could not. Status also depended upon the support of key institutional figures. David Livingstone's whole credibility as an explorer and his seeming clarity of vision was in large measure due to the promotional activities of Sir Roderick Murchison, who played up Livingstone's brief training in natural history by Richard Owen and generally magnified his persona as representing the fulfillment of Victorian ideals of Christian manliness.[80]

Powerful operators in the halls of science such as Murchison could just as swiftly deprive an individual of status gained over a number of years. When the artist Thomas Baines was dismissed from the Zambesi Expedition in 1859 on what appear to have been highly spurious charges leveled by Charles Livingstone, his standing as an explorer and career as scientific observer was effectively ruined. Despite valiant attempts over a number of years to clear his name and gain credit for his work as official artist, Baines found that his reputation as a man of character and artistic and scientific talent had been destroyed. He was denied the opportunity for a public hearing at the RGS and refused work on further society expeditions. Consequently, although Baines's paintings of the Zambesi were exhibited at the society, he was given far less recognition for his artistic and scientific work than was his due. This is not because Baines's pictures were less detailed, accurate, or of value to science than Charles Livingstone's photographs. In fact, the opposite is true, and Baines's paintings have proved far more durable and reliable visual legacies of the Zambesi Expedition than have Charles's photographs, which have all but vanished. Rather, the reputation and scientific standing of Baines and his visual work were ruined because he found himself on the wrong side of the explorer-hero David Livingstone and his powerful allies.[81]

John Kirk, by contrast, remained on good terms with David Livingstone. He worked at photography under his own auspices, made photographs for his own private use and in the interests of public science, particularly of botany and geography. Kirk's photographs were also exhibited in London, most probably at the RGS, in the late 1860s. Somewhat ironically, given his own difficulties, it was Thomas Baines who praised the aesthetic qualities of Kirk's photographs. In his 1871 coauthored handbook, *Shifts and Expedients of Camp Life*, while advising explorers on the uses of photography, Baines noted: "We have lately seen in London many most beautiful pictures taken by Dr. Kirk on the Zambesi with a small and inexpensive camera."[82]

Until the development and widespread adoption of halftone reproduction processes in the 1880s, the reproduction of photographs in publications was costly and therefore generally limited to specialist productions. A few notable individuals, such as John Thomson, managed to persuade publishers and patrons to produce books with photographic illustrations, either printed with text or "tipped-in" photographs.[83] However, until the 1880s, most travelers' photographs reached their widest audience only by being transposed into a different medium. This was certainly the case with popular exploration narratives such as those produced by David Livingstone. David and Charles Livingstone thus credited photographs as having "materially assisted in the illustrations" in their published narrative.[84] This material "assistance" is partly evidenced by similarities between Charles Livingstone's photograph of women and children (see above, figure 8.4) and various wood-block engravings in the volume, including one on the title page of a woman grinding corn in a granite block and another showing women tilling the soil. The fact that this stereoscopic photograph survives in the archives of Livingstone's publisher, John Murray, also suggests that the engravers of the woodcuts saw, if not emulated, its content and composition.

With its many woodblock engravings, David and Charles Livingstone's narrative was itself a highly visual text. However, as suggested above, images of men and women at work reproduced throughout the volume were placed there less as scientific records of ethnological data than as rhetorical devices to support the assertions of the text that productive labor offered the best strategy of improvement and civilization for Africans. Although a few images drew on photographs for detail and composition, most were constructed either from existing collections of drawings or from a negotiation between the textual description of David Livingstone and the imagination of the illustrator, Josiah Wood Whymper.[85] The latter process repeated a formula adopted with David Livingstone's earlier bestseller *Missionary Travels and Researches in South Africa* (1857), which described his early travels and coast-to-coast crossing of the continent.

Livingstone showed particular concern for the veracity of woodblock engravings in his books, most famously in *Missionary Travels*, when he wrote to John Murray complaining about the inaccuracy of Woolf's engraving of the explorer being attacked by a lion. Similarly, in preparing his and Charles's 1865 *Narrative*, David Livingstone went to some length to achieve his idea of "exactness" in an illustration of his wife's grave. His annotations on the engraver's watercolor sketch of the scene ask, among other details, that the thickness of the baobab tree be increased, so as to make it "more like [the] photo," referring to John Kirk's photograph of the scene.[86]

Notwithstanding Livingstone's concern that the illustrations be as true to life as possible, it is questionable whether he would have sanctioned the use of actual photographs in the book, had it been within the technical and financial means of John Murray to have published them. Photographs potentially obviated the need for middlemen such as illustrators and engravers. However, authors and publishers soon realized that photographs imposed their own kind of evidential limitations. Unlike photographs, woodblock engravings allowed the easy exercise of artistic license: the incorporation of a variety of subjects into a single scene or the editing of unwanted detail and a closer correlation with a dramatic narrative and assertions of the text. Whereas photographs, in common with other techniques of observation and collection in the field, were tied to the onward march of the expedition, woodblocks could be made in retrospect. Despite the disadvantages of protracted negotiations between engraver and author-explorer, the flexibility was useful, since it was only once explorers returned home and began to complete their travel accounts that it became apparent what visual representations were actually required.

PHOTOGRAPHY AND *HINTS TO TRAVELLERS*

The revolutionary impact of photography in the science of geography may also be measured in its absorption into everyday practices within geographical societies. By 1864, when the Zambesi Expedition was recalled, photographs were a familiar sight at the RGS. Since the mid-1850s, photographs had been exhibited at meetings alongside maps, paintings, specimens, and instruments. Photographs of the scenery, plants, animals, and "races" of previously little-known portions of the world were also being added to the society's collections, as a complement to its maps. By the late 1880s, the society was regularly hosting evening meetings illustrated by lantern slides, publishing photographs in its *Proceedings,* and had appointed an official instructor in photography to train explorers. Many provincial geographical societies established in these decades also built up collections of photographs and relied upon the use of lantern slides to enliven evening meetings.

It is no accident that photography was promoted at the RGS at the same time that many geographers were advocating more precise modes of travel observation. The increasing emphasis on scientific exploration was reflected in the changing nature of publications such as *Hints to Travellers.* This highly influential and long-running series (first published in 1854) evolved from the efforts of influential geographers to provide advice to explorers on equipment and observational techniques in order to improve the applicability of their labors to geographical science.[87] The second edition of *Hints to Travellers* (1865) contained, for the first time, a section on photography, consisting of an account by John Kirk of the photographic equipment used on the lower Zambesi and an essay by one Professor Pole, who argued that, with more commercially accessible dry plates, "every traveller and tourist may be his own photographer, with much less trouble and difficulty than is generally supposed." The inclusion of photography in this widely read guide signaled an important recognition of the medium's application in exploration.[88]

Moreover, the volume was part of a wider concern with training explorers to observe more fully, clearly, and accurately. Indeed, this concern might be thought of as an attempt to make explorers more like photographic cameras, encouraging them to capture visual facts in the field and then bring them home for scrutiny and reordering. A more significant indication of the importance that was being attached to photography, particularly in exploration, can be found in the subsequent efforts of important RGS fellows such as Francis Galton to encourage the society to actively train travelers in photography. A central figure on the RGS Council from the mid-1850s, Galton played a significant role in the society throughout the second half of the nineteenth century, particularly in its publications, finances, library, and expeditions. As I have noted, Galton had undertaken his own experiments into the geographical uses of photography since 1865. He remained a prime agitator for the RGS to employ photography more formally. In 1877, Galton's circular *The Promotion of Scientific Branches of Geography* identified the main scientific branches of geography that should be promoted. As well as special scientific expeditions and lectures on scientific geography, the RGS Council wished to encourage the collection and publication of geographical data and improvements in apparatus useful for geographical instruction or scientific research by travelers. The society's Scientific Purposes Committee, formed in January 1878, which Galton chaired regularly over the next decade, set about such work. Galton's circular was sent to the Councils of the Royal, the Anthropological, the Statistical, the Linnean, the Geological, and the Zoological Societies. The Scientific Purposes Committee also organized the printing of a new edition of *Hints to Travellers*.[89] In 1879, on the basis of Clements Markham's scheme for the training and instruction of travelers, the Scientific Purposes Committee was successful in getting the council's approval for initial instruction in surveying and mapping by John Coles. By the following year, the committee reported the completion of the society's rooftop observatory and new instrument room.

In 1880 Francis Galton wrote to the RGS secretary, Henry Bates, to back the idea—apparently already under consideration—that instruction be given to travelers in dry-plate photography, "which would imply a dark room and a glass house wherein they could practice photographing weapons, curiosities, portraits etc. & of course all matters connected with the focusing & manipulation of the cameras & lenses."[90] Galton envisioned a wealth of opportunities, from reproducing maps and diagrams for the RGS's *Proceedings* to making facsimiles of travelers' maps, which might stem from engaging "the occasional attendance of a professional photographer," and he asked Bates if the society would not generally "find much convenience in having adequate photographic facilities under our own roof?"[91] At the Scientific Purposes Committee meeting two days later, with Galton (at that time vice-president of the society) as chairman and Bates on the committee,

> it was directed that the London Stereoscopic Company and some of the leading photographers be addressed with a view to ascertaining what arrangements could be made for instructing travellers in photography.[92]

After some years' delay, due largely to financial constraints, the RGS duly appointed John Thomson as its official instructor in photography in 1886. Though it had taken some time, photography was now fully incorporated into the scientific training schemes of the society. This institutional arrangement confirmed much general practice whereby, as Thomson himself put it in 1885, "[n]o expedition, indeed, now-a-days, can be considered complete without photography to place on record the geographical and ethnological features of the journey."[93] With this in mind, Thomson gave lessons at his studio at 70a Grosvenor Street, only a short walk from 1 Savile Row, the then home of the RGS.[94] He reported to Henry Bates in 1886 that he had been "very successful with [his] pupils."[95] Years later he claimed "from Sir H. Stanley onwards the leading explorers sought my instruction so enabling them to secure excellent photographic records of their routes."[96] Thomson certainly worked with a range of key geographical figures, including Henry Stanley and Halford Mackinder. Thomson's help with color photography on Halford Mackinder's 1899 expedition to Mount Kenya was even acknowledged with a geographical landmark on the mountain.[97] Thomson continued his work for some twenty years; the RGS was still advertising his training sessions in 1905.[98]

Precisely how Thomson taught photography is not known. It is likely that he followed contemporary training guides that stressed the exercise of training and imagination in both science and art. In an 1871 book of instruction in photography for use at the School of Military Education, Chatham, Lieutenant William Abney asserted that "to become a good photographer it is necessary to turn to it with an *artistic* and *scientific* mind." Abney went on to stress that only by learning scientific skills through experience and manuals could an operator practice "clean manipulation" of the medium and ensure against becoming "photographically degenerate."[99] The moral tone of such training manuals was not unusual. For example, Lowis D'A. Jackson's influential *Aid to Survey-Practice* (1889), used in the RGS's own instruction program, emphasized the masculine demands on a surveyor, noting that he must be capable of good management as well as rational judgment. Jackson argued that knowledge of surveying, like drawing, "should be invariably comprised in the ordinary education of an English Gentleman as it may be useful at anytime." The manual goes on to note that management of "survey work of the highest and most perfect type," such as the Great Trigonometrical Survey, should be "entrusted to scientific men, with whom rest the initiation or adoption of improved methods, instruments, and appliances; while the carrying out of the detailed routine work, its checking and superintendence, is usually delegated to inferior men."[100] The establishment of hierarchies of social and scientific respectability was clearly essential in the training and practice of geographical survey and description. Such training in observation was also informed by a strong sense of improved masculinity. This is evident in the claims made for the RGS's training program by people like Lord Aberdare, its president, in 1882:

> [I]ntending travellers are every season trained in the use of instruments for geographical observation: who, fired with a true chivalrous spirit, set forth armed at all points for scientific adventure in various regions of the earth.[101]

Geography was not the only science in which attempts were made to improve observational and photographic techniques. Other sciences, such as anthropology and meteorology, also sought to enhance the scientific value of amateur, camera-wielding fact finders.[102] Moreover, it is no accident that the incorporation and refining of photography into scientific practices occurred during the decades that witnessed a popular revolution in photography. What had been at the end of the 1870s a relatively elite pursuit became, by 1890, a phenomenon of the mass market. With such a change, "professional" photographers became increasingly keen to distinguish their work, perhaps as "art," from the mere "snap-shooting" of amateurs. Similarly, the popularity of photography and the evolution of its technological and commercial basis was a double-edged sword for institutions such as the RGS forging the shape of professional science. Geographers recognized that increasingly portable and inexpensive photographic equipment had huge potential to enhance their scientific enterprise. However, if it was now possible for almost anyone to take and transport photographs of his or her travels, what was to distinguish a proper explorer from a mere tourist? As the 1865 edition of *Hints to Travellers* had pointed out, photography was increasingly available to tourist and traveler, men and women.[103] Given the heated debates throughout the 1890s over the admission of women to the RGS, it is not surprising that many of its fellows remained deeply ambivalent about embracing a medium of such popular appeal to both sexes.[104]

* * *

I have attempted in this chapter to outline some of the revolutionary impacts of photography within Victorian geography. In so doing, I have drawn a necessary distinction between the impact of photography on the making of geographical observations in the field and its influence on the making of geographical knowledge through reproduction.[105] At a practical level, measured in terms of expenditure and investment at the Royal Geographical Society, for example, photography did not seriously threaten geography's reliance on maps and mapping. However, as I have argued, photography was embraced by explorers as a means of providing objective evidence of their travels and discoveries. As photographic equipment became more easily adaptable to the demands of travel, the medium became, by the 1890s, an indispensable part of the apparatus of geographical observation in the field. Nevertheless, as I have stressed, the use of photography on expeditions did not result in the abandonment of sketching, painting, or mapping. Photographs might be unassailable for the fidelity and detail of their imagery, but scale, color, comparative detail, and contextual information were often better provided by other means. Indeed, many geographers considered field sketching superior to photography because the former involves prior analysis and selection.[106] However, as I have suggested, the camera facilitated a mode of geographical observation whereby complex environments were visualized in terms of discrete categories, such as "scenery" or "human types." Photography was thus quickly accommodated into geographical science's overarching theory of knowledge based upon creating an ever expanding and comprehensive visualization of the world. In an age of positivism and empiricism, photography seemed to fit perfectly into such a mission, since it produced naturalistic representations of the world and seemingly incontrovertible visual, geographical facts. The scientific utility of the medium to mapping was anticipated from the 1850s, and by the 1890s advances allowed the more effective combination of photographic and survey instruments.[107]

The significance of photography in Victorian geography lay not only in its use by travelers to witness their explorations and record their observations, but through its use to reproduce scenes of unfamiliar places and people for an increasing popular audience at home. Thus, the truly revolutionary impact of photography on Victorian geography came less from its practical use as a scientific tool of observation, survey, and measurement than from its symbolic effectiveness, in an age of positivism and empiricism, as a means of collecting visual facts to authenticate written and spoken accounts of exploration and to reproduce the sights of global exploration. In this way, photographs in all their guises—from stereoscopes to lantern slides—made travel a virtual possibility for almost all. Photography revolutionized the geographical imagination of Victorians by collapsing space and by bringing new audiences face to face with distant lands and peoples.

As I have shown, the evidential quality of a photograph as a record of geographical fact did not reside merely in the photograph itself. This evidence had to be corroborated through a variety of additional means, including accompanying text and spoken accounts. Such circumstantial evidence was invariably conveyed within particular spaces of knowledge, including journals, the lecture room, map room, and library. The credibility of photographic evidence within such spaces depended not only upon the context of image making and display, but also on the networks of power and patronage that enabled certain individuals to claim legitimacy as trustworthy observers.

Much of the employment of photography in Victorian geography, whether through stereoscopic views, lantern slides, or photographically illustrated books, served to produce a sense of an expanding and all-encompassing global vision. However, as the capacity for vision was being ever extended in science, notably through developments in optics and theories of light, so it became apparent how much lay beyond the field of vision. Thus, at the turn of the nineteenth century, it was no longer possible to entertain the idea, as it had been in the heyday of Victorian exploration, that if the globe could be surveyed comprehensively on maps of a uniform scale then a parallel project might accomplish the same task with photographic views. Photographs not only collected portions of the geography of the world, they ensured its endless proliferation.

It is also worth remembering, as I have noted, that important currents of ambivalence existed among British geographers throughout the 1870s, 1880s, and 1890s concerning the incorporation of photography into geographical orthodoxy and methodological handbooks of geographical science such as *Hints to Travellers*.[108] For while photography was a potentially useful tool of popular geography, as evidenced in the work of commercial travel photographers and the increasing portability and affordability of cameras, its ability to be practiced *by anyone,* with little or no training, threatened the Royal Geographical Society's institutional dominion over geographical expertise. The introduction of training courses and an official instructor in photography in 1886 was thus partly a response to such concerns, a means of securing institutional hold over a medium and distinction between, on the one hand, properly trained scientific explorers and, on the other, mere tourists.

Photography has long been caught up in disputes over the proper place of the popular and the visual within science. Studies of the uses of photography in astronomy, meteorology, and bacteriology in the nineteenth century show how the medium, with its apparent objectivity, was equally celebrated and mistrusted as a form of scientific observation and reproduction.[109] Similarly, in geography the accuracy and increasing ubiquity of photography was regarded with both praise and suspicion. Although the Royal Geographical Society had long courted popular prestige, many of its fellows in the 1890s also resisted what they saw as popularizing tendencies, such as the introduction of magic-lantern slides and the admission of women as fellows. Indeed, it is ironic that while photography in the field was practiced within existing notions of heroic, manly observation, its use as a means of popular display was linked in the minds of many men to feminine and nonscientific entertainment. This attitude did not disappear easily. In his *Record of the Royal Geographical Society,* published at the society's centenary in 1930, H. R. Mill praised the purpose-built projection room and screen of the new lecture theater recently completed at Lowther Lodge, the society's Kensington Gore site, but noted indignantly that the London County Council "does not discriminate between the private hall of a scientific society composed of rational persons and the public theatre of any degree."[110] Of course, performance and entertainment had always been part of the process through which geographical knowledge and scientific authority was produced. Moreover, a key aspect of such performances was their reliance on spectacle and technologies of visualization. It is not, therefore, surprising to find, in the early twentieth century, photographic projection built into the very architecture of an institution such as the Royal Geographical Society. Indeed, it is a measure of the extent to which, despite occasional protestations by some geographers, photography had revolutionized geography, helping to transform its popular face and evolving professional status.

...

Notes

1. David N. Livingstone, *The Geographical Tradition: Episodes in the History of a Contested Enterprise* (Oxford: Blackwell, 1992), 99.

2. Felix Driver, "Visualizing Geography: A Journey to the Heart of the Discipline," *Progress in Human Geography* 19 (1995): 123–34.

3. George Perkins Marsh, *Man and Nature*, ed. David Lowenthal (1864; Cambridge: Harvard University Press, 1965), 10; quoted in H. C. Darby, "The Problem of Geographical Description," *Transactions of the Institute of British Geographers* 30 (1962): 5. See also Yi-Fu Tuan, "Sight and Pictures," *Geographical Review* 69 (1979): 413–22; Douglas C. D. Pocock, "Sight and Knowledge," *Transactions of the Institute of British Geographers* 6 (1981): 385–93.

4. See, for example, Denis Cosgrove, *Social Formation and Symbolic Landscape* (London: Croom Helm, 1984); Stephen Daniels, *Fields of Vision: Landscape Imagery and National Identity in England and the United States* (Cambridge: Polity, 1993); Denis Cosgrove and Stephen Daniels, eds., *The Iconography of Landscape: Essays on the Symbolic Representation, Design and Use of Past Environments* (Cambridge: Cambridge University Press, 1988); David Matless, *Landscape and Englishness* (London: Reaktion, 1998); Nicholas Alfrey and Stephen Daniels, eds., *Mapping the Landscape: Essays on Art and Cartography* (Nottingham: Nottingham University Art Gallery and Castle Museum, 1990).

5. See, for example, J. Brian Harley, "Deconstructing the Map," in *Writing Worlds: Discourse, Text and Metaphor in the Representation of Landscape,* ed. Trevor J. Barnes and James S. Duncan (London: Routledge, 1992), 231–47; David Turnbull, *Maps Are Territories: Science Is an Atlas* (Chicago: University of Chicago Press, 1993); Matthew H. Edney, *Mapping an Empire: The Geographical Construction of British India, 1765–1843* (Chicago: University of Chicago Press, 1997); Denis Cosgrove, ed., *Mappings* (London: Reaktion, 1999); Charles W. J. Withers, "Authorizing Landscape: 'Authority,' Naming and the Ordnance Survey's Mapping of the Scottish Highlands in the Nineteenth Century," *Journal of Historical Geography* 26 (2000): 532–54.

6. See, for example, Joan M. Schwartz, "The Geography Lesson: Photographs and the Construction of Imaginative Geographies," *Journal of Historical Geography* 22 (1996): 16–45; James R. Ryan, *Picturing Empire: Photography and the Visualization of the British Empire* (London: Reaktion; Chicago: University of Chicago Press, 1997); Joan M. Schwartz and James R. Ryan, eds., *Picturing Place: Photography and the Geographical Imagination* (London: I. B. Tauris, 2003).

7. See, for example, Martin Rudwick, "The Emergence of a Visual Language for Geological Science, 1760–1840," *History of Science* 14 (1976): 149–95; Martin Rudwick, *Scenes from Deep Time: Early Pictorial Representations of the Prehistoric World* (Chicago: University of Chicago Press, 1992); Alex Soojung-Kim Pang, "Visual Representation and Post-Constructivist History of Science," *Historical Studies in the Physical and Biological Sciences* 28 (1997): 139–71; Bernard Lightman, "The Visual Theology of Victorian Popularizers of Science: From Reverent Eye to Chemical Retina," *Isis* 91 (2000): 651–80; Anne Secord, "Botany on a Plate: Pleasure and the Power of Pictures in Promoting Early Nineteenth-Century Scientific Knowledge," *Isis* 93 (2002): 28–57; Isobel Armstrong, "The Microscope: Mediations of the Sub-Visible World," in *Transactions and Encounters: Science and Culture in the Nineteenth Century,* ed. Roger Luckhurst and Josephine McDonagh (Manchester: Manchester University Press, 2002), 30–54; Simon Naylor, "The Field, the Museum and the Lecture Hall: The Spaces of Natural History in Victorian Cornwall," *Transactions of the Institute of British Geographers* 27 (2002): 494–513.

8. Jennifer Tucker, "Photography as Witness, Detective, and Imposter: Visual Representation in Victorian Science," in *Victorian Science in Context,* ed. Bernard Lightman (Chicago: University of Chicago Press, 1997), 378–408; Susan Gamble, "An Appealing Case of Spectra: Pho-

tographs on Display at the Royal Society, London, 1891," *Nuncius* 17 (2002): 635–51; Michael Lynch, "Science in the Age of Mechanical Reproduction: Moral and Epistemic Relations between Diagram and Photographs," *Biology and Philosophy* 6 (1991): 205–26; and Alex Soojung-Kim Pang, "Victorian Observing Practices, Printing Technology, and Representations of the Solar Corona," part 2, "The Age of Photochemical Reproduction," *Journal for the History of Astronomy* 26 (1995): 63–75.

9. Théophile Gautier, *L'univers illustré* (1858); quoted in Michel F. Braive, *The Photograph: A Social History,* trans. David Britt (London: Thames and Hudson, 1966), 186.

10. Raymond Williams, *Keywords* (London: Fontana, 1976).

11. As quoted in Ann Thomas, "The Search for Pattern," in *Beauty of Another Order: Photography in Science,* ed. Ann Thomas (New Haven: Yale University Press, 1997), 76.

12. Walter Benjamin, "The Work of Art in the Age of Mechanical Reproduction" (1936), in *Visual Culture: The Reader,* ed. Jessica Evans and Stuart Hall (London: Sage, 1999), 72–79.

13. William Mills Ivins, *Prints and Visual Communication* (1953; Cambridge: MIT Press, 1980), 94; quoted in Tucker, "Photography as Witness, Detective, and Imposter," 378.

14. A point made by William Mills Ivins, "Photography and the Modern Point of View: A Speculation in the History of Taste," *Metropolitan Museum* 1 (1928): 16–24; and Peter Galassi, *Before Photography: Painting and the Invention of Photography* (New York: Museum of Modern Art, 1981).

15. Jonathan Crary, *Techniques of the Observer: On Vision and Modernity in the Nineteenth Century* (Cambridge: MIT Press, 1996).

16. [William] Lake Price, *A Manual of Photographic Manipulation, Treating of the Practice of the Art; and Its Various Applications to Nature,* 2nd ed. (1858; New York: Arno, 1973), 1–2.

17. Joan M. Schwartz, "Agent of Sight, Site of Agency: The Photograph in the Geographical Imagination" (PhD diss., Queen's University, Kingston, Canada, 1998), 99.

18. Felix Driver, *Geography Militant: Cultures of Exploration and Empire* (Oxford: Blackwell, 2001); and Robert Stafford, *Scientist of Empire: Sir Roderick Murchison, Scientific Exploration and Victorian Imperialism* (Cambridge: Cambridge University Press, 1989).

19. See Schwartz, "Agent of Sight, Site of Agency," 101.

20. George Bellas Greenough, "Anniversary Meeting Presidential Address," *Journal of the Royal Geographical Society* 11 (1841): lxxvii.

21. Paul Carter, *On Living in a New Country: History, Travelling and Language* (London: Faber and Faber, 1992), 33.

22. Roderick Murchison, "Presidential Address," *Journal of the Royal Geographical Society* 28 (1858): 155; and "Presidential Address," *Journal of the Royal Geographical Society* 29 (1859): 152.

23. Sir John F. W. Herschel, *Physical Geography: From the Encyclopaedia Britannica* (Edinburgh: Adam and Charles Black, 1861), 2.

24. Rosalind Krauss, "Photography's Discursive Spaces: Landscape/View," *Art Journal* 42 (1982): 311–20. See also Elizabeth Edwards, "Photographs as Objects of Memory," in *Material Memories: Design and Evocation,* ed. Marius Kwint, Christopher Breward, and Jeremy Aynsley, (Oxford: Berg, 1999), 221–36.

25. John Thomson, "Photography and Exploration," *Proceedings of the Royal Geographical Society* 13 (1891): 673.

26. David R. Stoddart, *On Geography and Its History* (Oxford: Blackwell, 1986), 180.

27. Quoted in Andrew J. Birrell, "The North American Boundary Commission: Three Photographic Expeditions, 1872–74," *History of Photography* 20 (1996): 120 n. 5.

28. Livingstone, *The Geographical Tradition,* 171.

29. See, for example, Anne McClintock, *Imperial Leather: Race, Gender and Sexuality in the Colonial Contest* (Routledge: London, 1995), 122–25; and Timothy Mitchell, "The World-as-Exhibition," *Comparative Studies of Society and History* 31 (1989): 230.

30. Key studies of anthropology and photography include Elizabeth Edwards, ed., *Anthropology and Photography, 1860–1920* (London: Yale University Press, 1992); Elizabeth Edwards, *Raw Histories: Photographs, Anthropology and Museums* (Oxford: Berg, 2001); and Christopher Pinney, *Camera Indica: The Social Life of Indian Photographs* (London: Reaktion, 1997).

31. Tucker, "Photography as Witness, Detective, and Imposter," 380.

32. "The Application of the Talbotype," *Art Union* 8 (1846): 195. The "Talbotype," named after the negative photographic process invented by William Henry Fox Talbot in 1840, and also known as the Calotype, used sensitized paper negatives that were easier to transport than glass. The technique remained popular until the 1850s and was thought by many to be ideal for use by travelers and explorers.

33. The image is taken from an album of twenty-seven photographs of Zanzibar, most of which were made by Grant (Royal Geographical Society Photos X73/018784–018810).

34. Jonathan Crary, *Techniques of the Observer.*

35. Francis Galton, "On Stereoscopic Maps, Taken from Models of Mountainous Countries," *Journal of the Royal Geographical Society* 35 (1865): 102. (Additional notes by Robert Cameron Galton appeared in the same issue of the journal, pages 105–6.)

36. On Western imagery of the earth and space photography, see Denis Cosgrove, "Contested Global Visions: One-World, Whole-Earth, and the Apollo Space Photographs," *Annals of the Association of American Geographers* 84 (1994): 270–94; and Denis Cosgrove, *Apollo's Eye: A Cartographic Genealogy of the Earth in the Western Imagination* (Baltimore: Johns Hopkins University Press, 2001).

37. Thomson, "Photography and Exploration," 674.

38. Schwartz, "The Geography Lesson."

39. Frederick Scott Archer's "wet-plate" collodion process (1851) produced much finer images than Calotypes; required an exposure time of a few seconds as opposed to a few minutes, and was free from English patent restrictions which continued to limit the use of Talbot's Calotype until the mid-1850s.

40. Waxed-paper negatives, developed by Gustave Le Gray in 1850, gave finer quality negatives than the normal Calotype paper, but were not as detailed or as sensitive as collodion. A total of twenty-nine of Kirk's Zambesi photographs, as paper and glass negatives and prints, survive in a private collection on loan to the National Library of Scotland, Acc. 9942/40 and 41.

41. David Livingstone to Charles Livingstone, 10 May 1858, in *The Zambezi Expedition of David Livingstone, 1858–1863: The Journal Continued with Letters and Dispatches Therefrom*, ed. John P. R. Wallis (London: Chatto and Windus, 1956), 2:431.

42. Major studies of graphic illustration of scientific travel include Bernard Smith, *European Vision and the South Pacific*, 2nd ed. (New Haven: Yale University Press, 1985); Barbara Maria Stafford, *Voyage into Substance: Art, Science, Nature, and the Illustrated Travel Account, 1760–1840* (Cambridge: MIT Press, 1984); and James Krasner, *The Entangled Eye: Visual Perception and the Representation of Nature in Post-Darwinian Narrative* (Oxford: Oxford University Press, 1992).

43. David Livingstone, "Extracts from the Despatches of Dr David Livingstone to the Right Honourable Lord Malmesbury," *Journal of the Royal Geographical Society* 31 (1861): 256–96.

44. David Livingstone to Lord Malmesbury, 17 December 1858, in *Zambezi Expedition*, 2:294.

45. Livingstone, *Zambesi Expedition*, 2:299.

46. See Tim Jeal, *Livingstone* (London: Heinemann, 1973), 202–14; Reginald Coupland, *Kirk on the Zambesi: A Chapter of African History* (Oxford: Clarendon, 1928), 136.

47. Elizabeth Edwards, "Photographic 'Types': The Pursuit of Method," *Visual Anthropology* 3 (1990): 235–58; and Ryan, *Picturing Empire*, 140–82.

48. David Livingstone to Charles Livingstone, 10 May 1858, in *Zambezi Expedition*, 2 : 431.

49. Mary Cowling, *The Artist as Anthropologist* (Cambridge: Cambridge University Press, 1989).

50. A second photograph, of a baobab tree, survives in the collection of the National Museum, Livingstone, Zambia.

51. David Livingstone to Charles Livingstone, 10 May 1858, in *Zambezi Expedition*, 2 : 431.

52. Although no documentation survives with this particular stereoscopic photograph, circumstantial evidence suggests that it was made when the expedition explored the western side of Lake Nyasa (today's Lake Malawi) toward the end of the expedition and that it shows people known to Livingstone as the "Mang'anja." The depiction of African women at work is of particular significance, not least because it combines the Victorian male fascination—notable particularly in medicine and ethnology—with the bodies of black women, with the conventional prominence of work within Victorian art. As a sign of potential Christian moral virtue, the capacity for hard work was always at the forefront of David Livingstone's mind when observing people, not least himself and his own expedition officers but, more specifically here, Africans. Indeed, Livingstone regarded the "Mang'anja" highly for their agricultural skill, industry, and ability to work, virtues he thought would make them a suitable labor force for a future European colony in the Shire Highlands.

53. Charles Livingstone to Hariette Livingstone, 14 Sept–21 Dec 1858, G5/10 National Museum, Livingstone, Zambia; quoted in Gary W. Clendennen, "Charles Livingstone: A Biographical Study, with Emphasis on His Accomplishments on the Zambesi Expedition, 1858–1863" (PhD diss., University of Edinburgh, 1978), 251.

54. Ibid.

55. Mary-Louise Pratt, *Imperial Eyes: Travel Writing and Transculturation* (London: Routledge, 1992), 61.

56. James Chapman, *Travels in the Interior of South Africa, 1849–1863*, ed. Edward C. Tabler, 2 vols. (Cape Town: A. A. Balkema, 1971).

57. James Chapman, "Notes on South Africa," *Journal of the Royal Geographical Society* 30 (1860): 17–18.

58. Ibid.

59. James Chapman, *Travels in the Interior of South Africa*, 2 : 211.

60. Thomson, "Photography and Exploration," 5.

61. G. A. Farini, *Through the Kalahari Desert: A Narrative of a Journey with Gun, Camera, and Note-book to Lake N'gami and Back* (London: Sampson Low, Marston, Searle, and Rivington, 1886), vii.

62. Ibid., 42–44, 73–78, 223–24.

63. G. A. Farini, "A Recent Journey in the Kalahari," *Journal of the Royal Geographical Society* 8 (1886): 437–53.

64. Farini, *Through the Kalahari Desert*, ix.

65. Ibid., 164–66.

66. Samuel Bourne, "Narrative of a Photographic Trip to Kashmir (Cashmere) and Adjacent Districts," *British Journal of Photography* 13 (1866): 474; see also 474–75, 498–99, 524–25, 559–60, 583–84, 617–19, and the continuation of the narrative in *British Journal of Photography* 14 (1877): 4–5, 38–39, 63–64.

67. John Thomson and Adolphe Smith, *Street Life in London* (London: Sampson Low, Marston, Searle, and Rivington, 1878); John Thomson, *Illustrations of China and Its People, a Series of Two Hundred Photographs with Letterpress Description of the Places and People Represented,* 4 vols. (London: Sampson Low, Marston, Low, and Searle, 1873–74); John Thomson, *Through Cyprus with the Camera, in the Autumn of 1878* (London: Sampson Low, Marston, Searle and Rivington, 1879).

68. Thomson, introduction to *Illustrations of China,* vol. 1.

69. Hugh Robert Mill, *The Record of the Royal Geographical Society, 1830–1930* (London: Royal Geographical Society, 1930), 86–87.

70. Thomson, introduction to *Illustrations of China,* vol. 3, plates 17–24.

71. Rosalind Krauss, "Photography's Discursive Spaces." See also Estelle Jussim and Elizabeth Lindquist-Cock, *Landscape as Photograph* (New Haven: Yale University Press, 1985).

72. Luciana L. Martins, "A Naturalist's Vision of the Tropics: Charles Darwin and the Brazilian Landscape," *Singapore Journal of Tropical Geography* 21 (2000): 19–33. See also Smith, *European Vision;* and Stafford, *Voyage into Substance.*

73. William Allen, *Picturesque Views on the River Niger, Sketched during Lander's Last Visit in 1832–33* (London: John Murray, 1840), preface.

74. Theodore Hoffman, "Exploration in Sikkim: To the North-East of Kanchinjinga," *Proceedings of the Royal Geographical Society* 14 (1892): 613–18.

75. "Twenty-three Photographs of Mountain Scenery in Sikkim," *Proceedings of the Royal Geographical Society* 15 (1893): 288.

76. Charles Livingstone to Austin Layard, n.d., FO97/322, fol. 156; quoted in Clendennen, *Charles Livingstone,* 255.

77. Richard Owen to Charles Spring-Rice, 9 December 1863, FO97/322, fol. 158; quoted in ibid., 256.

78. Tucker, "Photography as Witness, Detective, and Imposter," 389.

79. Morag Bell and Cheryl McEwan, "The Admission of Women Fellows to the Royal Geographical Society, 1892–1914: The Controversy and the Outcome," *Geographical Journal* 162 (1996): 295–312.

80. Norman Vance, *The Sinews of the Spirit: The Ideal of Christian Manliness in Victorian Literature and Religious Thought* (Cambridge: Cambridge University Press, 1985). 2–3; see also Driver, *Geography Militant.*

81. Jane Carruthers and Marion Arnold, *The Life and Works of Thomas Baines* (Cape Town: Fernwood, 1995), 54–65; and Stafford, *Scientist of Empire,* 181–82.

82. Thomas Baines and William Barry Lord, *Shifts and Expedients of Camp Life* (London: Horace Cox, 1871).

83. For a notable example of the latter, see H. B. George, *The Oberland and Its Glaciers: Explored and Illustrated with Ice Axe and Camera* (London: Alfred W. Bennett, 1866).

84. David and Charles Livingstone, *Narrative of an Expedition to the Zambesi and Its Tributaries and of the Discovery of the Lakes Shirwa and Nyassa, 1858–1864* (London: John Murray, 1865), vii.

85. See Tim Barringer, "Fabricating Africa: Livingstone and the Visual Image, 1850–1874," in *David Livingstone and the Victorian Encounter with Africa,* ed. John M. MacKenzie (London: National Portrait Gallery, 1996), 183.

86. See Oliver Ransford, *David Livingstone: The Dark Interior* (London: John Murray, 1978), 232; MacKenzie, *David Livingstone,* 51.

87. Francis Galton, "Hints to Travellers," *Journal of the Royal Geographical Society* 24 (1854): 345–58.

88. John Kirk, "Extracts from a Letter," *Journal of the Royal Geographical Society* 34 (1865): 290–92; Pole, "Photography for Travellers and Tourists," *Journal of the Royal Geographical Society* 34 (1865): 295.

89. Francis Galton, ed., *Hints to Travellers* (London: Royal Geographical Society, 1878).

90. Francis Galton to H. W. Bates, 10 November 1880, Royal Geographical Society Archives, London (henceforth RGS Archives).

91. F. Galton to H. W. Bates, 10 November 1880, RGS Archives.

92. Scientific Purposes Committee, 12 November 1880, Royal Geographical Society Committee Minute Book, RGS Archives, 177.

93. John Thomson, "Exploration with the Camera," *British Journal of Photography* 32 (1885): 373.

94. Thomson had moved his studio from Buckingham Palace Road, Belgravia, to Mayfair in the early 1880s and remained at 70a Grosvenor Street until 1905, when the studio moved to nearby New Bond Street. One Savile Row was the headquarters of the Royal Geographical Society from 1870 until 1913, when it moved to its present home at 1 Kensington Gore.

95. John Thomson to H. W. Bates, 28 December 1886, RGS Archives. Thomson initially charged ten shillings for an hour's teaching, half of this amount covering the cost of instruments, chemicals, and all other materials. See John Thomson to H. W. Bates, 25 January 1886, RGS Archives.

96. John Thomson to A. R. Hinks, 7 November 1917, RGS Archives. In 1886 Thomson offered Henry Stanley free training in photography for any of his officers on his Emin Pasha expedition. It is uncertain if his offer was taken up. See J. Thomson to H. W. Bates, 28 December 1886, RGS Archives.

97. It seems that Thomson found out about this monument to himself only by chance in 1921, the year he died. John Thomson to A. R. Hinks, 10 February 1921; A. H. Hinks to John Thomson, 15 February 1921, RGS Archives.

98. "Instruction for Intending Travellers," *Journal of the Royal Geographical Society* 26 (1905): viii.

99. William de W. Abney, *Instruction in Photography: for use at the S.M.E. Chatham* (Chatham: SME, Printed for Private Circulation, 1871), 1–2.

100. Lowis D'a Jackson, *Aid to Survey-Practice: For Reference in Surveying, Levelling, and Setting-out; and in Route-Surveys of Travellers by Land and Sea* (London: Crosby Lockwood and Son, 1889), xii, 1, 2.

101. Lord Aberdare, "The Annual Address on the Progress of Geography," *Proceedings of the Royal Geographical Society* 4 (1882): 329–39.

102. See Elizabeth Edwards, "Photographic 'Types'"; and Tucker, "Photography as Witness, Detective, and Imposter."

103. Pole, "Photography for Travellers and Tourists."

104. See Bell and McEwan, "Admission of Women Fellows."

105. The importance of this distinction is also noted in Ann Shelby Blum, *Picturing Nature: American Nineteenth-Century Zoological Illustration* (Princeton: Princeton University Press, 1993); and Alex Soojung-Kim Pang, *Empire and the Sun: Victorian Solar Eclipse Expeditions* (Stanford: Stanford University Press, 2002), 113.

106. For a later assertion of this view, see Sidney W. Wooldridge, "The Status of Geography and the Role of Field Work," *Geography* 40 (1955): 73–83.

107. See, for example, H. Schlichter, "Celestial Photography as a Handmaid to Geography," *Proceedings of the Royal Geographical Society* 14 (1892): 714–15; H. Schlichter, "The Determination of Geographical Longitudes by Photography," *Geographical Journal* 2 (1893): 423–29;

and E. G. Ravenstein, "Correspondence on 'The Determination of Longitudes by Photography,'" *Geographical Journal* 2 (1893): 557–58.

108. See also Driver, *Geography Militant,* 49–67.

109. For debates in astronomy where photography was seen from the 1890s as an "authoritive chemical retina" that could provide detailed accurate records of the stars, see Lightman, "Visual Theology," 675. For discussion of debates over the meaning of photographs within meteorology, bacteriology and spiritualism, see Tucker, "Photography as Witness, Detective, and Imposter." See also Lorraine Daston and Peter Galison, "The Image of Objectivity," *Representations* 40 (1992): 81–128.

110. Mill, *Record of the Royal Geographical Society,* 229–30.

PART III

GEOGRAPHY AND POLITICAL REVOLUTION

Geography and State Governance

Political revolutions regularly come with geographical qualifiers: English, Chinese, Cuban, French, Iranian, Mexican, Russian, to name only a few. In such contexts, questions of geography are often taken to be synonymous with matters of scale—revolution as local "rebellion," a national issue, as a shared international feature—or they are understood to follow from the spatially significant unequal distribution of resources within a given society. There is, too, a sense that through political revolution, a new and better political state can be attained, a new political geography *and* a new geography of politics if you will. Future political well-being can result in, and be the result of, greater geographical equity. In these several senses, there is, in short, a geography *of* and *to* political revolution.

The four chapters making up this part do draw upon such questions. For Mayhew in chapter 9, for example, part of the difficulty of understanding the nature of the "English Revolution" of the 1640s lies in knowing its regional expression and causes while recognizing that it also had British and European dimensions. In France, the Revolution of 1789 and thereafter was rooted in a disaffection for the monarchy more evident in some places than in others and may even have been prompted by local geographies of harvest failure and consequent price rises following an unseasonably wet summer. As Heffernan notes in chapter 10, one of the outcomes of the French Revolution was a new administrative structure for France, a geography rearranged to match the nation's new forms of governance. The chapters in part 3 are rather more concerned, however, with the place of geography *in* various revolutions and revolutionary contexts than they are with advancing spatial explanations for the nature of political revolutions. For some authors more than others, attention is also directed at the nature of revolutionary change in what geography was in periods of intellectual and political turmoil.

Mayhew explores the institutional sites of Oxford University and the textual spaces of English geography books in order to address the connections between geography and political revolution in mid-seventeenth-century England. Geography, a long-term adjunct to history in humanist pedagogy, was in this period anyway undergoing a "revolution" in its method and purpose, a revolution which, Mayhew demonstrates, allowed geography books to become sites for political and religious debate. Such textual revolution did not begin in Oxford, but it was certainly much evident there, in books of systematic geography especially. As geography's texts became sites in a war of words over the nature of Protestant theology, so geography was differently called upon according to one's political and religious allegiance. Political revolution was thus served by the appeals made in and to the generic revolution in geography's books.

If geography's textual conventions changed as the world turned "upside down" in mid-seventeenth-century England, geography in late-eighteenth-century France provided a means both to national order and, for Edme Mentelle, a means of survival in revolutionary times. Here, too, complex connections can be discerned: between a revolution in geography in regard to conventions concerning method, the place of geography in an age of revolution as the subject was differently taught in a variety of pre- and postrevolutionary institutions, and in terms of the social geography of revolutionary sites and spaces. As Heffernan shows, Edme Mentelle's geography text of 1758 provided a passport to teach at the École Royale Militaire in Paris, a site that afforded him a base for many more works of geography and, latterly, facilitated his introduction into courtly society. But the later Mentelle was no die-hard royalist. New nations demand new geographies, if not always new geographers. From 1791, Mentelle began a program of republican geography books. Lectures given by him at the École Normale show him using geography as factual knowledge for a new citizenry. Shortly before his death, Mentelle, the by then thrice self-fashioned geographer-cum-political survivor, was honored by the king, Louis XVIII. In this account of Mentelle and of his works, we may see how geography, principally as an educational discourse, was used in and through a revolutionary age to serve different political agendas and, if not always tellingly, ultimately to serve France.

Elsewhere, geography was likewise being called upon to serve the needs of a new nation. As Livingstone shows in chapter 11, for Jedidiah Morse in particular, geography *in* America in the early Republic was vital *for* America, a means both to dispel the continuing misperceptions of Europeans and to shape persistent and potentially divisive localisms into one nation united under God. Morse, like Jefferson, assumed the role of patriot geographer. Yet different geographies were appealed to in shaping postrevolutionary America. Where Jefferson's 1787 *Notes on the State of Virginia* was at once "a regional inventory" of that state and a defense of America's identity, Morse's *American Geography* (1789) looked to provide a "moral geography of the nation" cast in the mold of New England. So, too, for Timothy Dwight, whose vision for America as a Christian Republic was rooted in the landscapes of New England's virtuous pastoralism. As Livingstone demonstrates, here, clearly, certain sorts of geography, even certain geographical conditions, were being looked upon to provide for a postrevolutionary new world *and* a new moral order rooted in civic virtue, and the means to the deliverance of both was through geography's books.

Like Heffernan in part, Rupke deals in chapter 12 with the theme of geographers in an age of revolution. But whereas Mentelle's tale is one of living through revolution, of keeping his head while the world revolved about him, Humboldt's reworking as a revolutionary was a posthumous affair. In death, Humboldt the geographer was "revolutionized," as Rupke has it, to become Humboldt the revolutionary. Noting this fact matters less, perhaps, than knowing how and why. Humboldt's published private correspondence to a German democrat friend was used to read the public and international Humboldt differently in death. As was the case for his active geographical life, the geographies of the reception of Humboldt after his death had different national and political significance. Images in Britain of Humboldt the cosmopolitan geographer—images useful to the promotion of certain visions for science—could be sullied by Humboldt's seeming association with revolutionary politics in general and with German nationalism in particular.

Certain important themes are thus highlighted in this section. In at least three different geographical and revolutionary contexts—postrevolutionary America, revolutionary France, and Civil War England—geography was called upon to help understand political revolution, to know what "revolution" there meant, and to assist in managing its consequences. Geography was used to "map" revolution actually and conceptually. Similarly, the idea of revolution is useful in understanding changes in what geography was and what it did, notably in the genres of geography's books and in respect of method. Methodological questions have in turn been raised here. For while none of the indi-viduals here discussed was a "geographer-revolutionary"—despite Humboldt's detractors' trying to make him so in death—documenting an individual's life path can help throw a fuller light on the place of geography and geographers in relation to wider social and political concerns.

9

GEOGRAPHY'S ENGLISH REVOLUTIONS

Oxford Geography and the War of Ideas, 1600–1660

...

Robert J. Mayhew

GEOGRAPHY AND THE ENGLISH REVOLUTION: HOBBESIAN RUBRICS

L ooking back from the 1660s, Thomas Hobbes's history of the "English Revolution," *Behemoth,* pointed to the dominant rubric by which geography and the "English Revolution" have been conjoined in recent historiography.[1] That rubric concerns the spatial scale at which the events of 1640 to 1660 can be understood. Hobbes points to three scales of analysis. First, he suggests there is a *European* context in which the English Revolution must be understood. For Hobbes, Charles I was too close to the Catholic powers of France and Spain, which led to the distrust of Parliament and of Calvinists.[2] Second, Hobbes framed the Civil War in the context of a *nation of multiple kingdoms,* suggesting that tensions between England, Scotland, and Ireland were central in the breakdown of government.[3] Third, Hobbes looks to the *local scale* in two ways. He discusses the geography of allegiance, arguing that towns were Presbyterian and, therefore, Parliamentarian strongholds, implicitly placing the countryside in Charles's corner. Later, Hobbes's narrative sketches the geography of the unfolding conflict.[4]

Debates about these three scales and their relative merits as explanatory contexts have fueled the main connections that historians have forged between geography and the English Revolution. Reversing our order and giving a mere flavor of the work produced, debates on the local scale about allegiance have canvassed whether allegiance was geographically more "spotty" than neat formulae such as Hobbes's town-country divide suggest.[5] These debates are more than merely empirical: there has been considerable questioning of the extent to which talk of county identities makes sense when gentry lifestyles were so metrocentric.[6] In other words, the geographical concept of "localism" as an explanatory tool has come under scrutiny. Similar comments apply to the more sophisticated "localist" thesis of David Underdown, who suggested allegiance might vary by ecological region, pastoral areas being less subject to the structures of government and more likely than arable areas to ally with Parliament. Underdown, then, sought to move from a mere spatial pattern of allegiances to an explanation determined by regional ecology. This model has also met with serious objections, concerning its general applicability and even its purchase on the experience of Wiltshire, Underdown's own case study.[7] In short, how to conceive of the local and how, indeed whether, localism affected the English Civil War, are issues where no consensus has been reached.

Hobbes's second scale, the British nation and its constituent kingdoms, has been perhaps the most discussed historiographical issue connecting geography and the English Revolution. There is not space to go into detail here, but the "English" Revolution is now recognized, in both its causes and its unfolding, to have been an unavoidably British affair rooted in the tensions, misunderstandings, and alliances that the differing politico-theological regimes of England, Scotland, and Ireland could facilitate. As Russell summarized matters:

> [T]he English Civil war is the name we give to that part of the British Civil Wars which was fought by Englishmen on English soil. That part is rather a small one. It is not even the first part of the British Civil Wars. . . . If we look at the British Civil Wars as a whole, it is clear that they began and ended as a struggle between England and the rest for supremacy over the British Isles.[8]

If this contention has received widespread support, it has also been questioned from two angles. On the one hand, Hirst has argued that at the time "Britain" as a concept meant very little compared to England or county affiliation.[9] This point seems to recapitulate at a different spatial scale the arguments made about the lack of reality of county communities, but to less effect, for the multiple-kingdoms argument is not that allegiance was driven by spatial scale per se (unlike that for the county scale), but that identifiable political, theological, and historical differences between the kingdoms of Britain led them into conflict regardless of their geographical self-images.

The second questioning of the centrality of the "British question" to the Civil War opens up Hobbes's third spatial scale. It is argued, notably by Scott, that the category "British" is a retrospective construction not part of the conceptual lexicon of those who fought in the Civil War.[10] For Scott, we need to turn to the scale at which the English understood their troubles, that being not the nation, but the European stage. At one level, Scott's argument has the same weakness as Hirst's: the fact (and it is debatable) that the English did not think in British terms does not mean they did not act in ways determined by a logic whose spatial scale was that of the nation.[11] Equally, there is no need to see these scales as mutually exclusive: multiple kingdom situations were rife across Europe, and the political and religious tensions that they sparked were common currency in British debates.[12]

Yet debates about the spatial scales on which to understand events do not address the key rubric by which, on Hobbes's account, geography and the English Civil War should be connected. Hobbes's interpretive key for understanding the events of 1640–60 is not a revolutionary *scale,* but a revolutionary *space,* the space of the universities, which for him were "the core of the rebellion" for two reasons. First, universities fomented republican values by their preoccupation with ancient commonwealths. Second, universities bred debates within Protestant theology between Calvinism and Arminianism, these debates spilling out into the public realm and ushering in a war of ideas.[13]

The universities have been intensively studied by historians of the English Revolution. Christopher Hill and Charles Webster have argued that Puritanism was conducive to scientific endeavor, as it questioned received authority.[14] There are important differences between the two, Hill seeing the universities as, Gibbonian avant la lettre, places of torpor and inactivity, looking to merchants, instrument makers, explorers, and Gresham College for a vibrant culture of learning. By contrast, Webster's Puritan revolution is one that had far more purchase in and indeed dependence on the universities. In both accounts, the culture of geographical learning has received attention, Hill seeing the Puritan/Parliamentarian circle of the Earl of Leicester as supporting geography and exploration, and Webster looking at Nathanael Carpenter's Oxonian *Geography* (1625), Lewis Roberts's commercial geography, and Gerald Boate's *Natural History of Ireland.*[15]

The Hill/Webster analysis needs qualification on three grounds. First, both historians tend to treat all forms of Calvinism as "Puritan," where others, as we shall see, have shown that orthodox Calvinism was the theological "mainstream" in early Stuart England. For this reason, the lines of force between Puritanism and scientific revolution are far less clear than has been suggested. Second, it is not clear precisely what either Hill or Webster takes "science" to mean. As historians of science recover Renaissance and late Scholastic natural philosophy,[16] so the simple division between "modern" science and the hidebound Scholastic science it succeeded, a division Hill and Webster draw on, ceases to make sense. Finally, and as a corollary to the preceding point, both Hill and Webster treat geography as a timeless and self-evident inquiry, not a historically variable one. Thus Webster refers to Boate's work as "a major development in economic geography," where it is not clear what this division of knowledge could have meant in the seventeenth century.[17] Similarly, Hill's geography is by and large exploration and its narration, which excludes the textual tradition that dominated contemporary definitions of geography (see also Withers's arguments to this effect in chapter 4).[18]

The Hill/Webster position has been strongly criticized by Nicholas Tyacke and Mordechai Feingold, both of whom point to university cultures as anything but stagnant and as anything but reliant on Puritanism or even Calvinism for their vigor.[19] Yet neither moves the debate forward in terms of the two other weaknesses in the Hill/Webster argument, namely, the definition of science and of geography. Both mention geography as sketchily as do Hill and Webster, and they do so in a way that accords with the picture they are opposing more than that they propose, simply pointing to Carpenter and a group of Calvinist geographers at Exeter College, Oxford. As we shall see, there is grist to their mill in Oxford's geographical culture, even if they have not picked up on it.

If geography is incidental to this debate on the culture of university learning, that is because the debate, true to Hill's initial terms of reference, has covered a gamut of inquiries labeled, somewhat anachronistically, "science," and geography's relationship with this label has been opaque (as Withers also points out above). The historians of geography on whom historians of the English Revolution have drawn have done little to clarify the situation for several reasons. First, the dominant image of geography's history has suggested that as a university subject it did not come into being in the British universities until 1887.[20] What work has been done on geography at the English universities before Mackinder has focused on matters other than its political content and context, which is the point at issue among historians of the Civil War. Thus, Taylor's survey (on which Hill draws extensively) provides little insight into politics, being an enumerative bibliography whose anachronistic conception of "real" geography as exploration was followed by Hill,[21] while Cormack's fine study of English university geography to 1625 focuses on matters of genre, book availability, and reading, not on politics beyond issues of patriotism and empire.[22]

My argument in this chapter takes up Hobbes's revolutionary space, the university, and looks at the geographical culture of Oxford University in the period 1600–60. On Hobbes's own argument, there are prima facie reasons to look at geography, as it participated in the two aspects of university learning he argued had spread the canker of rebellion. First, humanist pedagogic strictures repeatedly aligned geography as an adjunct to history. Geography was the "eye" by which scholars could read aright the histories of ancient commonwealths that Hobbes found so dangerous. Hobbes himself recognized the link in his translation of Thucydides, to which he appended two maps on the entirely conventional grounds that without this geographical information the history could "neither patiently be read over, perfectly understood, nor easily remembered."[23] Second, geography books were sites for theological debate, and conflicts between Arminianism and Calvinism were aired in this context. If Hobbes portrayed the English Revolution as a war sparked by ideas, some of the first salvoes were fired in geography.

This chapter has three parts, following Hobbes's structure of argument. First, it argues that there was a revolution in the generic form of geography in the period 1600–25, inspired by late humanist pedagogic theories. Second, this newly codified genre allowed for politico-theological debates, and in the period 1600–60, those debates were keyed to a division between Calvinism and Arminianism. Finally, moving out from geography books, we will see that pamphlet debates surrounding the Civil War drew on geographical information.

THE REVOLUTION IN ENGLISH GEOGRAPHY

The nature of English geography as a textual genre underwent a revolutionary change in the years after 1600, and the impetus for this change came in a series of works produced in the context of Oxford University.

Before this time, the main medieval contexts in which geographical knowledge had been conveyed were the chronicle, which tended to preface its history with a geographical treatment, and the encyclopedia, which in providing a summation of all knowledge treated knowledge of the globe as a standard element.[24] Taking Britain's most popular chronicle, Ranulph Higden's fourteenth-century *Polychronicon*, a few features stand out in contradistinction to geography books produced after 1600. Higden's geographical description of the world forms book 1 of the *Polychronicon*, prefaced to his history.[25] Each nation is described briefly, but there is little by way of systematic organization, different nations being described in different ways despite Higden's opening claim that "hit shalle be expressede by ordre."[26] Thus chapters on India and France start with toponymic etymologies and go on to the bounds of each nation, where the descriptions of Greece and Rome see a different structure dictated by Higden's desire to cite and marshal classical sources. The complications and inconsistencies are still furthered by the Welsh section, which in Caxton's 1480 English edition was a versified rendering of information from Giraldus Cambrensis. There was, then, no standardized format by which information was ordered or presented.

In the half century before 1600, geography was emerging as a separate genre as the medieval chronicle lost favor, a process of generic reordering that has been charted more thoroughly for history than geography.[27] Yet the early geography texts produced in this reordering were hardly systematic. Thus Roger Barlow's *Briefe Summe of Geography* (1541) took paragraphs wholesale from ancient texts and had long sections of heterogeneous material. Holinshed's chronicle similarly had a geographical section drawn up by William Harrison, which frequently juxtaposed diverse material.[28]

So what revolution took place in geography texts after 1600? We can gain an initial insight by looking at the image that prefatory material gave of the task of writing geography. All the major Oxford geographers at this time speak of their ambition to write by "method." It was in this regard that they claimed, despite cribbing material from other sources, to be authorially creative. Thus Peter Heylyn argued his *Microcosmus* was "properly called the issue of mine owne braine," adding that this meant "the matter I derive from others, the wordes for the most part are mine owne, the method totallie."[29] Robert Stafforde in his *Geographicall Description* could likewise say the information in his book was taken from others, but "the Methode I had from my Tutor," the rector of Exeter College, John Prideaux.[30] Furthermore, those collecting travel accounts began to describe their difference from geography in terms of a lack of method. Thus Samuel Purchas opined that *Hakluytus Posthumus* was arranged "not by one professing Methodically to deliver the Historie of nature according to the rules of Art, nor Philosophically to discusse and dispute."[31]

What did this appeal to method mean? Did it entail a genuinely more systematic approach than had Higden's claim to express himself by order? All the Oxford geographers made clear divisions of their material between "general" considerations and "special" issues particular to specific examples. This is particularly apparent in Carpenter's *Geography* (1625), where layer upon layer of division is created in terms of a general/special binary, this being set out in introductory charts. Further, the two books of the title deal respectively with general issues concerning the universe (book 1) and with what he calls "topical" matters (book 2). Even where this technique is less foregrounded, it is still present. For example, George Abbot's *Briefe Description of the Whole Worlde* (1599) sets out the (general) terms of geography before dealing at greater length with their (special) instantiations around the world. Similar divisions can be found in Heylyn's *Microcosmus* and in William Pemble's *Briefe Introduction to Geography* (1630).[32]

"Method" had an important effect on the ways in which nations were described, one that did amount to a systematization when compared to the work of geographical predecessors such as Higden or Barlow. Geographers now followed a rigid order in the presentation of information. Even Abbot's *Briefe Description,* which true to its name is a short octavo of about sixty pages, showed a clear system, each nation being described sequentially in terms of its boundaries, the kingdoms of which it was composed, nations it abutted, and finally a section covering religion, history, and curiosities. Abbot had, then, a set order in which he presented material, and this can also be seen in more expansive geographical works. Content did not differ wildly from medieval precedents: etymologies, boundaries, religion, and curiosities still featured, but in a regularized sequence. Compared with medieval chronicles, there was a reduction in references to Providence, but this was not a secularization of geography. Rather, as we shall see, in methodized geographical writings ecclesiology and theology came to have a central place.[33] Claims to a methodical presentation are made good by all these authors. Moreover, they all seem to follow versions, with varying degrees of elaboration, of the schema described for Abbot, where earlier geographers had followed very different plans from one another, there being a presentational vacuum as the generic norms of the chronicle collapsed. Geography books from 1600 thus started to take on a certain method within themselves, and a certain family resemblance between themselves, which justifies seeing them as forming a coherent genre.

The edges of this generic consolidation were still ragged in the early seventeenth century. Carpenter, praised for his scientific approach by historians of geography, has a long poetic lament about his lack of rewards at Oxford, while Thomas May's 1631 translation of Barclay's global survey of cultures, *Icon Animorum,* inhabits a mental world little altered by appeals to method.[34] Yet overall, it is the construction of an agreed genre of geography, centered on a method of general/special binaries, that impresses the reader when compared with earlier geographical texts.

What were the intellectual origins of this remarkable generic consolidation around "method"? We can discern two origins, appropriately a general and a special one. In general, European humanism demonstrated counterpoised ambitions both to enjoy the plentitude of information available and to control and organize that copiousness into some form of order.[35] Both ambitions could be pursued through the technique of "commonplacing," collecting information gleaned from reading under a range of headings.[36] There can be little doubt that such commonplacing was the raw material for geographical surveys, but this does not on its own explain their appeal to method. It has been argued that the balance between the desire for plenitude and that for organization changed over time, the later humanism of the late sixteenth and early seventeenth centuries witnessing a drive for organization in the face of a flood of information.[37] It is this shift that can be seen in geography books and is the "special" origin of the drive for method. The geographical writings of Barlow in the 1540s and Harrison in the 1570s show a Rabelaisian joy in copiousness, whereas with George Abbot some twenty years after Harrison, we see the pendulum swing toward organization.

The shift to a desire for method is particularly associated with two movements. First were the pedagogic reforms of Petrus Ramus in Paris. Ramus's mantra was "method" as he sought to overturn the edifice of Aristotelian Scholasticism. Ramus deployed branching diagrams of the relation between inquiries and showed an organizational obsession with binaries.[38] Second, the work of the "systematics" was a more direct influence on the geography texts produced in Oxford. German pedagogues—the most important of whom were Bartholomew Keckermann and Johann Alsted—sought to take on some of Ramus's innovations, but to reconcile them with Aristotelianism; to move away from Ramus's preoccupation with binary structures in favor of more general branching organizations of knowledge; and to take the drive for method into new realms of inquiry in their encyclopedic approach to knowledge. Notably, both extended their remit to include geography, Keckermann writing a small treatise on geography and Alsted including it in his encyclopedic projects.[39]

The importance of the systematics (and, to a lesser extent, Ramus) to Oxonian geography and its drive to method is made explicit at a number of points. Carpenter displayed a thoroughly Ramist hostility to Aristotelian natural science in his *Geography,* but like Ramus himself remained "totally imprisoned by scholastic concepts and terminology."[40] Carpenter cited Keckermann at one point, but his influence is more apparent in Heylyn's *Microcosmus,* the preliminaries to which were constructed with frequent marginal references to Keckermann's 1612 geographical treatise, which clearly provided Heylyn's method regardless of his aforementioned claim to have devised it for himself.[41] Heylyn's textual format of marginal references strongly suggests that he was ordering the fruits of his reading and commonplacing under continental and national headings, the result being a "systematic" form of geography book. Beyond these explicit references, it stretches credibility not to see the structures of Abbot, Pemble, and Stafforde as deriving from the same sources, Keckermann and Alsted's manuals being highly popular in early-seventeenth-century Oxford.[42] Oxonian geographers seized upon the method advocated by the systematics as a way to organize commonplace material under a structure of spatial headings, which then amounted to a genre called geography.

Two points of great salience to our project connecting geography and the English Revolution emerge from this analysis. First, the context from which the pedagogic projects of Ramus and the systematics emerged was the so-called Second Reformation, in which European Protestants sought to come to terms with their embattled status in the Wars of Religion. In other words, just as historians of the English Revolution have sought to recover the affinities between that event and its European context, so we can see that the generic revolution in English geography stems from the context of European confessional conflicts. But where accepted wisdom sees the transference of German encyclopedism to England as a Puritan, or at least Calvinist, phenomenon, in geography its methodizing habits of thought were taken up across theological divides by Calvinists such as Abbot and Carpenter, but also and indeed most transparently by the trenchantly Arminian Peter Heylyn. This reinforces a suggestion ventured by Howard Hotson: because the systematics argued against Calvinism that we could come to know God through our philosophic efforts, theirs was an attitude congenial to Arminians.[43]

The second point is that this generic revolution allowed for the participation of geographical texts in politico-theological debates. One of the main preoccupations of methodized encyclopedic knowledge was to develop a map of the disciplines. In such maps, geography's status as the "eye" of history was reinforced, and its place in an arts curriculum designed to create politically astute actors was solidified. Furthermore, if geography's disciplinary location in the arts curriculum politicized it at a general level, the generic revolution created a methodized geography in which the convention was to include information concerning the politics, religion, and history of each country. In other words, the systematic geography books that emerged in the early seventeenth century had clear sites in which to engage with potentially divisive political issues. How this played itself out in the English Revolution is the subject of the next section.

THE ENGLISH REVOLUTION IN GEOGRAPHY, 1600–1660

Contention and Context

The era of the English Revolution saw a series of ideological conflicts, which were played out in systematic geography books produced in Oxford. The key to those conflicts was a clash between two rival Protestant theologies, Calvinist and Arminian, as Hobbes argued in *Behemoth*. This debate was explicitly aired in geography books and drove differing geopolitical representations of European political and religious establishments. Before we can make good this series of contentions, two sets of contexts need to be established. First, that of the theological clash between Calvinism and Arminianism, and second, the Oxford geographers involved in this clash.

There has been an enormous shift, indeed a reversal, in the accepted view of the theological dynamics that precipitated the breakdown of English government in the mid-seventeenth century. In Christopher Hill and others' rubric of the English Revolution, a staid establishment centered on Charles's trusted archbishop of Canterbury, William Laud, was depicted as being undermined by the incipient radicalism of Puritan doctrine, so that the English Revolution was in good part a Puritan Revolution. This view was carried forward to explain the culture of science within the universities. Yet Tyacke's previously noted attack on this conception of university scientific culture is part of a broader revision of our understanding of religious culture.[44] Taking a perspective from 1600, mainstream churchmanship was Calvinist: "'Orthodox' meant Calvinist."[45] This statement applies as late as 1620, but in the following decade a group of churchmen—the Arminians—began to attack Calvinist doctrine. On the revisionist view, it is Laud and the Arminians who, anything but staid, led a revolution in ideas that was a major trigger to the traumatic breakdown in English governance.

The debate between Calvinism and Arminianism convulsed European Protestant thought as a whole, leading to the reproof of Arminian doctrines at the Synod of Dort in 1618. In England it was at just this time that the doctrine started to gain adherents, the process centering on Oxford. This theological war of ideas had a major impact on the ways in which the first generations of methodical geography books were constructed as politico-theological statements. The contending theological parties in Oxford saw a number of their key members write geography books in the period between Abbot's Briefe Description in 1599 and Heylyn's Cosmographie in 1652. On the Calvinist side, George Abbot was the main defender of a pan-European Calvinism as archbishop of Canterbury, his Briefe Description, a work of his youth, demonstrating a Calvinist approach.[46] Similarly, Exeter College under Prideaux produced a number of geography books that articulated a Calvinist viewpoint, notably Carpenter's and Stafforde's.[47] William Pemble, at Magdalen Hall—the "nest of Puritans"[48]—was similarly a Calvinist theologian as well as the author of a mathematical geography. The Arminian side in Oxford geography is principally represented by one figure, Peter Heylyn of Hart Hall, whose editions of Microcosmus in 1621 and 1625 pioneered an

Arminian geography, which his *Cosmographie,* written in the heart of the Interregnum, confirmed even in the midst of its defeat. Yet Heylyn was not a lone voice in geography. Robert Johnson's embellished translations of Giovanni Botero's geographical surveys were Arminian in their content, and in the 1630s as Laudianism became the official position of the English church, geography books as of standard incorporated positions that would have shocked the Calvinist mainstream a generation before.

Contending Orthodoxies

Geography books were one site in which differing perspectives, Calvinist and Arminian, about the nature of a true English Protestant church were presented. Calvinist geographers developed a position consonant with their ecclesiological views more generally. The first edition of Abbot's *Briefe Description* was too brief to go into any detail, but the extended 1605 edition did adopt a position. In material newly added to the English section, Abbot argued that the British had converted to Christianity by the time of Tertullian (ca. 200 AD) and that the mission sent by Pope Gregory the Great under Augustine was not, therefore, pivotal. In other words, British Christianity was primitive not papal.[49] It was only under the Saxons that Romish corruptions entered the British church.[50] Abbot added a strong attack on monasteries before the Reformation for "giving themselves to much filthyness, and dyvers sorts of uncleannesse."[51]

Twenty years after Abbot, Carpenter's *Geography* also adopted a Calvinist position. There is no doubt of Carpenter's personal Calvinism: a number of his sermons show his hostility to Arminianism, which he lumped together with Roman Catholicism.[52] Yet Carpenter's geographical work has always been seen in terms of its mathematical and organizational rigor by historians of geography, such that its theological interventions have been ignored. The first book was dedicated to the Earl of Pembroke for his "Zeale to *Religion,*" and it was Pembroke who, as chancellor of Oxford University, "delayed" the "full impact of the Arminian revolution on Oxford."[53] In terms of the "politics of scholarship," Carpenter is critical of the trustworthiness of Jesuits, who in science are "a combined faction of their owne society, unwilling to contradict" each other.[54] He also criticized humanist textual approaches to scientific questions, such as those adopted by Joseph Scaliger (whom Heylyn cited admiringly in *Microcosmus*) for their theological resonances:

the *Crickes* . . . of our Age, who like *Popes* or *Dictatours,* have taken upon them an Universal authority to censure that which they never understood. . . . To

seeke for a determination of a *Cosmographicall* doubt in the Grammaticall resolution of two or three Hebrew wordes . . . were to neglect the kernall, and make
a banquet on the shells.[55]

In short, if Carpenter is mathematical or scientific in his geography to the
modern eye, there is a clear politics to his method, distancing him from the
authoritarianism of modes of scholarship that he associated with "Popery."
 Yet Carpenter's *Geography* is more explicit in its Calvinism, for he also
makes direct theological comments. Contrary to the tide of Arminianism, already on the rise in Oxford as he wrote, Carpenter opines that *"externall rites*
and *Ecclesiasticall discipline* . . . by wise men have bin esteemed no other,
then matters *indifferent* which may admit of *change."*[56] More striking, however, is Carpenter's geographical refutation of Sabbatarianism. As an orthodox
Calvinist, Carpenter was incensed by the Arminian equation of Calvinism
and Puritanism:

[H]ow much the odious name of a *Puritane* hath abused many a sincere Christian . . . [such] that a *Protestant* must make a hard shift, either by *Popery* or
Arminianism to save himself.[57]

In this context, he wanted to distance himself from extreme Protestantism,
including Sabbatarianism, the doctrine of rigorous observance of the Sabbath.
Here, his geographical description of time zones allowed him to support orthodox Calvinism. Carpenter asks us to imagine a Christian traveling around
the globe westward and losing a day, an Islamic worshiper traveling eastward,
gaining a day, and a Jew remaining at home (on the broader "revolutions in
the times," see Glennie and Thrift in chapter 7 above). The result would be

the *Sarazen* according to the Law of *Mahomet,* shall observe his Friday, the *Jew*
his Saturday, being his Sabboth; and the *Christian* the Lords day, being the Sun-
day; yet so, as all shall happen on the same day; all of them excluding any errour
in their calculation.

From this Carpenter drew his moral:

Methinkes this . . . were a reason sufficient to convince some strait-laced men,
who rigidly contend our Lords day (which they erroniously tearme the Sabboth)
to be meerely morall. . . . If it were so, according to our Premises before demonstrated, this absurditie would ensure necessarily: That the Morall Law, which

they call also in a sort the Law of nature, is subject to manifold mutation, which by our best Diuines is utterly denied.

Carpenter concluded that the observance of the Lord's Day was connected to the "*Ecclesiasticall* constitution" not moral law, such that each "should celebrate the . . . Lords day according to the institution of their owne Church."[58]

Turning to Arminian geography books, Heylyn's *Microcosmus* agreed with those just canvassed concerning the primitive origins of English Christianity. Yet even in this agreement, there was a subtle difference, Heylyn refusing to dismiss the role of Augustine as "a fabulous vanity" cooked up by Rome: "[T]o say that *Austin* first preached the Gospell here, . . . is not to be understood absolutely, that he first preached it; but that he first preached it to the *Saxons.*"[59] As we have already seen, this is just the sort of grammatical sleight of hand that Carpenter associated with popery. On monasteries, Heylyn was far more qualified in his condemnation than Abbot had been, lamenting as an Arminian believer in the beauty of holiness the dissolution as "spoyling the Church ornaments [which] were most exquisite."[60] Most important, Heylyn made the English Reformation a via media between Calvinism and popery, thereby implicitly labeling Calvinism extremist: "[T]he English bearing respect neither to *Luther, Zwinglius,* nor *Calvin;* but abolishing such things as were dissonant to Gods word retained such ceremonies as without offence the liberty of the Church might establish." He concluded that "had the reformed part continued an allowed correspondency in some circumstances with the *Romish* Church, as the Church of England doth now: it had beene farre better."[61]

Other geography books developed this Arminian conception of the English Reformation. Notable here was Johnson's translation of Botero's *Relations of the Most Famous Kingdomes and Common-wealths thorowout the World,* which has aptly been described as a Counter-Reformation political cosmography.[62] The first edition of Johnson's translation was published in 1608, and marginal notes show that Heylyn drew on this geopolitical survey in constructing *Microcosmus.* The expanded editions of 1616 and especially 1630 show Johnson's already rather free translation moving in an Arminian direction. The preface to the 1630 edition pointed out how much "all the Writers of Geographie" since Botero's time had cribbed from him, notably Petrus Bertius, who was lambasted as "that Turncoat Apostatazing Plagiarie, that Enemie and Threatner of our *English* Nation."[63] Bertius was "Arminius's leading Dutch disciple" and had written a number of works of mathematical geography before moving to Paris and converting to Catholicism.[64] Bertius's conversion was potentially awkward to Arminians, given the Calvinist equation of

Arminian doctrine with popery. Johnson's Botero adopted a strongly Arminian line in this outspoken criticism, a move in line with the dominance Arminianism had established by 1630. The text was in accord with these prefatory comments, the depiction of the English church as a via media and Calvinism as an extreme being all but identical with Heylyn's.[65] Lest the point was not clear enough, it was reinforced in Botero's section on the Netherlands, resolving back from a description of Dutch Calvinism to England: "[B]ut our men at home (zealous ones of the *Geneva* discipline) are much deceived if they looke for such a face of a Church, such decent Service of God, such devotion, or strict observation of the Lords day, in any of the *Calvinist* Churches, as in the Church of *England.*"[66] This material was new to the 1630 edition, which strengthens my point that Johnson's Botero, sailing with the wind, became progressively more Arminian in successive editions. It should be added that by 1630 the translation was no longer Johnson's alone, for the simple reason that he died in 1625. The text does not make it clear who updated the translation, but it is notable that it was printed by John Haviland. Haviland printed a range of authors, including the most celebrated Arminian controversialist of the 1620s, Richard Montagu.[67] It is of a piece, then, that Haviland printed a decidedly Arminian reworking of Botero's *Relations* in 1630. The Jesuit's political cosmography, read aright, could be reworked to speak in an Arminian language.

Arminian geography books equated Calvinism with insurrection. Heylyn in *Microcosmus* contrasted the English Reformation, taken on "with mature deliberation" with the actions of Luther and Calvin, "receaued tumultuously."[68] The equation of Calvinism with tumult became, unsurprisingly, a more insistent feature in geography books written after the Bishops' Wars of 1639–40, wherein Scottish Calvinism had been defended against the interventions of Charles I and Laud. Ephraim Pagitt's geographical survey of Christianity, the *Christianography,* in its third, 1640 edition, dealt with the events of the Bishops' War head on, attacking the Scottish church:

> But why do I terme this their doing a Reformation? is this a Reformation, in which the subjects are armed, against their most gracious and religious Sovereigne . . . who sincerely professeth, and maintaineth the blessedly reformed Religion.[69]

Finally, Heylyn, who wrote both the first great Arminian geography, *Microcosmus,* and the last, *Cosmographie,* thought his fears about Calvinism had come true by the time he wrote the latter in the depths of the Protectorate. He argued in *Cosmographie* that the gentry had sent their children to Geneva to be educated, with fatal consequences:

> the *frie* or *seminary* of our Gentry being seasoned in their youth with Genevan Principles; have many times proved disaffected to the forms of Government (as well *Monarchicall* as *Episcopall*) which they found established here at home: to the great imbroilment of the State, in matters of most near concernment.[70]

CONTENDING MAPS OF CONFESSIONAL ALLEGIANCE

We have seen how Arminian geographers' discussions of the English church spilled out into non-British sections, with Johnson's Botero, Pagitt, and Heylyn all reflecting back on the English situation from their discussions of other Calvinist nations. More generally, "it is in English Protestant divines' perceptions of . . . foreign churches—Roman and Reformed—that their different images of the nature of the English church come into clearer perspective."[71] The same point can be made the other way around: different images of the English church—Calvinist and Arminian—fed into discernibly different patternings of sympathy, allegiance, and fear when treating European nations in the geography books of the English Revolution.

Calvinists supported a pan-European alliance of Protestant nations in the face of resurgent Catholicism, and Calvinist geographies reflected this. In fact, the English only adopted such a Protestant policy from 1609 to 1616,[72] and a key supporter was George Abbot. Abbot's "pathological" fear of popery is amply represented in his geographical work, the *Briefe Description*.[73] It is noticeable, for example, that between the 1605 and 1620 editions of this work, Abbot saw fit to add a new section expressing his fear of Spanish ambitions to achieve a universal monarchy that would subject all to the pope.[74] Between 1605 and 1620, Abbot fell out of favor at court, in good part for his obsession with a confessional foreign policy, which did not accord with James I's plans.[75] It could only have added to Abbot's marginalization that he expressed such views at a time when James was trying to negotiate a marriage between his son Charles and the Spanish Infanta. Abbot also, in a formula that was common currency to Calvinists but that came under increasing attack from Arminians, saw the pope as the antichrist and Rome as prefigured in the book of Revelation: "Who so looketh on the description layde downe by the Holy ghost in the *Revelation,* shall see that the *Whore of Babylon* there mentioned, can be understood of no place, but the *Cittie of Rome.*"[76]

The other, positive, side of the Calvinist confessional map of Europe was its support for those nations that embraced Protestantism. Further, Calvinists supported all Christian churches that lay outside the ambit of Rome, suggesting an affinity in opposition to the pope. In this connection, Abbot supported the Greek Orthodox Church, and his *Briefe Description* made the same point about the Coptic Church of Abyssinia: "[T]he people therefore are Christians, as is also their prince: but differing in many things from the West Church: and in no sorte acknowledging any supreame prerogative of the B. of Rome."[77] A number of highly geographical treatises surveyed the faiths of the world or, more specifically, the spatial distribution of the Christian faith, like Pagitt's previously cited *Christianography* (1640) and Edwin Sandys's *Europae Speculum* (1629), with similar ambitions to trace opponents of papal supremacy.

As a footnote to this Calvinist confessional map, Tomasso Campanella's millenarian geopolitical survey, *De Monarchia Hispania,* was translated into English in 1650 and 1659. This work, written around 1600, surveyed the nations of the world, looking to their governmental structure and suggesting how Spain under Philip II could pursue a policy of universal monarchy. It was translated, in the words of the title of the 1659 edition, "for awakening the English to prevent the approaching ruine of their Nation."[78] A preface by William Prynne advised the reader to look to Campanella's chapters on England, which argued that England could be ruined by "intestine wars between *England, Scotland, Ireland* and the *Netherlands.*" Prynne pointed to "wars with the *Scots* and *Hollanders* . . . subverting our ancient *Kingly Government* to metamorphose us into a *Commonwealth.*" Prynne had been a staunch but independent Parliamentarian under Charles and a violent critic of Arminianism, but he attacked Parliament from 1640 on. As a consequence, his view of the Civil War during the Interregnum period was very different from Heylyn's. Where Heylyn, as we have seen, saw Calvinist principles as the heart of the trouble, for Prynne it was precisely schism from England's Calvinist neighbors, the Netherlands and Scotland, that had been the problem. The confessional map of allegiances that Abbot had advocated as England's salvation, then, was still being invoked in geographical contexts some sixty years later.

Turning to Arminian geography books, a very different map of confessional allegiances emerges. Arminians were reluctant to cast Rome as the antichrist and shifted away from advocacy of a confessional foreign policy.

Thus Heylyn's *Microcosmus,* penned in 1621, referred in its dedication to Charles, Prince of Wales (the future Charles I), to his "serious negotiations," meaning the protracted efforts to secure his marriage to the Spanish Infanta.[79] Indeed, it is no exaggeration to see *Microcosmus* as a defense of the Spanish match (to which, as we have seen, Calvinists like Abbot were opposed). In this way the work functioned, true to the emergent Arminian position, by down-playing the threat posed by the Catholic nations of France and Spain. Heylyn did retain the traditional Calvinist equation of the pope as antichrist and feared Spanish universal empire in ways that make *Microcosmus* a liminal document in his oeuvre.[80] However, he also noted Spanish qualities in church and state in a way Calvinists did not countenance: "[I]n offices of pietie very devout, to their King very obedient, and of their civill duties to their betters not unmindfull."[81] More remarkably, Heylyn did not simply paint the Spanish Inquisition as a force for evil, as it had been founded to enforce Christian conformity on the Moorish converts, which "custome in it selfe was wondrous tollerable and laudable."[82] Similarly, although Heylyn retained the equation of the pope with the antichrist, he was prepared to admit that the popes up to Nicholas III had been true to Christian doctrine.[83]

If Heylyn toned down the critique of Roman Catholics, he also offered criticism of the European Reformed tradition, the result being, as in Arminianism more generally, that popery and Puritanism were figured as equal and opposite dangers: "I have heard a worthy Gentleman . . . say; till the Jesuites were taken from the Church of Rome, and the peevish Puritan Preachers out of the Churches of Great Brittaine, hee thought there would never be any peace in Christendome."[84] Indeed, in one case Heylyn could even (implicitly) tip the balance against the Reformed churches, for while he praised the Spanish for devoutness and obedience, he exempted Biscay:

> They admit no Bishops to come amongst them, and when *Fernand* the Catholique came in progresse hither, accompanied amongst others, by the Bishop of *Pampelune*, the people arose in armes, drove back the Bishop, and gathering all the dust on which they thought he had trodden, flunge it into the Sea.

A marginal note to this anecdote added that Biscay was "[a] good place for Puritanes to dwell in."[85] For Heylyn, Puritans and by extension Calvinists threatened stability in church and state and could be more problematic than Catholics.

The Arminian map of confessional sympathies and fears envisaged a catholic, that is, universal, church comprising several independent and equal branches, Rome being one of them. For this reason, the Arminian view was positive about the French Catholic Church, which had retained more ecclesiological independence from Rome than other churches.[86] Pagitt embodied this sanguine attitude in his survey of Christian churches: condemning (as cited earlier) the Scottish Calvinists for arming against Charles I, he went on "in these moderne popular Reformations, I doe commend the Church

Gallican, who neither have dispoiled their Bishops, nor robbed the Church,"[87] a commendation that implicitly condemned the Calvinist churches for doing just that.

Heylyn's *Cosmographie* followed the same line on France. Heylyn attacked the fanatical rigors demanded by French Protestants and their attempt to create a power base separate from the king:

> Being grown too insolent by reason of so great a strength, and standing upon terms with the King as a *Free Estate* (the *Common-wealth of Rochell,* as King *Henry* the fourth was used to call it) they drew upon themselves the jealousie and fury of King *Lewis* the thirteenth.

Writing under the English Commonwealth, Heylyn seems once more to intimate that Calvinism is the root of the problem of insubordination toward monarchs. By contrast, he praises the French Catholic Church for standing "stoutly to their naturall rights, against the usurpations and encroachments of the see of *Rome.*"[88] A similar picture of more than ambivalence toward a confessional map of allegiances emerges in *Cosmographie's* treatment of the Low Countries, divided into the United Provinces and those areas remaining under Spanish control. Heylyn refuses to support the former merely for their Protestantism, arguing that the Spanish are "the true Proprietary of the whole," and adding that the citizens of the United Provinces have hardly benefited from their change of government instead having "drawn upon themselves more arbitrary and illegal payments, than any Nation in the World."[89] Again, parallels with the English Interregnum government seem close to the surface, especially in his conclusion on the Low Countries:

> [T]here is nothing wanting to these Countries, wherewith the God of all blessings doth enrich a Nation, but a *gracious Prince, unitie of Religion,* and *a quiet Government:* which if it pleased the Almighty to confer upon them, they would surpass all neighbouring States in treasure, potency, content, and all worldly happiness.[90]

In summary, there were two clearly distinguishable ways in which the newly methodized geography book was politicized in the half century after its inception. These two modes were primarily theological in their grounding, being Calvinist and Arminian. Differences between the two were manifest both in their depictions of the English church and state and in their networks of sympathies and hostilities across Europe. Calvinist geographies stressed the corruptions of the English church wrought by Roman Catholicism and the continuing depravity of Rome, Spain, and, to a lesser extent, France. They supported a web of confessional alliances among the Calvinist nations of Europe and beyond to Christian countries outside Rome's ambit. By contrast, Arminian geographies emphasized the continuities between the English Reformed church and the Roman church, criticizing in contradistinction the tumultuous Calvinist reformations that, thanks to her Scottish neighbors, created a model of sedition that had cost England dear. Spain was seen as less of a threat than it had been before the Armada, and France's established church was praised quite as much if not more than its Reformed church. From a Calvinist perspective, Arminian geography was dangerously popish, and this division of Britain from her natural confessional allies had wrought catastrophe. To Arminians, Calvinism was a pattern for sedition, the importation of whose ideas from Geneva, the Netherlands, and Scotland had precipitated crisis. The evidence of English geography books bespeaks a war of religious ideas fomented over a long time span.

"Directed to Farther Action": Geography in English Revolutionary Debates, 1640–1660

If geography in its newly methodized form was engaged in political and theological debates, lines of force also worked in the opposite direction, geographical information and imagery being deployed in political debates as England descended into civil war. The reason for this reverse influence is that throughout the early modern period, reading was closely linked to political action. Anthony Grafton and Lisa Jardine have uncovered how Gabriel Harvey read Livy's histories in varying ways to understand and justify political action: he "studied for action."[91] Harvey was not the exception but the rule, following conventional wisdom about the aims of the university arts curriculum. Nathanael Carpenter identified the arts as "not contented with a bare knowledge or speculation, but . . . directed to some farther work or action," but opined that geography was outside this rationale, being a science.[92] Against this, two points must be made. First, as we have seen, Carpenter, contrary to his rhetoric here, did in fact make political points in his *Geography* such that it could be studied for action. Second, most who read Carpenter would have been partaking of the arts curriculum or more generally studying with practical ends in mind.

Historians of the book have shown the longevity of this approach and that it applied to geography. Grafton and Jardine noted that Sir Philip Sidney's advice on how to read for political advice subsumed geography as the eye of history: "For historicall maters, I woold wish you before you began reed a little of Sacroboscus Sphaere, & the Geography of some moderne writer."[93] It is worth noting here with reference to Carpenter's claims that mathematical geography is as much politicized as special geography (in this case through that university standard for the subject, Sacrobosco). Beyond this, Sir William Drake's reading throughout the course of the English Revolution shows him using geographical material to understand events. Drake, like many of his contemporaries, read by noting down salient points in a commonplace book under themed headings.[94] Commonplace books, as we have seen, were vital to the construction of a methodical form of geography, and once constructed, of course, these books themselves became available for later generations to mine for useful political information.

Geography books, then, were used in the creation of commonplace books, on which scholars seeking to defend and debate political action drew, this being a feature of early modern intellectual life spanning a far longer period than the English Revolution. A full study of this process is beyond my remit: I seek, instead of showing the process in action by a study of commonplace books, to give a brief glimpse of the end product, namely, the ways in which geographical information and imagery were deployed in debates surrounding the English Revolution.

As a starting point, it is worth noting that all our Oxonian "geographers" were also important discussants of political and theological matters. Perhaps unsurprisingly, there is evidence of them drawing on their authority as geographers to make polemical points in pamphlet debates. This is most apparent in a tract written by George Abbot. Abbot attacked a Catholic tract on the grounds that it falsely sought "to extend the territories of the Pope" and thus to claim that the Roman Catholic Church was "Catholic," that is, universal, in a geographical sense. He rebutted the claim that Roman Catholicism had spread to the farthest western regions of the Old World: "[W]here by the way you faile a litle in your Geography, as wel as in your Divinity: for . . . in Africa, the partes about Morocco doe without controversie exceed them all [in westward longitude]" yet are not under Rome's sway. Abbot reverted to this argument—his opponent's "Geography is just as sound as your Divinity"—at a later point, to refute the claim that Roman Catholicism spread over the whole world:

> Untill that of late the Portingales attempting to goe to Calicut, found the Cape of *Buona speranza* . . . all of which was but a little before the going out of Columbus; what was there within the whole compasse of Africa, which knew ought of the Romishe doctrine? . . . And as you spedde in Africa, so did you in Asia,

the whole compasse of that huge region taking no notice of your Pope & of his Idolatry.[95]

Abbot, then, used his authority as a geographer to rebut his opponent: geography could be studied for anti-Catholic action. In this Abbot was by no means alone. On the contrary, his argumentation was closely akin to that of Paggit's *Christianography* and Sandys's *Europae Speculum,* both of which also surveyed the globe to scrutinize the geographical meaning of claims to Catholicity. The Catholicity question was key to the debate between Rome and the Reformed churches and by definition could be won and lost in the geographical arena.

More broadly, the pamphlet debates that surrounded the English Revolution drew on geography for two types of material. First, specific geographical information was used to defend or rebut an argument. Second, generalized geographical imagery was deployed: widely purveyed and accepted commonplaces about different nations were key currency in debates about relations between the people, Parliament, and the king.

Looking at the debates that led to Charles raising his standard at Edgehill in 1642, Parliament's defenders saw Charles as having followed models of absolutism derived from Continental, particularly Roman Catholic, nations, and expressed this by deploying geographical imagery. Thus Henry Parker saw the levying of ship money as negating the English constitution and used a range of geographical images to make his point:

> [T]his invention of ship-money makes us as servile as the Turkes. . . . If we shall examine why the Mohametan slaves are more miserably treated, than the Germans, or why the French Pesants are so beggarly, wretched, and bestially used more than the Hollanders, or why the people of Milan, Naples, Sicily are more oppressed, trampled upon, and inthralled than the Natives of Spaine? there is no other reason will appeare but that they are subject to more immoderate power, and have lesse benefit of law to releeve them.[96]

Parker suggests that the non-Christian and the non-Protestant tend toward tyranny owing to their disregard for the law. It was the sort of message any conscientious reader of geography books could have constructed. For Charles I, the whole tenor of such criticisms was to undermine the constitution and remodel it on foreign premises incompatible with Englishness: "We are resolved not to quit . . . the ancient, equall, happy, well-posed, and never-enough commended Constitution of the Government of this Kingdom, nor to make

Ourself of a King of *England* a Duke of *Venice,* and this of a Kingdom a Republique." [97]

Where for Parker the king had moved England away from her legal foundations toward Turkish tyranny, on Charles's reading Parliamentary interference tipped matters in the other direction, toward a Venetian republic. Clearly, as constitutional positions became polarized, so competing geographical representations could be drawn upon to suggest that one's opponents were outlandish and un-English. These representations were repeated on both sides throughout the course of events. Charles could thus be depicted by Charles Herle as making English laws "the same with those of *France* [and] *Turkie,*" and asked "[H]ow else were the *Monarchy*" on his conception "*mixt* more than that of *Turkie?*" [98] On the other side, the tract by Henry Ferne that aroused Herle's ire had argued Parliament sought to make "the Crown . . . altogether conditionall, as in the meerly elective kingdoms of Polonia, Swedeland, &c." [99] Poland, the Netherlands, and Venice were all depicted as republics with a welter of seditious sects by Royalists in ways closely related to geography books. Thus Heylyn in *Microcosmus* and *Cosmographie* depicts Poland as a "Babel of Religions" and links Poland to the Netherlands in this regard. [100]

This process of debating through stock geographical images continued in the run up to and aftermath of Charles's execution. As English political thinkers sought to fathom uncharted political waters, they looked across time and space for buoys to guide them, and this by definition meant looking to history and geography. A pamphlet published on the eve of Charles's execution justified the people's right to try their king by a range of geographical and historical examples, drawing particularly on French and Scottish evidence. [101] Importantly, Parliament's own justification for Charles's execution drew on a range of historical-geographical examples:

> Parliament received encouragement, by their observation of the *Blessing* of *God* upon other States; The *Romans,* after their *Regifugium* of many hundred years together, prospered far more than under any of their *Kings* or *Emperors.* The State of *Venice* hath flourished for One thousand three hundred years. How much do the Commons in *Switzerland,* and other *Free States,* exceed those who are not so, in *Riches, Freedom, Peace,* and all *Happiness?* Our *Neighbors* in the *United-Provinces,* since their change of Government, have wonderfully increased in *Wealth, Freedom, Trade,* and *Strength.* [102]

Clearly, rather than being a Royalist bogey, the Venetian example could now be embraced positively by Parliament.

John Milton was one of the most distinguished political writers in this period, and in him we find a thinker who persistently drew on catalogs of geographical examples in making his points. Milton justified cashiering a king on the grounds of Parliament's supreme authority, "appealing to the known constitutions of both the latest Christian Empires in Europe, the Greek and the German, besides the French, Italian, Arragonian, English and not least the Scottish Histories."[103] In Milton's case, it is easy to trace much of this catalog back to his commonplace book, which shows him collecting information about the present state of Europe from De Thou's *Historia Sui Temporis* (1620), but also gathering more exotic examples concerning China and Africa from Samuel Purchas.[104] Even more impressive was the geographical catalog Milton assembled to refute Salmasius in his *Defence of the People of England.* In proving his point that "it is very particularly in accordance with nature that tyrants should be punished in some way . . . all nations have over and over again done this," Milton ranges for supporting evidence from Europe to Egypt, from Egypt to Ethiopia and then back to Gaul. Clearly, Milton's commonplaces had been studied for action, and arranged by him in a geographical framework.[105]

On the eve of the Restoration, Milton was still deploying the same geographical forms of argumentation. Under two months before Charles II was invited back to England, Milton warned the people not to be seduced by the examples of Venice and the United Provinces into creating even a constitutional monarchy:

> [T]his facilitie we shall have above our next neighbouring Commonwealth, (if we can keep us from the fond conceit of something like a duke of *Venice,* put lately into many men's heads, by some one or other subtly driving on under that prettie notion his own ambitious ends to a crown) that our liberty shall not be hampered or hovered over by any ingagement to such a potent family as the house of *Nassaw* . . . but we shall live the cleerest and absolutest free nation in the world.[106]

Just as it had for Charles I, then, Venice formed a feared geographical example, but the image was invoked for very different reasons.

RESTORATION AFTERMATHS: GEOGRAPHICAL REVOLUTIONS ACCOMPLISHED AND EFFACED

English geographical culture in the period 1600–60 intersected with two revolutions. The first was a generic revolution that saw a postmedieval or early modern textual format developed for geography books. This revolution centered on late humanist—specifically Ramist and systematic—attempts to create methods by which copious knowledge could be organized, using binary or branching categories of general and special heads. In a culture of commonplacing, geography was created by collecting materials under spatial headings and was then presented in a form dictated by the notion of method. This generic revolution demanded that geography books contain material on the political and theological state of nations of the world and it was this dictate that facilitated geography's participation in the war of ideas surrounding the English Revolution. This change seems to have some claim to the title of a "revolution," in that it marked a clear break with long-established practice, occurred in a compressed space of time (ca. 1600–25), and was to dominate geographical writing for a very long time to come. The second—the window that geography books provide on the English Revolution—suggests that it was a war of ideas in which theological differences between Calvinist and Arminian doctrines were a fuel for conflict, this conflict being pictured on a global, but more particularly on a European, stage. Further, the culture of commonplacing ensured that geographical information was mobilized in the pamphlet exchanges through which varied conceptions of the causes and consequences of the English Revolution were debated.

Yet what happened at the Restoration? Simply put, both the geographical revolutions I have depicted were simultaneously effaced and accomplished. The generic revolution was, it might seem, simply accomplished, in that the basic structures of methodized geography remained intact and would guide the writing of geography books until the nineteenth century. Yet the generic revolution's ideological origins were effaced. The routes by which (largely) Calvinist pedagogic theory forged a new type of geography book at the breakdown of the medieval chronicle have passed all but unnoticed. Its origins in Protestant responses to confessional conflict were rapidly lost as the genre spread to all nations and denominations. The contention that geography's generic revolution was accomplished and effaced is supported, respectively, by Heffernan's and Livingstone's studies of the French and American Revolutions (chapters 10 and 11 below). Both Edme Mentelle and Jedidiah Morse clearly deployed trees of knowledge and forms of textual organization in their geography books that derived from the context outlined in this chapter. Indeed, the continuity in this regard is remarkable, but neither shows any awareness of where their geographical formats came from.

A reverse pattern might be depicted for the second revolution we have discussed. At the Restoration, as Collinson and Tyacke in particular have noted, the sheer centrality of Calvinism in early Stuart England was rendered invisible. One result was that Calvinism ceased to be considered as English in

English-language geography books. Only with the influence of the Scottish Enlightenment did the perception of Calvinism as anything but an "other" to the Anglican "self" reemerge as a possibility in English geography books.[107] At a simple level, then, Calvinism and its formative influence in early modern geography was effaced. Yet, at another level, the revolution politicizing geography was accomplished: the culture of politico-theological debate in geography books that had been forged in the English Revolution continued to exist under the later Stuarts and long beyond. The arguments and language changed, but the locus of ideological contestation remained.

On the eve of the Restoration, the otherwise unnamed "R.P." penned *A Geographicall Description of the World,* which spoke of "the *English* . . . now having changed the name of Kingdom into a Protectorship."[108] In the copy in Cambridge University Library, this clause is inked out, presumably a Restoration nicety. One does not have to be too steeped in Derridean ideas to accept that this erasure not only conceals but also emphasizes both the change in government and what went before. If the Restoration did in many ways ink out geography's English revolutions, generic and political, it also could not but acknowledge them whenever a geographer put pen to paper.

...

Notes

For comments and information that have greatly improved this chapter, I wish to thank David Armitage, Howard Hotson, Anthony Milton, Jonathan Scott, and Quentin Skinner. My thanks for further stimulating comments also go to the editors of this volume and to participants at the conference on which this book is based.

1. Where once there was a debate between proponents of the terms "English Civil War" and "English Revolution," occurring down largely party-political lines, today there are numerous competing representations of the events of 1640–60. In this context the nonaligned can use the terminology with a certain irenicism. For helpful introductions to the historiography of the English Revolution, see Alastair MacLachlan, *The Rise and Fall of Revolutionary England: An Essay on the Fabrication of Seventeenth-Century England* (London: Macmillan, 1996); and Ralph C. Richardson, *The Debate on the English Revolution,* 3rd ed. (Manchester: Manchester University Press, 1998).

2. Thomas Hobbes, *Behemoth; or the Long Parliament,* ed. Ferdinand Tönnies (Chicago: University of Chicago Press, 1990), 3–4.

3. Ibid., 33.

4. Ibid., 22–23, 122–23.

5. See Ann Hughes, "King, Parliament and the Localities during the English Civil War," *Journal of British Studies* 24 (1985): 236–63; Ralph C. Richardson, ed., *Town and Country in the English Civil War* (Manchester: Manchester University Press, 1992); and John Morrill, *Revolt in*

the *Provinces: The People of England and the Tragedies of War, 1630–1648,* 2nd ed. (London: Longmans, 1999).

6. Clive Holmes, "The County Community in Stuart Historiography," *Journal of British Studies* 19 (1980): 54–73; and Richard Cust, "News and Politics in Early-Seventeenth Century England," *Past and Present* 112 (1986): 60–90.

7. David Underdown, *Revel, Riot and Rebellion: Popular Politics and Culture in England, 1603–60* (Oxford: Oxford University Press, 1985); John Morrill, "The Ecology of Allegiance in the English Civil War," *Journal of British Studies* 26 (1987): 451–67; and Ann Hughes, "Local History and the Origins of the Civil War," in *Conflict in Early Stuart England: Studies in Religion and Politics, 1602–1642,* ed. Richard Cust and Ann Hughes (London: Longmans, 1989), 224–53.

8. Conrad Russell, *The Causes of the English Civil War* (Oxford: Clarendon, 1990), 217–18.

9. Derek Hirst, "The English Republic and the Meaning of Britain," in *The British Problem, c. 1534–1707: State Formation in the Atlantic Archipelago,* ed. Brendan Bradshaw and John Morrill (London: Macmillan, 1996), 192–219.

10. Jonathan Scott, *England's Troubles: Seventeenth-Century English Political Instability in European Context* (Cambridge: Cambridge University Press, 2000).

11. At a simple level, John Speed's *Theatre of the Empire of Great Britain* (1627) includes maps of England, Wales, Scotland, and Ireland, as well as a map of the four together. On the other hand, Camden's *Britannia* (1586), written before the Union of Crowns in 1603, does not include Scotland (although the 1607 edition does). Yet both projects clearly can draw on the conceptual category of "Britain," however much their usages might vary.

12. J. H. Elliott, "A Europe of Composite Monarchies," *Past and Present* 137 (1992): 48–71; and J. H. Elliott, "England and Europe: A Common Malady?" in *The Origins of the English Civil War,* ed. Conrad Russell (London: Macmillan, 1973), 246–57.

13. Hobbes, *Behemoth,* 3–4, 58, 62, 73, and 90.

14. Christopher Hill, *The Intellectual Origins of the English Revolution Revisited,* 2nd ed. (Oxford: Clarendon, 1997), 15–76; and Charles Webster, *The Great Instauration: Science, Medicine and Reform, 1626–1660* (London: Duckworth, 1975).

15. Hill, *Intellectual Origins,* 122–47; and Webster, *Great Instauration,* 118, 129, 356, and 427–42.

16. Dennis Des Chene, *Physiologia: Natural Philosophy in Late Aristotelian and Cartesian Thought* (Ithaca: Cornell University Press, 1996); and Anthony Grafton and Nancy Siraisi, eds., *Natural Particulars: Nature and the Disciplines in Renaissance Europe* (Cambridge: MIT Press, 2000).

17. Webster, *Great Instauration,* 429.

18. Lesley Cormack, "'Good Fences Make Good Neighbours': Geography as Self-Definition in Early Modern England," *Isis* 82 (1991): 639–61; and Robert J. Mayhew, "The Character of English Geography, c. 1660–1800: A Textual Approach," *Journal of Historical Geography* 24 (1998): 385–412.

19. Nicholas Tyacke, "Science and Religion at Oxford before the Civil War," in *Puritans and Revolutionaries: Essays in Seventeenth Century History Presented to Christopher Hill,* ed. Donald Pennington and Keith Thomas (Oxford: Clarendon, 1978), 73–93; and Mordechai Feingold, *The Mathematician's Apprenticeship: Science, Universities and Society in England, 1560–1640* (Cambridge: Cambridge University Press, 1984).

20. Charles W. J. Withers and Robert J. Mayhew, "Rethinking 'Disciplinary' History: Geography in British Universities, c. 1580–1887," *Transactions of the Institute of British Geographers* 27 (2002): 11–29.

21. Eva G. R. Taylor, *Late Tudor and Early Stuart Geography, 1583–1650* (London: Methuen, 1934).

22. Lesley Cormack, *Charting an Empire: Geography at the English Universities, 1580–1625* (Chicago: University of Chicago Press, 1997).

23. Thomas Hobbes, *English Works*, ed. William Molesworth (London: John Bohn, 1843), 8:x.

24. Natalia Lozovsky, *"The Earth Is Our Book": Geographical Knowledge in the Latin West, c. 400–1000* (Ann Arbor: University of Michigan Press, 2000).

25. Churchill Babington, ed., *Polychronicon Ranulphi Higden Monachi Cestrensis; together with the English Translations of John Trevisa and of an Unknown Fifteenth Century Writer* (London: Longmans, 1865–86), 1:1–2:174.

26. Ibid., 1:79.

27. Daniel R. Woolf, *Reading History in Early Modern England* (Cambridge: Cambridge University Press, 2000).

28. Roger Barlow, *A Briefe Summe of Geography*, ed. Eva G. R. Taylor (1541; London: Hakluyt Society, 1932); William Harrison, *The Description of England*, ed. Georges Edelen (1578; New York: Dover, 1994).

29. Peter Heylyn, *Microcosmus; or, A Little Description of the Great World: A Treatise Historicall, Geographicall, Politicall, Theologicall* (Oxford: Oxford University Press, 1621), sig. ¶¶r. An augmented and revised edition of the *Microcosmus* was published by John Lichfield and William Turner in 1625.

30. Robert Stafforde, *A Geographicall and Anthologicall Description of All the Empires and Kingdomes, both of Continents and Ilands in this Terrestriall Globe* (London: Simon Waterson, 1634), sig. A2v.

31. Samuel Purchas, *Hakluytus Posthumus; or, Purchas His Pilgrimes: Contayning a History of the World in Sea Voyages and Lande Travells* (Glasgow: James MacLehose, 1905–8), 1:xl. This does not mean, however, that the drive to method did not influence travel accounts more generally. On this point see Justin Stagl, "The Methodising of Travel in the Sixteenth Century: A Tale of Three Cities," *History and Anthropology* 4 (1990): 303–38.

32. Heylyn, *Microcosmus* (1621 ed.), 2–4; and William Pemble, *A Briefe Introduction to Geography: Containing a Description of the Grounds and General Part Thereof,* 5th ed. (Oxford: Oxford University Press, 1675). Pemble's subtitle obviously invokes this division.

33. The argument that method "liberated" geography from religion is made in a series of articles by Manfred Büttner, the most easily accessible to the English reader being "Bartolomaus Keckermann, 1572–1609," *Geographers: Biobibliographical Studies* 2 (1978): 73–79.

34. See Margarita Bowen, *Empiricism and Geographical Thought: From Francis Bacon to Alexander von Humboldt* (Cambridge: Cambridge University Press, 1981), 72–75; Nathanael Carpenter, *Geography Delineated Forth in Two Books* (Oxford: Oxford University Press, 1625), 2:267–72; T[homas] M[ay], *The Mirrour of Mindes; or, Barclay's Icon Animorum Englished* (London: Thomas Walkley, 1631).

35. Terence Cave, *The Cornucopian Text: Problems of Writing in the French Renaissance* (Oxford: Clarendon, 1979); Neil Gilbert, *Renaissance Concepts of Method* (New York: Columbia University Press, 1965); and Neil Kenny, *The Palace of Secrets: Béroalde de Veville and Renaissance Conceptions of Knowledge* (Oxford: Clarendon, 1994).

36. Ann Moss, *Printed Commonplace Books and the Structuring of Renaissance Thought* (Oxford: Clarendon, 1996).

37. William Bouwsma, *The Waning of the Renaissance, 1550–1640* (New Haven: Yale University Press, 2000).

38. Wilbur Howell, *Logic and Rhetoric in England, 1500–1700* (Princeton: Princeton University Press, 1956); and Walter Ong, *Ramus, Method and the Decay of Dialogue: From the Art of Discourse to the Art of Reason* (Cambridge: Harvard University Press, 1954).

39. Howard Hotson, *Johann Heinrich Alsted: Between Renaissance, Reformation and Universal Reform* (Oxford: Clarendon, 2000).

40. Carpenter, *Geography Delineated Forth,* 1 : 17 and 2 151–52; Webster, *Great Instauration,* 118.

41. Carpenter, *Geography Delineated Forth,* 2 : 110; Bartholomew Keckermann, *Systema Geographicum Duobus Libris* (Hanover: Petrus Janichius, 1612) Heylyn in his *Microcosmus,* cited Keckermann seven times in his brief "General Praecognita of Geography," more than any other author.

42. Margot Todd, *Christian Humanism and the Puritan Social Order* (Cambridge: Cambridge University Press, 1987), 79; Feingold, *Mathematician's Apprenticeship,* 12, 97; and Ong, *Ramus,* 298–300.

43. Hotson, *Johann Heinrich Alsted,* 228–29.

44. Patrick Collinson, *The Religion of Protestants: The Church in English Society, 1559–1625* (Oxford: Oxford University Press, 1982); Nicholas Tyacke, *Anti-Calvinists: The Rise of English Arminianism, c. 1590–1640* (Oxford: Clarendon, 1987); and Anthony Milton, *Catholic and Reformed: The Roman and Protestant Churches in English Protestant Thought, 1600–1640* (Cambridge: Cambridge University Press, 1995).

45. Collinson, *Religion of Protestants,* 82.

46. Paul Welsby, *George Abbot: The Unwanted Bishop, 1562–1633* (London: SPCK, 1962); Susan Holland, "Archbishop Abbot and the Problem of 'Puritanism,'" *Historical Journal* 37 (1994): 23–43; and Kenneth Fincham, "Prelacy and Politics: Archbishop Abbot's Defence of Protestant Orthodoxy," *Historical Research* 61 (1988): 36–64.

47. Feingold, *Mathematician's Apprenticeship,* 58; Webster, *Great Instauration,* 19–20.

48. Hill, *Intellectual Origins,* 278.

49. George Abbot, *A Briefe Description of the Whole Worlde* (London: John Browne, 1605), sig. M4r. A still later edition of Abbot's work was published by J. Marriott in 1620.

50. Abbot, *Briefe Description* (1605 ed.), sig. M3v.

51. Ibid., sig. N2r.

52. Nathanael Carpenter, *Achitophel; or, The Picture of a Wicked Politician* (London: J. Okes, 1633), sig. Er.

53. Carpenter, *Geography Delineated Forth,* 1: sig. ¶3v; Tyacke, "Science and Religion," 81.

54. Carpenter, *Geography Delineated Forth,* 1 : 74.

55. Ibid., 93. Heylyn draws on Scaliger's work on historical chronology, the sister "eye" of history to geography. On this point, see Anthony Grafton, *Joseph Scaliger: A Study in the History of Classical Scholarship,* vol. 2, *Historical Chronology* (Oxford: Warburg Institute, 1993).

56. Carpenter, *Geography Delineated Forth,* 2 : 243.

57. Carpenter, *Achitophel,* sig. C3v.

58. Carpenter, *Geography Delineated Forth,* 1 : 238–39.

59. Heylyn, *Microcosmus* (1621 ed.), 248.

60. Ibid., 243.

61. Ibid., 249.

62. Joan-Pau Rubiés, *Travel and Ethnology in the Renaissance: South India through European Eyes, 1250–1650* (Cambridge: Cambridge University Press, 2000), 294–300.

63. Giovanni Botero, *Relations of the Most Famous Kingdomes and Common-wealths thorowout the World,* 3rd ed., trans. Robert Johnson (London: John Haviland, 1630), sig. A2r–v.

64. Tyacke, *Anti-Calvinists*, 72; cf. Petrus Bertius, *Tabularium Geographicam* (Amsterdam: J. Hondius, 1616).

65. Botero, *Relations of the Most Famous Kingdomes*, 66.

66. Ibid., 207–8. The grammatical construction of this sentence, "our men," makes it transparent that this is written by an Englishman, that is, by Johnson the translator, not by the Italian Botero, and, indeed, Johnson said as much (Ibid., sig. A4v).

67. Milton, *Catholic and Reformed*; and Johan P. Sommerville, *Politics and Ideology in England, 1603–1640* (London: Longmans, 1986).

68. Heylyn, *Microcosmus* (1621 ed.), 249.

69. Ephraim Pagitt, *Christianography; or, The Description of the Multitude and Sundry Sorts of Christians in the World, Not Subject to the Pope*, 3rd ed. (London: J. Okes, 1640), 186. I am not suggesting that Pagitt was himself an Arminian, only that his work tended toward sympathy with that position under the pressure of events.

70. Peter Heylyn, *Cosmographie in Four Bookes*, 2nd ed. (London: Henry Seile, 1657), 159.

71. Milton, *Catholic and Reformed*, 10.

72. Simon Adams, "Spain or the Netherlands? The Dilemmas of Early Stuart Foreign Policy," in *Before the Civil War: Essays on Early Stuart Politics and Government*, ed. Howard Tomlinson (London: Macmillan, 1983), 79–101.

73. Holland, "Archbishop Abbot," 24; and Fincham, "Prelacy and Politics," 37.

74. George Abbot, *Briefe Description* (1620 ed.), sig. B2r.

75. Fincham, "Prelacy and Politics," 52.

76. Abbot, *Briefe Description* (1605 ed.), sig. C4v.

77. Fincham, "Prelacy and Politics," 50; Abbot, *Briefe Description* (1605 ed.), sig. Cviiir.

78. Tomassso Campanella, *Thomas Campanella: An Italian Friar and Second Machiavel; His Advice to the King of Spain for Attaining the Universal Monarchy of the World*, trans. Edmund Chilmead (London: Philemon Stephens, 1659), sig. A2v.

79. Heylyn, *Microcosmus* (1621 ed.), sig. ¶2v; Thomas Cogswell, "England and the Spanish Match," in Cust and Hughes, *Conflict in Early Stuart England*, 107–33.

80. Heylyn, *Microcosmus* (1625 ed.), 175–77, 34, and 71–72. For *Microcosmus* as a less than fully Arminian tract, see Anthony Milton, "The Creation of Laudianism: A New Approach," in *Politics, Religion and Popularity in Early Stuart England: Essays in Honour of Conrad Russell*, ed. Thomas Cogswell et al. (Cambridge: Cambridge University Press, 2002), 165–68.

81. Heylyn, *Microcosmus* (1625 ed.), 31.

82. Heylyn, *Microcosmus* (1621 ed.), 35.

83. Heylyn, *Microcosmus* (1625 ed.), 105.

84. Ibid., 196.

85. Heylyn, *Microcosmus* (1621 ed.), 36.

86. Milton, *Catholic and Reformed*, 265–66.

87. Pagitt, *Christianography*, 187–88.

88. Heylyn, *Cosmographie*, 176.

89. Ibid., 393 and 395.

90. Ibid., 396.

91. Anthony Grafton and Lisa Jardine, "'Studied for Action': How Gabriel Harvey Read His Livy," *Past and Present* 129 (1990): 30–78.

92. Carpenter, *Geography Delineated Forth*, 1 : 4.

93. Grafton and Jardine, "Studied for Action," 39.

94. Kevin Sharpe, *Reading Revolutions: The Politics of Reading in Early Modern England* (New Haven: Yale University Press, 2000).

95. George Abbot, *The Reasons which Doctour Hill Hath Brought, for the Upholding of Papistry, which Is Falselie Termed the Catholicke Religion: Unmasked* (Oxford: Oxford University Press, 1604), 4–5, 89, and 181–82.

96. Henry Parker, "The Case of Shipmony Briefly Discoursed" (1640), in *The Struggle for Sovereignty: Seventeenth-Century English Political Tracts,* ed. Joyce Lee Malcolm (Indianapolis: Liberty Fund, 1999) 1 : 109–10.

97. Charles I, "His Majesties Answer" (1642), in Malcolm, *Struggle for Sovereignty,* 1 : 167.

98. [Charles Herle], "A Fuller Answer to a Treatise Written by Doctor Ferne" (1642), in Malcolm, *Struggle for Sovereignty,* 1 : 254 and 229.

99. Henry Ferne, "The Resolving of Conscience" (1642), in Malcolm, *Struggle for Sovereignty,* 1 : 198.

100. Heylyn, *Microcosmus* (1621 ed.), 189, and *Microcosmus* (1625 ed.), 525.

101. "The People's Right Briefly Asserted" (1649), in Malcolm, *Struggle for Sovereignty,* 1 : 362–68.

102. "A Declaration of the Parliament of England" (1649), in Malcolm, *Struggle for Sovereignty,* 1 : 380.

103. John Milton, "The Tenure of Kings and Magistrates" (1650), in *John Milton: Political Writings,* ed. Martin Dzelzainis (Cambridge: Cambridge University Press, 1991), 10.

104. John Milton, "Commonplace Book," in *The Complete Prose Works of John Milton,* ed. Don Wolfe (New York: Columbia University Press, 1953–82), 1 : 362–513

105. John Milton, "A Defence of the English People" (1658), in Dzelzainis, *John Milton: Political Writings,* 157.

106. John Milton, "The Readie and Easie Way to Establish a Free Commonwealth" (1660), in Malcolm, *Struggle for Sovereignty,* 1 : 518.

107. Robert J. Mayhew, *Enlightenment Geography: The Political Languages of British Geography, c. 1650–1850* (London: Macmillan, 2000).

108. R.P., *A Geographicall Description of the World* (London: John Streater, 1659), 3.

10

EDME MENTELLE'S GEOGRAPHIES
AND THE FRENCH REVOLUTION

...

Michael Heffernan

[D]ans cette grande tempête de la Révolution . . . il y a eu des jours où les âmes les plus ordinaires se sont senties soulevées et ont eu leur heure d'héroïsme.

Claude Perroud, 1896

In this chapter, I examine the life and works of Edme Mentelle (1730–1815), a prominent French geographer of the revolutionary era. It must immediately be acknowledged that Mentelle is a very minor figure in the history of European geography. Despite his impressive output, most of his geographical writings seem to have been ignored by subsequent generations. Neither did he distinguish himself as a traveler or explorer. While other scientists of his era sought fame, fortune, and credibility through overseas adventure, Mentelle remained the most sedentary of armchair geographers. He seems rarely to have strayed beyond his beloved Paris.

The few historians of geography who have examined Mentelle's works are understandably perplexed by them. Leslie Marchant's outstanding account of Mentelle's life and works concludes with a long list of the errors, redundancies, and conceptual confusions that litter his books.[1] Mentelle is likewise the bête noire of Anne Godlewska's audacious survey of French geography from Cassini to Humboldt.[2] According to Godlewska, Mentelle's arid writings and rigid teaching methods were partially responsible for the "loss of direction and status" she feels geography experienced during and after the Revolution.[3]

This begs an obvious question: why bother with such an intellectually insignificant individual? More specifically, why use Mentelle as the centerpiece of a discussion that purports to deal, at least in some degree, with the impact of the French Revolution on the theory and practice of geography? The answer to these questions lies in the intriguing discrepancy between the harsh criticisms that have been leveled against Mentelle by historians of geography and the remarkable prestige he achieved and sustained during his lifetime. Despite his shortcomings as an intellectual, Mentelle was an extraordinarily successful man of letters. Not only was he one of the most prolific geographers of his generation, he also possessed a chameleonlike ability to reinvent himself continuously as the political climate demanded. During his long life, he secured generous patronage from a bewildering variety of personalities and political leaders under the ancien régime, the various phases of the Republic, and the Napoleonic First Empire. Throughout these upheavals, Mentelle retained prominent positions in the country's leading academic institutions, despite the regular, sweeping reorganizations to which these institutions were subjected.

It must be emphasized that this chapter is *not* an attempt to rescue Mentelle from the calumny heaped upon him by posterity; nor is it an attempt to rehabilitate his status as a geographer of note. Rather, the objective is to understand how a geographer as limited as Mentelle was able to chart such a successful route, despite the all-too-apparent weaknesses of his work, through the minefield of the French political landscape as the country lurched from monarchy to republic to empire. At one level, the story recounted below can be read as a commentary on the politics of scientific survival during the French Revolution. But there is a wider, methodological purpose here. The underlying argument of this chapter is that historians of geography can gain fresh insights into the changing nature of their subject by looking beyond its more obviously distinguished representatives. The life and work of a geographer as intellectually undistinguished as Mentelle, whose fame was nevertheless as great as any of his contemporaries, arguably tells us more about the place of geography in French political and intellectual life during the Revolution than the career of more illustrious figures whose achievements transcended their historical context.

The belief that the life and works of minor writers can be used to say something significant about large overarching themes—in this case the changing character of French geography during the Revolution—has been inspired by the work of the American cultural historian Robert Darnton on what he has called the "the literary underground" of the ancien régime, the seditious and scandal-mongering literature of "Grub Street."[4] The thousands of sensationalist and often pornographic pamphlets produced by hack writers in the dying years of the ancien régime were, Darnton insists, as significant to the maelstrom of ideas that billowed forth after 1789 as the more elevated philosophical inquiries of the "High Enlightenment."[5] As Darnton further demonstrates, it is in any case extremely difficult to draw a clear distinction between these apparently very different forms of eighteenth-century writing. Many of the desperate and disaffected hacks scribbling away on Grub Street were also to be found rubbing shoulders with the great and the good in the leading literary salons where the weighty issues of the day were discussed. In the vivid portrait of late-eighteenth-century Paris crafted by Darnton, the same writer can be found dashing off a quasi-pornographic scandal sheet in the morning, usually to satisfy a creditor, before settling down for an afternoon's serious writing on a major work of political economy.[6] A surprising number of these "high brow–low life" writers ended up as leading figures in the Revolution; men such as Jacques-Pierre Brissot de Warville, the philosopher turned revolutionary who led the Girondins before his execution during the Terror. According to Darnton, Brissot combined his pre-1789 career as a high-minded author of treatises on penal reform and the theory of criminal law, with a secret life as a police spy and writer of anonymous pamphlets alleging corruption and scandal in the Paris Bourse, the latter produced to order by a Swiss financier who paid off the debts that had landed Brissot in jail in both Paris and London.[7] As we shall see, the unusual career of Edme Mentelle (who was a close friend and correspondent of Brissot) suggests that Darnton's arguments about the complex relationship between the High Enlightenment and the "low life" of literature might be as useful to the historian of geography as they have been to the cultural historian of the Revolution.[8]

GEOGRAPHIES IN AND OF THE FRENCH REVOLUTION

Before analyzing Mentelle's career, it is important to situate his work in the context of French geography at the time of the Revolution. As we have seen, the best account of French "geographic science" in this period, Anne Godlewska's *Geography Unbound,* paints a gloomy picture of an uncertain protodiscipline suffering a crisis of identity about its purpose and core principles.[9] In Godlewska's view, the leading French geographers of the revolutionary period, notably Mentelle, remained stubbornly wedded to a sterile, descriptive mode of inquiry and consciously rejected the very possibility of explanation or theoretical argument. Geography, in this form, was simply the presentation of unadulterated, usually unconnected facts about different places, the meaning and interpretation of which were left to representatives of other disciplines.

The intellectual weakness of French geography during the 1790s was surprising, insists Godlewska, because the conditions were highly favorable for the development of a new and more powerful geography that might have brought together the human and natural sciences. Rapid advances in cartographic techniques and the emergence of explicitly scientific forms of exploration during the eighteenth century had provided unprecedented quantities of factual information, as well as accurate maps covering an impressive part of the globe.[10] Contemporaries were well aware of the opportunities: Didier Robert de Vaugondy's entry "Géographie" for Denis Diderot and Jean Le Rond d'Alembert's *Encyclopédie* in the 1750s had insisted that the eighteenth century would be seen as a golden age of French geography.[11] According to Godlewska, however, the discipline failed to develop a language adequate to the task of explaining the mountain of new evidence it had been bequeathed. The few genuinely important intellectuals whose explorations and writings pointed the way forward (men such as Alexander von Humboldt, Jean-Antoine Letronne, and the comte de Volney) were sadly ignored by most of their colleagues.[12]

Other historians of French geography during the revolutionary period have been less judgmental, without displaying much enthusiasm for the discipline's achievements. Numa Broc, in his splendid review of *La géographie des philosophes,* agrees that "la Révolution marque un ralentissement très net de la recherche géographique" and offers a simple explanation: "[L]es troubles intérieurs, les guerres continuelles, la désorganisation de la marine, [et] la perte des colonies" interrupted the flow of new information and the general sense of excitement that had accompanied the preceding era of exploration, particularly in the Pacific. Between the ill-fated expeditions of the late 1780s and early 1790s by Lapérouse and d'Entrecasteaux (the latter dispatched in a doomed attempt to find the former) and the subsequent explorations in the early 1800s by Baudin in Australasia and the East Indies, there stretched what Broc calls a "vide à peu près complèt."[13] While this hiatus might have provided an opportunity for geographers "s'interroger sur les bases méthodologiques de leur science," the few pages Broc devotes to their efforts in an otherwise extensive review are an eloquent testimony of the discipline's inability to function at a conceptual or theoretical level when deprived of new raw materials.[14]

Martin Staum's more sharply focused study of geography in the Classe des Sciences Morales et Politiques in the Institut National des Sciences et Arts between 1796 and 1803 provides a more generous review of the discipline's track record, but even he concludes that the revolutionary era failed to consolidate France's eighteenth-century preeminence as a center of geographical investigation. By the mid-nineteenth century, Staum suggests, "the lead in geographical studies had undeniably passed to the German speaking world."[15]

Insofar as these negative assessments are valid, they suggest a further irony. The Revolution was, after all, a quintessentially geographical process, which might logically have sparked off debate about the new geographies, both real and imagined, that emerged after 1789. At the very core of the revolutionary mythology was the conviction that the Revolution represented a fundamental break with the past and had ushered in entirely new conceptions of space and time. Jules Michelet, the most sympathetic nineteenth-century historian of the Revolution, perfectly encapsulated the revolutionary ideals a generation later when he insisted that 1789 had "annihilated" the old geographies and temporalities: "There were no longer any mountains, rivers, or barriers between men. . . . Time and space, those natural conditions to which life is subject, were no more."[16]

For the revolutionaries seeking to build their new republican order, this meant forcibly imposing new attitudes to time, notably through the republican calendar,[17] and reorganizing the country's geographies at all scales from the local to the national. The new revolutionary spaces they created—the restructured administrative geographies, the reordered and rededicated public (and especially religious) spaces of the towns and cities, the new institutional spaces created in these same towns and cities (including the educational and scientific institutions), and the microgeographies of newly controlled domestic and personal environments—have all been extensively studied in recent years.[18] This research has demonstrated that these new spaces were not merely products of the Revolution. They were also the locales wherein a revolutionary political culture was made, challenged, and remade after 1789.[19] As Mona Ozouf has observed in her fine work on the ephemeral spaces of revolutionary festivals:

> There is no end to the list of spatial metaphors associated with the Republic or with Revolutionary France: from the beginning of the Revolution a native connivance linked rediscovered liberty with reconquered space. The beating down of gates, the crossing of castle moats, walking at one's ease in places where one was once forbidden to enter: the appropriation of a certain space, which had to be opened and broken into, was the first delight of the Revolution.[20]

When one considers the remarkable geographical changes that occurred as a result of the Revolution, the apparently lifeless geographical descriptions produced at the time, which have been rightly lamented by historians of geography, seem all the more unaccountable. These descriptions were written in the midst of a seismic upheaval that was transforming the country's geography. The fact that the Revolution also swept away the moral, religious, and political constraints under which authors had labored during the ancien régime makes the timidity of French geographers seem stranger still.

This chapter offers some explanations for this surprising discrepancy between text and context. The complex life and works of Edme Mentelle is considered here in an attempt to understand the relationship between the new geographies constructed textually and cartographically during the French Revolution and the new geographies that emerged on the ground; between the geographies in and the geographies of the French Revolution.

Geography and the End of the Ancien Régime: Edme Mentelle, 1760–1789

Edme Mentelle was born in Paris into a solid bourgeois family, on 11 October 1730. Little is known of Edme's early life. His younger sibling, François-Simon Mentelle, enjoyed a more noteworthy education in Paris, partly under the guidance of Philippe Buache, the Geographer to the King from 1726 to 1773. This was a promising start for a would-be geographer and paved the way for François-Simon's career as a cartographic assistant to César François Cassini de Thury (Cassini III) in the 1750s and as an explorer and astronomer in French Guiana, where he settled in 1763 and eventually died, aged of sixty-eight, in 1799.[21]

Edme Mentelle seems to have been educated at less expense than his brother in Beauvais by a professor of rhetoric and ancient history, Jean Baptiste Louis Crévier. Mentelle then embarked on an ill-fated career in commerce, before finally gaining employment as a minor government functionary in Paris. Nothing in Mentelle's early years suggests a career as a professional geographer, and as a young man, he seemed more interested in literature and poetry. His first publication, printed in 1751, was a poetic reworking of a classical saga, *La mort de Polieucte,* and this was followed by other short poems, at least two of which appeared in the widely read *Mercure de France.*[22]

In the late 1750s, while still employed as a clerk in the finance ministry, Mentelle tried his luck as a writer of satire. He published a pseudonymous letter to an imaginary foreign visitor in 1757 satirizing the French periodical press of the time.[23] This was followed by a full-length play in 1758, *L'amour libérateur,* which was performed in Bordeaux. Mentelle's play was an energetic farce involving some familiar, off-the-peg characters: the illegitimate son of a local aristocrat, a beautiful peasant girl, and a scheming curé, whose adventures end in the happy union of the hero and heroine.

Mentelle's less than flattering treatment of the curé in *L'amour libérateur* provides a clue to his emerging political views. These seem to have been shaped by a Voltairean anticlericalism that revealed itself more openly in a subsequent pseudonymous work, *Le porte-feuille du R. F. Gillet,* in which Mentelle poked fun at the attempts by the Catholic Church and the monarchy to interrupt the publication of the early volumes of the *Encyclopédie* in the late 1750s.[24] In Mentelle's satire, an ironic and rather rude "A to Z" of "new" definitions is presented as an entrée to a mock-heroic poem about the descent into hell of a Jesuit priest who, having forged a pact with Pluto, the god of hell, returns to earth to continue his wicked ways. Mentelle's poem appeared two years after d'Alembert's *Sur la destruction des Jésuites en France, par un auteur désintéressé* (1765) and five years after the expulsion of the Jesuit order from the country, so his target was neither original nor controversial.[25] But unlike Catholic critics of the Jesuits, who were responsible for the order's removal from France, Mentelle clearly saw the Jesuitical tradition as a sinister manifestation of a wider Catholic malaise.

Given this less than prolific start in the literary world, Mentelle's decision to write a geography school textbook as he experimented with these other genres cannot easily be explained, though he was almost certainly influenced by his younger brother's cartographic work on the survey of France with Cassini III. Mentelle's *Élémens de géographie,* which appeared in 1758, was a simple text, divided into two sections, the first containing basic facts about the four "continents'" of Europe, the Americas, Africa, and Asia; the second containing similar details on France itself; and the whole prefaced by introductory remarks about the need to move from the general to the particular. On the face of it, this innocuous textbook bears little relationship to the stridently anticlerical (and more specifically anti-Jesuit) ideas Mentelle revealed in his other writings, but it may have been influenced by a desire to challenge the hegemony of Catholic, and particularly Jesuit, teaching methods.

According to Marcel Grandière, eighteenth-century French schooling can be divided into three periods: 1715–46, when Catholic educational techniques retained a virtual monopoly; 1746–62, when alternative, more secular ideas gathered momentum; and 1762–88, when the influence of Jean-Jacques Rousseau's *Émile* recast the entire educational debate and opened the way for an explicitly national education that prefigured developments under the Revolution.[26] Mentelle's modest textbook appeared as the church's educational monopoly was being challenged for the first time, and its distinctly secular tone may have reflected that fact. It was certainly written just as the clash between the *philosophes* and the church over the nature and content of the *Encyclopédie* was reaching its climax. As we have seen, this conflict directly informed Mentelle's satirical works, and it is possible that the same topic influenced his decision to write a geography textbook that might serve as a modest alternative to the works used in Catholic schools.[27] In these institutions, geography was taught both as evidence of a divine order and as a practical, universalizing subject that underpinned the expansion of the faith. In Mentelle's anticlerical vision, geography was a straightforward textual and cartographic technique to communicate resolutely secular facts about the world. In the Catholic worldview, the study of geography legitimized and sustained the church's role; from Mentelle's Voltairean perspective, the study of geography implicitly questioned the church's role, without explicitly challenging the idea of a God-created world.

Mentelle's textbook probably explains his appointment in 1760 to teach geography and history at the École Royale Militaire, the officer training institution recently established in Paris through which the young Napoleon Bonaparte would soon pass. Precisely how Mentelle's appointment came about remains unclear, but the ambitious young professor seized his opportunity, offering a series of lectures for the officer cadets. Geography and history had been taught at the École Militaire since its inception, but as separate programs.[28] Mentelle drew them together into a series of geohistorical narratives about warfare, politics, and the rise and fall of civilizations, with examples drawn mainly from the classical world.[29] These ideas were developed in Mentelle's first work during his tenure at the École Militaire, the *Manuel géographique, chronologique, et historique,* which was accompanied with numerous fine maps. Mentelle dedicated this volume to Mlle de Fitzjames, granddaughter of the Anglo-Irish Duke of Brunswick, founder of what had become, by the mid-eighteenth century, an important military dynasty in the service of France.[30]

Secure in his position, Mentelle produced volume after volume of geography and history textbooks for use in secondary schools and colleges through the 1760s, 1770s, and 1780s, together with dozens of new maps and atlases.[31] Many of these were multivolume works, published at Mentelle's own expense, though not unprofitably, if the number of reissues can be taken as a guide. His most successful prerevolutionary titles appear to have been *Élémens de l'histoire romaine* (1766, reissued in two volumes in 1773–74) and *Cosmographie élémentaire* (1781, reissued in 1782, 1795, and 1798–99), though his three-volume contribution *Géographie ancienne* for Charles-Joseph Panckoucke's *Encyclopédie méthodique* (1787–93) and the associated three volumes of his *Dictionnaire de géographie moderne* (1784–89) for the same publisher were probably his most prestigious commissions.[32]

The diagram Mentelle constructed to indicate the different domains of geographical knowledge at the end of his *avertissement* in the first volume of his

Géographie ancienne reveals the rigidly structured approach he developed in his teaching, though it must be emphasized his own writings were rarely as disciplined as this schema implies (fig. 10.1). Mentelle's desire to set down a branching "map" of the domains covered by geography may be taken to imply that French Renaissance concerns with systematic "method" retained some influence on Enlightenment educational practice.[33] The fact that a belief in "method" as the only way to communicate potentially limitless information had been forged in earlier battles between a Catholic and Aristotelian orthodoxy and an emerging Protestant critique may also have resonated with Mentelle's by now pronounced anticlericalism.[34] That said, Mentelle's repeatedly published schema of geographical domains also reflected a quintessentially eighteenth-century preoccupation with classificatory systems, revealed most obviously in the *Encyclopédie* and in the taxonomic systems of Carolus Linnaeus and other natural scientists.

Mentelle's activities at the École Militaire continued peacefully enough until the accession of Louis XVI in 1774. Mentelle had taught the king's youngest brother, the comte d'Artois, at the École Militaire, and this louche young aristocrat became the sober geographer's rather unlikely patron. By 1778, Mentelle had been appointed to the entirely notional position of d'Artois's official historiographer.[35] Under the protection of the comte, Mentelle gained an entrée into court society, where he assumed a leading role in the geographical education of the royal household during the 1780s, alongside Jean Nicolas Buache de la Neuville, who had recently been appointed Geographer to the King. Mentelle's role as a geography tutor in the royal household was a significant one. Philippe Buache, the long-serving Geographer to the King through the middle decades of the eighteenth century and an uncle of the aforementioned Buache de Neuville, had established a pedagogic regime for young princes based on a thorough grounding in geography. In Buache's view: "La géographie surtout est l'étude qui embrasse le plus de notions à l'usage des rois et des grands princes."[36]

Mentelle's most enduring contribution to the geographical education of the royal household in the 1780s was the remarkable globe he designed to teach the dauphin, Louis-Joseph, some basic geographical facts. At Mentelle's suggestion, the king commanded the engineer Jean-Tobie Mercklein to construct a new globe in March 1786. The original design, proposed by Mentelle, envisioned a globe 1.5 meters in diameter, with three encasing layers, each made of removable, interlocking sections like pieces in a jigsaw puzzle. The outer encasing was to display in relief the physical features of the globe, and could be dismantled to reveal a second encasing underneath showing the political geography of the world in the classical era, which could be removed

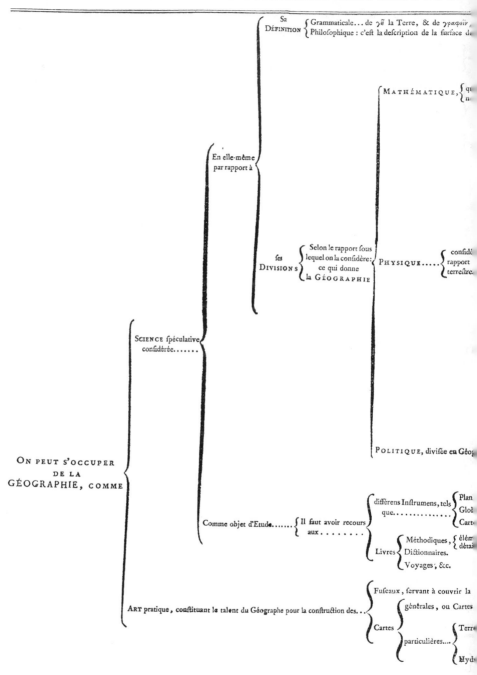

Figure 10.1. Mentelle's organization of geographical knowledge, 1787. From Edme Mentelle, *Encyclopédie méthodique* (Paris: Panckoucke), 1:xvi. Reproduced with the permission of the British Library.

ils fur les Peuples qui l'habitent.

a { circonférence, furface, folidité.

étendue en.. { dégrés de... { latitude. longitude. mefures géodéfiques..... { lieues.... milles.... } &c.

fituation par rapport aux Climats de...... { demi-heure mois..... } d'où réfulte la connoiffance de leurs plus { longs...... courts...... } jours, foit { d'Hiver. d'Eté.

{ Villes........ Obfervatoires, } leur pofition { propre..... relative.... } par les Obfervations aftronomiques des Eclipfes.... { de Soleil,.. de Satellites, } &c.

ion de fa furface divifée en..... {

TERRES..... { élevées, ou Montagnes divifées en { plateaux...... grandes chaînes } qui forment la carcaffe du Globe.

inclinées, ou Baffins { généraux, au fond de chacun defquels eft une Mer recevant toutes les eaux du Baffin. particuliers, au fond defquels coulent des Fleuves ou de grandes Rivières.

EAUX.... { permanentes...... { Mers. Lacs. Fontaines.

coulantes........ { Fleuves...... Rivières..... } &c.

ons........ { animales..... végétales... minérales... } afin d'indiquer en quels lieux elles fe trouvent.

HOMMES, confidérés...... { en eux-mêmes.... { configuration extérieure, couleur.... caractère.... } &c.

relativement aux lieux qu'ils habitent, dont il faut connoître...., ... { la fituation... les faifons.... } &c.

l'habitent, les... {

ANIMAUX { terreftres...... aquatiques... } que la Géographie doit faire connoître pour.......... { les avantages que l'on peut en retirer. les dangers que l'on court, foit à caufe de leur férocité, de leur venin, &c.

confidérant les {

PAYS, comme habitations d'Hommes réunis en Corps politiques, { Empires..... Royaumes... Républiques.. } dont elle décrit les Provinces, les Villes, &c.

PEUPLES, dont elle fait connoître................ { l'Origine............ la Langue........... la Religion.......... le Gouvernement....... les Sciences.......... le Commerce......... les Mœurs & ufages...... } & les révolutions hiftoriques.

raphiques.

ues.

ios.

{ planes. réduites.

in turn to reveal a third encasing displaying the political geography of the world in the late 1780s. The lavishly decorated version of this globe that was eventually presented to the king in 1788 was a modification of Mentelle's original idea, and still stands in the Musée du Château de Versailles. The Mentelle globe sits on a solid, gold-embossed base, supported by the upwardly curved bodies of three dolphins. It is operated by lifting the upper hemisphere of the outer globe, which depicts the political geography of the world of the 1780s, along four iron stands, thus revealing an inner globe, showing the earth's physical geography. In return for his labors, Mercklein received the princely sum of 8,000 livres, nearly 4,000 livres less than Mentelle was paid. Sadly, the young dauphin died a few months after his new acquisition arrived, in June 1789, at the age of eight.[37]

Mentelle's anticlericalism notwithstanding, he would appear to have been a loyal servant of the crown, though hints that his political views were changing are suggested by the new relationships he fostered with reform-minded intellectuals in the last years of the ancien régime. His most inspirational new associate was the comte de Mirabeau, the charismatic opponent of an unreformed French monarchy and the principal advocate of a British-style constitutional regime in the early revolutionary period.[38] The two men met in the mid-1780s, shortly before Mirabeau embarked on a tour of Prussia in 1786, ostensibly as a private individual but in reality at the behest of the authorities in Versailles who hoped to secure an alliance with Prussia to counterbalance the recent Austro-Russian entente. This excursion was to draw Mentelle into controversy, but also propel him into the circle of radical intellectuals associated with Mirabeau.

Mirabeau's task in Berlin, which was to be kept secret from his Prussian hosts, was to win over the court advisers of Frederick the Great, who unfortunately died soon after his arrival, and those of his successor, Frederick-William. While in Berlin, Mirabeau wrote sixty-six secret reports, from July 1786 to January 1788, many dealing with the scandals and intrigues of the Prussian court that had fallen under the influence of a cabalistic sect of illuminati associated with the gullible Frederick-William.[39] Mirabeau drew on his Berlin experience to produce a seven-volume account of the geographical, economic, and political conditions in Prussia, prepared with the assistance of Mentelle and Jakob Mauvillon, a liberal economist and military strategist.[40] The work was accompanied by a large-format atlas, comprising ten foldout maps, all by Mentelle, together with dozens of statistical tables and diagrams of the Prussian armed forces, mostly the work of Mauville. Mentelle also assisted Mirabeau in the lengthy geographical descriptions of the Prussian territories in Central Europe that appeared in volumes 1 and 2 of the main text.[41]

Mentelle's professional association with the immensely popular Mirabeau added to the former's prestige, despite the controversy that subsequently erupted when the real purpose behind Mirabeau's mission to Berlin became clear. During his tempestuous 1788 campaign to find a fief from which he could be elected to the Estates General as a representative of the nobility, Mirabeau gave permission to a London publishing house to issue the unexpurgated reports he had dispatched to Paris from Berlin. The resulting two-volume work, complete with a characteristically impertinent letter Mirabeau had written to Frederick-William on the day of his coronation, created a storm of controversy that affected many of Mirabeau's collaborators, including Mentelle.[42] Despite this setback, Mirabeau was elected to the Estates General, though as a representative of the Third Estate for Aix-en-Provence rather than as a noble.

Mentelle's newfound political confidence, born of his association with Mirabeau, was reinforced by his own activities. By the eve of the Revolution, Mentelle's modest apartments on the rue de Seine, where he maintained his library and extensive private collection of maps and globes, had become an important gathering point for other leading geographers, notably Jean-Baptiste Bourguignon d'Anville and Buache de Neuville, as well as prominent scientists like Pierre-Simon Laplace, Joseph-Louis Lagrange, Antoine-Laurent de Lavoisier, and Gaspard Monge.[43] The most noteworthy political visitor to Mentelle's *salon géographique* was probably Jean-Baptiste Brissot de Warville, the aforementioned radical journalist, whose most recent activities had centered on the French antislavery society, the Société des Amis des Noirs, which he had founded in February 1788. Mentelle's name does not appear alongside those of Condorcet, Mirabeau, the duc de Rochefoucauld, and the other 130 original members of the Société des Amis des Noirs, but there is clear evidence that Brissot relied on the geographer's expertise in developing the case against the French-controlled slave trade.[44] It seems reasonable to assume from this that Mentelle shared at least some of Brissot's liberal and republican views, and certainly the journalist greatly appreciated the geographer's assistance and advice. Brissot spoke warmly in his *Mémoires* of his friend's gifts as a teacher and scholar, and even asked Mentelle to act as godfather to one of his sons.[45]

GEOGRAPHY, REVOLUTION, AND EMPIRE: EDME MENTELLE, 1789–1815

The storming of the Bastille took place a few months before Mentelle's fifty-ninth birthday. Mentelle continued teaching much as before through the uncertain early months of the Revolution, while other members of his immediate family, including his son, contemplated and eventually opted for emigration. Mentelle's patron, the comte d'Artois, made a similar calculation for more pressing reasons. Mentelle had no intention of leaving his native city, however, safe in the knowledge that his friend Mirabeau was at the center of national decision-making. But by the spring of 1791, Mirabeau was dead, having expired at the early age of forty-two, worn down by chronic ill health and the endless political maneuvering that had eventually secured his presidency of the National Assembly.[46] Deprived of his ally at the heart of government and no longer protected by a powerful royal patron, Mentelle seems to have decided his interests would best be served by a public display of his liberal republican credentials.[47]

This took the form of letter to the city authorities in Paris, published in 1791 as a seven-page pamphlet under the title *Lettre d'un auteur citoyen à la Commune de Paris en faveur de la liberté de la presse.* Mentelle's pamphlet, which bore all the hallmarks of Mirabeau, was a carefully argued plea to continue the liberal policy of press freedom, enshrined in Article 11 of the Déclaration des Droits de l'Homme et du Citoyen, which had proclaimed the complete freedom of expression in 1789. The pamphlet seems to have been prompted by a decision in late 1791 to suppress Jean-Paul Marat's campaigning radical newspaper, *L'Ami du Peuple,* and by the ill-judged attempts to limit the number of *colporteurs* who sold the more popular forms of literature on the streets of the city.[48]

The impact of Mentelle's self-important letter was almost certainly negligible, but it served a wider purpose in demonstrating its author's commitment to the revolutionary ideals.[49] By the end of 1791, Mentelle was firmly associated with the group of radical republican deputies, journalists, and intellectuals who wrote for Brissot's newspaper, *Le Patriot Français.* This community, dubbed (more in contempt than affection) the "Brissotins," included the newly installed mayor of Paris, Jérome Pétion, and the liberal writer and scientist Condorcet.[50] The Brissotins assembled regularly at the salon of Jean-Marie Roland de la Platière and his formidable wife, Manon Jeanne Phlipon, on the rue Guénégaud, and here Mentelle received a warm welcome, secured by his friendship with the group's eponymous leader.[51]

The Rolands had recently moved to Paris from Lyons at the behest of Brissot, who had been impressed by the articles that appeared under Roland's name in the *Courrier de Lyon.* Most of these pieces had been written by Mme Roland, and merely signed by her husband, who was a factory inspector and economist.[52] The Roland salon, which initially welcomed Robespierre, subsequently the group's mortal enemy, provided a launching pad for Roland's bid to take control of the Ministry of the Interior, which was eventually forced upon a reluctant and beleaguered monarch in March 1792. Throughout the following year, Brissot, Roland, and their supporters in the press and the Legislative Assembly were the most articulate and impassioned advocates of a radical assault on the remaining monarchical powers and an aggressive policy of republican expansionism abroad, the latter aspect of their design culminating with the French declaration of war against Austria in 1792.

Mentelle can scarcely be regarded as a prominent figure within the Brissotin ranks, but he was closely associated with the ideals they espoused, a fact revealed by the warmth of his relationship with Brissot and Mme Roland, with whom he corresponded intermittently.[53] Mentelle certainly benefited from the Rolands' generosity. Although he was still officially employed by the École Militaire, the Revolution had raised serious doubts about the future of this academy, and the National Convention (established in September 1792 after the overthrow of the monarchy) subsequently closed the institution. Roland, who controlled the former royal palaces in his capacity as minister of the interior, made available a suite of rooms in the Louvre to allow Mentelle to pursue his calling, alongside other artists and writers, as a "professeur public de géographie."[54] In return, Mentelle was required to provide classes to adults and children alike on the new, republican geographies that the Revolution sought to construct, using maps, books and equipment from his offices in the École Militaire, as well as materials gathered from the royal palaces, including the globe designed for the dauphin a few years earlier.[55]

The nature of the classes provided by Mentelle from his offices in the Louvre will never be known, but it is unlikely that he radically changed the methods that had served him well in the past and that he would reprise in his later appointments. But the changed conditions required a new body of geographical literature, and Mentelle was only too keen to oblige by writing this himself. His first explicitly republican geography as an *auteur citoyen* appeared as early as 1791 under the title *Méthode courte et facile pour apprendre aisément et retenir sans peine la nouvelle géographie de la France.* This was designed to educate children about the names, *chefs-lieux*, and borders of the newly created French *départements.* It was followed in 1792 by two textbooks, *Tableau élémentaire de géographie de la République française* and *Tableau raisonné de la nouvelle division économico-politique*, which also dealt with the recent changes in the country's political geography. In the *Tableau élémentaire*, readers were encouraged to reject the petty provincialism that had defined the prerevolutionary era in favor of the uniform fraternity of the new republican age:

N'était-il pas ridicule de répondre étant à la porte presque de Dunkerque, quand on demandait aux gens: vous êtes Flamands? Non Monsieur, nous sommes Wallons. A Dieppe, à Caudebec, vous êtes de vrais Normands? Non Monsieur, nous sommes Cauchois. Actuellement, nous disons: nous sommes frères.[56]

In early 1793, Mentelle's career as "professeur public de géographie" was threatened by the larger crises of the time, into which he was drawn as much by his loyalty to his friends as by his political sympathies. By the beginning of that year, Brissot's supporters (now known as Girondins in recognition of their provincial origins) formed a majority in the National Convention, but they proved incapable of halting the slide toward anarchy, particularly during the trial of the king. Radicals during the Legislative Assembly, the Girondins had become the conservatives of the National Convention, opposing regicide and generally seeking to maintain order in the gathering chaos. Hemmed in by his Montagnard opponents, Roland resigned on 23 January 1793, two days after the king's execution. Roland's resignation unleashed a campaign of vitriolic abuse from the radical, left-wing press, with Marat's *L'Ami du Peuple*, the very paper that Mentelle had previously defended, leading the assault. The Girondins, screamed Marat's editorials, were seeking to destroy the Revolution and reinstate a federal monarchy.

By the summer of 1793, the Girondin power base in Paris had evaporated and the Montagnards, led by Robespierre, seized the initiative. Under threat of imprisonment, Roland escaped to the relative safety of Rouen, leaving his wife, at her insistence, in Paris. Brissot was captured as he attempted to leave the country, and the unfortunate Mme Roland was also arrested on 1 June 1793, two days before the remaining Girondin deputies were taken into custody. The Girondin leadership, including Mme Roland, were found guilty of treason and guillotined in the autumn of 1793, the latter's execution on 8 November recalled in the mythology of the Revolution by her famous remark directed toward the clay statue of the goddess Liberty that had been erected alongside the guillotine in the place de la Révolution: "O Liberté! Que de crimes on commet en ton nom!" On hearing the news of his wife's execution, her devoted husband committed suicide, as did other Girondins, inspired in their acts of self-destruction by the defiant words of Brissot and his colleagues on the steps of the guillotine: "Plutôt la mort que l'esclavage!"[57]

Incarcerated in the Conciergerie prison, Mme Roland had been allowed occasional visitors and limited correspondence. The last six letters she wrote from her cell, composed from late September to late October 1793 as she agonized about whether to starve herself to death, are among the most poignant documents of the Terror. These letters were written to a mysterious "Jany," her "unique consolateur," whom Mme Roland entreated to gather together her papers for safekeeping, including the documents she had composed in prison, and to carry messages to friends and supporters still at liberty. She feared her husband was already dead and her letters to "Jany," a code name selected to avoid incriminating her correspondent, spilled forth the anger, fear, and physical pain she was experiencing. "Quant à moi, Jany, tout est fini," she wrote in her fourth letter around 14 October 1793:

> Vous savez la maladie que les Anglais appellent heart-breaken [*sic*]? J'en suis at- teinte sans rèmede, et je n'ai nulle envie d'en retarder les effets; la fièvre com- mence à se developper, j'espère que ce ne sera pas très long. C'est un bien; ja- mais ma liberté ne me serait rendue; le Ciel m'est temoin que je la consacrerais à mon époux![58]

According to Claude Perroud, who edited Mme Roland's correspondence for publication just over a century ago, "Jany" was none other Mentelle, whose role in this affair was confirmed in a letter dated March 1800 that Mentelle sent to the Rolands' then editor, along with the bundle of documents he had kept under lock and key since Mme Roland's execution.[59] From the one-sided correspondence between Mme Roland and "Jany," it is clear that Mentelle vis- ited Roland and other imprisoned Girondin leaders, including Brissot; smug- gled documents and memoirs from prison cells; gathered together Mme Ro- land's other papers; and carried messages to Girondin supporters.[60]

Mentelle's actions on behalf of the doomed Girondins in late 1793 de- manded great personal courage, but he urgently needed to reposition himself once again if he was to avoid their fate. On 1 October 1793, while he was still receiving letters from the desperate Mme Roland, he wrote to the president of the Comité d'Instruction Publique in the National Convention, announc- ing his intention to offer a new course of public lectures at his atelier in the Louvre on what he called "la géographie comparée," based in part on his ear- lier book of that title but updated to consider the changes brought about by the Revolution.[61] The committee was asked to visit his salon to approve his methods. Two members of the committee duly visited him and submitted a favorable report, commenting positively on the globe Mentelle would be using to illustrate his teaching. Whether the inspectors realized this globe had been designed for the instruction of the dauphin is not recorded.[62] Men- telle sought to underline his commitment to the more radical forms of re- publicanism now in the ascendant in other ways as well, most dramatically through his toe-curlingly awful musical version of the Déclaration des Droits de l'Homme et du Citoyen, which he dedicated "à tous les SANS-CULOTTES de la République" (fig. 10.2).⁶³

The fall of Robespierre in July 1794 rehabilitated the Girondin "martyrs" and removed any lingering stigma clinging to Mentelle as a result of his association with Brissot and the Rolands. In the autumn of that year, Mentelle was offered a permanent post at the École Normale as one of two professors of geography, alongside Buache de la Neuville. The École Normale had recently been established to train a scientifically informed republican citizenship for the new revolutionary era,⁶⁴ and the thirteen lectures and seminars on geography, delivered by Mentelle and Buache between 22 January and 12 May 1795, provide a fascinating insight into the theory and practice of geography during the Revolution.⁶⁵ In their opening "*programme*," Mentelle and Buache divided geography into three branches: mathematical or astronomical (which studies the movements of the earth and planets), physical (which considers the shape of the earth and its physical products), and political (which considers the political divisions of the earth into states and the relations between these states). Taking their cue from the Lockean "sensationalism" of Condillac and Rousseau, Mentelle and Buache insisted that geographical education rested ultimately on information acquired through the sensory faculties, particularly the eyes, either through observing the environment itself or through studying scientifically formulated representations such as maps. "[L]a géographie", they insisted, "est une science qui ne s'apprend bien qu'avec les yeux."⁶⁶ Geography was not, in this sense, a theoretical or conceptual subject that was equipped to deal with the intangible or the abstract. It was a simple observational and factual subject. The task for the teacher of geography was merely to order and catalog facts into sensible systems.⁶⁷

As Godlewska has noted, the geography lectures provided by Mentelle and Buache at the École Normale were not universally appreciated at the time. An anonymous reviewer in the *Décade Philosophique,* the journal of the revolutionary *idéologues,* lambasted the course as boring, pointless, and irrelevant.⁶⁸ But when one considers the circumstances in which Mentelle and Buache were working, as well as Mentelle's recent experiences of the regime's brutal intolerance of its enemies, the decision to emphasize simple, apparently irrefutable facts, while studiously avoiding any of the controversies associated with abstract or theoretical reasoning, must have seemed an extremely sensible choice. In the absence of clear or consistent guidelines from an unstable and fickle polity, and confronted by students of widely varying abilities, Mentelle's emphasis on uncontroversial, if occasionally uninspiring, facts was entirely understandable.

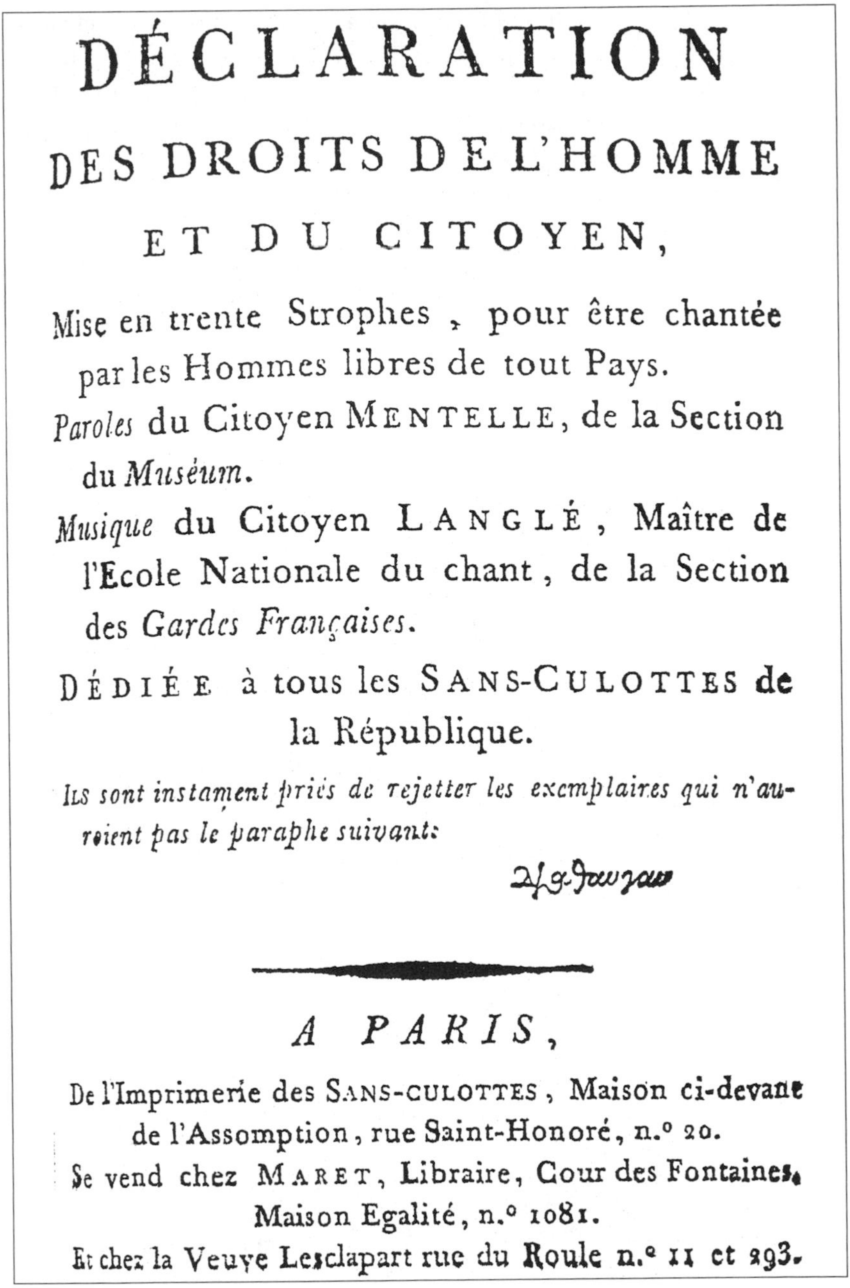

DÉCLARATION

DES DROITS DE L'HOMME

ET DU CITOYEN,

Mise en trente Strophes , pour être chantée
par les Hommes libres de tout Pays.
Paroles du Citoyen MENTELLE, de la Section
du Muséum.
Musique du Citoyen LANGLÉ , Maître de
l'Ecole Nationale du chant , de la Section
des Gardes Françaises.
DÉDIÉE à tous les SANS-CULOTTES de
la République.

Ils sont instament priés de rejetter les exemplaires qui n'au-
roient pas le paraphe suivant:

A PARIS,

De l'Imprimerie des SANS-CULOTTES , Maison ci-devant
de l'Assomption, rue Saint-Honoré, n.º 20.
Se vend chez MARET, Libraire, Cour des Fontaines,
Maison Egalité, n.º 1081.
Et chez la Veuve Lesclapart rue du Roule n.º 11 et 293.

Figure 10.2. Title page of Mentelle's musical version of the Déclaration des Droits de l'Homme, 1794. Source: Edme Mentelle, *Déclaration des droits de l'homme et du citoyen* (Paris: Imprimérie des Sans-Culottes / Maret). Reproduced with the permission of the British Library.

This was clearly what the authorities wanted, and Mentelle became, in effect, the official geographer of the Republic during the latter phase of the National Convention and through the Directory, from late 1794 to 1799. He acquired a variety of new responsibilities as a result, delivering geography and history classes at the new republican lycées in Paris in 1794, and at the Écoles Centrales in the Panthéon and Quatre-Nations from 1796 to 1804. He eventually joined the Conseil d'Administration of the latter institution. He was also invited to sit on various government commissions dealing with the arts and school textbooks.[69] In 1795, Mentelle was nominated by a National Convention decree as one of only a handful of scientists deemed to be worthy of official encouragement and financial support, and in the same year, he became involved in serious, if ultimately fruitless discussions to establish a Musée de Géographie, de Topographie Militaire, et d'Hydrographie in Paris, which would have brought together the disparate resources of the Dépôt des Cartes et Plans de la Marine and the Dépôt de la Guerre.[70]

Most important, Mentelle was elected to the Classe des Sciences Morales et Politiques in the Institut National, the new republican institution of higher learning established in October 1795 by one of the last acts of the National Convention. The Institut National was to replace the defunct *académies* that had been suppressed two years earlier.[71] The Classe des Sciences Morales et Politiques was initially divided into six sections devoted to the study and teaching of ideas, ethics, social science and legislation, political economy, history, and geography. The principal members of the sixth (geography) section were Mentelle, Buache, Carl-Friedrich Reinhard, Charles-Pierre d'Eveux Claret de Fleurieu, Paschal-François-Joseph Gosselin, and Louis-Antoine de Bougainville. During the seven years of the class's existence, Mentelle presented thirteen different memoirs, more than any of the other geographers except Buache. Indeed, only seven other members of the Institut National were more prolific than Mentelle.

The Classe des Sciences Morales et Politiques quickly became the institutional focus of the *idéologues,* a disparate group of antireligious intellectuals who sought to uncover the scientific principles underlying the formation and development of ideas, part of an ambitious plan to engineer a new, more rational, society.[72] As we have seen, Mentelle shared some of these underlying views, but his resolutely empirical approach rarely embraced this more abstract philosophical terrain. According to Martin Staum, however, the work of the geography section reflected and shaped many of the larger debates of the day. The ongoing discussions about Rousseau, particularly the question of "noble" versus "ignoble" savagery, the controversies about the nature and potential of human progress in different parts of the world,[73] the role of climate in the formation of different cultures, the relative importance of the physical and cultural environments in determining racial characteristics, and the implications these questions carried for the burning issue of slavery, were all themes Staum detects in the geographical teaching developed within the Classe des Sciences Morales et Politiques. Equally significant, claims Staum, was Mentelle's use of statistical techniques to back up his geographical presentations.[74]

Mentelle converted much of his new teaching into a range of additional books. His 1795 text, *La géographie enseigné par une méthode nouvelle; ou, Application de la synthèse à l'étude de la géographie* was especially commissioned by the Comité d'Instruction Publique as a geography textbook for republican primary schools, and it went through six editions down to 1813. Following an introductory section in which readers are introduced to the basics of map reading and orientation, the book considers the detailed geography of the town of Bourges in the Loire valley. The reader then progresses to consider the geography of the new *département* of the Cher, of which Bourges was *chef-lieu*, and so on through a series of expanding concentric circles ending with a consideration of the geography of France as a whole.[75] Additional historical works also appeared, notably a history of the Jews from Moses to the Romans; an abridged translation, the first into French, of William Guthrie's *Universal Geography;* and another three-volume work on cosmography, geography, and chronology.[76]

By the early years of the nineteenth century, Mentelle's prestige was such that Nicolas Baudin had named Cape Mentelle, in southwest Australia, in honor of his former geography teacher during his Pacific explorations of 1800 to 1804. Mentelle, by then in his early seventies but newly married for a second time to the daughter of the comte de Lanoue, might easily have slipped into a peaceful retirement, had not the drama of Napoleon's rise to power proved such an irresistible topic for him to explore. Unfortunately, Mentelle's first publication of the Napoleonic Consulate proved to be his most controversial. The *Précis de l'histoire universelle,* which appeared in 1801, marked the high point of Mentelle's antireligious writing. This odd review of early Christian Europe insisted that Jesus Christ was an impostor and that the early church had actively hindered the development of a peaceful coexistence between different peoples. These views would have seemed extreme, even at the height of antireligious purges in the mid-1790s. By 1801, they were wildly at odds with a rapidly changing public mood, and the *Précis de l'histoire universelle* dealt a significant blow to Mentelle's carefully cultivated reputation.

Napoleon's decision to reorganize the Institut National in 1803, and to dismiss the academics in the Classe des Sciences Morales et Politiques, including the geographers, was a further blow to Mentelle. Napoleon had no time for the abstract academic theorizing of most academics at the institute, an assessment that must have been especially galling for Mentelle, who took pride in the factual and educational nature of his work, far removed, he believed, from the vague ideas of the *idéologues*. This may explain why Mentelle tried so hard to convince the new Napoleonic regime of his practical support.

Following the Peace of Amiens in 1802, which held out at least the prospect of a lasting settlement of Anglo-French rivalries, Mentelle decided that the time was right for a general review of the world's political geography in the light of the preceding decades of more or less permanent warfare. Wrongly anticipating that the emergence of Napoleon would spell the end of revolutionary uncertainty in France and, as a result, the end of warfare in Europe, Mentelle conceived an ambitious multivolume political geography of the world. With the assistance of a younger colleague, Conrad Malte Brun, a political refugee from Denmark who had settled in revolutionary Paris during the 1790s and who later helped to establish the Société de Géographie de Paris in 1821, Mentelle threw himself into this new project with his usual energy. The first few volumes of what became a sixteen-volume work, entitled *Géographie, mathématique, physique et politique de toutes les parties du monde*, were published in 1803, and the entire work had appeared by the end of 1805.

This was a remarkable production, a work inspired by the same reasoning that would motivate Isaiah Bowman's *New World* (1921) more than a century later at the end of World War One. Like Bowman, Mentelle sought to describe the world as it emerged from an era of uncertainty and warfare in the hope and anticipation that the picture thus presented would remain in place for the foreseeable future. Mentelle's account would be overtaken by events even more rapidly than Bowman's once Napoleon's imperial adventure began to transform the political landscape of Europe, but Mentelle and Malte Brun's *Géographie* was a hugely successful publication in its day, not least because the authors were able to update the relevant parts after Napoleon's initial fall in 1813 for what would become a posthumous edition in 1816 entitled *Géographie universelle ancienne et moderne*.

Both editions of this enormous work were extremely complimentary toward Napoleon, as were the other works of history and geography Mentelle wrote in the last ten years of his life, particularly his 1804 *Géographie physique, historique, statistique et topographique de la France en cent huit départements*.[77] As the title suggests, this volume was a revision of Mentelle's earlier school geographies of France's revolutionary administrative structure, updated to take into consideration the territorial expansion of the country in the imperial age. This was subsequently revised, in its turn, into a posthumous, post-Napoleonic volume, published as late as 1821.

Even into his dotage, Mentelle remained an active and surprisingly passionate correspondent, particularly with women. His 1809 guide to the most appropriate teaching methods for young girls was an unusual reflection on Rousseau's educational ideas that seems to have been inspired by his correspondence with the pioneering feminist and educationalist, Stéphanie de Genlis, an inspector of schools during the First Empire.[78] He also corresponded with Constance de Salm, a prominent feminist playwright, on a range of topics, his letters dotted with exclamations of devotion and affection. Salm would return Mentelle's compliments in due course, by writing a short biography of him some years after his death.[79]

The books Mentelle wrote in the final years of his life, notably his *Exercices chronologiques* (1810), were among his most successful, being reissued several times down to the 1860s. But despite the praise Mentelle lavished on the emperor, his final honor—the Légion d'Honneur—was awarded, with supreme irony, just a few months before his death on 28 December 1815, not by Napoleon but by the newly restored monarch, Louis XVIII, elder brother of the comte d'Artois, Mentelle's prerevolutionary patron.[80]

GEOGRAPHY AND THE POLITICS OF SURVIVAL IN REVOLUTIONARY FRANCE

Edme Mentelle was a particular kind of geographer—he was not an explorer who wrote of his adventures, nor was he a natural scientist who reported on his experiments and observations. He was, rather, an educationalist, whose geographies were designed to communicate as many useful facts about the world as could readily be assimilated, particularly by the young. He saw himself as a writer rather than a scientist, and his most sustained exchanges and influences came from the world of letters and politics, particularly on educational matters, rather than the world of science.

Mentelle believed that geography had a very specific, and rather limited educational purpose, but his views on this matter were shaped by political expediency as well as intellectual conviction. He lived through an era of exceptional political upheaval, yet he remained successful under the monarchy, the various revolutionary regimes, and the imperial military dictatorship. He achieved this partly by reinventing himself politically and culturally, but also by maintaining a limited, uncontroversial, and inherently conservative perspective in his professional writings, which sought, to borrow Martin Staum's phrase, to "stabilize" the Revolution, at least in the pages of his textbooks. The trick was always to anticipate and never to overcommit; to do what was required rather than what was possible. To seek for larger explanations and greater truths might bring intellectual satisfaction, and even fleeting fame; but in the context of the French Revolution, fame was both fickle and dangerous.

The discrepancy identified at the start of our discussion, between Mentelle's resolutely antitheoretical, fact-laden texts, so often reviled by subsequent historians, and the ideologically charged context in which they were written, may partially be explained by his understandable desire to avoid potentially controversial theoretical discussion in an era when political and intellectual shifts were unpredictable, bewilderingly frequent, and occasionally life threatening. For long periods during the 1790s, and particularly during the Terror, when so many of Mentelle's friends and political allies were executed, the official, Jacobin view of science and education was defined in strictly utilitarian terms. Abstract, theoretical discussion was seen as unnecessary, elitist and ideologically suspect.[81] In this sense, Mentelle's pallid geographies simply reflected what the regime required, and therefore lend support to the contention developed by historians of other sciences who have pointed out the disastrous consequences of the unprecedented attempt to shape and control the scientific process to meet ideological needs during the Revolution.[82]

Viewed in this way, it is scarcely surprising that Mentelle's extensive writings have not stood the test of time. But the twists and turns of Mentelle's complex geobiography—his boundless energy, his endless quest for influence and prestige, his repeated compromises and reversals, and his occasional acts of defiance and even heroism—tell us a great deal about the highly personalized politics of geographical writing at a pivotal moment in European history.

...

Notes

The epigraph to this chapter is taken from Claude Perroud's description of Edme Mentelle's actions on behalf of Jeanne-Marie Roland de la Platière (Mme Roland) during the latter's imprisonment prior to her execution, on trumped-up charges of harboring monarchist sympathies, in October 1793. See Claude Perroud, "Jany, le dernier correspondant de Madame Roland," *La Révolution Française* 30 (1896): 37. I wish to thank Roselyn du Perray of the Musée du Château de Versailles for her assistance with Mentelle's globe for the education of the dauphin; Professor Randolph Runyon of Miami University, Ohio, who is a descendant of Mentelle, for his most generous advice; and the editors of this volume for their remarkable patience.

1. Leslie R. Marchant, "Edmunde Mentelle, 1730–1815, and François-Simon Mentelle, 1731–1799," *Geographers: Biobibliographical Studies* 11 (1987): 93–103.

2. Anne Godlewska, *Geography Unbound: French Geographic Science from Cassini to Humboldt* (Chicago: University of Chicago Press, 1999).

3. Ibid., 57–66.

4. The main works by Robert Darnton are *Mesmerism and the End of the Enlightenment in France* (Cambridge: Harvard University Press, 1968); *The Business of Enlightenment: A Publishing History of the Encyclopédie, 1775–1800* (Cambridge: Harvard University Press, 1979); *The Literary Underworld of the Old Regime* (Cambridge: Harvard University Press, 1982); *The Great Cat Massacre and Other Episodes in French Cultural History* (Harmondsworth: Penguin, 1984); *The Forbidden Best-Sellers of Pre-Revolutionary France* (New York: Norton, 1995); and *The Corpus of Clandestine Literature in France, 1765–1789* (New York: Norton, 1995). See also Robert Darnton and Daniel Roche, eds., *Revolution in Print: The Press in France, 1775–1800* (Los Angeles and Berkeley: University of California Press, 1989).

5. Darnton, *Literary Underworld*, 1–40. See also Roger Chartier, *The Cultural Origins of the French Revolution* (Durham: Duke University Press, 1991).

6. Darnton, *Literary Underworld*, 1–40.

7. Jean-Baptiste Brissot de Warville, *Théorie des lois criminelles*, 2 vols. (1781; Paris: J.-P. Aillaud, 1836); and Darnton, *Literary Underworld*, 1–40; though see also Simon Burrows, "The Innocence of Jacques-Pierre Brissot," *Historical Journal* 46 (2003): 843–71.

8. The idea that biography, or "life paths," can be used to illuminate historical geographies of period and place is the starting point for Stephen Daniels, *Humphry Repton: Landscape Gardening and the Geography of Georgian England* (New Haven: Yale University Press, 1999), and is further explored in the essays coedited by Stephen Daniels and Catherine Nash, eds., *Lifepaths: Geography and Biography*, special issue of *Journal of Historical Geography* 30 (2004). The conventional biography has always been an important format for the history of geography, of course, but more complex forms of biographical investigation are also suggested by Felix Driver's work on Winwood Reade, a minor Victorian writer, Darwinist, and would-be explorer. Driver's analysis of Reade's life and works goes beyond the conventions of a normal biography to examine how Reade's struggles to reinvent himself as an explorer can illuminate Victorian attitudes to exploration. See Felix Driver, *Geography Militant: Cultures of Exploration and Empire* (Oxford: Blackwell, 2001), 96–116.

9. Godlewska, *Geography Unbound*, 18–86, esp. 57–86.

10. Josef Konvitz, *Cartography in France, 1660–1848: Science, Engineering, and Statecraft* (Chicago: University of Chicago Press, 1987); Numa Broc, *La géographie des philosophes: Géographes et voyageurs français au XVIIIe siècle* (Paris: Ophrys, 1974).

11. Frank A. Kafker and Serena L. Kafker, *The Encyclopedists as Individuals: A Biographical Dictionary of the Authors of the Encyclopédie*, Studies on Voltaire and the Eighteenth Century (Oxford: Voltaire Foundation, 1988), 330–33.

12. Godlewska, *Geography Unbound*, 191–303; on Volney, see Michael Heffernan, "Historical Geographies of the Future: Three Perspectives from France, 1750–1825," in *Geography and Enlightenment*, ed. David N. Livingstone and Charles W. J. Withers (Chicago: University of Chicago Press, 1999), 125–64, esp. 136–46.

13. Broc, *La géographie des philosophes*, 460; see also John Dunmore, *French Explorers in the Pacific*, 2 vols. (Oxford: Oxford University Press, 1965); Catherine Gaziello, *L'Expédition de Lapérouse, 1785–1788: Réplique française aux voyages de Cook* (Paris: C.T.H.S., 1984); and Hélène Richard, *Une grande expédition scientifique au temps de la Révolution française: Le voyage de d'Entrecasteaux à la recherche de Lapérouse* (Paris: C.T.H.S., 1986).

14. Broc, *La géographie des philosophes*, 460–74.

15. Martin Staum, *Minerva's Message: Stabilizing the French Revolution* (Montreal and Kingston: McGill–Queen's University Press, 1996), 159–60, 171.

16. Quoted in Hayden White, *Metahistory: The Historical Imagination in Nineteenth-Century Europe* (Baltimore: Johns Hopkins University Press, 1973), 151.

17. On the republican calendar, which replaced the Gregorian calendar with a new sequence of months and years commencing with the establishment of the Republic rather than the death of Christ, see Bronislaw Baczko, "Le calendrier républicain," in *Les lieux de mémoire,* vol. 1, *La République,* ed. Pierre Nora (Paris: Gallimard, 1984), 38–82. Popular conceptions of a speeded-up time have also been detected by some historians in autobiographical writings from the 1790s. See Lynn Hunt, "The World We Have Gained: The Future of the French Revolution," *American Historical Review* 108 (2003): 1–19; and Dorinda Outram, "Life Paths: Autobiography, Science and the French Revolution," in *Telling Lives in Science: Essays on Scientific Biography,* ed. Michael Shortland and Richard Yeo (Cambridge: Cambridge University Press, 1996), 85–102.

18. A concern with the spaces of the Revolution has emerged as a key characteristic of the so-called new cultural history, heralded in the late 1970s by François Furet's stridently anti-Marxist reading of the revolutionary period in *Penser la Révolution française* (Paris: Maspéro, 1978), and developed by such general works as Keith Baker, *Inventing the French Revolution: Essays on French Political Culture* (Cambridge: Cambridge University Press, 1990); Chartier, *Cultural Origins;* Lynn Hunt, *Politics, Culture, and Class in the French Revolution* (Berkeley and Los Angeles: University of California Press, 1984); Lynn Hunt, *The Family Romance of the French Revolution* (London: Routledge, 1992); and Emmet Kennedy, *A Cultural History of the French Revolution* (New Haven: Yale University Press, 1989). Historical and cultural geographies of the Revolution can be uncovered from the work on cartographic and other representations of the Republic's national space by Josef Konvitz, *Cartography in France,* 32–55; Josef Konvitz, "The Nation-State, Paris and Cartography in Eighteenth- and Nineteenth-Century France," *Journal of Historical Geography* 16 (1990): 3–16; and Daniel Nordman, "La pédagogie du territoire, 1793–1814," in *Atlas de la Révolution française,* vol. 4, *Le territoire: Réalités et représentations,* ed. Daniel Nordman and Marie Vic-Ozouf Marignier (Paris: Gallimard, 1989), 62–64; on regional scale geographical tensions and representations by Ted Margadant, *Urban Rivalries in the French Revolution* (Princeton: Princeton University Press, 1992); Mona Ozouf, "La Révolution française et la perception de l'espace national: Fédérations, fédéralisme, et stéréotypes régionaux," in *L'Ecole de la France: Essais sur la Révolution, l'utopie et l'enseignement,* ed. Mona Ozouf (Paris: Gallimard, 1984), 27–54; Marie Vic-Ozouf Marignier, *La formation des départements et la réprésentation du territoire français à la fin du XVIIIe. siècle* (Paris: Gallimard, 1989); and Michel Vovelle, *La découverte de la politique: Géopolitique de la Révolution française* (Paris: La Découverte, 1993); on the ideas behind local urban reconstruction by Richard A. Eltin, *Symbolic Space: French Enlightenment Architecture and Its Legacy* (Chicago: University of Chicago Press, 1994), 30–47; and Benjamin Nathans, "Habermas's 'Public Sphere' in the Era of the French Revolution," *French Historical Studies* 16 (1990): 620–44; and on personal, corporeal spaces by Dorinda Outram, *The Body and the French Revolution: Sex, Class, and Political Culture* (New Haven: Yale University Press, 1989); and Richard Wrigley, *The Politics of Appearances: Representations of Dress in Revolutionary France* (Oxford: Berg, 2002). Mona Ozouf's work *Festivals and the French Revolution* (Cambridge: Harvard University Press) is a key statement on the spatial performance of the Revolution, and from the history of science, Emma Spary's *Utopia's Garden: French Natural History from Old Regime to Revolution* (Chicago: University of Chicago Press, 2000) reveals how a specific scientific site, the Muséum d'Histoire Naturelle, was implicated in the formation of revolutionary culture.

19. For a fascinating study of the constitutive power of space, in this case the "spaces of modernity" in eighteenth-century London, see Miles Ogborn, *Spaces of Modernity: London's Geographies, 1680–1780* (New York: Guilford, 1998).

20. Ozouf, *Festivals and the French Revolution,* 126.

21. Broc, *La géographie des philosophes*; Marchant, "Edmunde Mentelle," 93–4.

22. Edme Mentelle, *La mort de Polieucte* (Paris, 1751).

23. Edme Mentelle, *Lettre à un seigneur étranger sur les ouvrages périodiques de France, par M. l'abbé D. C. d'H**** (Paris, 1757).

24. Darnton, *Business of Enlightenment*, 9–14; Frank A. Kafker, *The Encyclopedists as a Group: A Collective Biography of the Authors of the Encyclopédie*, Studies on Voltaire and the Eighteenth Century (Oxford: Voltaire Foundation, 1996), 17–120.

25. Catherine M. Northeast, *The Parisian Jesuits and the Enlightenment, 1700–1762*, Studies on Voltaire and the Eighteenth Century (Oxford: Voltaire Foundation, 1991); and Dale van Kley, *The Jansenists and the Expulsion of the Jesuits from France, 1757–1765* (New Haven: Yale University Press, 1975).

26. Marcel Grandière, *L'idéal pédagogique en France au XVIIIe siècle*, Studies on Voltaire and the Eighteenth Century (Oxford: Voltaire Foundation, 1998); and Gilbert Py, *Rousseau et les éducateurs: Étude sur la fortune des idées pédagogiques de Jean-Jacques Rousseau en France et en Europe au XVIIIe siècle*, Studies on Voltaire and the Eighteenth Century (Oxford: Voltaire Foundation, 1997).

27. On the Jesuits and geography, see François de Dainville, *La géographie des humanistes* (Paris: Beauchesnes, 1940). It should be emphasized that Mentelle's textbook was remarkably similar to others produced around the same time. See Yves Saint-Hilaire, *Élémens de géographie; ou, Nouvelle méthode simple et abrégée, pour apprendre en peu de temps et sans peine la géographie* (Lyons, 1750); and M. de Leris, *La géographie rendue aisée; ou, Traité méthodique pour apprendre la géographie* (Paris, 1753). Bizarrely, Mentelle's *Élémens de géographie* was reissued in 1783 with no significant alteration, despite the new discoveries in the Pacific. See Edme Mentelle, *Élémens de géographie, contenant: 1er, Les principales divisions des quatre parties du monde; 2ème, Une description abrégée de la France, à l'usage des commerçans, avec des cartes* (Paris: By the author, 1783).

28. R. Laulan, "L'enseignement à l'École Royale Militaire de Paris de l'origine à la réforme du comte de Saint-Germain," *L'Information Historique* 19 (1957): 152–58.

29. There are some scattered documents on Mentelle's teaching at the École Royale Militaire in the Archives Nationales MM679, Mémoires, 1760–65, and O1 1605 Mémoire, 1764; cited in François Labourie and Daniel Nordman, "Leçons de géographie du Buache et Mentelle," introduction to *L'École Normale de l'An II: Leçons d'histoire, de géographie, d'economie politique; Édition annoté des cours de Volney, Buache de la Neuville, Mentelle, et Vandermonde avec introductions et notes*, ed. Daniel Nordman (Paris: Dunod), 142. See also Grandière, *L'idéal pédagogique*, 173–82.

30. Edme Mentelle, *Manuel géographique, chronologique et historique, par M***, professeur d'histoire et de géographie, dédié à Mlle. de Fitz-James* (Paris: Dufour, 1761).

31. Mentelle's major works in this period were *Éléments de l'histoire romaine, divisée en deux parties; avec des cartes et un table* (Paris: A. Delalain, 1766); *Géographie abrégée de la Grèce ancienne, par un professeur d'histoire et de géographie* (Paris: Barbou, 1772); *Géographie comparée; ou, Analyse de la géographie ancienne et moderne des peuples de tous les pays et de tous les âges; Accompagnée de tableaux analytiques et d'un grand nombre de cartes*, 7 vols. in 8 books, plus atlas (Paris: By the author, 1778–84); *Cosmographie élémentaire, divisée en parties astronomique et géographique: Ouvrage dans lequel on a tâché de mettre les vérités les plus intéressantes de la physique céleste à la portée de ceux même qui n'ont aucune notion de mathématiques* (Paris: By the author, 1781); *Choix de lectures géographiques et historiques, présentées dans l'ordre qui a paru le plus propre à faciliter l'étude de la géographie de l'Asie, de l'Afrique et de l'Amérique, précédé d'un abrégé de géographie, avec des cartes* (Paris: By the author, 1783); *Géographie; ou, Annonce de Quelques ouvrages relatifs à cette science, avec quelques vues sur la manière de l'enseigner* (Paris: P.-G. Simon et N.-H. Nyon, 1784); *Dictionnaire de géographie*

moderne, 3 vols (Paris: Panckoucke, 1784–89); and *Encyclopédie méthodique: Géographie anci-enne,* 3 vols. (Paris: Panckoucke, 1787–93).

32. On the *Encyclopédie méthodique,* see Darnton, *The Business of Enlightenment,* 460–519. Labourie and Nordman, "Leçons de géographie," 144, claim some Mentelle titles sold as many as four thousand copies.

33. Terence Cave, *The Cornucopian Text: Problems of Writing in the French Renaissance* (Oxford: Clarendon, 1979); and Ann Moss, *Printed Commonplace Books and the Structure of Renaissance Thought* (Oxford: Clarendon, 1996).

34. As Robert Mayhew shows in his contribution to this volume (chapter 9), the Renais-sance concern with "method" emerged from the critique of Aristotle developed in Paris by the Protestant convert Pierre de la Ramée (Petrus Ramus) in the early sixteenth century, and had important implications for the organization of seventeenth-century English geographies.

35. The comte d'Artois has been described as "the most dashing of the Versailles bloods: a notorious rakehell at the hunt and in the ballroom and bed" (Simon Schama, *Citizens: A Chron-icle of the French Revolution* [London: Viking, 1989], 9). The comte became far more pious and serious-minded during his exile through the revolutionary period, after which he returned to France under the Restoration monarchy. He was eventually crowned King Charles X, in a quasi-medieval ceremony, in Rheims in 1824.

36. Ludovic Drapeyron, "L'éducation géographique de trois princes français au XVIIIe. siè-cle—le duc de Berry et les comtes de Provence et d'Artois [Louis XVI, Louis XVIII, Charles X]," *Revue de Géographie* 11 (1887): 244; quoted in Labourie and Nordman, "Leçons de géogra-phie," 138.

37. Christian Baulez, "Notes sur quelques meubles et objets d'art des appartements inté-rieurs de Louis XVI and Marie-Antoinette," *Revue du Louvre* 5/6 (1978): 360–73, esp. 364–65. Mentelle's globe has had several homes since 1789. It was moved to the Tuileries with the royal household immediately after the Revolution, and was then used by Mentelle himself in classes he delivered from an apartment in the Louvre during the mid-1790s. Napoleon moved it again, together with a host of other educational equipment, to the Château Meudon, where he planned to create an imperial institute wherein his own son, the king of Rome, and other princes of the imperial dynasty might be educated together. The globe was apparently sold to a private citi-zen in 1826 and was later reacquired by the Bibliothèque Nationale in 1877, where it was dis-played in the Salle des Cartes et Plans. In 1931, it was displayed in the entrance hall of the Ex-position Coloniale in Vincennes, the building which today houses the Musée des Arts d'Afrique et d'Océanie. It was returned to Versailles in the 1960s.

38. Barbara Luttrell, *Mirabeau* (New York: Harvester-Wheatsheaf, 1990).

39. Fernando Salleo, "Mirabeau en Prussie, 1786–1787: Diplomat parallèle ou agent sècret?" *Revue d'Histoire Diplomatique* 3/4 (1977): 346–56; and Henri Welshinger, ed., *La mission se-crète de Mirabeau à Berlin, 1786–1787; d'après les documents originaux des Archives des Af-faires Étrangères* (Paris: Plon, Nourrit et Cie., 1900).

40. Mirabeau, comte de [Riquetti, H.-G], *De la monarchie prussienne sous Frédéric le Grand, avec un appendice contenant des recherches sur la situation actuelle des principales contrées de l'Allemagne,* 7 vols., plus atlas (London, 1788). Mirabeau dedicated this work to his formidable father, the marquis de Mirabeau, who was a noted writer on economics, in the hope of repair-ing their troubled relationship, one of the many casualties of Mirabeau's youthful ill-discipline, which had seen him imprisoned, pursued by creditors, and condemned to death for seducing and "abducting" the (by no means unwilling) wife of another man.

41. Ibid., 1:241–431; 2:1–47.

42. Mirabeau, comte de [Riquetti, H.-G], *Histoire secrète de la cour de Berlin; ou, Correspon-dence d'un voyageur françois, depuis le mois de juillet 1786 jusqu'au 19 janvier 1787; Ouvrage posthume avec une lettre remise au roi de prusse regnant, le jour de son avènement au trône,* 2 vols.

(London: S. Bladon, 1789). See also M. de la Haye de Launay, *Justification du système d'économie politique et financière de Frédéric II, roi de Prusse, pour servir de réfutation à tout ce que le comte de Mirabeau a hazardé à ce sujet dans son ouvrage de la monarchie prussienne* (Paris, 1789).

43. Labourie and Nordman, "Leçons de géographie," 142.

44. The papers of Thomas Clarkson, the campaigning young member of the British Society for Effecting the Abolition of the Slave Trade, established a few months earlier than its French equivalent in April 1787, indicate that Mentelle's maps and other information were circulating among antislavery activists in both Paris and London. Clarkson's papers include thirteen letters in French addressed to Mirabeau in late 1789 and early 1790 designed to assist the French campaign. They deal with various aspects of the slave trade through the French-controlled West African enclaves of Gorée and St. Louis, based on maps and other statistics derived (either directly or indirectly) from Mentelle. See Thomas Clarkson Manuscripts, AMS, Lettres nouvelles sur le commerce de la Côte de Guinée, December 1789–January 1790, William L. Clements Library, University of Michigan. On Clarkson and French antislavery debates before and during the Revolution, see Robin Blackburn, *The Overthrow of Colonial Slavery, 1776–1848* (London: Verso), 138–40 and 161–264.

45. Jean-Baptiste Brissot, *Mémoires, 1754–1793*, [ed. C. Perroud], (Paris: Picard, [1912?]), 1:178; Labourie and Nordman, "Leçons de géographie," 142.

46. Mirabeau's funeral produced an outpouring of popular grief as his body was laid to rest in the newly established Panthéon, though this turned to bitter anger when documents were subsequently discovered proving his secret negotiations with the king to protect the monarchy in return for cash. See Luttrell, *Mirabeau*, 321–32.

47. It should be noted that Mentelle continued to style himself as historiographer of the comte d'Artois until his patron's exile in 1791.

48. Michael Heffernan, "Rogues, Rascals and Rude Books: Policing the Popular Book Trade in Early Nineteenth-Century France," *Journal of Historical Geography* 16 (1990): 90–107, esp. 93–98; and more generally, Jack R. Censer, *The Prelude to Power: The Parisian Radical Press, 1789–1791* (Baltimore: Johns Hopkins University Press, 1976); Hugh Gough, *The Newspaper Press in the French Revolution* (Oxford: Oxford University Press, 1988); Jeremy D. Popkin, *The Right-Wing Press in France, 1792–1800* (Chapel Hill: University of North Carolina Press, 1980); Jeremy D. Popkin, *Revolutionary News: The Press in France, 1789–1799* (Durham: Duke University Press, 1990).

49. For a brilliant analysis of the formation of a revolutionary identity, see Timothy Tallack, *Becoming a Revolutionary: The Deputies of the French Assembly and the Emergence of a Revolutionary Culture* (Princeton: Princeton University Press, 1996).

50. Keith M. Baker, *Condorcet: From Natural Philosophy to Social Mathematics* (Chicago: University of Chicago Press, 1975).

51. Claude Perroud, ed., *Lettres de Madame Roland (Jeanne-Marie Roland), 1788–1793* (Paris: Imprimérie Nationale, 1900–2), 2:767.

52. Simon Dalton, "Gender and the Shifting Ground of Revolutionary Politics: The Case of Madame Roland [Jeanne-Marie Roland]," *Canadian Journal of History* 36 (2001): 259–82; and Gita May, *Madame Roland and the Age of Revolution* (New York: Columbia University Press, 1970).

53. Perroud, *Lettres de Madame Roland*, 2:767–77.

54. Ibid., 769–70.

55. Baulez, "Notes sur quelques meubles," 364; Labourie and Nordman, "Leçons de géographie," 143.

56. Edme Mentelle, *Tableau élémentaire de la géographie de la République française, à l'usage des écoles du 1er et 2ème âge* (Paris: F. Hocquet, 1792), 13; also quoted in Ozouf, "La Révolution française," 33.

57. On the complex politics leading to the purge of the Girondins, see Morris Slavin, *The Making of an Insurrection: Parisian Sections and the Gironde* (Cambridge: Harvard University Press, 1986).

58. Claude Perroud, "Jany, le dernier correspondant de Madame Roland," *La Révolution Française* 30 (1896): 15.

59. In the intervening years, Mentelle apparently forgot the code name he had used seven years earlier, recalling this as "Betzy" rather than "Jany" in his 1800 letter. According to Perroud, this was either "une lamentable défaillance de mémoire" or "une énorme distraction de plume" ("Jany," 35).

60. Ibid.; Claude Perroud, "Un dernier mot sur Jany-Mentelle," *La Révolution Française* 30 (1896): 227–28.

61. See Mentelle, *Géographie comparée*, published in seven volumes between 1778 and 1784 (see note 31 above).

62. Mentelle's letter and the subsequent report can be found in James Guillaume, ed., *Procès-verbaux du Comité d'Instruction Publique de la Convention Nationale* (Paris: Ministère de l'Instruction Publique, 1893), 2 : 136, 526–27, and 599.

63. Emphasis in the original. See Edme Mentelle, *Déclaration des droits de l'homme et du citoyen, mise en trente strophes, pour être chantée par les hommes libres de tout pays* (Paris: Imprimérie des Sans-Culottes / Maret, 1794).

64. Robert R. Palmer, *The Improvement of Humanity: Education and the French Revolution* (Princeton: Princeton University Press, 1985).

65. Nordman, *L'École Normale de l'An II*, 163–336.

66. Quoted in Labourie and Nordman, "Leçons de géographie," 145; and in Nordman, *L'École Normale de l'An II*, 169.

67. Py, *Rousseau et les éducateurs*, 512–20.

68. Godlewska, *Geography Unbound*, 57–86. As we have noted, Godlewska is entirely in sympathy with the anonymous reviewer, but for more positive assessments of Mentelle's École Normale lectures, see Labourie and Nordman, "Leçons de géographie," and Sergio Moravia, "Philosophie et géographie à la fin du XVIIIe. siècle," *Studies on Voltaire and the Eighteenth Century* 57 (1967): 937–1011, esp. 949–53.

69. Labourie and Nordman, "Leçons de géographie," 144; and Louis Trenard, "Manuels scolaires au XVIIIe. siècle et sous la Révolution," *Revue du Nord* 55 (1973): 99–111.

70. Numa Broc, "Une Musée de Géographie en 1795," *Revue d'Histoire des Sciences* 27 (1974): 37–43.

71. Maurice Crosland, *Science in France in the Revolutionary Era* (Cambridge: Harvard University Press, 1969); Roger Hahn, *The Anatomy of a Scientific Institution: The Paris Academy of Science, 1666–1803* (Berkeley and Los Angeles: University of California Press, 1971); Dorinda Outram, "The Ordeal of Vocation: The Paris Academy of Sciences and the Terror, 1793–95," *History of Science* 21 (1983): 257–73; Daniel Roche, *Le siècle des lumières en province: Académies et académiciens provincaux, 1680–1789* (Paris and the Hague: Mouton, 1978).

72. Cheryl B. Welch, *Liberty and Utility: The French Idéologues and the Transformation of Liberalism* (New York: Columbia University Press, 1984).

73. Heffernan, "Historical Geographies of the Future."

74. Staum, *Minerva's Message*, 154–71. See also Martin Staum, "Human Geography in the French Institute: New Discipline or Missed Opportunity?" *Journal of the History of the Behavioural Sciences* 23 (1987): 332–40. According to Larry Wolff's analysis of French attitudes to eastern Europe in the eighteenth century, Mentelle's reliance on stereotypical views of Russia merely restated ideas previously expressed by Voltaire and other leading figures of the Enlight-

enment. See Larry Wolff, *Inventing Eastern Europe: The Map of Civilization on the Mind of the Enlightenment* (Stanford: Stanford University Press, 1994), 195–234.

75. Nordman, "La pèdagogie du territoire." According to Labourie and Nordman, "Leçons de géographie, 145–46, the local archives in Seine-et-Marne indicate that Mentelle's text was used in the many disputes between different towns hoping to be given *chef lieu* status.

76. Edme Mentelle, *Précis de l'histoire des Hébreux depuis Moyse jusqu'à la prise de Jérusalem par les Romains: Ouvrage dans lequel on a tâché de concilier l'exactitude des faits avec les sains lumières de la raison, à l'usage des écoles primaires et centrales de la République française* (Paris: By the author, 1797–98); *Abrégé de la Géographie universelle de William Guthrie* (Paris: Bernard, 1799); *Cours de cosmographie, de géographie, de chronologie et d'histoire ancienne et moderne, divisé en cent vingt–cinq leçons,* 3 vols. (Paris: Bernard, 1800–1801).

77. See also Edme Mentelle, *Abrégé élémentaire de géographie ancienne et moderne* (Paris: Bernard, 1804), and *Géographie classique et élémentaire,* 2 vols. (Paris: Germain-Mathiot, 1813).

78. Edme Mentelle, *Études convenables aux demoiselles, à l'usage des écoles et des pensions* (Paris: Bossange, 1809); and Jennifer Birkett, "Madame de Genlis: The New Man and the Old Eve," *French Studies* 42 (1988): 150–64; Penny Brown, "'La femme enseignante': Mme de Genlis and the Moral and Didactic Tale in France," *Bulletin of the John Rylands University Library* 76 (1994): 23–42; Marie-Emmanuelle Plagnol-Diéval, *Madame de Genlis et le Théâtre d'Éducation au XVIIIe. siècle,* Studies on Voltaire and the Eighteenth Century (Oxford: Voltaire Foundation, 1997); Judith Still, "Genlis's *Mademoiselle de Clermont:* A Textual and Intertextual Reading," *Australian Journal of French Studies* 37 (2000): 331–47.

79. See Elizabeth Colwill, "Epistolary Passions: Friendship and the Literary Public of Constance de Salm, 1767–1845," *Journal of Women's History* 12 (2000): 41; and Constance de Salm, *Notice sur la vie et les ouvrages de Mentelle* (Paris: F. Didot, 1839).

80. Commentaries on Mentelle's life soon after his death include Bon-Joseph Dacier, "Notice historique sur la vie et les ouvrages de M. Mentelle," *Histoire et Mémoires de l'Institut Royal de France: Académie des Inscriptions et Belles-Lettres* 7 (1824): 212–222; and P.-J. Larche, "Notice sur Mentelle," *Magazin Encyclopédie* 1 (1816): 359–71.

81. Patrice Higonnet, *Goodness beyond Virtue: Jacobins during the French Revolution* (Cambridge: Harvard University Press, 1998).

82. Crosland, *Science in France;* Nicole and Jean Dhombres, *Naissance du pouvoir: Sciences et savants en France, 1793–1824* (Paris: Payot, 1989); and Charles Coulson Gillispie, *Science and Polity in France at the End of the Old Regime* (Princeton: Princeton University Press, 1981).

11

"RISEN INTO EMPIRE"

Moral Geographies of the American Republic

...

David N. Livingstone

I t was precisely because the United States had recently "risen into Empire," that the Reverend Jedidiah Morse felt compelled to embark on the business of geography. Hitherto, he lamented, "Europeans have been the sole writers of American Geography, and have too often suffered fancy to supply the place of facts." Now, "since the United States have become an independent nation," Morse insisted, "it would be reproachful for them to suffer this ignorance to continue."[1]

Morse's sentiments, expressed in what his biographer Richard Moss calls "the book that made him America's leading geographer," advertise the intimate connection between geographical writing and questions of national identity.[2] This is a theme that has, of late, begun to receive critical attention by historians of geography, not least because of the power both history and geography have to foster what David Hooson calls "identity-consciousness."[3] Now, thanks to a raft of specialist studies, we are beginning to appreciate something—to take a random sample—of how geographical performances like surveying and public masques were instrumental in national "self-fashioning" in mid-seventeenth-century Britain, of the role played by mapping in the manufacture of a sense of national identity in a variety of regional settings, of how in Ireland school geography was "taught as an *instrumentum regni*."[4] Such cases could doubtless be elaborated in extenso.[5]

Here I want to turn attention to the production of geographies in the period of the American Revolution and to draw upon several different species of geographical text in order to ascertain how they were mobilized in different ways to support conceptions of the new Republic. To be sure, all of the authors were united in the conviction that the facts of geography underwrote the moral and political legitimacy of the Revolution. But beyond that, these works of geographical scholarship disclosed different strategies for justifying how best to regulate a novel set of governmental arrangements, for determining what should be learned from the moral economy of the different states, for figuring out how to shore up public virtue in an era jettisoning inherited tradition, monarchy, social hierarchy, and established religion as the grounds of civic authority, and for instilling in the new nation a sense of its own identity. All this means that these texts have to be read differently from how historians of geography have traditionally approached them. Despite the popularity of Morse's geographies in the early Republic, for instance, his significance was relegated to a mere footnote in Preston James's compendious survey.[6] William Warntz called attention to "inaccuracies resulting from scanty information or uncritical acceptance of material sent to him from the various localities he described" and remarked that "his books suffered further from bias in favour of New England, religious orthodoxy, and of extreme conservatism in morals."[7] Margarita Bowen observes that the "long popularity of Morse's work is indicative of the lack of vigour in American geography at the this time."[8] John Greene finds Morse's endeavors wanting on account of his failure to keep "abreast of the latest researches in American geography, geology, and natural history," so much so that, Greene surmises, if "Jefferson ever read the account of American geography and natural history in Morse's 1819 edition, he must have been deeply discouraged and not a little disgusted."[9] Such evaluations may or may not approach accuracy. My concerns are simply different. I want to read geographical texts of this sort as active political interventions at a time when the newly born United States of America was struggling to establish itself. For, as has recently been argued, "the culture of geographic letters . . . [was] integral to the early republican culture of letters in the United States."[10]

What further supports scrutinizing geographical texts in the period along these lines, is the realization that during the 1780s and 1790s, genre painters like Ralph Earl routinely depicted human subjects in the different regions— a Massachusetts matron, a Connecticut shopkeeper, a Vermont tavern owner— using "the discursive materials of geography." These citizens, Martin Brückner tells us, "frequently posed with geographic instruments"—globes, charts, atlases, and the like—and in so doing rhetorically called upon geographical discourse to "negotiate and transform the representation of personal, regional, and political difference into material figures of national consent." Indeed, Brückner goes so far as to suggest that it was through textual geography that "Americans invented their variant of modern nationalism" because in such texts they found it possible to reconcile "regional diversity and geopolitical unity" and thus to produce "the nation as a material and inherently readable form."[11] If this analysis approaches accuracy, the advent of print culture and the generation of geographical works constituted "a basic structure and imaginary form through which Americans effectively gave themselves the official imprimatur of a national identity."[12] Geoliteracy provided the ideological cement that held together Massachusetts farmers and Virginia lawyers, Connecticut merchants and Georgia plantation owners. Through geography texts, not least school textbooks, disparate regional peoples and cultures were fashioned into local varieties of a generic *Homo americanus.*

At the same time, these texts promised to deliver Noah Webster's dream of a geoliterate nation inhabited by children who, as he put it in his essay *"On the Education of Youth"* in 1788, were well "acquainted with [their] own country" and formed "attachments to it."[13] Webster, after all, had planned to cooperate with Morse on a geography treatise, and as Robert Lawson-Peebles observes, the hopes Morse "nursed for American Geography were echoed by Webster in relation to American language."[14] Geography books, then, were seen "as the material conduit through which to achieve this imaginary fusion between the concrete individual and the abstract figure of the nation."[15] All of this, moreover, was in keeping with Webster's vision of national standards of word pronunciation in order to facilitate communication between regions and classes, and as he himself wrote in the introduction to his 1783 *Grammatical Institute* (commonly known as Webster's *Speller*), "to promote the honour and prosperity of the confederated republics of America; and cheerfully [to throw] his mite into the common treasure of patriotic exertions."[16] And it clearly resonated with John Jay's comment in the second *Federalist* that Providence had delivered a common geographic space to a single people: "Providence has been pleased to give this one connected country to one united people," he remarked. Indeed, it had given him much pleasure to reflect "that independent America was not composed of detached and distant territories, but that one connected, fertile, wide-spreading country, was the portion of our western sons of liberty."[17] Any text that impressed on the new nation's citizens a shared sense of geographical identity played a constitutive role in the construction of national self-consciousness.

THOMAS JEFFERSON'S VIRGINIA AND THE QUERELLE D'AMERIQUE

Thomas Jefferson's *Notes on the State of Virginia,* of course, is the natural point of departure. First published in English in 1787, this classic of the early American canon was a stimulus to the growth of scientific learning in the early Republic. In the present context, detailed exegesis is unnecessary. Jefferson's more general scientific interests and the role that scientific inquiry played in his political thinking are not germane to my intentions here.[18] Nor is there any need to dwell on the contradictions that Jefferson's personal life discloses—an American railing against the corruptions of commercial individualism yet enjoying the epicurean delights of Monticello, a slaveholding landowner championing inalienable natural rights.[19] My purpose in referring to Jefferson's most celebrated publication (the Declaration of Independence passing as his magnum opus) is simply to provide something of the context within which the moral geographies to which I shall presently turn need to be domesticated.

Composed in response to the comprehensive questionnaire on the geography, products, institutions, religion, and finances of Virginia that François Marbois, the secretary of the French Legation in Philadelphia, had presented to Jefferson, *Notes on the State of Virginia* was in essence a regional inventory.[20] It surveyed everything from the state's boundaries, rivers, mountains, and climate to seaports, population, laws, manners, and much else besides. But it was withal an imagined inventory of an envisioned Virginia. Take, for example, his gloss on the moral virtues of agrarian laborers in Query 19 dealing with Virginian "Manufactures":

> Those who labour in the earth are the chosen people of God, if ever he had a chosen people, whose breasts he has made his peculiar deposit for substantial and genuine virtue. It is the focus in which he keeps alive that sacred fire, which otherwise might escape from the face of the earth. Corruption of morals in the mass of cultivators is a phaenomenon of which no age nor nation has furnished an example. It is the mark set on those, who not looking up to heaven, to their own soil and industry, as does the husbandman, for their subsistance, depend for it on the casualties and caprice of customers. . . . The mobs of great cities add just so much to the support of pure government, as sores do to the strength of the human body. It is the manners and spirit of a people which preserve a republic in vigour. A degeneracy in these is a canker which soon eats to the heart of its laws and constitution.[21]

Of course Jefferson's was a *re*imagined Virginia. The seemingly dry rehearsal of the state's "mines and other subterraneous riches; its trees, plants, fruits, &c," for example, was designed to counter its earlier projection as either El Dorado or Gehenna, as heaven or hell. In contrast to these metaphorical extremes, Jefferson's Virginia was a real place, measurable, fathomable, and thus conceivable as a site for a new form of political being. Such declarations confirm, as Robert Dawidoff remarks, that the "Jeffersonian imagination envisioned the possible, the desirable, and the hopeful, set in a pervasive future that might invade the present and surely colored the past. . . . The *Notes* abounds in this special sort of Jeffersonian vision, which is his way of reading the scientific tea leaves and picturing the future. In the book Virginia becomes a scientifically delineated and republican paradise, the right setting for the free government Jefferson has in mind."[22] This vision, moreover, was all of a piece with Jefferson's understanding of the imagination as an organ of moral conveyance. Reflecting on the uses of fiction, he observed that here the "spacious field of imagination is . . . laid open to our use, and lessons may be formed to illustrate and carry home to the mind every moral rule of life."[23]

At the same time, *Notes on the State of Virginia* was also designed to defend Jefferson's own state from the calumnies to which it had been subjected by European writers, the French naturalist Buffon in particular. It was thus a key text in the "dispute of the New World," albeit structured according to the norms of European regionalism as set forth in Robert Boyle's *General Heads for a Natural History of a Country* of 1692. Because Buffon had expanded on the zoological inferiority of the American continent, Jefferson elaborated in extenso on the greater size and volume of animals in the New World. As Antonello Gerbi tellingly put it: "Jefferson gets carried away with enthusiasm at seeing so many of his champions, his reindeer and bears and wolves, defeating or at least equaling in weight the European champions; he is determined to rout his opponent completely, and onto the metaphorical scales he throws the vast American mammoth, bringing the balance crashing down on the side of the New World."[24]

At the same time, Jefferson's zoogeographical philippic was an anthropological apologia for the state of *human* nature in the New World. Here his target was Guillaume Thomas François Raynal, the French abbé who had presumed to inform European readers that America was immature and its people decrepit. When Raynal ventured the opinion that "America has not yet produced . . . one man of genius in a single art or a single science," Jefferson rose to the bait. Had not America produced a Washington ("whose memory will be adored while liberty shall have votaries"), a Franklin ("than whom no one of the present age has made more important discoveries"), and a Rittenhouse ("second to no astronomer living")?[25] Jefferson's sense of moral outrage at Raynal's slur can be gleaned from his vituperative quip that, even if France could achieve a quota of genius per head of population comparable to America, the glory of Britain was "fast descending to the horizon. Her philosophy," he went on, "has crossed the Channel, her freedom the Atlantic, and herself seems passing to that awful dissolution, whose issue is not given human foresight to scan"[26]—and this, despite his assigning to the gallery of immortals on Monticello's hallowed walls portraits of Isaac Newton, Francis Bacon, and John Locke. Indeed, Jefferson's case for the anthropological defense extended to the native peoples of America, though—as is well known—he urged, on putatively scientific grounds, that whites and blacks should be permanently segregated.[27] Nevertheless, he used the celebrated 1774 speech of Chief Logan to the governor of Virginia, Lord Dunmore, to combat what he described as "the contumelious theory of certain European writers, whose celebrity gave currency and weight to their opinions, that our country, from the combined effects of soil and climate, degenerated animal nature, in the general, and particularly the moral faculties of man."[28] Here we find laid bare the motive forces that underlay Jefferson's penning of the *Notes*. For, as he explained in a letter to Governor Henry of Maryland in 1797, his treatise "had called to the bar of fact and reason" the repugnant "theory, so unfounded and degrading to one third of the globe," that "supposed that there is something in the soil, climate and other circumstances of America, which occasions animal nature to degenerate, not excepting even the man, native or adoptive, physical or moral."[29] *Notes on the State of Virginia* was a geographical apologia for the moral integrity at once of American nature and of American human nature. America was no place of moral peril; its physical, human, and moral geography was well suited to the needs of a new republic.

JEDIDIAH MORSE AND THE MORAL TOPOGRAPHY OF THE NEW REPUBLIC

In the early decades of the American Republic, thousands of Americans learned about "their" nation through the writings of a Congregationalist clergyman, the Reverend Jedidiah Morse (1761–1826). His various geographies "were dominant in the nation's schools and a staple on family bookshelves" for decades.[30] Indeed, these works were so popular that only the Bible and Webster's spelling books outsold them.[31] By Morse's own estimate, 20,600 copies of his *American Geography* had been sold by 1794.[32] And his writings exerted significant influence. As Madison Kuhn demonstrated, "Michigan owed some part of its rapid settlement to the Rev. Jedidiah Morse, father of American Geography."[33] Throughout his multiple careers, as minister, geographer, and campaigner for doctrinal orthodoxy, Morse saw himself as guardian of the new Republic's morality. And his geographical works were thus, fundamentally, exercises in moral topography. They were projects too in the justification of the American Revolution.[34] Lester Cohen, in his analysis of the ways in which historians legitimized the Revolutionary War, speaks of how "patriot historians" deployed ideas of natural law and of nature's God to justify

"American separation from Britain on the grounds of historical necessity."[35] Indeed, he calls upon Morse's 1824 *Annals of the American Revolution* to demonstrate how questions of political expediency and what Morse called "metaphysical disquisitions about abstract rights" were deployed in the arsenal of historical justification.[36] To this, we may surely add a geographical correlate. For Morse was equally insistent in his geographical writings that

> [t]he God of nature never intended that some of the best part of the earth should be inhabited by the subjects of a monarch 4000 miles away from them. And may we not venture to predict, that, when the rights of mankind shall be more fully known, and the knowledge of them is fast increasing both in Europe and America, the power of European potentates will be confined to Europe, and their present American dominions, become like the United States, free, sovereign, and independant [*sic*] empires.[37]

I want to begin with some remarks on the *American Geography,* which first appeared in 1789—just as the Constitution was being ratified—and subsequently came out in many different editions and versions.[38] It was "the most widely read description of the nation created by the passage of the Constitution."[39] A second London edition appeared from John Stockdale in 1792 in octavo, and a further new edition appeared there two years later in large quarto style with three folding maps and twenty-five regular maps, including "[a] map of Kentucky drawn from actual observations by John Filson." An Edinburgh edition, with seven folded maps, made its appearance in 1795. Other editions, with differing map complements, different pagination, and other alterations, were also to be found. All this reminds us of the textual fluidity that characterized books in the period, and therefore that different audiences were encountering different renditions of the treatise.[40]

The early part of the book was organized around traditional geographical themes—astronomical principles (including an elucidation of the planets, comets, and fixed stars), guidance on the use of globes, and the nature of atmospheric circulation. Typically these subjects were domesticated within a natural theology framework, which accounted for the "rich and beautiful canopy" of the stars by reference to "the wisdom of the creator." At the same time, Morse alluded to the possibility of the plurality of worlds by observing that all these planetary "systems . . . are filled with inhabitants suited to their respective climes; and are so many theatres, on which the Great Creator and governor of the Universe displays his infinite power, wisdom and goodness" (3). This format, of course, was entirely in keeping with geographical works of the day. As his later collaborator Elijah Parish observed in his own 1810

New System of Modern Geography, "Though geography is an earthly subject, it has been just denominated 'a heavenly study.' It includes many of those subjects which enlarge the mind, and improve the heart, which give just views to Providence, and of human nature."[41]

Shortly, Morse turned to the geography of America, describing the salient features of the continent, recording its European discovery and settlement, delineating both the physical and cultural circumstances of the new United States, and after reprinting in full the new Constitution proposed to replace the Articles of Confederation, concluding with brief laudatory epitaphs of such key figures in the recent drama of the Republic as Washington, Montgomery, Greene, and Lafayette. Writing in the flush of republican zeal, Morse's own pro-French sentiments (though he would later shift his ground) were clearly discernible, as were his frustrations with the British. In contrast to our "good and faithful allies and friends, the French," whose "newly established Federal Government" had been "liberal" in its commercial policies regarding the United States, the British had sought "to shackle our trade with every possible embarrassment" (86, 84). Indeed, in several sermons preached during the mid-1790s, Morse celebrated the way the French had "burst the chains of civil and ecclesiastical tyranny" and "espoused the case of LIBERTY, which is the birthright of mankind."[42] In this he was certainly not a lone voice among the clergy; his mentor Ezra Stiles, Congregationalist president of Yale, for example, was a zealous supporter of the French Revolution.[43]

With such conventional subjects treated, Morse next turned to a regional portrait of the several states of the Union, and to what was in effect a moral geography of the nation. It was, at base, an apologia for New England as the mold in which every other state of the Union should be cast. New Englanders were "almost universally of English descent," and it was on account of their enthusiasm for education that "the English language has been preserved among them so free of corruption." New Englanders were "tall, stout, and well-built," and possessed in excelsis that "spirit of freedom" that manifested itself as the "essence of true liberty." New Englanders were characterized by "industry and frugality" and what Morse called "that happy mediocrity . . . which, by inducing oeconomy and industry, removes from them temptations to luxury, and forms them to habits of sobriety and temperance" (145). New England women were "handsome"; New England sports were "healthy"; New England gaming was "practised by none but those who cannot, or rather will not find a respectable employment" (148). Of course, the portrait was not entirely blemish-free. Morse thought New Englanders too prone to litigation, which, to be sure, was the "genuine fruit of republicanism," but at the same time evidence of the "corruption of virtue" (146). Still, New England was a place of industry, frugality, and charity. The people obtain "their estates by hard and persevering labour. They of consequence know their value, and spend with frugality. Yet in no other country do the indigent and unfortunate fare better" (147). Not quite utopia, to be sure, but pretty close. Thus Massachusetts, with its impressive list of charitable societies and academies, was "in point of military, political, and literary importance . . . inferior to none, and superior to most states in the union" (193). Rhode Island was "exceedingly pleasant and healthful," was "celebrated for its fine women," and was therefore known to travelers as the "*Eden* of America" (202). And supremely, Connecticut—Morse's own state—was, in religious terms, the "best in the world, perhaps, for a republican government," so much so that he felt its mode of exercising church government and discipline "might not improperly be called a republican religion" (219). As for intellectual attainments, "in no part of the world is the education of all ranks of people more attended to than in Connecticut" (225). Not surprisingly, Morse considered it "the *Athens of America*" (214). In all, Connecticut was characterized by an austere virtue whose "industrious sagacious husbandmen" managed farms that furnished them "with all the necessaries, most of the conveniences, and but few of the luxuries of life" (240). It was "perhaps as perfect and happy a republic as has ever existed" (241).

Elsewhere conditions were less rosy. In Virginia, good character was restricted to "a few eminent men" (387). More generally, quoting a certain Rev. Andrew Burnaby, Morse depicted the Virginians as "indolent," "given to convivial pleasures," and chary of "expos[ing] themselves to fatigue"; in consequence, they displayed "extravagance, ostentation, and a disregard of oeconomy," frequently "out-ran their incomes," and showed little aptitude for business (388, 389). And according to "another discerning traveller," they were "addicted to gaming, drinking, swearing, horse-racing, cock-fighting, and most kinds of dissipation" (390). Conditions elsewhere were no better. Conversational life in North Carolina, for example, revolved around "negroes, the prices of indigo, rice, tobacco, &c." Here there was "as little taste for the sciences as for religion," with the result that "[p]olitical enquiries, and philosophical disquisitions, are attended to but by a few men of genius and industry, and are too laborious for the indolent minds of the people at large" (417). Morally, too, things were at a low ebb: "Temperance and industry," Morse reported, "are not to be reckoned among the virtues of the North Carolinians. The time which they waste in drinking, idling, and gambling, leaves them very little opportunity to improve their plantations or their minds" (417). As for South Carolina, the evils of slavery, "by exempting great numbers from the necessities of labour," had led to "luxury, dissipation and extravagance"—

a particularly unfortunate set of social circumstances that were only reinforced by "a climate which favours indulgence, ease and a disposition for convivial pleasures" that left the state's inhabitants "with barely knowledge enough to transact the common affairs of life" (423). In these circumstances, it was hardly surprising that South Carolina had accumulated very considerable debts. In Georgia too, climatic conditions—in good Neo-Hippocratic style— conspired to crush virtue. There, conditions were far from "salubrious" and the "disorders" that the climate induced included "indolence," "disease," and a love of "immense quantities of spirituous liquors" that disclosed "a species of intemperance, which too often proves ruinous to the constitution" (445). In these circumstances, it was no surprise that many took "a fancied pleasure at the gaming table, which . . . frequently terminates in the ruin of their happiness, fortunes, and constitutions," and that in "regard to religion, politics and literature, this state is yet in its infancy" (451). Taken overall, as Moss nicely characterizes it, "everywhere that Morse found slavery and warm weather, he found a culture he did not like."[44]

Further testimony is not needed to appreciate the degree to which Morse's *American Geography* can be considered "an extended and complex jeremiad" in which he "lamented the sins of the nation and exulted its virtues."[45] Virtue and vice, to Morse, had a distinctive geographical distribution. The former was prevalent in New England, the latter everywhere else. But his jeremiad was not intended as apocalyptic; his aim was to engender repentance and renewal so that the whole nation might eventually come to own "the honourable name of AMERICANS" (67). At the same time, Morse was forced to finesse his account in more or less imaginative ways. Having already expressed himself on the "injustice and iniquity of enslaving the Africans," on slavery's "pernicious" influence on both "manners and morals," on his hope that "all slaves in the united States will in time be emancipated," and on the value of societies "for the manumission of slaves" in northern states (65, 67), for example, it was clearly tricky to hold up the slaveholding George Washington as a paragon of "virtuous simplicity" (130). When he rode out "to his different farms," Morse recorded, Washington remained with his "labourers until a little past two o'clock." On paper, though hardly in practice, Washington's slaves quietly transmuted into "labourers" (133).

Still, the moral suture lines of the new nation largely followed attitudes to racial difference. Take the matter of America's native peoples. In both the *American Geography* and the *Compendious History,* Morse foregrounded the contrast between "the justice of New England's dealings with the Indians and the injustice of their fate in the western states."[46] These judgments prefigured his later *Report to the Secretary of War of the U.S. on Indian Affairs,* which he composed in the wake of a tour he undertook during the summer of 1820. To be sure, Morse's hope was that the Indians would become hardworking, Christian yeoman-farmers, and to that degree their only prospect of survival was in cultural extinction. If Native Americans could only adopt the village life of the idealized New England that Morse celebrated in his geographies, they could find salvation in this world and the next. Nevertheless, he admired the nonacquisitive character of their society, deplored their exploitation at the hands of traders, lamented their destruction through a decadent capitalism, and spoke of their "heart-breaking situation" in which they were "left miserably to waste away for a few generations, and then to become extinct for ever." He had, too, a sanguine view of intermarriage, claiming that when they were educated and Christianized "the principal obstacles . . . would be removed." After all, he insisted like Stanhope Smith that the Indians were "of one blood, with ourselves."[47] In any event, treatment of the Indian was a marker in the moral landscape of the Republic, though it must have disturbed his moral cartographics to read—if he ever did—the remarks of Edward Everett, president of Harvard, secretary of state, and later governor of Massachusetts on his *Report.* For Everett happily watched with the coldest possible eyes the inevitability of white conquest. To him, it was entirely plain that the Indians "have no just cause of murmuring at the progress of the whites, who, had they from their first landing used no other means of extending themselves and extinguishing the Indian claims than the sword, would have stood on as good a right as the Indians themselves."[48]

While the *American Geography* was widely read and strongly recommended in such serials as the *Farmer's Almanac,* Morse's portrayal—not least of the supposedly exemplary Connecticut clergy—was not universally welcomed. In his three-volume *History of the Anti-Christ,* the maverick populist Elias Smith—sometime minister, physician, dentist, publisher, and merchant—lampooned Morse's Federalist narrative.[49] An antielitist democrat, Smith's outlook was moored to the conviction that radical Jeffersonian egalitarianism was entirely compatible with biblical religion, and he used every opportunity to attack traditional denominations and established clerical authority. Clergymen like Morse, Smith reckoned, were smug supporters of the status quo who studied "*human divinity,* enough to learn several prayers by rote," enjoyed a college education, and "*preserved an aristocratical balance over democracy in Connecticut.*" Small surprise that the ranks of the Connecticut clergy were so numerous! Smith did not mince his words: "[T]here is no way that a young gentleman can be in business with a *small stock* so easy as to be a *clergyman—a suit of black, a few notes,* something of a share of hypocrisy; he can have a parsonage house and land, twenty cords of wood per year, several hundred dollars to begin with, several more hundred to continue with, an obligation for life, by which he is sure if the town is sick of him, that he shall have several hundred more to leave off with. No wonder they are numerous—who would not be a *clergyman*!!" At bottom Smith was sure that the clergy "are a dangerous lot of men, because they are opposed to the *Republican Government of our Country.*"[50] Along with figures like Lorenzo Dow and William Miller, Smith was one of a group of what Nathan Hatch describes as "upstarts" who were "dismissed as mindless demagogues" by a spooked traditional clergy who nevertheless "could not restrain people from flocking to their cause."[51]

Smith's portrayal of the Connecticut clergy was, at least as far as Morse was concerned, not far off the mark. For Morse was an enthusiast for the Federalist Party—partly out of personal friendship with, and loyalty to, George Washington—and he was profoundly troubled by Democratic-Republicans of Jeffersonian stripe. The Federalists, until the party's demise in the wake of the War of 1812, had a more elitist view of government and a preference for strong central authority. This viewpoint, according to Richard Buel, "was the choice of those who felt insecure as leaders because of the changes wrought by the Revolution."[52] Of course the Democratic-Republicans and Federalists were not diametrically opposed on every issue, not least since both were convinced of the rightness of the Revolution itself. Accordingly, Morse's Federalist sympathies cannot be construed as a rejection of Jefferson in toto. The *American Geography* was entirely supportive of Jefferson's anti-Buffonianism, though Morse ascribed the American inferiority thesis to "want of information" rather than "prejudice" on the part of "some ingenious and eloquent European writers." Either way, he rejoiced that "Mr Jefferson has confuted this theory; and by the ingenuity and abilities he has shewn in doing it, has exhibited an instance of its falsehood" (63). For all that, Morse found democratic doctrines on the Jefferson model little short of an abomination. Morse, indeed, was always an "anxious republican" fearful of radical democracy and religious pluralism.[53] The Democratic-Republican rallying cry of "equality" unnerved him because he believed it spawned individualism and a subversion of cultural authority. Indeed it was for this reason that he portrayed the Connecticut clergy as the guardians of "aristocratical balance in the very democratical government of the state" and as a check on "the overbearing spirit of republicanism" (219).

Jeffersonian republicanism, Morse feared, would unleash infidelity, individualism, and indulgence. Subsequently—after he soured on the French Revolution—he came to refer to Jefferson's party as "Jacobins" embodying French irreligion. His own vision was different. America could only flourish in the soil of simplicity and self-sacrifice, virtue, and communalism. Freedom, without virtue, as he said of New Englanders, is "licentious" (146). To Morse, genuine republics needed sumptuary laws that would restrict superfluous consumption and excesses in diet and dress alike.[54] High-priced buttons, balloon hats, and meals of more than two courses were just some of the things he believed should be prohibited by law.[55] For as Montesquieu had already argued, it was lust after luxury that had brought about the ruin of the Greek and Roman republics. Thus, in the *American Geography*, Morse railed against the "immorality of importing and consuming such amazing quantities of spirituous liquors," which "impair the estates, debilitate the bodies, and occasion the ruin of the morals of thousands of the citizens of America" (89).[56] This was indeed a common Federalist theme. As John Adams later put it to Jefferson himself in 1819, the question was "how to prevent riches from producing luxury . . . [and] how to prevent luxury from producing effeminacy, intoxication, extravagance, vice and folly."[57]

In the years following the first appearance of the *American Geography*, Morse sensed an increasing corruption of American character. Locally he feared an invasion of newcomers to his parish in Charlestown, Massachusetts (where he had moved in 1789), and he urged a policy of barring "as far as possible the settlement of the idle, the intemperate, the unprincipled, and the poor. . . . Our civil Fathers whose business it is to manage the policy of the town from their known regard for its moral, religious as well as temporal welfare will it is hoped pay all needful attention to this important business."[58] At the global level, he came to believe that "a vast and insidious conspiracy," emanating from certain radical Enlightenment philosophers, was threatening the very fabric of the new Republic. Thus, during 1798 and 1799, he delivered three sermons intended to expose what he referred to as the Bavarian "Illuminati," which catapulted him to national prominence. Here he based his analysis on a book by John Robison, published in Edinburgh in 1797 and entitled *Proofs of a Conspiracy against all the Religions and Governments of Europe, Carried on in the Secret Meetings of the Free Masons, Illuminati, and Reading Societies.*[59] To Morse such analyses impressed on Americans the need to hold fast to their traditions, to do all they could to resist the erosion of patriotism and loyalty, to spurn sexual promiscuity and epicurean pleasures, and to tenaciously cling to the doctrines of Providence.

In large measure, Morse had been encouraged in this venture by Oliver Wolcott, a national figure in the Federalist movement, who, during correspondence with Morse during the mid-1790s, had spoken of "a mental epidemic" spreading through the nation like wildfire and had urged that only New Englanders possessed the capacity to resist the onslaught. Of course there were many factors involved in the controversy—Morse's innate conservatism, a passion to curry favor with public figures, an attempt to regain the legitimacy he was losing in his own parish on account of preferring geography writing to fulfilling his pastoral duties, and so on.[60] But in the present context, the affair confirms the fundamentally moral character of his undertakings and an anxiety about the direction in which the new Republic seemed to be traveling. To this end, he founded in 1800 a Federalist newspaper in Boston, the *New England Palladium,* with precisely the aim of exposing Jacobinism. Similarly inspired was his later role in the establishment of the *Panoplist* as a literary vehicle for the cultivation and maintenance of an orthodox religious community over against what he saw as the liberalizing trends in Massachusetts symbolized in the replacement of the late David Tappan by the Unitarian Henry Ware at Harvard.[61] What bothered him, as a Calvinist, with such allies as Timothy Dwight at Yale and Ashbel Green at Princeton, was that he considered "Unitarianism as the *democracy* of Christianity. It dissolves all the bonds of Christian union and deprives religion of all its efficacy and influence upon Society. Our ecclesiastical affairs are fast assuming the portentous aspect and convulsed state of our political affairs."[62]

Morse's endeavors to deliver textual constructions of the new Republic, of course, were not restricted to the *American Geography.* No less significant were the suite of geography textbooks he produced over the years in ever newer editions to inculcate in the Republic's newest citizens a sense of their own identity, among them *Geography Made Easy* (1784) and *Elements of Geography* (1795). The tactics he deployed in order to effect this vision were manifold. Typically he inverted the conventional ordering of subjects, for example, by providing a geographical inventory of America before portraying Europe, Africa, and Asia. (The *American Geography,* we should note in passing, devoted only about fifty of its more than five hundred pages to the rest of the world.) And then, as a final response in his catechetical *Elements of Geography,* students were required to answer, "I am truly delighted, Sir, with the account you have given of my country, and I am sure I shall love it more than I ever did before."[63] Commenting on the wider implications of such submission to geographical authority, Brückner observes that "at the same time when the American community negotiated issues of nationality through strategies of cultural mediation, primarily vacillating between forms of orality and print, Americans resorted to the geographic print characters as the most flexible, orally rehearsed, typographically set, and economically viable vehicle through which to fuse the abstract ideology of nationalism to a material form."[64] To this his *Compendious History of New England* (1804), coauthored with Elijah Parish, could be added, for here too, as William Gribbin writes, the aim was "to counteract the mockery of New England's past and present and thus to hold young readers loyal to their heritage. It was unabashed filiopietism."[65]

Timothy Dwight, *Travels in New England*, and the Virtues of Pastoral Orderliness

Timothy Dwight (1752–1817), Morse's friend and fellow Congregationalist clergyman, president of Yale College since 1795, found in New England the cultural moderation, or "mediocrity" as he himself styled it, that best expressed what American identity at its best was all about. To Dwight, such virtues took visible form in the cultural landscape, and his posthumously published *Travels in New England and New York* (1821–22) gave voice to his republican reading of the social relations and material features of the region. The four-volume work took the form of a fictive set of letters to "an English gentleman," recording Dwight's travel experiences begun in September 1796. Its twofold aim was to present a snapshot of a rapidly changing New England "in a manner resembling that in which a painter would depict a cloud," and to correct "the misrepresentation which foreigners, either through error or design," had circulated.[66]

In an era wondering whether the new nation would model itself on the agrarian unpretentiousness of New England, the plantation gentry of the South, or the radical democratic egalitarianism of France, Dwight's *Travels* constituted a strategic Federalist intervention in the debate. To be sure, as Jane Kamensky has shown, Dwight's landscape vision was shot through with ambivalences.[67] Sometimes wild land was beatific, disclosing traces of a benevolent divine; sometimes it stood as an impediment to the moral progress of civilization. Sometimes in new towns Dwight read the signs of national well-being—enterprise, industry, and civility; sometimes they signified godlessness and moral jeopardy. Sometimes the primeval forest announced God's presence; sometimes its conversion into fruitful fields signaled the coming of the New Jerusalem heralded in the American Revolution.[68] Either way, the material landscape was a text in which the moral caliber of the new nation's regions could be read.

And yet for all this seeming equivocation, Dwight's landscape hermeneutics remained crystal clear. What Dwight, as a Federalist partisan, most valued was a society in which tradition and deference took precedence over Jeffersonian modernity and egalitarianism, and he thus valorized those landscapes in which these virtues were crystallized. The pastoral New England town was Dwight's ideal, for it was a virtuous landscape composed of "discrete settlements focused about single centers." The "country," he plainly asserted, "is more virtuous than the city" (2:17). And as Kamensky observes, the "nearest earthly incarnation of this vision of perfection could be found in the Connecticut River Valley, where Dwight's family had farmed for generations."[69] Read in this register, the very ordinariness of the pastoral New England townscape signaled events of epochal significance. From "the plains of Concord," he wrote, "will henceforth be dated a change in human affairs; an alteration in the balance of human power; and a new direction to the course of human improvement" (2:387). It was not, as Tichi has observed, that the American Revolution was a "mere *sign* of the beginning of the millennial Age of Liberty"; rather "it *was* the beginning."[70] And it was in the landscape that the qualities of the New Earth were to be seen. As Dwight made clear in the general preface to the work: "A forest, changed within a short period into fruitful fields, covered with houses, schools, churches, and filled with inhabitants, possessing not only the necessaries and comforts, but also the conveniences of life, and devoted to the worship of JEHOVAH, when seen only in prophetic vision, enraptured the mind even of Isaiah; and, when realised, can hardly fail to delight that of a spectator. At least it may compensate the want of ancient castles, ruined abbeys, and fine pictures" (1:7–8). To Dwight, then, such landscapes were valued, as John Sears notes, "not for their picturesque effect, but because they add up to the pious, educated, and prosperous New England village that in Dwight's view was the cornerstone of America's well-being."[71]

A synthesis of American experimentalism and ethical continuity was what Dwight most cherished, and it was out of the tension between constancy and change that his landscape compositions took shape.[72] And compositions they assuredly were, for Dwight was acutely conscious of the need to regulate landscape vision and educate sensibility in order to "fix" a vista and render it subject to the "composing eye." No doubt this reflected his own exceptionally poor eyesight—a condition, it has recently been suggested, that imposed itself on the very homiletic structure and rhetorical cadence of his myriad sermons.[73] But it also sprang from his synthesis of Locke's empiricism, Reid's Scottish philosophy of common sense, and his grandfather Jonathan Edwards' Calvinist theology. In his scientific investigation "On Light," for example, Dwight grounded reasoning in experience and observation, and it is thus no surprise that the operations of visibility had to be cultivated with extreme care.[74] To him "observation" was different from mere "looking" because—in keeping with Webster—it meant (as Timothy Spears remarks) "a thinking process which, in a steady, methodical, even ritualistic fashion, led . . . from specific, discernible aspects of the landscape to broader, universal laws."[75]

Consider in this connection his deliberations when visiting the White Mountains of northern New Hampshire:

> In the first view of a new region, the mind, from its absolute want of knowledge concerning everything within its grasp, finds not a little difficulty in settling upon certain capital points as stations from which it may proceed in all its inquiries, and to which everything of inferior importance might be reduced in order to fix the proportions and relations of the subordinate parts, and thus correctly adjust the situation of the whole. In this case, the imagination is equally at a loss with the intellectual, and until it has fixed its own stations, drawn its outline, and referred its inferior images to this scheme will find all its succeeding efforts to form the perfect picture for which it labors in vain. (2:93–94)

Plainly, Dwight's was a reading that fully acknowledged the positionality of the travel eye. His observations on his first visit to the Lake George region in 1802—a spot to which he returned nine years later—bear ample testimony. From the outset he explained how certain features were discernible only "at certain times" from "particular positions," and how the variegated play of light and shade on the water was subject to "the progressive change which the traveler sailing on its bosom perpetually finds in his position" (3:247, 250–51). But Dwight's visual sensibilities were modulated in two further ways—by eschatological anticipation and historical recollection. Contemplating a millennial future, Dwight looked for the time when the "hand of the husbandman," already active in the business of clearance, would adorn the landscape with "all the smiling scenes of agriculture" (3:252). Such pastoral transformations presaged the postmillennial world for which he hoped. Retrospectively, the fact that it was on this very site that some of the most stirring events of the War of Independence—on which he elaborated in detail in two further letters (3:252–72)—had taken place shaped his visual hermeneutics. At Ticonderoga, scene of the 1758 battle, for example, Dwight found himself confronted by "a prospect superior to any which I ever beheld" (3:251). To a considerable degree, Dwight's basking in the glorious scenery of Lake George was because here, as John Stilgoe observes, "occurred glorious events which presaged a glorious future." His description, Stilgoe goes on, "glows with the energy of mated religion and patriotism." Here, "out of the darkness, out of the forest of bewilderment and struggle, the republic emerges into the promised land ripe for improvement, a land washed in the light of Protestantism, ordered liberty, and the sun."[76]

Whereas Morse eschewed panoramic vistas, preferring to move over the terrain, as Tichi tellingly puts it, "like an inchworm on a relief map,"[77] Dwight relished precisely expansive visibility. Thus, in Gilpinesque tones, his portrayal of the Connecticut River valley was replete with the vocabulary of sublimity: ridges were "magnificent," summits "finely figured," surfaces "graceful," the whole bound by mountains of "grandeur indescribable" (2:128). *Travels in New England* thus offered a reading of landscape, rather than a regional inventory. And yet it did not fail to record, in empirical detail, statistical data that advanced Dwight's republican politics. Culling from a catalog that appeared in Williams's *History of Vermont*, for example, Dwight compared American and European animal weights—for bears, wolves, deer, foxes, beaver, otters, and so on—to refute yet again Buffonian claims about American inferiority. Thus, while the European bear weighed in at around 153 pounds, its Vermont equivalent recorded 456 pounds; for the wolf, the respective figures were 69 and 92 pounds (1:32).[78] And again, Dwight used available vital statistics to confirm Connecticut's demographic superiority over Virginia, the Carolinas, and Georgia, in order to support the salubrious character of the state.

The significance of landscape in Dwight's *mentalité* was not restricted to his posthumous classic. To the contrary, as Kamensky writes, "Dwight nearly always placed landscape at the center of his work: the biblical landscape in sermons and early epic poems, the idealized pastoral in later poetic works, and the varied terrain of New England in his travel narratives."[79] Take, for example, his book-length poetic pastoral, *Greenfield Hill* (1794), in which the merits of the rural village were extolled. This idealized village, with every house on its own land, disclosed order, improvement, and that "bless'd" unity derived from variegated integration. Here was a landscape neither too manicured nor too savage, neither too commercial nor too wild. Here was a landscape of moderation bearing a moral community. And here was the way to build the New Jerusalem—a landscape escaping the corruption and debauchery of its Old Testament equivalent. Like Penniman's Meetinghouse Hill in Roxbury, Massachusetts, Dwight's Greenfield Hill "brushes upon Eden," as Stilgoe puts it, "uniting agricultural and domestic economy with the disciplined worship of the Almighty."[80] In what Dwight called this "flourishing village":

No griping landlord here alarms the door,
To halve, for rent, the poor man's little store.
No haughty owner drives the human swain
To some far refuge from his dread domain;
Nor wastes, upon his robe of useless pride,
The wealth, which shivering thousands want beside;

Nor in one palace sinks a hundred cots;
Nor in one manor drowns a thousand lots.[81]

Or as he put it in the 1795 sermon *The True Means of Establishing Public Happiness:* "[N]either ... splendour and luxury on the one hand, nor ... suffering and meanness on the other" could deliver the kind of civic virtue that the new nation needed.[82] To quote Kamensky once more, "[I]n virtually every landscape he described, real or imagined, Dwight promoted an ideal of moderation, an ideal based on a profound awareness of natural limits, an ideal that looked as much to the republican future as to the Puritan past and strove for the best each had to offer."[83] All in all, Dwight's vision of the Republic was mediated through the vista of a pastoral aesthetic most conspicuously crystallized in the tamed landscape of the Connecticut River town. It was here that postrevolutionary America assumed the form of "Canaan's promised shores" and heralded a time when the land would be "stripped of the forests ... [and] measured out into farms, enlivened with all the beauties of cultivation" (2 : 140).

Dwight's landscape evocations were rooted in the moral economy of a Christian republicanism. But it would be mistaken to think that, even in *Travels in New England and New York,* scenic panoramas and pastoral poetics were the sole focus of attention. In letter 13 of volume 1, he advertised a moral register that later, in volume 4, he would expand into a three-hundred-odd page apologia for New England character. Enterprise and industry, a love of science and learning, a passion for liberty, upright morality, and an earnest piety were all typical of the resourceful New Englander. The early establishment of schools and colleges, the relatively low level of capital crimes, and the numerous churches that dotted the landscape, all bore testimony to New England's excellence. As for political ideology, New Englanders were "the only people on this continent who originally understood, and have ever since maintained, the inseparable connection between liberty and good order, or who practically knew that genuine freedom is found only beneath the undisturbed domination of equitable laws" (1 : 123). Here was the template from which the new nation should construct its political identity. In consummate detail, and at considerable length, Dwight delineated these threads in the final two thirds of the fourth volume of the enterprise. One or two emblematic excerpts must suffice to provide some sense of the whole.

Take, first, the matter of language. Contrary to popular opinion, Dwight insisted that, taking the pronunciation of "well-bred people in London" as the standard, the English language was more correctly pronounced in New England than in most of England. (4 : 195). Indeed, even within London itself,

Dwight could itemize a lengthy list of mispronunciations, which, he believed, was vastly more extensive than anything that could be drawn up for New England. Second, the diffusion of literary and scientific knowledge through "the multitude of schoolhouses appearing everywhere at little distances" disclosed a people given to intellectual enlightenment. The political implications were enormous. "In a republic," Dwight mused, where the "complicated affairs of school districts, parishes, townships and counties demand a considerable share of intelligence, and the agency of a great number of persons," where the "debates of a town meeting" were intricate and weighty, and where large numbers of citizens held "some public office or other," governance without education was unthinkable (4:211–12). In New England, the benefits of the diffusion of knowledge were enormous; here "religious novelties operate with less fascination and are more reluctantly received than among those who are ignorant," and the region enjoyed "in a more desirable manner, a good deal of that stability which is elsewhere produced by energy in the government" (4:212). Third, Dwight mustered the forces of moral geography to refute the charge printed in the November 1809 issue of the *Quarterly Review* that "a trace of savage character" produced "by the circumstances of society and of external nature" was evident in the American disposition (4:235). In disputing this characterization, Dwight remarked that in New England "half to two thirds of the inhabitants sleep round the year without bolting or locking their doors" (4:235), that elections were carried out without the "scenes of riot, tumult, and violence" reportedly common in England (4:236), that the Sabbath was observed "with a greater degree of sobriety and strictness than in any other part of the world" (4:255), and that those "barbarous and profligate sports" that prevailed in Britain—horse racing and cockfighting—were virtually unknown (4:238).

With these sentiments it comes as no surprise that, like Morse, Dwight was distressed by infidel philosophy and was convinced that Jeffersonian Jacobinism possessed no means of "restraining vice, or promoting virtue."[84] To be sure, Dwight was an enthusiastic advocate of the Revolution. He was allergic to the idea of hereditary aristocracy; he believed high office had to be earned; he insisted leadership was bestowed by the community. He was suspicious of mere tradition and, in the *Duty of Americans,* warned his fellow countrymen that "one of the first political errors" was to hold "too high a respect for the state of society in Greece and Rome."[85] But he was withal suspicious about the power of unlimited democracy to deliver the kind of synthesis of piety and knowledge that was essential to public virtue. The uneducated crowd was not a good basis for government. Jeffersonian republicanism simply could not deliver the moderation and restraint needed to curb extrava-

gance, consumerism, and overindulgence. For its own political health, the nation needed to look to the moral landscapes of New England and pattern itself on the virtues made manifest there.

SAMUEL STANHOPE SMITH, ENVIRONMENT, AND THE FOUNDATIONS OF PUBLIC VIRTUE

Between the first appearance of Morse's *American Geography* and the posthumous publication of Dwight's *Travels*, the second and much enlarged edition
of Samuel Stanhope Smith's *Essay on the Causes of the Variety of Complexion
and Figure in the Human Species* became available in 1810.[86] Its prior incarnation almost a quarter of a century earlier in 1787, established it—according to John Greene—as "the first American treatise devoted to the causes of
racial variation in the human species" and secured for its author an inaugurat-
ing presence in the annals of American anthropology.[87] At that stage, Smith
had been professor of moral philosophy at the College of New Jersey (Princeton) for some eight years and later, in 1795, would succeed his father-in-law,
John Witherspoon, as president.[88] In that environment, it is entirely unsurprising that he had immersed himself in the Scottish Common Sense school
of philosophy that Witherspoon championed and that was to inform his political, theological, and scientific outlook, as indeed it dominated early American college life more generally.

Because Smith's text dealt with questions of racial difference, it has routinely been examined in the context of ethnological history and its climatic
account of human racial variation given appropriate prominence.[89] Marvin
Harris, indeed, claimed that the "ultimate pitch of environmentalism" had
been reached in Smith's influential narrative.[90] Here, however, I want to mobilize Smith's fundamentally geographical account for different purposes. For
what we encounter here is not a regional portrait of one particular state as
with Jefferson, a geographical inventory of the state of the Union as with
Morse, or a traveler's reading of landscape as with Dwight. Instead we find
a treatise in the natural philosophy of the human species in which the physical environment is called upon to do both scientific and—crucially—ideological work.

Scientifically, Smith's intervention was simply to urge, over against the
polygenists, that climate, migration, and social environment were sufficient
explanatory mechanisms to make sense of human variation. By such geographical factors, Smith believed that observable differences in the races
could be explained, and this implied that some kind of evolutionary account

of race history—involving the mechanism of the inheritance of acquired characters—must be involved. A further implication was human cosmopolitanism. Humankind was evidently capable of acclimatizing to new climatic regimes, and Smith called upon the medical testimony of Dr. Carl Strack, a German physician at Mainz, and data from experiments in animal acclimatization, to support the case.[91] Human hair, skeletal stature, forehead shape, physiognomic features, even mental powers, were thus all correlated with "climates . . . some states of society, and modes of living."[92] More locally, Smith's mechanistic environmental determinism—thoroughly characteristic as it was of Newtonian science—seemed confirmed in what Winthrop Jordan referred to as the "ethnological laboratory" of America itself.[93]

Smith's inclination to fix upon environmental conditions as the cause of racial variation, however, is less interesting than what he hoped such a mechanism could deliver politically. By turning to geographical factors, he believed he had preserved the reality of a single human constitution, flexible to be sure, but common nonetheless. In a new republic Smith felt this was politically foundational. To glean some sense of its significance, we need to be clear about what such a doctrine did, and did not, entail. A shared origin and a shared nature, for example, did not imply ethnic or cultural uniformity. Evidently as accepting of cultural relativism as Montesquieu, Smith accounted for "savage life" in terms of degeneration. Nevertheless, he remained convinced that differences between the races were literally only skin deep. He was as certain as Morse and Dwight, for example, that "if the Anglo-American, and the indian [sic] were placed from infancy in the same state of society, in this climate which is common to them both, the principal differences which now subsists between the two races, would, in a great measure, be removed when they should arrive at the period of puberty."[94] To be sure, this did not amount to modern egalitarianism, but it did confirm the unity of the human species. Right from the very first page of the treatise, Smith announced: "The unity of the human race, notwithstanding the diversity of colour, and form under which it appears in different portions of the globe, is a doctrine, independently of the authority of divine revelation, much more consistent with the principles of sound philosophy, than any of those numerous hypotheses which have referred its varieties to a radical and original diversity of species."[95]

The political implications for the Republic were plain. A common human nature was fundamental to political stability. For if human nature was inherently unstable and if there were a range of essentially different constitutions, then general principles of morality and duty—and hence of the regulation and government of the polity—could not be promoted:

> I must repeat here an observation which I made in the beginning of this essay, and which I trust I am now entitled to make with more confidence, that the denial of the unity of the human species tends to impair, if not entirely destroy, the foundations of duty and morals, and, in a word, of the whole science of human nature. No general principles of conduct, or religion, or even of civil policy could be derived from natures originally and essentially different from one another, and, afterwards, in the perpetual changes of the world, infinitely mixed and compounded. The principles and rules which a philosopher might derive from a study of his own nature, could not be applied with certainty to regulate the conduct of other men, and other nations, who might be of totally different species. The terms which one man would frame to express the ideas and emotions of his own mind must convey to another a meaning as different as the organization of their respective natures. But when the whole human race is known to compose only one species, this confusion and uncertainty is removed, and the science of human nature, in all its relations, becomes susceptible of system. The principles of morals rest on sure and immutable foundations.[96]

If the recent, and to Smith obnoxious, speculations about polygenism with which Scotland's Lord Kames had been flirting in his *Sketches of the History of Man* were to be confirmed, and the human race was actually composed of a variety of distinct species, Smith was sure that—as he put it in the first edition of the work—any "science of morals would be absurd; the law of nature and nations would be annihilated."[97] Preserving a universal human nature was thus far more important than any seeming materialism that his siding with Montesquieu's dictum that "the empire of climate is the first of all empires" might seem to imply. For that charge was a good deal less worrisome than anthropology à la Kames. Recall, as Mark Noll has put it, that the new Republic was "busily repudiating the props upon which virtue had traditionally rested—tradition itself, divine revelation, history, social hierarchy, an inherited government, and the authority of religious denominations."[98] In such circumstances, the common human nature that was part and parcel of Scottish moral philosophy was precisely suited, as Norman Fiering writes, "to the needs of an era still strongly committed to traditional religious values and yet searching for alternative modes of justification for those values."[99] A single human constitution, albeit molded into different shapes by the power of geography, gave foundations to the very possibility of public virtue.

What Smith's account also reveals is just how central a role moral philosophy played in the political thought of the early Republic. For the constitutional unity of the human species ensured the possibility of an inductive mental science. It was precisely because there *was* a common human nature that it could be the object of empirical scrutiny. In this way Christian moralists could co-opt scientific method and make human ethics a science even while remaining convinced that the findings of this new Baconian venture would confirm Christian virtue. If, as Witherspoon was certain, a republic was right to break both with its monarchical past and with religious establishment; and if republics, dependent as they were on the performance of public virtue, were the fragile thing Montesquieu had shown them to be, then morality could only be preserved in the public square if a universal ethical sense could be extracted from human nature by the methods of secular science. To this Stanhope Smith remained passionately dedicated. And he was not alone. As Roger Smith has recently reminded us, such language was central to eighteenth-century moral philosophy and "had the status of being the shared ground on which writers of many persuasions erected moral standards of universal validity."[100]

Take James Madison—himself a student of Witherspoon at Princeton—for instance. A devotee of the Scottish moralists, Madison drew inspiration from David Hume's social theory in his political formulations about the nature of the Constitution.[101] As Hume put had put it:

[T]here is a great uniformity among the actions of men, in all nations and ages, and . . . human nature remains still the same, in its principles and operations. The same motives always produce the same actions; the same events follow from the same causes. . . . Would you know the sentiments, inclinations, and course of life of the Greeks and Romans? Study well the temper and actions of the French and English. Mankind are so much the same, in all times and places, that history informs us of nothing new or strange, in this particular. *Its chief use is only to discover the constant and universal principles of human nature,* by showing men in all varieties of circumstances and situations, and furnishing us with materials, from which we may form our observations and become acquainted with the regular springs of human action and behaviour.[102]

It was with these Humean sentiments echoing in his mind, side by side with Locke's conviction that there were universal and immutable laws of human nature to be uncovered, that Madison composed his famous tenth *Federalist.* It was precisely because of the immutability of the human constitution that, as Douglass Adair has observed, he "quite deliberately . . . set his limited personal experience in the context of the experience of men in other ages and times."[103] When Madison wryly commented on the "propensity of mankind to fall into mutual animosities, that where no substantial occasion presents itself, the most frivolous and fanciful distinctions have been suffi-cient to kindle their unfriendly passions and excite their most violent con-flicts," he was revisiting Hume's observation that "[m]en have such a pro-pensity to divide into personal factions, that the smallest appearance of real differences will produce them."[104] More generally, it was this conviction that encouraged the framers of the Constitution to look to historical experience in order to glean lessons about how to mix ideas of monarchy, aristocracy, and democracy in the new American Republic.[105] Thus John Adams, for example, observed in 1786: "The History of Greece should be to our countrymen what is called in many families on the Continent, a *boudoir,* an octagonal apart-ment in a house, with a full-length mirror on every side, and another on the ceiling."[106] What sustained this turning to the repository of historical experi-ence for inspiration, instruction, and upholding the political experiment that was the United States Constitution was the assurance that the human condi-tion was unchanging through time. And it was this that Stanhope Smith sought to preserve in his espousal of environmental influence. Recall the very first sentence of his treatise: "The unity of the human race, notwithstanding the diversity of colour, and form under which it appears in different portions of the globe, is a doctrine . . . much more consistent with the principles of sound philosophy, than any of those numerous hypotheses which have referred its varieties to a radical and original diversity of species."[107]

* * *

Whatever their particularisms, Jefferson, Morse, Smith, and Dwight were all engaged in one form or another of geographical apologetics for the rightness of the Revolution, for the integrity of the new Republic, and for the refutation of that irritating French witness for the prosecution, Buffon. Their conjoint aim was to vindicate America, as Robert Beverley had sought to do in 1705, when in the preface to his *History and Present State of Virginia* he com-plained: "There are no books so stuffed with Poetical Stories, as Voyages; and the more distant the countries lie, which they pretend to describe, the greater License those priviledg'd Authors take, in imposing upon the world."[108] Now, in the aftermath of the Revolution, the need to speak up for America was more pressing than ever. But vindicating America required more than accu-rate "earth-writing."

To these writers, issues of civic virtue and moral sensibility were at the heart of their geographical texts, because they were at the heart of republican ideology. To be sure, different writers had different views on the role of re-ligion and orthodoxy in the new Republic. Some, like John Adams and to some degree Washington, retained a hankering after religious establishment;

others, like Jefferson and Madison, tended toward a minimalist natural religion and to absolute freedom of religious persuasion from state power. Yet all knew that, as Polybius had made clear so many centuries before, Rome was beset with corruption and decline; that Adam Ferguson attributed the dissolution of the Roman Empire to the influx of obnoxious nonprovidentialist Greek philosophy; that Francis Hutcheson, the influential Hiberno-Scottish Enlightenment moral philosopher, was sure that the "denial of a moral providence, or of the obligations of the moral or social virtues . . . directly tend to hurt the state in its most important interests."[109] Whatever their different takes on the subject, as Colin Kidd has put it, the "central dilemma for proponents of religious freedom was the perceived need for a mechanism of moral obligation which would be effective in maintaining virtue." As he goes on: "The nascent American republic was wracked by worry about the dangers which might befall its experiment in government. Financial speculation, standing armies, and luxurious consumption all threatened to undermine the moral supports of republic society."[110] In this context, the fact that the new geographies of the Republic were *moral* geographies is hardly surprising; they are precisely what geography had to deliver.

...

Notes

I am most grateful to David Lowenthal, Mark Noll, and Charles Withers for insightful comments on an earlier version of this paper.

1. Jedidiah Morse, *The American Geography; or, A View of the Present Situation of the United States of America,* 2nd ed. (London. John Stockdale, 1792), iii.

2. Richard J. Moss, *The Life of Jedidiah Morse: A Station of Peculiar Exposure* (Knoxville: University of Tennessee Press, 1995), 35. See also John Kirtland Wright, "Some British 'Grandfathers' of American Geography," in *Geographical Essays in Memory of Alan G. Ogilvie,* ed. R. Miller and J. Wreford Watson (London: Thomas Nelson, 1959), 144–65; William Buell Sprague, *The Life of Jedidiah Morse, D.D.* (New York: Anson D.F. Randolph, 1874); John Rennie Short, "A New Mode of Thinking: Creating a National Geography in the Early Republic," in *Surveying the Record: North American Scientific Exploration to 1930,* ed. Edward C. Carter II (Philadelphia: American Philosophical Society, 1999), 19–50.

3. David M. Hooson, introduction to *Geography and National Identity,* ed. David M. Hooson (Oxford: Blackwell, 1994), 2.

4. On national self-fashioning, see Charles W. J. Withers, "Geography, Royalty and Empire: Scotland and the Making of Great Britain, 1603–1661," *Scottish Geographical Magazine* 113 (1997): 22–32. On the role of mapping in national identity, see Josef W. Konvitz, *Cartography in France, 1660–1848: Science, Engineering, and Statecraft* (Chicago: University of Chicago Press, 1987); D. Graham Burnett, *Masters of All They Surveyed: Exploration, Geography, and a*

British El Dorado (Chicago: University of Chicago Press, 2000); and Thongchai Winichakul, *Siam Mapped: A History of the Geo-body of a Nation* (Honolulu: University of Hawai'i Press, 1994). On school geography in Ireland, see Anne Buttimer and Gerard Fahy, "Imaging Ireland through Geography Texts," in *Text and Image: Social Construction of Regional Knowledges*, ed. Anne Buttimer, Stanley D. Brunn, and Ute Wardenga, special issue of *Beiträge zur Regionalen Geographie* (Leipzig), 49 (1999): 179–91. See also John A. Campbell, "Modernisation and the Beginnings of Geography in Ireland," in *Geography and Professional Practice*, ed. Vincent Berdoulay and H. van Ginkel (Utrecht: Nederlands Geographical Studies, 1996), 125–38.

5. The most sustained treatment of the theme is Charles W. J. Withers, *Geography, Science and National Identity: Scotland since 1520* (Cambridge: Cambridge University Press, 2001).

6. Preston James, *All Possible Worlds: A History of Geographical Ideas* (Indianapolis: Bobbs-Merrill, 1972), 193.

7. William Warntz, *Geography Then and Now,* American Geographical Society Research Series, no. 25 (New York: American Geographical Society, 1964), 136.

8. Margarita Bowen, *Empiricism and Geographical Thought* (Cambridge: Cambridge University Press, 1981), 171.

9. John C. Greene, *American Science in the Age of Jefferson* (Ames: Iowa State University Press, 1984), 214–15.

10. Martin Brückner discusses Morse's influence on the early American novel in "Geography, Reading, and the World of Novels in the Early Republic," in *Early America Re-explored: New Readings in Colonial, Early National, and Antebellum Culture,* ed. Klaus H. Schmidt and Fritz Fleischmann (New York: Lang, 2000), 387.

11. Martin Brückner, "Lessons in Geography: Maps, Spellers, and Other Grammars of Nationalism in the Early Republic," *American Quarterly* 51 (1999): 311, 313–15.

12. Ibid., 316. This was not the only way in which a national culture was constructed. Important too was a common-sense republicanism informed by Christian faith and mediated in part via the Methodist connexional system. See Mark A. Noll, *America's God: From Jonathan Edwards to Abraham Lincoln* (New York: Oxford University Press, 2002).

13. Noah Webster, "On the Education of Youth in America" (1788), in *Essays on Education in the Early Republic,* ed. Frederick Rudolph (Cambridge: Harvard University Press, 1965), 65.

14. Robert Lawson-Peebles, *Landscape and Written Expression in Revolutionary America* (Cambridge: Cambridge University Press, 1988), 74.

15. Brückner, "Lessons in Geography," 318.

16. Noah Webster, *A Grammatical Institute of the English Language: Part I* (1783; Menston: Scolar Press, 1968), 13.

17. John Jay, "Federalist Number II," in *The Federalist; or, The New Constitution,* 2nd ed., ed. Max Beloff (Oxford: Blackwell, 1987), 5–6.

18. On Jefferson's scientific accomplishments see Silvio A. Bedini, *Thomas Jefferson: Statesman of Science* (New York: Macmillan, 1990); and Greene, *American Science in the Age of Jefferson.* On the influence of science in Jefferson's political theory, see Garry Wills, *Inventing America: Jefferson's Declaration of Independence* (Garden City: Doubleday, 1978); the critique by Ronald Hamowy, "Jefferson and the Scottish Enlightenment: A Critique of Gary Wills' *Inventing America*," *William and Mary Quarterly* 36 (1979): 502–23; and I. Bernard Cohen, *Science and the Founding Fathers: Science in the Political Thought of Thomas Jefferson, Benjamin Franklin, John Adams and James Madison* (New York: Norton, 1995).

19. On this later topic, see especially Michael P. Zuckert, *The Natural Rights Republic: Studies in the Foundation of the American Political Tradition* (Notre Dame: University of Notre Dame Press, 1996); Michael P. Zuckert, "Founder of the Natural Rights Republic," in *Thomas Jefferson and the Politics of Nature,* ed. Thomas S. Engeman (Notre Dame: University of Notre Dame

Press, 2000), 11–58; Joseph J. Ellis, *American Sphinx: The Character of Thomas Jefferson* (New York: Knopf, 1997).

20. On Jefferson as a geographer, see George T. Surface, "Thomas Jefferson: A Pioneer Student of American Geography," *Bulletin of the American Geographical Society* 41 (1909): 743–50; A. W. Greely, "Jefferson as a Geographer," *National Geographic Magazine* 7 (1896): 269–71; Gary S. Dunbar, "Thomas Jefferson, Geographer," *Special Libraries Association, Geography and Map Division, Bulletin* 40 (April 1960): 11–16.

21. Thomas Jefferson, *Notes on the State of Virginia*, ed. William Peden (1784; Chapel Hill: University of North Carolina Press, 1955), query 19, 165.

22. Robert Dawidoff, "Rhetoric of Democracy," in Engeman, *Thomas Jefferson and the Politics of Nature*, 109.

23. Cited in Lawson-Peebles, "Thomas Jefferson and the Spacious Field of Imagination," in *Landscape and Written Expression*, 175.

24. Antonello Gerbi, *The Dispute of the New World: The History of a Polemic, 1750–1900*, rev. ed., trans. Jeremy Moyle (Pittsburgh: University of Pittsburgh Press, 1973), 257.

25. Jefferson, *Notes on the State of Virginia*, 64.

26. Ibid., 65.

27. See the discussion in James W. Caeser, "Natural Rights and Scientific Racism," in Engeman, *Thomas Jefferson and the Politics of Nature*, 165–89.

28. Jefferson, *Notes on the State of Virginia*, app. 4, 227.

29. Ibid., 230.

30. Moss, *Life of Jedidiah Morse*, ix.

31. William J. Gilmore, *Reading Becomes a Necessity of Life: Material and Cultural Life in Rural New England, 1780–1835* (Knoxville: University of Tennessee Press, 1989), 64.

32. Cited in Moss, *Life of Jedidiah Morse*, 49.

33. Madison Kuhn, "Tiffin, Morse, and the Reluctant Pioneer," *Michigan History* 30 (1966): 126.

34. Wright makes the comment that Morse's works may be "deemed American with regard to their authorship, subject matter, and sources; as americanistic where they reflect an ardent and characteristically American patriotism in passages extolling our republican institutions and defending American against European disparagement; and as subamericanistic where they reveal strong sectional or denominational prejudices" (John Kirtland Wright, "What's 'American' about American Geography?" in *Human Nature in Geography: Fourteen Papers, 1925–1965* [Cambridge: Harvard University Press, 1966], 129).

35. Lester H. Cohen, *The Revolutionary Histories: Contemporary Narratives of the American Revolution* (Ithaca: Cornell University Press, 1980), 131.

36. Jedidiah Morse, *Annals of the American Revolution; or, A Record of the Causes and Events which Produced, and Terminated in the Establishment and Independence of the American Republic . . .* (Hartford, 1824), 109.

37. Morse, *American Geography*, 469. On the widespread use of providential language in support of the Revolution, see Catherine L. Albanese, *Sons of the Fathers: The Civil Religion of the American Revolution* (Philadelphia: Temple University Press, 1976); John F. Berens, *Providence and Patriotism in Early America, 1640–1815* (Charlottesville: University of Virginia Press, 1978).

38. Hereafter quotations from the *American Geography* are cited in the text by page number.

39. Moss, *Life of Jedidiah Morse*, 38.

40. Ralph Hall Brown, "The American Geographies of Jedidiah Morse," *Annals of the Association of American Geographers* 31 (1941): 145–217.

41. Elijah Parish, *A New System of Modern Geography,* 2nd ed. (Newburyport, MA: E. Little, 1812,), iii.

42. Quoted in K. Alan Snyder, "Foundations of Liberty: The Christian Republicanism of Timothy Dwight and Jedidiah Morse," *New England Quarterly* 61 (1983): 387.

43. For the attitude of Morse, Stiles, and others, see Gary B. Nash, "The American Clergy and the French Revolution," *William and Mary Quarterly* 22 (1965): 392–412.

44. Moss, *Life of Jedidiah Morse,* 41.

45. Ibid., 38.

46. William Gribbin, "A Mirror to New England: The *Compendious History* of Jedidiah Morse and Elijah Parish," *The New England Quarterly* 45 (1972): 349. See also Ralph Ketcham, *From Colony to Country: The Revolution in American Thought, 1750–1820* (New York: Macmillan, 1974), esp. chap. 16, "Indians, Blacks, Race, and Slavery."

47. Jedidiah Morse, *A Report to the Secretary of War of the U.S. on Indian Affairs, Comprising a Narrative of a Tour Performed, in the Summer of 1820, under a Commission from the President of the U.S., for the Purpose of Ascertaining, for the Use of the Government, the Actual State of the Indian Tribes, in Our Country* (Newhaven, 1822), 81.

48. Edward Everett, "On the State of the Indians" (1823), *North American Review* 258 (1973): 11.

49. Michael G. Kenny, *The Perfect Law of Liberty: Elias Smith and the Providential History of America* (Washington, DC: Smithsonian Institution Press, 1994).

50. Elias Smith, *The History of Anti-Christ: In Three Books, Written in Scripture Stile, in Chapters and Verses; For the Use of Schools* (Portland, 1811), 116–20; emphasis in original.

51. Nathan O. Hatch, *The Democratization of American Christianity* (New Haven: Yale University Press, 1989), 134–35.

52. Richard Buel, Jr., *Securing the Revolution* (Ithaca: Cornell University Press, 1972), 85.

53. Moss, *Life of Jedidiah Morse,* xi.

54. On such issues more generally, see Joseph J. Ellis, *After the Revolution: Profiles of Early American Culture* (New York: Norton, 1979), chap. 2, "Paradoxes: Culture and Capitalism."

55. See Richard J. Moss, "Republicanism, Liberalism, and Identity: The Case of Jedidiah Morse," *Essex Institute Historical Collections* 126 (1990): 209–36.

56. It should be noted that Morse added: "In a pint of beer, or half a pint of Malaga or Tenerife wine, there is more strength than in a quart of rum. The beer and the wine abound with nourishment, whereas the rum has no more nourishment in it than a pound of air." He then proceeded to harness this argument in the cause of American independence: "These considerations point out the utility, may I not add, the necessity of confining ourselves to the use of our own home made liquors, that it might encourage our own manufactures, promote industry, preserve the morals and lives of our citizens, and save our country from the enormous annual expence [*sic*] of four millions of dollars" (Morse, *American* Geography, 90).

57. Cited in Peter Marshall and Ian Walker, "The First New Nation," in *Introduction to American Studies,* ed. Malcolm Bradbury and Howard Temperley (London: Longman, 1981), 57.

58. From an untitled sermon quoted in Moss, *Life of Jedidiah Morse,* 71.

59. The standard account is Vernon Stauffer, *New England and the Bavarian Illuminati* (New York: Columbia University Press, 1918).

60. See Richard J. Moss, "Jedidiah Morse and the Illuminati Affair: A Re-Reading," *Historical Journal of Massachusetts* 16 (1988): 141–53. The quotations from the Morse-Wolcott correspondence are taken from this article. See also Conrad Wright, "The Controversial Career of Jedidiah Morse," *Harvard Library Bulletin* 31 (1983): 64–87.

61. See Elizabeth Barnes, "The 'Panoplist': Nineteenth-Century Religious Magazine," *Journalism Quarterly* 36 (1959): 321–25.

62. Letter to Joseph Lyman, June 15, 1805, in Moss, *Life of Jedidiah Morse,* 87.

63. Jedidiah Morse, *Elements of Geography* (Boston, 1795), 121.

64. Brückner, "Lessons in Geography," 334. See also note 10 above.

65. Gribbin, "A Mirror to New England," 342–43. Gribbin reviews some of the strategies employed in successive editions of the *Compendious History* to mobilize history for contemporary purposes.

66. Timothy Dwight, *Travels in New England and New York,* ed. Barbara Miller Solomon (1821–22; Cambridge: Harvard University Press, 1969), 1 : 1. We know from the final volume of the *Travels* that Dwight had figures like Volney, Weld, la Rochefoucauld, and John Lambert in mind. Hereafter, quotations from this edition of the *Travels* are cited in the text by volume and page number.

67. Jane Kamensky, "'In These Contrasted Climes, How Chang'd the Scene': Progress, Declension, and Balance in the Landscapes of Timothy Dwight," *American Quarterly* 43 (1990): 80–108.

68. See Cecilia Tichi, *New World, New Earth: Environmental Reform in American Literature from the Puritans through Whitman* (New Haven: Yale University Press, 1979).

69. Kamensky, "In These Contrasted Climes," 102.

70. Tichi, *New World, New Earth,* 70.

71. John F. Sears, "Timothy Dwight and the American Landscape: The Composing Eye in Dwight's *Travels in New England and New York,*" *Early American Literature* 9 (1976–77): 317.

72. Dwight's compositional techniques are treated in Peter M. Briggs, "Timothy Dwight 'Composes' a Landscape for New England," *American Quarterly* 40 (1988): 359–77.

73. See John R. Fitzmier, *New England's Moral Legislator: Timothy Dwight, 1752–1817* (Bloomington: Indiana University Press, 1998), 81ff. This work is now the standard biography.

74. Dwight's scientific contributions more generally are the subject of Kathryn and Philip Whitford, "Timothy Dwight's Place in Eighteenth-Century American Science," *Proceedings of the American Philosophical Society* 114 (1970): 60–71.

75. Timothy B. Spears, "Common Observations: Timothy Dwight's *Travels in New England and New York,*" *American Studies* 30 (1989): 39.

76. John R. Stilgoe, "Smiling Scenes," in *Views and Visions: American Landscapes before 1830,* ed. Edward J. Nygren (Washington, DC: Corcoran Gallery of Art, 1986), 214.

77. Tichi, *New World, New Earth,* 112.

78. Dwight also rejected Buffon's claims about the inherent intellectual inferiority of the American Indian urging that Indian "degradation" was attributable to social conditions. As Spears puts it: "Not through violence or genocide, but through education and conversion, Dwight . . . proposed to regenerate the Indian, and swell the ranks of useful citizenry" (Sears, "Dwight and the American Landscape," 46).

79. Kamensky, "In These Contrasted Climes," 86.

80. Stilgoe, "Smiling Scenes," 222.

81. Timothy Dwight, *Greenfield Hill,* lines 81–88, in *The Major Poems,* ed. William J. McTaggart and William K. Bottorff (Gainesville: Scholars' Facsimiles and Reprints, 1969).

82. Timothy Dwight, *The True Means of Establishing Public Happiness* (New Haven: T. S. Green, 1795), 36.

83. Kamensky, "In These Contrasted Climes," 106.

84. Timothy Dwight, "The Nature and Danger of Infidel Philosophy," in *Sermons* (New Haven, 1828), 1 : 340.

85. Timothy Dwight, *The Duty of Americans* (New Haven: Greens, 1789), 16

86. I have discussed Smith elsewhere and what follows draws on material published in David N. Livingstone, "Geographical Inquiry, Rational Religion and Moral Philosophy: Enlight-

enment Discourses on the Human Condition," in *Geography and Enlightenment,* ed. David N. Livingstone and Charles W. J. Withers (Chicago: University of Chicago Press, 1999), 93–119.

87. Greene, *American Science in the Age of Jefferson,* 323.

88. See Mark A. Noll, *Princeton and the Republic, 1768–1822: The Search for a Christian Enlightenment in the Era of Samuel Stanhope Smith* (Princeton: Princeton University Press, 1989).

89. So, for instance, Thomas F. Gossett, *Race: The History of an Idea in America* (Dallas: Southern Methodist University Press, 1963); William Stanton, *The Leopard's Spots: Scientific Attitudes toward Race in America, 1815–59* (Chicago: University of Chicago Press, 1960).

90. Marvin Harris, *The Rise of Anthropological Theory: A History of Theories of Culture* (New York: Thomas Y. Crowell, 1968), 86.

91. See Samuel Stanhope Smith, *Essay on the Causes of the Variety of Complexion and Figure in the Human Species* (Philadelphia: Robert Aitkin, 1787), 34, 48. For a general analysis, see David N. Livingstone, "Human Acclimatization: Perspectives on a Contested Field of Inquiry in Science, Medicine and Geography," *History of Science* 25 (1987): 359–94.

92. Samuel Stanhope Smith, *Essay on the Causes of the Variety of Complexion and Figure in the Human Species* (1810; reprint ed., Cambridge: Belknap Press of Harvard University Press, 1965,), 78.

93. Winthrop Jordan, introduction to Smith, *Essay on the Causes of the Variety of Complexion and Figure* ([1810] 1965 ed.), xxviii.

94. Smith, *Essay on the Causes of the Variety of Complexion and Figure* ([1810] 1965 ed.), 107.

95. Ibid., 7.

96. Ibid., 149.

97. Smith, *Essay on the Causes of the Variety of Complexion and Figure* (1787 ed.), 48. On Kames, see Ian Simpson Ross, *Lord Kames and the Scotland of His Day* (Oxford: Clarendon, 1972); and William C. Lehmann, *Henry Home, Lord Kames, and the Scottish Enlightenment: A Study in National Character and in the History of Ideas* (The Hague: Martinus Nijhoff, 1971). Paul Wood comments, however, that Kames's contemporaries among the Scottish men of letters rejected his polygenetic inclinations "as subverting the foundations of religion and morality" (Paul B. Wood, "The Science of Man," in *Culture of Natural History,* ed. Nicholas Jardine, James A. Secord, and Emma C. Spary [Cambridge: Cambridge University Press, 1996], 204).

98. Mark A. Noll, "The Rise and Long Life of the Protestant Enlightenment in America," in *Knowledge and Belief in America: Enlightenment Traditions and Modern Religious Thought,* ed. William M. Shea and Peter A. Huff (New York: Cambridge University Press, 1995), 100.

99. Norman Fiering, *Moral Philosophy at Seventeenth-Century Harvard: A Discipline in Transition* (Chapel Hill: University of North Carolina Press, 1981), 300.

100. Roger Smith, "The Language of Human Nature," in *Inventing Human Science: Eighteenth-Century Domains,* ed. Christopher Fox, Roy Porter, and Robert Wokler (Berkeley and Los Angeles: University of California Press, 1996), 101.

101. See, for example, Roy Branson, "James Madison and the Scottish Enlightenment," *Journal of the History of Ideas* 40 (1979): 235–50.

102. David Hume, *Enquiry Concerning Human Understanding* (London, 1748), sec. 8, "Of Liberty and Necessity."

103. Douglass Adair, "'That Politics May Be Reduced to a Science': David Hume, James Madison, and the Tenth Federalist," in *Fame and the Founding Fathers: Essays by Douglass Adair,* ed. Trevor Colbourn (Toronto: Norton, 1974), 97.

104. Quoted in ibid., 103.

105. Douglass Adair, "'Experience Must be Our Only Guide': History, Democratic Theory, and the United States Constitution," in Colbourn, *Fame and the Founding Fathers,* 107–23.

106. John Adams, *A Defence of the Constitution of the United States of America,* in *The Works of John Adams, Second President of the United States,* ed. Charles Francis Adams (Boston, 1850–1856), 4:469.

107. Smith, *Essay on the Causes of the Variety of Complexion and Figure* ([1810] 1965 ed.), 7.

108. Robert Beverley, preface to *The History and Present State of Virginia, in Four Parts* (London, 1705).

109. Francis Hutcheson, *A System of Moral Philosophy* (London, 1755), 2:310.

110. Colin Kidd, "Civil Theology and Church Establishment in Revolutionary America," *Historical Journal* 42 (1999): 1016, 1018.

ALEXANDER VON HUMBOLDT AND REVOLUTION

A Geography of Reception of the Varnhagen von Ense Correspondence

...

Nicolaas Rupke

Alexander von Humboldt (1769–1859) is not commonly depicted as a revolutionary—at least not in the Anglo-American literature on him. Some biographers of Humboldt may consider his contributions to plant geography or to meteorology of revolutionary importance.[1] But Humboldt is not seen as a political revolutionary in the way that, for example, Humboldt's younger contemporary Karl Marx was. Yet precisely this view of Humboldt as a supporter—even an instigator—of political revolutions was taken in the former East Germany, where the account of Humboldt's life and work was placed in the discursive space of Marxist historiography. There, a biographical narrative was constructed that highlighted Humboldt's friendships with revolutionaries, making him a revolutionary by association. One of these friendships had been with the "revolutionary democrat" Georg Forster, the "great champion of the idea of 1789" who had passionately supported the Jacobins in France and one of those who in 1793 in Mainz had established "the first democratic republic on German soil." Forster had exerted a profound influence on Humboldt, and as a result, Humboldt's ideas had developed along socialist lines, anticipating those of Friedrich Engels and coming close to Marx, by advocating proletarian internationalism and through active support of anticolonialism and antislavery. Another friendship of revolutionary purport had been with Simon Bolívar, whom Humboldt had influenced in the years leading up to the Bolívar-led wars of liberation of Latin America. Prominent among the German Democratic Republic's Marxist historians was the Leipzig expert on Latin America, Manfred Kossok. Kossok maintained that Humboldt had been the intellectual father of the wars of independence and that, with his writings on Cuba and Mexico, Humboldt had occupied a fixed place in the revolutions of the period 1810–26, when the yoke of Spanish colonial rule had been thrown off and various Latin American republics had come into existence.

This linkage of Humboldt to revolutionary causes was not new. It had in fact had been made by the earliest biographers of Humboldt in the German-speaking world, authors of the first distinct group of Humboldt biographies, published from around the time of the Revolution of 1848 until German uni fication under Bismarck in 1871. The Revolution of 1848 by and large failed, and accounts of Humboldt's life and work were produced in a context of frus-trated and redirected revolutionary fervor, Humboldt being depicted as a lib-eral democrat whose writings were a force of social emancipation and politi-cal unification of the German Volk. The early biographers of Humboldt were, for the most part, "Forty-eighters," opposed to absolutism in church and state, their left-leaning loyalties taking a variety of different forms: some Humboldt-ians participating in the Berlin battles of the barricades, others serving as can-didates for the Frankfurt and Stuttgart National Assembly, yet others advocat-ing radical antimonarchist and republican views.[2]

Particularly effective as a strategy in appropriating Humboldt on behalf of the revolutionary cause was the selective editing of his correspondence. In fact, no sooner had Humboldt died than his name was claimed on behalf of Berlin's revolutionary faction by means of an edition of Humboldt's letters to the notorious democrat Karl August Varnhagen von Ense (1785–1858). The volume's appearance was a bombshell, its impact being felt not just in Ber-lin and the wider German-speaking world, but across Europe. Publicly, Hum-boldt had never taken an unambiguous political stance, and he conducted a friendly correspondence with representatives of a wide political spectrum that included despots and reactionaries as well as liberals and radicals. Ad-mittedly, Humboldt had all along been known for the fact that he promoted liberal causes, and contemporary German Humboldtians had felt justified in claiming him on behalf of the elimination of feudal "Kleinstaaterei" and of the promotion of political freedom and the national unification of Germany. Yet to many outsiders, Humboldt seemed closely connected also with the Prussian and other European monarchical conservatives, and by the time of his death it was by no means self-evident that his great name should be so di-rectly usurped in the cause of revolutionary democracy. From an examination of Humboldt's correspondence with Varnhagen von Ense and how that corre-spondence was interpreted by contemporaries, I hope to illustrate how, after his death, Humboldt the geographer came in different ways to have a new worldly "afterlife" as Humboldt the revolutionary.

HUMBOLDT'S CORRESPONDENCE AND THE VARNHAGEN VON ENSE EDITION

Alexander von Humboldt died on 6 May 1859. Less than two months later, on 29 June 1859, while obituary accounts were still appearing in the periodical press, plans were set in motion to publish Humboldt's correspondence with Varnhagen von Ense.[3] During February 1860, the *Briefe von Alexander von Humboldt an Varnhagen von Ense aus den Jahren 1827 bis 1858* appeared, produced in Leipzig by the Brockhaus publishing company.

Humboldt had been a prodigious writer of letters. At the height of his fame, he wrote no fewer than two thousand letters per year, five or six a day. During the period 1789 to 1859, Humboldt wrote a total of some fifty thousand letters. Around thirteen thousand of these have been collected at the Alexander-von-Humboldt-Forschungsstelle in Berlin. Some thirty-five thousand are likely to have been lost. The number of letters Humboldt received may have amounted to a hundred thousand, yet of these a mere 4 percent has been preserved, as Humboldt habitually threw his correspondence away. With the increase in his fame, the flood of letters rose, and during the last few decades of his life perhaps three thousand a year would have been addressed to Humboldt. The names of over twenty-five hundred correspondents of Humboldt are known; the recipients of some seven hundred and fifty letters have as yet not been identified.[4]

This vast correspondence provides a picture of Humboldt's network of contacts. Biermann classified the correspondents in ten categories, ranging from those who received 10 to 24 letters from Humboldt, to those who received 350 or more. Additionally, Biermann grouped Humboldt's correspondents according to occupation, scientific and otherwise. Among his concluding observations were that the exchange of letters with astronomers was more frequent than that with geographers, which may have been because Humboldt was less familiar with astronomy than with geography and in greater need, therefore, of astronomers' advice.[5]

Of interest in this chapter is less the primary feature of letters to and from Humboldt, than the secondary one of posthumous editions of these letters and the question of why one part rather than another of Humboldt's many letters was selected for publication. Too often, volumes of letters are thought of as neutral, "primary sources." Editions of correspondence are no mere record of documentation. In many instances, they serve to prove—or can be made to so serve—a partisan point of ideology. For example, if one were to edit Humboldt's letters to and from the Jewish men and women from among his many contacts, this would strengthen the perception that Humboldt had been philo-Semitic. In late-nineteenth-century Germany, for example, such an edition could have served the cause of Jewish emancipation; in the period following World War II, it would have furthered post-Nazi reparation politics. And if one were to edit Humboldt's correspondence with the many women with whom he exchanged letters, the cause of feminism and especially of equal participation in science could be promoted. Work on Humboldt's correspondence with Helen Maria Williams, Caroline Herschel, Mary Somerville, and others is indeed in progress—women who significantly contributed to the translation into English of some of Humboldt's books.[6]

Editions of letters can therefore be highly contested, as was the case with the very first of such editions of Humboldt's letters, the Varnhagen von Ense correspondence. Varnhagen von Ense was a diplomat and writer on public affairs, who chronicled the events of his time in diaries and became known for biographical and autobiographical writings. His antiestablishmentarian political stance and his correspondence with various liberal figures have inspired a number of later studies of him.[7] Further editions of letters from his papers, such as with Thomas Carlyle, Richard Monckton Milnes, first Baron Houghton, the American socialist Albert Brisbane, and several others have also been the subject of study.[8] Varnhagen, Humboldt's close friend and confidant, carefully saved the letters he received from Humboldt and additionally faithfully recorded in his diary what his great scientific friend had said during the private visits he regularly paid. Moreover, Humboldt had presented Varnhagen with a selection of letters he received from eminent correspondents from across Europe, to be added to Varnhagen's vast but systematic collection of contemporary papers.

The published volume contained 225 letters from the private papers of Varnhagen von Ense, for the most part written by Humboldt to Varnhagen. Also included were several letters addressed to Humboldt by various famous people, among whom were King Christian VIII of Denmark, the Prince of Metternich, the British Tory prime minister Robert Peel, the Prince Consort to Queen Victoria, Albert, Dominique Francois Jean Arago, Friedrich Wilhelm Bessel, John Frederick William Herschel, Honoré de Balzac, and Victor Hugo. In general, Humboldt wrote far more letters to scientists than to figures from public life, but in this edition the latter category predominated.[9] Extracts from Varnhagen's diaries were added.

Varnhagen apparently planned to publish these letters after Humboldt's death, but although the younger of the two men, he himself died before Humboldt, having donated his papers and collections to his niece, Rosa Ludmilla Assing. It was she who, in the wake of Humboldt's death, proceeded with the publication of his correspondence. The public interest was enormous. In 1860 alone, no fewer than five German editions appeared in Leipzig, plus a further one in New York. Two English-language translations appeared, the first published in London, the other in New York. Three French translations were produced, one in Brussels and two different ones in Geneva. There was also a Danish translation.

CONTEMPORARY RECEPTION—SCANDAL

Let us now look at aspects of the geography of reception of this widely published volume of correspondence in the English-, French-, and German-speaking worlds. During 1860 alone, well over two hundred articles of varying length about the Varnhagen von Ense volume appeared in the European daily and periodical press.[10] Among these were lengthy and passionately formulated essays in leading periodicals, written by some of the best-known names in journalism. In Great Britain, substantial articles appeared in the *Athenaeum,* the *Edinburgh Review,* and *Fraser's Magazine.* Among the French-language periodicals, the *Revue des Deux Mondes,* the *Revue Contemporaine,* and the *Revue Germanique* all published major reviews of the volume of correspondence. In the German-speaking world, detailed discussions were printed in such varied periodicals as the *Zeitstimmen aus der Reformierten Kirche der Schweiz* and *Wolfgang Menzels Literaturblatt.* Hermann Marggraff, editor of the Brockhaus-owned weekly *Blätter für Literarische Unterhaltung,* observed in the issue of 9 August 1860—some six months after the volume had appeared—that the publication of Humboldt's correspondence still represented the "real event" in literature that year, and that not even writing better than of Homer or Shakespeare could have competed with a volume:

> The real "European event" in the world of literature is still the publication of the letters by Alexander von Humboldt to Varnhagen and of his oral communications that are contained in Varnhagen's diary pages. No epic, even if more important than Homer's, no drama, even if far superior to Shakespeare's, could at the moment compete with a publication such as this.[11]

One reason for the stir that this edition of Humboldt's correspondence caused was that it contained "delicious bits of scandal" as one contemporary put it.[12] The letters included wholly unexpected, disrespectful and sarcastic comments by Humboldt about several eminent people. There existed an obvious national-geographic spread of which instances were most prominently discussed, each national readership being most interested in Humboldt's indiscretions about famous men and women with whom they were especially familiar. The gossip aspect of the volume was highlighted in the *Edinburgh Review* article, written by Abraham Hayward, an essayist, frequent contributor to the quarterlies, and translator of Goethe's *Faust* (1833). Hayward criticized Humboldt for his snide remarks against Humboldt's royal patron, King Friedrich Wilhelm IV of Prussia and others at the Berlin court:

> Far be it from us to undervalue this book as a contribution to the history of science, literature, criticism, or society in Berlin, but its claims in this respect would have been very little, if at all, lessened by the omission of the most objectionable passages, and the proposition which we especially dispute is that the name and memory of Humboldt will derive fresh lustre from it as it stands. We knew already that he was endowed with many of the highest gifts of genius, that his energy was inexhaustible, his knowledge vast and varied, his intellect of the most comprehensive order, his imagination rich, his fancy versatile and lively, and his perceptions singularly quick. But we did not know, whatever we might suspect, that he had become envious and carping, wanting in charity and candour, faithless even to the royal friend with whom he sate at meat, a backbiter and a flatterer. In short, his bad qualities are now brought out in bold relief, to the (we hope) temporary obscuration of the good. (217–18)

Various instances of Humboldt's "irreverent comments" passed the review (215). One concerned Peel, whom Humboldt disliked. In January 1842, Friedrich Wilhelm visited England to attend the christening of the Prince of Wales. Humboldt accompanied him even though he objected to the visit, because he saw in it a conspiracy by Christian Karl Josiah, Freiherr von Bunsen, the Prussian ambassador to London, in support of the Anglican Church and the Tories. Varnhagen chronicled:

> Humboldt has given me a very favourable account of England. At court, great splendour, but a simple and natural mode of private life; conversation easy and friendly, and good-natured in its tone, even between the members of rival political factions. Peel he does not like, did not like him before, says that he looks like a Dutchman, is rather vain than ambitious, has narrow views. (230–31)

Further "mischievous innuendo" was directed against Dorothea Christophorana, Princess Lieven (née von Bekendorf), famous for her London and Paris salons, whose husband had been the Russian ambassador to London (215). Humboldt disrespectfully referred to her as "Madame de Quitzow," the Quitzows having been an infamous family of medieval robber barons in Brandenburg:

Madame de Quitzow, who has never written to me for the last twenty-five years, wants to know from me whether the Emperor Paul, during the epoch of his political insanity, had caused the proposal to be made by Kotzebue, that instead of the armies, the Ministers of Foreign Affairs should engage in single combat. I was at that time (1799 and 1800) traversing the Delta of South America, and had no knowledge whatever of the anecdote which the Russian princess (who, as it now appears to me, has a very strong leaning towards Western ideas and predilections), wishes to have authenticated. According to rather untrustworthy accounts which I have gathered, the proposal was, that the Monarchs themselves, not the Ministers, should enter the lists for the duel. (232–33)

Hayward defended the princess against this disparaging account, pointing out that her letter had been solicited by Humboldt and that "the notion of a combat between monarchs was too commonplace to have fastened on the imagination of the Czar" (233). In any case, "no amount of deference, and no height of celebrity, are sufficient to protect Humboldt's correspondents from his malice, if there is an ambiguous or infelicitous phrase to fasten upon" (232). This was true even for letters received from the Italian novelist and poet Alessandro Manzoni or from the Austrian statesman Metternich. Hayward was particularly irritated by hypercritical remarks about a letter written by Prince Albert. Humboldt's "malice" concerned a metaphor the Prince Consort had used in a letter of thanks for the receipt of a copy of Humboldt's *Kosmos*, having concluded the letter as follows: "May Heaven, of whose 'revolving Seas of Light and Terraces of Stars' you have given us so nobly a description, preserve you for many years to the Fatherland, the world, and 'Kosmos' itself, in unimpaired freshness of body and mind" (231). This missive was accompanied by the gift of a book, *Views of Ancient Monuments in Central America, Chiapas and Yucatan* (1844), written by one Frederick Catherwood. In one of his letters to Varnhagen, Humboldt criticized Albert, who was the younger son of the Duke of Saxe-Coburg-Gotha, for lack of politeness, the Prince Consort having waited too long before thanking Humboldt for the complimentary copy of *Kosmos*. Humboldt then managed to pack into the next few lines the following derogatory remarks:

He makes me speak of "revolving Seas of Light and *Terraces* of Stars"; a Coburg reading of my text, *quite English,* from Windsor, where all is full of terraces. In "Kosmos" there occurs once the expression *Star-carpet,* to explain the starless spots by *openings* in the firmament. The book on Mexican Monuments, which he *makes me a present of,* I bought two years ago. A fine illustrated edition of Lord Byron's works would have been a more delicate compliment. It is strange, too, that he never mentions Queen Victoria; who, perhaps, does not find my book on Nature sufficiently Christian. You see, I judge severely when princes write. (231)

Contemporary Reception—Politics

By focusing on "the delicious bits of scandal" and portraying Humboldt as a backbiter, attention was diverted from the left-wing political message of the Humboldt–Varnhagen von Ense edition. Accordingly, conservative reviewers highlighted the gossip-cum-scandal side of the correspondence, whereas liberal ones stressed the political content and implications of Humboldt's comments and criticisms. A lengthy essay in the *Athenaeum*, for example, at the time edited by the historian and traveler William Hepworth Dixon and known for occasionally functioning as a mouthpiece of democratic reform, discussed the volume as a serious document of antiestablishmentarian politics. Only at the very end did the reviewer turn his attention to the supposedly scandalous lines on Prince Albert, defending the Prince Consort against Humboldt's criticism, but softening its censure of Humboldt by stating that he had written the letter while "not being perhaps in the best of humours."[13]

Indeed, the letters revealed a very different Humboldt from the one many people believed themselves to have known: not a staunch monarchist and establishment pillar, but a Humboldt in whose character and actions there had existed a subversive streak and who had been more closely allied to the revolutionary politics of the time than many of his admirers liked. For many years—critics reminded their readers—Humboldt had been a courtier, a daily visitor at the palace of the king of Prussia, and a constant guest at the royal table. It thus might be expected that he was in general agreement with the policy of Friedrich Wilhelm IV and his ministers. Yet now the Varnhagen letters appeared to reveal the opposite, showing Humboldt to be a supporter of libertarianism and an opponent of the conservative pietists with whom the king had surrounded himself. Moreover, Humboldt seemed to have believed that even extreme forms of republicanism were far superior to the enlightened despotism that ruled in Berlin, Paris, and other parts of Europe. The *Athenaeum* quoted a letter demonstrating that Humboldt's hostility to the court was carried to such a pitch "that he sometimes runs into downright Republicanism."[14] The "Advertisement" to the London edition stated:

These letters have created the most lively sensation all over Germany, where, within a few weeks after their first publication, a fifth edition has already appeared. In the present eventful state of affairs they have been hailed as fresh and startling evidence of the fact, that liberal principles and a strong feeling of German nationality and unity have long been steadily gaining ground, even among the highest classes in Prussian society.[15]

This connection with the political left was apparent in a number of ways. The fact that Humboldt should have been so intimate with Varnhagen, who was an outspoken liberal democrat, appeared suggestive of Humboldt's private, political leanings.[16] Nor should we overlook the significance of the person who edited the letters, Ludmilla Assing, who at her uncle's home had acted as a salon hostess, continuing what her aunt, Rahel Levin, had done before her early death in 1833. Like the Jewish Levin, she was a committed radical sympathizer with the liberal ideals of the Revolution of 1848. Her choice of the Humboldt letters from the vast private archives of Varnhagen appeared to serve the family purpose of furthering liberal democracy in the German-speaking world. When she continued with her editorial work and published several volumes of Varnhagen von Ense's *Tagebücher,* the Berlin police seized the volumes and she was threatened with a political lawsuit, being sentenced in absentia in 1862 to an eight-month prison term, and in 1864 to a further two years.[17] Assing fled to Italy, where in Florence she associated with Italian as well as expatriate democrats, revolutionaries, and anarchists.[18]

Let me repeat this point. Conservative periodicals denounced the publication, whereas the more liberal ones hailed it. In the *Preußische Jahrbücher,* a conservative monthly of German capitalists and landowners published in Berlin from 1858 to 1935, Rudolf Haym attacked Assing. The book was a monument of impiety ("Denkmal der Impietät") to Humboldt, sullying his name and by that the honor of the nation and the glory of German science ("dem Ruhm deutscher Wissenschaft"). Haym adding: "Notions of glory and honour are understandably further removed from the conceptual sphere of a woman [than of a man]" ("Die Begriffe von Ruhm und Ehre liegen begreiflich dem Vorstellungskreise eines Weibes ferner").[19] The writer and editor of the Stuttgart *Literaturblatt,* Wolfgang Menzel, who had changed in his own political allegiance from moderate liberalism to a reactionary-conservative position, also condemned Humboldt and Assing, accusing the latter of lowly, pecuniary motives and of having sold Humboldt's and Varnhagen's sins to the bookseller ("Ludmilla Assing verkauft die geheimen Sünden Humboldts und Varnhagens an den Buchhändler").[20] Moreover, Menzel judged the book empty, worthless, and exhibiting a reprehensible hatred against Christianity and against the noble motives of the Prussian king.[21] The staunchly conservative

René-Gaspard-Ernest Taillandier, writing in the establishmentarian *Revue des Deux Mondes,* with its Orleanists and professors from the Collège de France as contributors, accused Humboldt of duplicity, regarded the correspondence as injurious to Berlin and Prussia, and emphatically condemned Assing and her motives, accusing her of feminine vanity and of improperly having usurped the right to make the letters public.[22] Hayward also condemned the volume, even though his Toryism was of the reformist, Peelite mode, and his essay appeared in the Whig *Edinburgh Review.* Hayward reproached Assing for bringing scandal and malice before the public:

> [T]here can be no doubt that she enjoyed the full confidence and esteem of both the eminent men who are so closely bound together in her book. This, in our opinion, aggravates her guilt in bringing them before the public in this fashion; and it is to be hoped that the merited censure she has incurred by her indiscretion will have some effect in preventing future offences of the sort. (236)

Someone who may well have hoped, too, that "future offences of the sort" be prevented, was Richard Moncton Milnes, first Baron Houghton. Milnes had good reason to be concerned. Like Varnhagen he was a sympathizer with liberal causes and himself had conducted a frank correspondence with him. These letters, covering the politically sensitive decade 1844 to 1854, were not published until 1922.[23] *Fraser's Magazine,* to which Milnes contributed, severely criticized the publication of Humboldt's private letters, accusing the person who published them of "treachery."[24] What most worried Milnes was that Humboldt's identity as a cosmopolitan scientist, who could be made to play a role in British scientific reform, was changed to that of a German nationalist by association with Berlin's dirty politics. Accordingly, Milnes made a valiant attempt to restore the image of his hero to that of a cosmopolitan genius of science who was above national politics:

> So noble indeed was the nature of Alexander von Humboldt, that it preserved, under an almost life-long weight of patronage, the elevation of his intellect and the integrity of his heart. His indefatigable industry was unimpeded by the constant round of small duties and vapid amusements, and the luxurious security of his official position never blunted his eager interest in the new acquisitions of all science, and in the fresh developments of literature. It was thus his signal good fortune to retain to the last, not only the wonderful stores of knowledge accumulated through so many years, but also the art to reproduce and dispose them for the delight and edification of mankind.[25]

By contrast, the *Athenaeum* saw in the volume an important commentary upon "the reactionary party in Prussia,"[26] causing chagrin to the aristocrats and delight to the democrats:

> Here is a book of wonders! Humboldt a democrat, a satirist—the philosopher of Berlin mocking and sporting in the garb of Pasquin! It sounds incredible; yet it seems most true. What will the illustrious sitters to this Prussian Gavarni say? Are not half the princes of Europe sending their subscriptions to Prussia in the name of Humboldt? Has not our own Prince Consort—has not Prince Frederick-William—have not the Emperors Francis Joseph and Louis Napoleon—given money, and time, and influence, to do honour to the memory of a philosopher, who was also believed to be a courtier—who appeared daily in royal palaces—who at table sat at the right hand of kings? Yet, here is evidence that, while bowing and smiling at the Schloss, Chamberlain Humboldt's heart was far away—that he looked on the court pageant as a comedy, on the princes and kings as merely players—that among the splendour of Sans Souci or Charlottenburg, he was mocking and railing with a Republican freedom more suited to the political atmosphere of New York. Here is a surprising revelation![27]

The *Athenaeum* critic cataloged instance after instance of Humboldt's antireactionary views. He had denounced King Ernest Augustus of Hanover, who as the fifth son of George III of England was also Duke of Cumberland, for his treatment of the "Göttingen Seven." The Hanover crown went through the male line only, and at the accession of his niece Queen Victoria in 1837, Ernst August, who had been associated with reactionary Tories and ultraconservatives, became Hanoverian king and rescinded the liberal constitution of 1833. When seven of Göttingen University's professors objected, they were summarily dismissed. Humboldt commented: "What a disgrace that such a man should pass for a German Prince!"[28] Further instances were cited of Humboldt's criticism of the Prussian king and his circle of reactionary and pietistic advisers. A passage from Varnhagen's diary of 18 March 1843 recorded Humboldt as having commented:

> The King does precisley what pleases him, whatever developes out of his early fixed notions; and any advice to him, even if he listens to it, is of no avail. . . . The King has given up none of his former plans, and may attempt at any moment to put them into execution; such as those relating to the Jews, the observance of the Sunday, the consecration of bishops after the English form, the new regulations touching the nobility, and so forth.[29]

A further instance of Humboldt's disaffection for his king concerned the confiscation by brute force of the Orleans property by Louis Napoleon during the coup d'état of 1851: the king approved; Humboldt was appalled.

Other liberal critics, too, welcomed the publication. A major article written by the liberal Protestant theologian Heinrich Lang appeared in the Zurich journal *Zeitstimmen aus der Reformierten Kirche der Schweiz*. Lang praised the volume as a contribution to the portrayal of a true and honest Humboldt: "Simple and true, like everything true and great, Humboldt's character shines forth from these free and easy letters" ("Einfach und wahr, wie alles Wahre und Große, leuchtet Humboldt's Charakter aus diesen zwanglosen Briefen hervor").[30] In a series of three consecutive articles, Lang lionized Humboldt for his liberal character and defended him against accusations of irreligiosity. Auguste Nefftzer, a French scion of the liberal press, and co-editor of the *Revue Germanique,* compared Humboldt's correspondence to that of Voltaire's, finding in the correspondence reasons to praise Humboldt for being non-nationalistic ("dénationalisé") and more French than German in matters of religion.[31] Edouard Simon, a Francophile German Jew who in the issue of the Decembrist *Revue Contemporaine* of February 1860 had just defended his friend, the radical revolutionary and materialist Carl Vogt, now in the June issue hailed Humboldt's letters to Varnhagen von Ense as a record of admirable liberalism. His lengthy article was a catalog of proofs concerning Humboldt's liberal views on politics and religion, at the Prussian court as well as in relation to national and international events, among which was the infamous case of the Göttingen Seven. He even defended Humboldt for his hypercriticism of Prince Albert, arguing that this had issued from a "noble sentiment" in that Humboldt's dislike of the Prince Consort stemmed from the latter's disparaging remarks made about Irish and Polish people.[32]

HUMBOLDT, GEOGRAPHY, AND REVOLUTION

Humboldt was one of the great geographers of his century. Yet "geography" here is not defined by his accomplishments in geography, but by the landscape of appropriations of Humboldt, staked out by the many and various reactions to his correspondence with Varnhagen von Ense. This approach is analogous to that of a previous study of the geography of reception of Humboldt's Mexican work.[33] Nikolaus Gatter has considered the responses of the periodical press to Assing's Humboldt volume in terms of the sociology of scandal.[34] Gatter has additionally recognized that approval or condemnation by critics may have amounted to an appropriation of Humboldt's renown on behalf of sociopolitical causes.

This chapter has developed and refined the latter point in particular by showing that the collection of letters of Alexander von Humboldt to Varnhagen von Ense was located in the discursive space of a sociopolitical ideology that harked back to that of the Revolution of 1848. On the Continent, those who had sympathized with the uprisings or with its objectives approved of the volume. Conservative reactionaries there objected to it. The situation in Britain was different. There, in 1848, the liberal-reformist periodical press had, by and large, not waved the banner of revolution. Upon the volume's appearance, that press, with rare exceptions, was condemnatory in its verdict. Thus, two points may be made from this exploration of the contemporary meanings of the first published selection of Humboldt's private correspondence, and from consideration of the reactions to that correspondence. The first concerns the evidence contained in the correspondence itself and in the reactions to it, evidence that reveals a shift in Humboldt's reputation toward a perception of him as a revolutionary. The second is that, as with the reception of Humboldt's work during his lifetime, there is a discernible geography to the readings of Humboldt the revolutionary.

...

Notes

1. On Humboldt and plant geography, see Malcolm Nicolson, "Alexander von Humboldt, Humboldtian Science and the Origins of the Study of Vegetation," *History of Science* 25 (1987): 167–94; Malcolm Nicolson, "Humboldtian Plant Geography after Humboldt: The Link to Ecology," *British Journal for the History of Science* 29 (1996): 289–310; Malcolm Nicolson, "Alexander von Humboldt and the Geography of Vegetation," in *Romanticism and the Sciences*, ed. Nicholas Jardine and Andrew Cunningham (Cambridge: Cambridge University Press, 1990), 169–88. For Humboldt and meteorology, see Michael Dettelbach, "Global Physics and Aesthetic Empire: Humboldt's Physical Portrait of the Tropics," in *Visions of Empire: Voyages, Botany and Representations of Nature*, ed. David Philip Miller and Peter Hanns Reill (Cambridge: Cambridge University Press, 1996), 258–92.

2. These paragraphs are based on my *Alexander von Humboldt: A Metabiography* (Bern: Peter Lang, in press).

3. Nikolaus Gatter, *"Gift, geradezu Gift für das unwissende Publicum": Der Diaristische Nachlaß von Karl August Varnhagen von Ense und die Polemik gegen Ludmilla Assings Editionen, 1860–1880* (Bielefeld: Aisthesis Verlag, 1996), 73.

4. Kurt-R. Biermann, "Wer Waren die wichtigsten Briefpartner Alexander von Humboldts?" *NTM-Schriftenr. Gesch. Naturwiss., Technik, Med.* 18 (1981): 34; Christian Suckow, "Die

Alexander-von-Humboldt-Edition: Ein Projekt der Berlin-Brandenburgischen Akademie der Wissenschaften," *Jahrbuch der historischen Forschung in der Bundesrepublick Deutschland,* Berichtsjahr 1995, 16–21; Christian Suckow and Ingo Schwartz, "Zur Problematik einer auswählenden Briefedition, Beispiel: Die Briefe Alexander von Humboldts," in *Wissenschaftliche Briefeditionen und ihre Probleme,* ed. Hans-Gert Roloff (Berlin: Weidler Buchverlag, 1998), 119–22.

5. Biermann "Wer Waren die wichtigsten Briefpartner," 36–38.

6. Petra Werner, *Casanova ohne Frauen? Bemerkungen zu Alexander von Humboldts Korrespondenzpartnerinnen* (Berlin: Alexander-von-Humboldt-Forschungsstelle, 2000).

7. Walther Fischer, *Die Persönlichen Beziehungen Richard Monckton Milnes', ersten Barons Houghton, zu Deutschland, unter besonderer Berücksichtigung seiner Freundschaft mit Varnhagen von Ense* (Würzburg: Buchdruckerei Konrad Triltsch, 1918); Carl Misch, *Varnhagen von Ense in Beruf und Politick* (Gotha and Stuttgart: Perthes Verlag, 1925); Konrad Feilchenfeldt, *Varnhagen von Ense als Historiker* (Amsterdam: Verlag der Erasmus Buchhandlung, 1970); Terry H. Picket, *The Unseasonable Democrat: Karl August Varnhagen von Ense, 1785–1858* (Bonn: Bouvier Verlag, 1985); Werner Greiling, *Varnhagen von Ense—Lebensweg eines Liberalen: Politisches Wirken zwischen Diplomatie und Revolution* (Cologne: Böhlau, 1993); Gatter *"Gift, geradezu Gift für das unwissende Publicum."*

8. Richard Preuss, ed., *Briefe Thomas Carlyle's an Varnhagen von Ense: Aus den Jahren 1837–1857* (Berlin: Paetel, 1892); Walther Fischer, ed., *Die Briefe Richard Monckton Milnes', ersten Barons Houghton, an Varnhagen von Ense, 1844–1854* (Heidelberg: Carl Winters Universitätsbuchhandlung, 1922); Terry H. Picket, *Letters of the American Socialist Albert Brisbane to K. A. Varnhagen von Ense* (Heidelberg: Winter, 1986).

9. Biermann "Wer Waren die wichtigsten Briefpartner," 36–37.

10. Gatter *"Gift, geradezu Gift für das unwissende Publicum,"* (397–409.

11. Hermann Marggraff, "Stimmen des Auslandes über Alexander von Humboldt's Briefe," *Blätter für Literarische Unterhaltung,* 9 August 1860, 590. See also Hermann Marggraff, "Die Englische Übersetzung der Briefe Alexander von Humboldt's," *Blätter für Literarische Unterhaltung,* 14 June 1860, 443; and Hermann Marggraff, "Alexander von Humboldt's Briefe in England und Frankreich," *Blätter für Literarische Unterhaltung,* 27 September 1860, 718–19.

12. Abraham Hayward, "Correspondence of Humboldt," *Edinburgh Review* 112 (1860): 217–18. Subsequent quotations from this article are cited by page number in the text.

13. "Literature: Letters from Alexander von Humboldt to Varnhagen von Ense," *Athenaeum,* 17 March 1860, 366.

14. Ibid., 366.

15. Ludmilla Assing, ed., *Letters of Alexander von Humboldt, Written between the Years 1827 and 1858, to Varnhagen von Ense: Together with Extracts from Varnhagen's Diaries, and Letters from Varnhagen and Others to Humboldt* (London: Trübner, 1860), vii.

16. Misch, *Varnhagen von Ense in Beruf und Politick;* Feilchenfeldt, *Varnhagen von Ense als Historiker;* Picket, *Unseasonable Democrat.*

17. Gatter *"Gift, geradezu Gift für das unwissende Publicum,"* 356–62.

18. Ludmilla Assing, *Piero Cironi: Ein Beitrag zur Geschichte der Revolution in Italien* (Leipzig: Matthes, 1867).

19. Rudolf Haym, "Notizen," *Preussische Jahrbücher* 5 (1860): 415.

20. Wolfgang Menzel, "Zeitgeschichte: Briefe von Alexander von Humboldt an Varnhagen von Ense," *Wolfgang Menzels Literaturblatt,* 4 April 1860, 105.

21. Menzel, "Zeitgeschichte," 105, 109.

22. René-Gaspard-Ernest (Saint-René) Taillandier, "Lettres Intimes et Entretiens Familiers de M. A. de Humboldt," *Revue des Deux Mondes* 28 (1860): 66, 67, 88, 89.

23. Fischer, *Die Briefe Richard Monckton Milnes'*; Fischer, *Die Persönlichen Beziehungen*; Philip Glander, *The Letters of Varnhagen von Ense to Richard Mockton Milnes* (Heidelberg: Carl Winter, 1965).

24. "The Publication-of-Letters Nuissance," *Fraser's Magazine* 61 (1860): 563.

25. Richard Monckton Milnes, "Alexander von Humboldt at the Court of Berlin," *Fraser's Magazine* 62 (1860): 597.

26. "Our Weekly Gossip," *Athenauem*, 10 March 1860, 343.

27. "Literature: Letters from Alexander von Humboldt," 365.

28. Ibid., 365.

29. Ibid., 366.

30. Heinrich Lang, "Humboldt und Unsere Zeit," *Zeitstimmin aus der Reformierten Kirche der Schweiz* 2 (1860): 233.

31. Auguste Nefftzer, "Correspondence d'Alexandre de Humboldt," *Revue Germanique* 9 (1860): 687.

32. Eduard Simon, "La correspondence d'Alexandre de Humboldt avec Varnhagen de Ense, 1827 à 1858," *Revue Contemporaine* 25 (1860): 148.

33. Nicolaas Rupke, "A Geography of Enlightenment: The Critical Reception of Alexander von Humboldt's Mexico Work," in *Geography and Enlightenment*, ed. David N. Livingstone and Charles W. J. Withers (Chicago: University of Chicago Press, 1999), 319–39.

34. Gallei *"Gift, geradezu Gift für das unwissende Publicum,"* 232

AFTERWORD

Revolutions and Their Geographies

...

Peter Burke

In this brief afterword, written from the point of view of a cultural historian, and deliberately raising more questions than it is possible to answer (as the proliferation of question marks will reveal), I should like to emphasize the variety of both revolutions and their geographies.

As the editors point out in their introductory chapter, the term "revolution" has changed its meaning in a fundamental way in the course of its long history.[1] *Revolutio* meant "revolving," in the case of wheels as in that of planets. Since the planets were believed to exert "influence" on human affairs, the term "revolution" was extended to include political upheavals, "earthquakes of state" as one seventeenth-century Italian writer described them. What Thomas Hobbes stated explicitly was often assumed by other writers, and that is that after the earthquake subsided or the wheel revolved, the political structure would return to its former state.

The years following 1789 marked a revolution in the idea of revolution, symbolized by the adoption of a new calendar, with "Year One" as a declaration of the intent to make all things new, to abolish what was becoming known as the ancien régime and to follow the road of progress. "Revolution" now came to signify an irreversible change. In other words, the concept reversed its original meaning in the course of its incorporation in a linear rather than a cyclical view of history. Only "counterrevolutionaries," as they came to be called, still believed in the possibility of restorations.

The new concept of revolution spread from politics to other domains, as in the famous case of the nineteenth-century idea of an "Industrial Revolution."

More recently, especially since the publication of Thomas Kuhn's *Structure of Scientific Revolutions* (1962), historians and others have accustomed themselves to working with the idea of intellectual revolutions, including revolutions in particular disciplines; astronomy, chemistry, anthropology, or history itself. As Peter Dear remarked recently, "All revolutions are revolutions against something."[2] In other words, the concept implies its complementary opposite, the idea of an old regime, or to employ Kuhn's language, "normal science." In France in particular, the idea of "regime" has considerable appeal, from Michel Foucault's "regimes of truth" to François Hartog's recent discussion of "regimes of historicity."[3] By contrast, scholars in the English-speaking world seem to be more comfortable with Kuhn's concept of the "paradigm."

In what follows I should like to reflect, first, on what has already been discovered about the geographies of revolutions, and second, about what might now be done, the direction that research might take in the future and the ideas linking the terms.

POLITICAL, ECONOMIC, AND SCIENTIFIC REVOLUTIONS

My own working assumption as a historian is that it is impossible to explain any changes in human affairs, revolutions included, without looking at three dimensions of these changes: the chronological, the spatial, and the social. The first dimension has long formed part of historical practice. The third has become increasingly important in the past fifty years, with the rise of social history and the establishment of closer relations with sociology and social anthropology.

The second, the geographical dimension, has not been totally neglected. When I was a student of history in Oxford in the late 1950s, historical geography was a compulsory course in the first year, and I still remember reading works by W. Gordon East on the subject. The association between the study of history and the study of geography was even closer in France from the late nineteenth century onward, and this association contributed to the rise of the so-called Annales school. One of the leaders of that school or movement, Lucien Febvre, was an admirer of Vidal de la Blache and himself the author of a book on historical geography, *La terre et l'évolution humaine* (1922). It might be argued, however, that historians used to define the geographical dimension in too narrow a manner. It is only recently that the "spatial turn" discussed in the introduction to this volume has become visible in historical studies, although this is perhaps less the case in the history of science. Urban historians, for instance, like their colleagues in the history of architecture, are beginning to write the history of neighborhoods, of squares, and even of the arrangement of domestic interiors, analyzing what they call "the politics of space" as well as its sociology. Italian examples are numerous and for different periods.[4]

Historical studies of revolutions, whether they are political, economic, or cultural, have also had something to say about their geography. Consider, for example, the example of the English Revolution, studied here by Robert Mayhew in chapter 9. Civil wars are usually, among other things, wars between regions—the Spanish Civil War makes an obvious example—and the English Civil War is no exception to this rule. As Christopher Hill pointed out over sixty years ago, the Civil War was a conflict "of north and west versus south and east." Hill interpreted this geographical divide in economic terms, noting that the "economically backward areas" supported the king, while "those districts influenced by the demands of the London market" favored Parliament.[5] More recently, David Underdown has studied "the geographical distribution of allegiance" in the Civil War in considerable detail, presenting the war as a clash between regional cultures which are characterized in essentially ecological terms, contrasting arable areas with pastoral ones, such as the "cheese country" in Wiltshire.[6] As Mayhew here argues, however, such regional considerations alone present difficulties without reference to religious and other distinctions.

Similar points might well be made about the French Revolution, discussed by Michael Heffernan in chapter 10. Brittany is an obvious example of what Hill would have called a "backward" or peripheral area, and it was one where the peasantry generally opposed the Revolution.[7] At a microlevel and more than seventy years ago, Georges Lefebvre studied the geography of rumor during the Revolution, notably the notorious "great fear" of 1789.[8] Other studies have also dealt with the impact of the Revolution and, more especially, the Terror on particular regions, such as the Loire valley, where "simple geographical facts" such as the contrast between the mountains and the plains "prevented adjacent areas from responding similarly to the same impulses."[9] The French Revolution is a particularly interesting revolution for geographers to study in the sense that it included an attempt to unify France more closely (by changes in its administrative geography not least)—an attempt that led to and at the same time was encouraged by a growing awareness of the cultural differences between different parts of the country.[10]

All the same, it is striking that general analyses, like Theda Skocpol's well-known comparative study of the French, Russian, and Chinese Revolutions, have little to say about local variations in the response to revolutionary messages, although the resistance of the Cossacks to the Bolsheviks is well known, like the outsize contribution of the province of Hunan to the movement led by Mao Zedong.[11] In the case of the Mexican Revolution, things are only a little better developed, perhaps, in respect of sensitivity to geographical difference. Eric Wolf, for example, stressed the contribution of two rural areas, Chihuahua in the north and Morelos in the south. In his classic general study, Alan Knight made a number of references to local conflicts and suggested that the Mexican Revolution provided the opportunity for the redress of many local grievances, but he did not provide any systematic analysis of the pattern of regional responses.[12] My own view is that a crucial factor in these responses was the local system of land tenure. In the case of the Loire valley in the late eighteenth century, the contrast between highlands and plains was also a contrast between smallholdings and large estates. In the case of Spain and Portugal, too, the relation between land and revolution is very clear. In the south of both countries, where *latifundia* were dominant, the agricultural workers supported revolution and, in the Spanish case, a number of them turned anarchist. By contrast, the rural north of both countries was dominated by smallholders and was socially conservative.

In the case of economic revolutions, commercial, "industrious," and industrial, the importance of regions such as Yorkshire or Friesland was pointed out long ago. In the past generation, however, under the influence of contemporary debates in development economics, the spatial concepts of center and periphery have become increasingly important—not to say central—in historical analyses. One of the best-known examples of the trend is the work of that sociologist of Africa turned economic historian, Immanuel Wallerstein. In his study of "world systems," Wallerstein explained the rise of capitalism in the West in early modern times in terms of the changing relation between a center and its peripheries and "semiperipheries." In other words, he linked the rise of capitalism to the rise of serfdom in eastern Europe as well as to the import of slaves to the New World, stressing the intercontinental division of labor in which the periphery produced raw materials, allowing the center to specialize in trade and manufacture.[13]

The concepts of center and periphery have been extended to science, and to an understanding of its history, by Bruno Latour. Latour has pointed out that raw data are taken from many places, and turned into knowledge in what he calls centers of calculation.[14] Questions of geography—expressed variously as "center-periphery," "centers of calculation," "national styles," or as local sites and the networks that connect them—are central to the case of the Scientific Revolution, as a number of chapters in this volume show. Yet what also deserve to be noted are the geographical implications of the debates about the compatibility of science with different religious attitudes, even if the implications of such matters have not always been made explicit.

A generation before Max Weber's famous observations about the sociology of capitalism, for example, the French historian of science Alphonse de Candolle had drawn attention to the fact that French Protestants performed better than their Catholic counterparts in the scientific domain. Inspired by Weber's famous discussion of why China did not develop capitalism, Joseph Needham among others has analyzed what he describes as the Chinese failure to achieve the "breakthrough" into modern science. Making more explicit the geographical point implied in Candolle's study (since Protestants were mainly to be found in certain French regions), later historians have noted the uneven performance of different European countries in the Scientific Revolution; the outsize contribution of the Dutch, for instance, and the relative insignificance of that made by the Spaniards. These differences in what has been called the "cultural topography" or the "geopolitics" of science have often been explained in religious terms, with the Catholic Galileo, given his conflict with the church, as the exception that proves the rule.[15]

More recently, historians of science (such as John Henry here, for example, in chapter 3) have been adding qualifications to the simple dichotomies between Catholic and Protestant, East and West, without abandoning the quest for the geography of scientific revolutions. Geoffrey Lloyd, for instance, has compared the social organizations underlying curiosity about the world of nature in ancient Greece and ancient China, contrasting competitive individualism in one culture with state support in the other. As David Goodman has shown, early modern Spain is, like China, no longer viewed purely in terms of scientific deficit.[16] Accordingly, it might be more useful to think in terms of alternative styles of research or "styles of scientific thinking," as Alistair Crombie called them, rather than of either the presence or absence of science.[17] These styles of thinking are sometimes national, but before what might be called the "nationalization of culture" in the nineteenth century, linked to the rise of more centralized states, they were more likely to be either regional or religious.

Revolutions in Art and in Geography

Turning from the present to the future and extending from the issues explored in this volume, I should like to suggest two possible directions for further work on the connections between geography and revolution, work in which historians, geographers, and others might participate. The first is the geography of artistic revolutions, the second the place of revolutions in geography itself.

Reference has already been made here, and by Henry in chapter 3, to

Crombie's idea of "styles" of scientific thinking. Study of the history of art may also reveal some fascinating examples of the geography of styles, including stylistic revolutions. As in the case of political scientists, art historians are ready to admit the importance of local geographical variations, but they rarely try to write systematically about artistic regions with, perhaps, two notable exceptions.[18] One of the few attempts to do so is the work neither of an art historian nor of a geographer, but of that of a general historian, Victor Tapié. In seventeenth-century Europe, baroque and classicism were rival international styles. Long ago, Tapié suggested that they had their own geographies, often parallel to the geography of religion and science. The inhabitants of Catholic regions (Italy, Spain, Portugal, and central Europe) generally chose baroque, while Protestant regions (Britain, the Dutch Republic, north Germany, and Scandinavia) generally chose classicism. France was more complicated, a mainly Catholic country in which a prolonged hesitation between the rival styles was resolved in favor of classicism after 1650.[19]

How might one explain this contrast? Classicism and baroque have often been defined as opposite as well as rival styles, in terms of regularity versus irregularity, simplicity versus complexity, restraint versus exuberance, clarity versus difficulty, repose versus movement, balance versus imbalance, reason versus unreason, and so on. In the case of at least some of these contrasts we might speak of affinities between classicism and Protestantism (especially Calvinism) and between baroque and Catholicism (especially Counter-Reformation Calvinism). A few bolder historians might go further still in this direction. The Czech historian Josef Polišenský has interpreted the Thirty Years' War as a clash between regions that was also a clash between two civilizations or two cultural models, the Spanish and the Dutch, in which Spain represented the old regime while the Dutch symbolized a new way of life.[20]

Let me turn, however, from artistic revolutions to revolutions in the discipline of geography itself. The idea of writing a Kuhnian or a Foucauldian history of geography, focused on changes of paradigm or epistemic "ruptures" is an attractive one. There is no such general history available, so far as I know, despite the existence of important studies of particular countries and centuries, of which the most explicitly Foucauldian is probably Anne Godlewska's study of French geography.[21] When such a general history comes to be written, however, it will surely need to take account of a number of suggestions expressed in the chapters making up this volume. Suppose that we treat the period running from Ptolemy and Strabo to the Renaissance revival of interest in these scholars as a prolonged old regime (implying not that the study of geography was static but that there was no structural change, no major or sudden breach of continuity). We may then ask, when, where, and why did revolutions against this regime occur?

One revolution in geography has been discussed here. Charles Withers argues in chapter 4 that—associated as it was with astronomy and mathematics—"geography was part of what we have come to term the 'Scientific Revolution,'" but that, for various reasons, this place has not always been recognized. How many more revolutions or paradigm changes has geography undergone? A short list compiled from other chapters would include at least five more candidates. In chronological order, they comprise the printing of maps, as discussed by Jerry Brotton in chapter 6; the Darwinian revolution as explored by James Moore in chapter 5; the "visual revolution" in association with the introduction of photography as discussed by James Ryan in chapter 8; and even geography's own quantitative revolution in the mid-twentieth century, mentioned briefly in chapter 1.

To this list a seventh candidate might be added. That is, the "cultural revolution" in geography, exemplified by many of the contributors themselves. This cultural turn, parallel to that of the neighboring discipline of history, is revealed by a new vocabulary, including terms such as "geographical culture," "geoliteracy," or "cultures of exploration." The cultural approach has grown out of the social in geography and history alike. In Britain, the Society for the Study of Social Geography added the word "cultural" to its title in 1988, while the Social History Society made a similar move a decade later.[22]

All seven putative revolutions raise, however, the problem of the criteria for describing them as revolutionary. In some cases, the principal criterion seems to be a change of method, either the turn to systematic fieldwork and to state mapping, for instance, to statistics and political economy, or to surveys of one sort or another.[23] In other cases, it is the rise of new concepts such as evolution, and in yet others, the institutionalization of the subject in schools and universities. Is it intellectually permissible to shift from one criterion to another in this way? It might be better to argue that from a methodological point of view there have been, say, two revolutions, from the conceptual point of view three, and so on. Or would these distinctions dilute the idea of revolution so much as to disable it? The problem of continuity also demands discussion here. In the case of the trajectory or career of history, I would certainly want to argue that its so-called revolutions (Rankean, Burckhardtian, Braudelian, or whatever), however significant they were, did not erase but, rather, supplemented the traditions of the craft. Could this be the case for geography too?

There is at least one more question to ask about geographical revolutions—

or is it the geography of revolutionary thinking?—and that concerns their own geographies. That is to say, are there national styles of geography as there are (so John Henry has it in chapter 3) in the sciences? At one point, about 1900, the differences between German and French approaches (that of Friedrich Ratzel, say, or of Vidal de la Blache) seemed clear enough. Are such issues visible any longer? Or does geography, in most of the academic environments in which it is practiced, now like history speak with an American accent? In the case of the cultural geography and geographies of science exemplified in this volume, there seem to be international networks linking Cornell and Syracuse in New York with Edinburgh, London, Oxford, Bristol, and Belfast. But is it significant that only one contributor to the volume, Nicolaas Rupke, comes from outside the Anglophone culture area?

Matters Arising

As a number of examples discussed above suggest, approaches to the geography of and in revolutions may be divided into two kinds. On one side there are macrogeographies contrasting, for example, the north of Europe with its south, or Europe as a whole with China. On the other side, there are microgeographies following the prompts of James Ryan's topography of the Royal Geographical Society in nineteenth-century London, for instance, or of David Underdown's mapping of regional cultures in early modern England.

In similar fashion, the approach to "revolutions in the times" followed by Paul Glennie and Nigel Thrift in chapter 7 above might be illustrated and extended by both macro- and microexamples. At the macrolevel, we have the standardization and westernization of world times in the later nineteenth century. Japan adopted the Gregorian calendar in 1873, for example, and Turkey, following Mustafa Kemal Atatürk's revolution, did so in 1925. Following the Washington Conference on World Standard Time in 1884, one country after another adopted Greenwich Time. At much the same time, electrical systems were devised to coordinate different clocks in the same city in Leipzig, Bern, Paris, and elsewhere.[24] At the microlevel, we might note, for instance, Thomas Hardy's vivid description of the variety of local times on Egdon Heath in the middle of the nineteenth century:

> On Egdon there was no absolute hour of the day. The time at any moment was a number of varying doctrines professed by the different hamlets, some of them having originally grown from a common root, and then become divided by secession, some having been alien from the beginning. West Egdon believed in Blooms-End time, East Egdon in the time of the Quiet Woman Inn. Grandfer

Cantle's watch had numbered many followers in years gone by, but since he had grown older faiths were shaken.[25]

Studies of the two levels raise the problem of the relation between the local and the global, a problem that is still far from being resolved, as the continuing debate about the value of microstudies shows. Some historians still dismiss microhistory as trivial. Others claim that the view through the historical microscope requires scholars to revise their notion of a plausible explanation by revealing the ways in which individuals or small groups find space in which to maneuver and escape the pressures of large-scale institutions such as states and churches. Again, it is argued that microstudies are a "strategy of knowledge" that has the advantage of keeping close to the experience of the agents. On the other hand, to keep one's eye glued to a microscope, so to speak, is to miss the big picture.[26]

What is to be done? It is generally agreed that a synthesis of the micro- and macrolevels is desirable. The problem is how to achieve this aim, or at least to facilitate it, in relation to geographical and historical scales alike. As several of the contributors here note, historians of science in particular and others have focused on this problem in recent years, discussing the process by which local spaces of local research and experimentation have created general knowledge.[27] Two more specific debates that have taken place in other disciplines, in folklore and in anthropology, are also highly pertinent here. One debate concerns "ecotypes," the other "brokers." "Ecotype" is a term that was long ago borrowed from botany by the Swedish folklorist Carl von Sydow, who was interested in the development of local variants of folktales that are known throughout Europe or even beyond, variants that he viewed as adaptations to the local milieu.[28] In similar fashion, and to return for a moment to my earlier remarks upon seventeenth-century European art, we might view Czech baroque architecture, say, as an ecotype of an international movement, the product of a conflict between centripetal and centrifugal forces. To take another example, the difference between the forms of Gothic architecture in France, Italy, and Scandinavia is even more visible.

Anthropologists, faced with the impact of globalization on the small-scale cultures they have traditionally studied, have made some of the most interesting contributions to the micro-macro debate, among them Eric Wolf, Fredrik Barth, Marshall Sahlins, and Ulf Hannerz. Half a century ago, in the course of his fieldwork in Mexico, Eric Wolf noted the importance of intermediaries, political "brokers," as he called them, between local communities and the wider world. He argued that just as anthropologists should not forget the state when studying villages, so political scientists should not overlook local communities when studying the nation. His advice is of obvious relevance to historians and geographers.[29]

The parallel concerns of historians and geographers, especially cultural historians and cultural geographers as well as historians of science interested in matters of geographical difference, are obvious enough in these pages. Let us hope that unlike parallel lines, the disciplines themselves as well as some of their practitioners will manage to meet more often in the future.

...

Notes

1. See also Karl Griewank, *Der Neuzeitliche Revolutionsbegriff* (Weimar: H Böhlaus Nachfolger, 1955); Karl-Heinz Bender, *Revolutionen: Die Entstehung des Politischen Revolutionsbegriffes in Frankreich zwischen Mittelalter und Aufklärung* (Munich: Wilhelm Fink, 1977).

2. Peter Dear, *Revolutionizing the Sciences: European Knowledge and Its Ambitions, 1500–1700* (Princeton: Princeton University Press, 2001), 3.

3. Michel Foucault, *Les mots et les choses* (Paris: Gallimard, 1968); François Hartog, *Régimes d'historicité* (Paris: La Découverte, 2003).

4. Dale V. Kent and F. William, *Neighbours and Neighbourhood in Renaissance Florence* (New York: Academic Press, 1982); Richard Krautheimer, *The Rome of Alexander VII* (Princeton: Princeton University Press, 1985); Joseph Connors, "Alliance and Enmity in Roman Baroque Urbanism," *Römisches Jahrbuch der Bibliotheca Hertziana* 25 (1989): 207–94; Edward Muir and Ronald Weissman, "Social and Symbolic Places in Renaissance Venice and Florence," in *The Power of Place: Bringing Together Geographical and Sociological Imaginations*, ed. John A. Agnew and James S. Duncan (Boston: Unwin Hyman, 1989), 81–103; Patricia Waddy, *Seventeenth-Century Roman Palaces* (New York: Architectural History Foundation and MIT Press, 1990); Laurie Nussdorfer, "The Politics of Space in Early Modern Rome," *Memoirs of the American Academy in Rome* 42 (1999): 161–86.

5. Christopher Hill, quoted in *Puritans and Revolutionaries: Essays in Seventeenth-Century History Presented to Christopher Hill*, ed. Donald Pennington and Keith Thomas (Oxford: Clarendon, 1978), 153–96.

6. David Underdown, "The Chalk and the Cheese," *Past and Present* 85 (1979): 25–48; David Underdown, *Revel, Riot and Rebellion: Popular Politics and Culture in England, 1603–1660* (Oxford: Oxford University Press, 1985).

7. Donald Sutherland, *The Chouans: The Social Origins of Popular Counter-Revolution in Upper Brittany, 1770–1796* (Oxford: Clarendon, 1982).

8. Georges Lefebvre, *La grande peur de 1789* (Paris: A. Colin, 1988).

9. Colin Lucas, *The Structure of the Terror: The Example of Javogues and the Loire* (Oxford: Oxford University Press, 1973).

10. Jacques Revel, "La région," in *Histoire de la France,* ed. Jacques Revel (Paris: Seuil, 1992), 1:851–83; Jacques Revel, "Knowledge of the Territory," *Science in Context* 4 (1991): 133–61; Daniel R. Headrick, *When Information Came of Age: Technologies of Knowledge in the Age of Reason and Revolution, 1700–1850* (Oxford: Oxford University Press, 2000).

11. Theda Skocpol, *States and Social Revolutions: A Comprehensive Analysis of France, Russia and China* (Cambridge: Cambridge University Press, 1979).

12. Eric Wolf, *Peasant Wars of the Twentieth Century* (New York: Harper and Row, 1969); Alan Knight, *The Mexican Revolution* (Cambridge: Cambridge University Press, 1986), 1:219–23.

13. Immanuel Wallerstein, *The Modern World System*, vol. 1, *Capitalist Agriculture and the Origins of the European World-Economy in the Sixteenth Century* (New York: Academic Press, 1974).

14. Bruno Latour, *Science in Action: How to Follow Scientists and Engineers through Society* (Milton Keynes: Open University Press, 1987). But see also Christian Jacob, *L'empire des cartes* (Paris: Gallimard, 1992); and Peter Burke, *A Social History of Knowledge from Gutenberg to Diderot* (Cambridge: Polity, 2000).

15. Alphonse de Candolle, *Histoire des sciences et des savants depuis deux siècles* (Geneva: H. Georg, 1873); Reijer Hooykaas, *Religion and the Rise of Modern Science* (Edinburgh: Scottish Academic Press, 1972), 98, 101; Roy Porter and Mikuláš Teich, eds., *The Scientific Revolution in National Context* (Cambridge: Cambridge University Press, 1992), 2. On China, see Joseph Needham, "Poverties and Triumphs of the Chinese Scientific Tradition," in *Scientific Change: Historical Studies in the Intellectual, Social and Technical Conditions for Scientific Discovery*, ed. Alistair Crombie (London: Routledge and Kegan Paul, 1963), 117–49.

16. Geoffrey Lloyd, *The Ambitions of Curiosity* (Cambridge: Cambridge University Press, 2002); David Goodman, "The Scientific Revolution in Spain and Portugal," in Porter and Teich, *Scientific Revolution*, 158–77.

17. Alistair Crombie, *Styles of National Thinking in the European Tradition*, 3 vols. (London: Duckworth, 1994).

18. Jan Białostocki, "The Baltic Area as an Artistic Region in the Sixteenth Century," *Hafnia* 6 (1976): 11–24; Thomas DaCosta Kaufmann, *Toward a Geography of Art* (Chicago: University of Chicago Press, 2004).

19. Victor-L. Tapié, *Baroque et classicisme* (Paris: A. Colin, 1957).

20. Josef Polišenský, *The Thirty Years War* (London: New English Library, 1974).

21. Anne Godlewska, *Geography Unbound: French Geographic Science from Cassini to Humboldt* (Chicago: University of Chicago Press, 1999). For other, less explicitly paradigmatic or Foucauldian studies of geography's history, see Robert J. Mayhew, *Enlightenment Geography: The Political Languages of British Geography, c. 1650–1850* (London: Macmillan, 2000); Charles W. J. Withers, *Geography, Science and National Identity: Scotland since 1520* (Cambridge: Cambridge University Press, 2001).

22. Christopher Philo, ed., *New Words, New Worlds: Reconceptualizing Social and Cultural Geography* (Lampeter: St. David's University College, 1991); Ian Cook et. al., eds., *Cultural Turn/ Geographical Turn: Perspectives on Cultural Geography* (Harlow: Prentice Hall, 2000).

23. On which, see respectively, Matthew H. Edney, *Mapping an Empire: The Geographical Construction of British India, 1765–1843* (Chicago: University of Chicago Press, 1997); Mayhew, *Enlightenment Geography*, 207–28; Withers, *Geography, Science and National Identity*, 210–24.

24. Peter Galison, *Einstein's Clocks, Poincaré's Maps: Empires of Time* (London: Sceptre, 2003), 13–47, 144–59.

25. Thomas Hardy, *The Return of the Native* (London: Macmillan, 1878), 137.

26. Giovanni Levi, "On Micro-History," in *New Perspectives on Historical Writing*, 2nd ed., ed. Peter Burke (Cambridge: Cambridge University Press, 2001), 97–119; James S. Amelang, "Micro-History and its Discontents: The View from Spain," in *Historia: A Debate*, ed. Carlos Barros (Santiago: Cosmos, 1995), 307–24; Jacques Revel, ed., *Jeux d'Échelle: La microanalyse à l'experience* (Paris: Gallimard, 1996).

27. See, for example, Crosbie Smith and Jon Agar, eds., *Making Space for Science: Territorial Themes in the Shaping of Knowledge* (London: Macmillan, 1998); Jan Golinski, *Making Natural Knowledge: Constructivism and the History of Science* (Cambridge: Cambridge University Press, 1998), esp. 79–102; David N. Livingstone, *Putting Science in Its Place* (Chicago: University of Chicago Press, 2003).

28. Carl von Sydow, *Selected Papers on Folklore* (Copenhagen: Rosenkilde and Bagger, 1948).

29. Eric Wolf, "Aspects of Group Relations in a Complex Society: Mexico," in *Peasants and Peasant Societies,* ed. Theodor Shanin (Harmondsworth: Penguin, 1975), 50–66; see also and compare the work of Fredrik Barth, *Process and Form in Social Life: Selected Essays of Fredrik Barth* (London: Routledge and Kegan Paul, 1981); Ulf Hannerz, "Theory in Anthropology: Small is Beautiful," *Comparative Studies in Society and History* 28 (1986): 362–67; Marshall Sahlins, "Cosmologies of Capitalism: The Trans-Pacific Sector of the 'World System,'" *Proceedings of the British Academy* 74 (1988): 1–52.

CONTRIBUTORS

Jerry Brotton is senior lecturer in Renaissance studies at Queen Mary, University of London. He is the author of *Trading Territories: Mapping the Early Modern World* (1997), *The Renaissance Bazaar* (2002), and with Lisa Jardine, *Global Interests: Renaissance Art between East and West* (2000). He is currently completing a book on the formation and dispersal of the art collection of King Charles I.

Peter Burke is professor of cultural history at the University of Cambridge and a fellow of Emmanuel College. A fellow of the British Academy, he is the author of more than twenty books, of which the latest, published by Cambridge University Press in 2004, is *Languages and Communities in Early Modern Europe*. He is now at work with his wife on an intellectual biography of the Brazilian sociologist-historian Gilberto Freyre.

Peter Dear is professor of history and of science and technology studies at Cornell University. He is the author of *Discipline and Experience: The Mathematical Way in the Scientific Revolution* (1995) and *Revolutionizing the Sciences: European Knowledge and Its Ambitions, 1500–1700* (2001). He is currently completing *The Intelligibility of Nature* for the University of Chicago Press.

Paul Glennie is senior lecturer in the School of Geographical Sciences at the University of Bristol. Among his recent publications is an essay with Nigel Thrift on the spaces of clock times in *History and the Social Sciences* (2002) and work on towns in an agrarian economy, 1500–1700, published in the *Cambridge Urban History of Britain* (2000). He has research interests in protoindustrialization, consumption, and consumerism in the early modern period and, with Nigel Thrift, is currently completing a book on the geographies and histories of timekeeping.

Michael Heffernan is professor of historical geography at the University of Nottingham. His research has focused on post-Enlightenment European and North American geographical thought, the political and cultural geography of Europe since the eighteenth century, and the cultural and intellectual implications of European imperial expansion and warfare in the nineteenth and twentieth centuries. Recent publications include *The Meaning of Europe: Geography and Geopolitics* (1998).

John Henry is a reader in the Science Studies Unit at the University of Edinburgh. He is the coauthor, with Barry Barnes and David Bloor, of *Scientific Knowledge: A Sociological Analysis* (1996). He has published widely on the history of science from the sixteenth to the nineteenth century, including *Moving Heaven and Earth: Copernicus and the Solar System* (2001) and *Knowledge is Power: Francis Bacon and the Method of Science* (2002). His textbook *The Scientific Revolution and the Origins of Modern Science* has recently appeared in a second edition (2002). He is currently engaged on editing the scientific works of Thomas Hobbes.

David Livingstone is professor of geography and intellectual history at the Queen's University of Belfast and a fellow of the British Academy. He is the author of numerous books, including *The Geographical Tradition* (1992) and *Putting Science in Its Place* (2003), and coeditor with Charles Withers of *Geography and Enlightenment* (1999). He is currently working on two projects: a book about the history of science, race, and religion entitled *Adam's Ancestors,* and a comparative study of the reception of Darwinism in different locations.

Robert Mayhew is reader in geography in the School of Geographical Sciences at the University of Bristol. He is the author of *Enlightenment Geography* (2000) and of *Landscape, Literature and English Religious Culture* (2004), as well as of numerous articles concerning the history of geographical thought, circa 1500–1900. He is currently working on a monograph tracing the interactions of geography and humanistic thought from ancient times to the present.

James Moore teaches history of science at the Open University and the University of Cambridge, and he has been a visiting professor at Harvard, Notre Dame, and McMaster Universities. He is coauthor with Adrian Desmond of *Darwin* (1991), and their edition of Darwin's *Descent of Man* appeared in 2004. His other publications include *The Post-Darwinian Controversies* (1979) and *The Darwin Legend* (1994). He is working on a biography of Alfred Russel Wallace and on a catalog of his correspondence.

Nicolaas Rupke is professor of the history of science at Göttingen University and director of the Institut für Wissenschaftsgeschichte. His areas of expertise are the late modern earth and life sciences. In exploring these, he has followed a biographical approach and has written on such nineteenth-century men of science as William Buckland in *The Great Chain of History* (1983), on Richard Owen in *Richard Owen* (1994), and most recently on Alexander von Humboldt, *Alexander von Humboldt: A Metabiography* (in press).

James R. Ryan is senior lecturer in geography at the University of Leicester. His publications include *Picturing Empire: Photography and the Visualisation of the British Empire* (1997) and, coedited with J. M. Schwartz, *Picturing Place: Photography and the Geographical Imagination* (2003). He is a member of the editorial board of the *Journal of Victorian Culture.*

Nigel Thrift is head of the Division of Life and Environmental Sciences and a student of Christ Church at the University of Oxford. His main research interests are in the history of timekeeping, international finance, the impact of information technologies, and nonrepresentational theory. His most recent publications include *Cities,* with Ash Amin (2002), *The Handbook of Cultural Geography,* coedited with Kay Anderson, Mona

Domosh, and Steve Pile (2003), *Patterned Ground*, coedited with Stephan Harrison and Steve Pile (2004), and *Knowing Capitalism* (2004).

Charles Withers is professor of historical geography at the University of Edinburgh. Recent coedited publications include *Geography and Enlightenment*, with David Livingstone (1999), *Science and Medicine in the Scottish Enlightenment*, with Paul Wood (2002), and *Georgian Geographies: Essays on Space, Place and Landscape in the Eighteenth Century*, with Miles Ogborn (2004). He is the author of *Geography, Science and National Identity: Scotland since 1520* (2001). He has research interests in the geographies of the Enlightenment and on memory, commemoration, and the history of geography.

BIBLIOGRAPHY

Abbot, George. 1604. *The Reasons which Doctour Hill hath Brought, for the Upholding of Papistry, which is Falselie termed the Catholicke Religion: Unmasked.* Oxford: Oxford University Press.

———. 1605. *A Briefe Description of the Whole Worlde.* London: John Browne.

———. 1620. *A Briefe Description of the Whole Worlde.* 5th ed. London: J. Marriott.

Aberdare, Lord. 1882. "The Annual Address on the Progress of Geography," *Proceedings of the Royal Geographical Society* 4:329–39.

Abney, William de W. 1871. *Instruction in Photography: for use at the S.M.E. Chatham.* Chatham: SME, Printed for Private Circulation.

Adair, Douglass. 1974. "'Experience Must be Our Only Guide': History, Democratic Theory, and the United States Constitution" In *Fame and the Founding Fathers: Essays by Douglass Adair,* ed. Trevor Colbourn, 107–23. Toronto: Norton.

———. 1974. "'That Politics May Be Reduced to a Science': David Hume, James Madison, and the Tenth Federalist." In *Fame and the Founding Fathers: Essays by Douglass Adair,* ed. Trevor Colbourn, 93–106. Toronto: Norton.

Adams, John. 1850–56. *A Defence of the Constitution of the United States of America.* In *The Works of John Adams, Second President of the United States,* vol. 4, ed. Charles Francis Adams. Boston.

Adams, Simon. 1983. "Spain or the Netherlands? The Dilemmas of Early Stuart Foreign Policy." In *Before the Civil War: Essays on Early Stuart Politics and Government,* ed. Howard Tomlinson, 79–101. London: Macmillan.

Albanese, Catherine L. 1976. *Sons of the Fathers: The Civil Religion of the American Revolution.* Philadelphia: Temple University Press.

Alder, Ken. 1995. "A Revolution to Measure: The Political Economy of the Metric System in France." In *The Values of Precision,* ed. M. Norton Wise, 39–71. Princeton: Princeton University Press.

Alfrey, Nicholas, and Stephen Daniels, eds. 1990. *Mapping the Landscape: Essays on Art and Cartography.* Nottingham: Nottingham University Art Gallery and Castle Museum.

Allen, William. 1840. *Picturesque Views on the River Niger, Sketched during Lander's Last Visit in 1832–33.* London: John Murray.

Alpert, Peter, Elizabeth Bone, and Claus Holzapfel. 2000. "Invasiveness, Invasibility and the Role of Environmental Stress in the Spread of Non-native Plants." *Perspectives in Plant Ecology, Evolution and Systematics* 3:52–56.

Amelang, James S. 1995. "Micro-history and its Discontents: The View from Spain." In *Historia: A Debate,* ed. Carlos Barros, 2:307–24. Santiago: Cosmos.

Anderson, Benedict. [1983] 1991. *Imagined Communities: Reflections on the Origin and Spread of Nationalism.* London: Verso.

Andrews, Kenneth. 1984. *Trade, Plunder and Settlement: Maritime Enterprise and the Genesis of the British Empire, 1480–1630.* Cambridge: Cambridge University Press.

Anstey, Peter. 2000. "The Christian Virtuoso and the Reformers: Are there Reformation Roots to Boyle's Natural Philosophy?" *Lucas: An Evangelical History Review* 27/28:5–40.

Applebaum, Wilbur, ed. 2000. *Encyclopaedia of the Scientific Revolution, from Copernicus to Newton.* New York: Garland.

"The Application of the Talbotype." 1846. *Art Union* 8:195.

Argles, Martin. 1990. "The Bark's Worse than the Bite," *Guardian* (London), 14 December, 31.

Armitage, Geoff. 1997. *The Shadow of the Moon: British Solar Eclipse Mapping in the Eighteenth Century.* Tring: Map Collector Publications.

Armstrong, Isobel. 2002. "The Microscope: Mediations of the Sub-Visible World." In *Transactions and Encounters: Science and Culture in the Nineteenth Century,* ed. Roger Luckhurst and Josephine McDonagh, 30–54. Manchester: Manchester University Press,

Armstrong, Patrick. 1992. "The Metaphors of Struggle, Conflict, Invasion and Explosion in Biogeography." *Ekológia* 11:437–45.

Aspinall, Arthur. 1949. *Politics and the Press c. 1780–1850.* London: Home and Van Thal.

Assing, Ludmilla. ed. 1860. *Letters of Alexander von Humboldt, Written between the Years 1827 and 1858, to Varnhagen von Ense; Together with Extracts from Varnhagen's Diaries, and Letters from Varnhagen and Others to Humboldt.* London: Trübner. Translation of *Briefe von Alexander von Humboldt an Varnhagen von Ense aus den Jahren 1827 bis 1858; Nebst Auszügen aus Varnhagen's Tagebüchern, und Briefen von Varnhagen und Andern an Humboldt* (Leipzig: Brockhaus, 1860).

———. 1867. *Piero Cironi: Ein Beitrag zur Geschichte der Revolution in Italien.* Leipzig: Matthes.

Ault, Warren O. 1972. *Open-Field Farming in Medieval England: A Study of Village Bye-Laws.* London: Allen and Unwin.

Axelson, Eric, ed. 1988. *Dias and His Successors.* Cape Town: Saayman and Weber.

Babington, Churchill, ed. 1865–86. *Polychronicon Ranulphi Higden Monachi Cestrensis; together with the English Translations of John Trevisa and of an Unknown Fifteenth Century Writer.* 9 Vols. London: Longmans.

Back, Vice-Admiral Sir George, Rear-Admiral Collinson, and Francis Galton. 1865. "Hints to Travellers." *Journal of the Royal Geographical Society* 34:272–308.

Bacon, Francis. 1857–61. *Works.* Ed. J. Spedding, R. L. Ellis, and D. D. Heath. 7 vols. London: Longmans.

Baillie, Granville H. 1951. *Clocks and Watches: An Historical Bibliography.* Vol. 1. London: N.A.G. Press.

Baines, Thomas, and William Barry Lord. 1871. *Shifts and Expedients of Camp Life, Travel and Exploration.* London: Horace Cox.

Baker, Keith Michael. 1975. *Condorcet: From Natural Philosophy to Social Mathematics.* Chicago: University of Chicago Press.

———. 1990. *Inventing the French Revolution: Essays on French Political Culture.* Cambridge: Cambridge University Press.

Barber, Richard, ed. 1982. *John Aubrey's Brief Lives.* Woodbridge: Boydell.

Barlow, Roger. [1541] 1932. *A Briefe Summe of Geography.* Ed. Eva G. R. Taylor, London: Hakluyt Society.

Barlow, Thomas. 1693. *The Genuine Remains of that Learned Prelate Dr. Thomas Barlow, Late Lord Bishop of Lincoln.* London: Printed for John Dunton.

Barnes, Elizabeth. 1959. "The 'Panoplist': Nineteenth-Century Religious Magazine." *Journalism Quarterly* 36:321–25.

Barrett, Paul H., ed. 1977. *The Collected Papers of Charles Darwin.* 2 vols. Chicago: University of Chicago Press.

Barrett, Paul H., Peter Gautrey, Sandra Herbert, David Kohn, and Sydney Smith, eds. 1987. *Charles Darwin's Notebooks, 1836–1844: Geology, Transmutation of Species, Metaphysical Enquiries.* London: British Museum (Natural History) / Cambridge University Press.

Barringer, Tim. 1996. "Fabricating Africa: Livingstone and the Visual Image, 1850–1874." In *David Livingstone and the Victorian Encounter with Africa,* ed. John M. MacKenzie, 169–200. London: National Portrait Gallery.

Barrotta, Pierluigi. 2000. "Scientific Dialectics in Action: The Case of Joseph Priestley." In *Scientific Controversies: Philosophical and Historical Perspectives,* ed. Peter Machamer, Marcello Pera, and Aristedes Baltas, 153–76. New York: Oxford University Press.

Barth, Fredrik. 1981. *Process and Form in Social Life: Selected Essays of Fredrik Barth.* London: Routledge and Kegan Paul.

Bataille, Georges. *Visions of Excess: Selected Writings, 1927–1939.* Trans. Allan Stoekl. Minneapolis: University of Minnesota Press.

Baulez, Christian. 1978. "Notes sur quelques meubles et objets d'art des appartements intérieurs de Louis XVI and Marie-Antoinette." *Revue du Louvre* 5/6:360–73.

Baxter, Stephen. 2004. *Revolutions in the Earth: James Hutton and the True Age of the World.* London: Phoenix.

Bechler, Zev, ed. 1982. *Contemporary Newtonian Research.* Dordrecht: Reidel.

Beddall, Barbara G. 1988. "Darwin and Divergence: The Wallace Connection." *Journal of the History of Biology* 21:1–68.

———. 1988. "Wallace's Annotated Copy of Darwin's 'Origin of Species.'" *Journal of the History of Biology* 21:265–89.

Bedini, Silvio A. 1990. *Thomas Jefferson: Statesman of Science.* New York: Macmillan.

Beer, Gillian. 1983. *Darwin's Plots: Evolutionary Narrative in Darwin, George Eliot and Nineteenth-Century Fiction.* London: Routledge and Kegan Paul.

Bell, Morag, and Cheryl McEwan. 1996. "The Admission of Women Fellows to the Royal Geographical Society, 1892–1914: The Controversy and the Outcome." *Geographical Journal* 162:295–312.

Bender, Karl-Heinz. 1977. *Revolutionen: Die Entstehung des Politischen Revolutionsbegriffes in Frankreich zwischen Mittelalter und Aufklärung.* Munich: Wilhelm Fink.

Benjamin, Walter. [1936] 1999. "The Work of Art in the Age of Mechanical Reproduction." In *Visual Culture: The Reader,* ed. Jessica Evans and Stuart Hall, 72–79. London: Sage.

Bennett, Jim. 1991. "The Challenge of Practical Mathematics." In *Science, Culture and Popular Belief in Renaissance Europe,* ed. Stephen Pumfrey, Paolo L. Rossi, and Maurice Slawinski, 176–90. Manchester: Manchester University Press.

———. 1998. "Practical Geometry and Operative Knowledge." *Configurations* 6:195–222.

———. 1998. "Projection and the Ubiquitous Virtue of Geometry in the Renaissance." In *Making Space for Science: Territorial Themes in the Shaping of Knowledge,* ed. Crosbie Smith and Jon Agar, 27–38. Basingstoke: Macmillan.

———. 2002. "The Travels and Trials of Mr Harrison's Timekeeper." In *Instruments, Travel and Science: Itineraries of Precision from the Seventeenth to the Twentieth Century,* ed. Marie-Nöelle Bourguet, Christian Licoppe, and H. Otto Sibum, 75–96. London: Routledge.

Berens, John F. 1978. *Providence and Patriotism in Early America, 1640–1815.* Charlottesville: University of Virginia Press.

Berg, Maxine, and Kristine Bruland, eds. 1998. *Technological Revolutions in Europe: Historical Perspectives.* Cheltenham: Edward Elgar.

Berggren, J. Lennart, and Alexander Jones, eds. 2000. *Ptolemy's Geography: An Annotated Translation of the Theoretical Chapters.* Princeton: Princeton University Press.

Berry, Andrew. 2002. *Infinite Tropics: An Alfred Russel Wallace Anthology.* London: Verso.

Bertius, Petrus. 1616. *Tabularium Geographicam.* Amsterdam: J. Hondius.

Bettey, J. H., ed. 1981. *The Casebook of Sir Francis Ashley JP, Recorder of Dorchester, 1614–35.* Dorset Record Society Publications 7. Dorchester.

Beverley, Robert. 1705. *The History and Present State of Virginia, in Four Parts.* London.

Biagioli, Mario, and Steven J. Harris, eds. 1998. *The Scientific Revolution as Narrative.* Special issue of *Configurations* 6.

Białostocki, Jan. 1976. "The Baltic Area as an Artistic Region in the Sixteenth Century." *Hafnia* 6:11–24.

Bibliotheca Symsoniana. 1712. *A Catalogue of the Vast Collection of Books, in the Library of the Late Reverend Learned Mr Andrew Symson.* Edinburgh.

Biermann, Kurt-R. 1981."Wer Waren die Wichtigsten Briefpartner Alexander von Humboldts?" *NTM-Schriftenr. Gesch. Naturwiss., Technik, Med.* 18:34–43.

Billinge, Mark D., Derek J. Gregory, and Ronald J. Martin, eds. 1984. *Recollections of a Revolution: Geography as Spatial Science.* London: Macmillan.

Binggeli, Pierre. 1994. "Misuse of Terminology and Anthropomorphic Concepts in the Description of Introduced Species." *Bulletin [of the British Ecological Society]* 25: 10–13.

Birkett, Jennifer. 1988. "Madame de Genlis: The New Man and the Old Eve." *French Studies* 42:150–64.

Birrell, Andrew J. 1996. "The North American Boundary Commission: Three Photographic Expeditions, 1872–74." *History of Photography* 20 : 113–21.

Blackburn, Robin. 1988. *The Overthrow of Colonial Slavery, 1776–1848*. London: Verso.

Blum, Ann Shelby. 1993. *Picturing Nature: American Nineteenth-Century Zoological Illustration*. Princeton: Princeton University Press.

Blunt, Alison, and Jane Wills. 2000. *Dissident Geographies: An Introduction to Radical Ideas and Practice*. Harlow: Prentice Hall.

Botero, Giovanni. 1630. *Relations of the Most Famous Kingdomes and Common-wealths thorowout the World*. 3rd ed. Trans. Robert Johnson. London: John Haviland.

Bourdieu, Pierre. 2000. *Pascalian Meditations*. Cambridge: Polity.

Bourguet, Marie-Noëlle, Christian Licoppe, and H. Otto Sibum, eds. 2002. *Instruments, Travel and Science: Itineraries of Precision from the Seventeenth to the Twentieth Century*. London: Routledge.

Bourne, Samuel. 1866–67. "Narrative of a Photographic Trip to Kashmir (Cashmere) and Adjacent Districts." *British Journal of Photography* 13 : 474–75, 498–99, 524–25, 559–60, 583–84, 617–19; 14 : 4–5, 38–39, 63–64.

Bouwsma, William. 2000. *The Waning of the Renaissance, 1550–1640*. New Haven: Yale University Press.

Bowen, Margarita. 1981. *Empiricism and Geographical Thought: From Francis Bacon to Alexander von Humboldt*. Cambridge: Cambridge University Press.

Bowker, Geoffrey C., and Susan L. Star. 1999. *Sorting Things Out*. Cambridge: MIT Press.

Bowler, Peter J. 1988. *The Non-Darwinian Revolution: Reinterpreting a Historical Myth*. Baltimore: John Hopkins University Press.

Bowman, Isaiah. 1921. *The New World: Problems of Political Geography*. Yonkers-on-Hudson: World Books.

Boxer, Charles R. 1969. *The Portuguese Seaborne Empire*. New York: Knopf.

Boyle, Robert. 1661. *Hydrostatical Paradoxes*. Oxford: Printed for W. Hall by R. Davis.

Brackman, Arnold C. 1980. *A Delicate Arrangement: The Strange Case of Charles Darwin and Alfred Russel Wallace*. New York: Times Books.

Bradshaw, Brendan, and John Morrill, eds. 1996. *The British Problem, c. 1534–1707: State Formation in the Atlantic Archipelago*. London: Macmillan.

Braive, Michel F. 1966. *The Photograph: A Social History,* trans. David Britt, London: Thames and Hudson.

Brand, Paul 2001. "Lawyers' Time in England in the Later Middle Ages." In *Time in the Medieval World,* ed. Chris Humphrey and W. M. Ormrod, 73–104. York: York Medieval Press.

Branson, Roy. 1979. "James Madison and the Scottish Enlightenment." *Journal of the History of Ideas* 40: 235–50.

Brayshay, Mark. 1991. "Royal Post-Horse Routes in England and Wales: The Evolution of the Network in the Late-Sixteenth and Early-Seventeenth Century." *Journal of Historical Geography* 17 : 373–89.

Brayshay, Mark, Philip Harrison, and Brian Chalkley. 1998. "Knowledge, Nationhood and Governance: The Speed of the Royal Post in Early-Modern England." *Journal of Historical Geography* 24 : 265–88.

Brewster, David. 1856. *The Stereoscope: Its History, Theory, Construction, and Application to the Arts and to Education.* London: John Murray.

Briggs, Peter M. 1988. "Timothy Dwight 'Composes' a Landscape for New England." *American Quarterly* 40:359–77.

Bright, Chris. 1998. *Life Out of Bounds: Bioinvasion in a Borderless World.* New York: Norton.

Brissot, J.-P. [1781] 1836. *Théorie des lois criminelles.* 2 vols. Paris: J.-P. Aillaud.

———. 1912. *Mémoires, 1754–1793,* [ed. C. Perroud]. 2 vols. Paris: Picard.

Broc, Numa. 1974. *La géographie des philosophes: Géographes et voyageurs française au XVIIIe siècle.* Paris: Ophrys.

———. 1974. "Une Musée de Géographie en 1795." *Revue d'Histoire des Sciences* 27:37–43.

Brockway, Lucile H. 1979. *Science and Colonial Expansion: The Role of the British Royal Botanic Gardens.* New York: Academic Press.

Brotton, Jerry. 1997. *Trading Territories: Mapping the Early Modern World.* London: Reaktion.

———. 2000. "Printing the World." In *Books and the Sciences in History,* ed. Marina Frasca-Spada and Nicholas Jardine, 35–48. Cambridge: Cambridge University Press.

Brown, Penny. 1994. "'La femme enseignante': Mme de Genlis and the Moral and Didactic Tale in France." *Bulletin of the John Rylands University Library* 76:23–42

Brown, Ralph Hall. 1941. "The American Geographies of Jedidiah Morse." *Annals of the Association of American Geographers* 31:145–217.

Browne, Janet. 1980. "Darwin's Botanical Arithmetic and the Principle of Divergence, 1854–1858." *Journal of the History of Biology* 13:53–89.

———. 1983. *The Secular Ark: Studies in the History of Biogeography.* New Haven: Yale University Press.

———. 1996. "Biogeography and Empire." In *Cultures of Natural History,* ed. Nicholas Jardine, James A. Secord, and Emma C. Spary, 305–21. Cambridge: Cambridge University Press.

Brückner, Martin. 1999. "Lessons in Geography: Maps, Spellers, and Other Grammars of Nationalism in the Early Republic." *American Quarterly* 51:311–43.

———. 2000. "Geography, Reading, and the World of Novels in the Early Republic." In *Early America Re-explored: New Readings in Colonial, Early National, and Antebellum Culture,* ed. Klaus H. Schmidt and Fritz Fleischmann, 385–410. New York: Lang.

Bryden, David J. 1972. *Scottish Scientific Instrument-Makers 1600–1900.* Edinburgh: Royal Scottish Museum.

Buel, Richard Jr. 1972. *Securing the Revolution.* Ithaca: Cornell University Press.

Buisseret, David. 2003. *The Mapmakers' Quest: Depicting New Worlds in Renaissance Europe.* Oxford: Oxford University Press.

Bullough, Geoffrey. 1963. *Luis de Camões: The Lusiads.* Trans. Richard Fanshawe. London: Penguin.

Burckhardt, Jacob. 1945. *The Civilization of the Renaissance in Italy.* London: Phaidon.

Burke, Peter. 2000. *A Social History of Knowledge from Gutenberg to Diderot.* Cambridge: Polity.

———, ed. 2001. *New Perspectives on Historical Writing.* 2nd ed. Cambridge: Cambridge University Press.

Burkhardt, Frederick, Sydney Smith, D. M. Porter et al., eds. 1985–. *The Correspondence of Charles Darwin*. Cambridge: Cambridge University Press.

Burnell, A. C., and P. A. Thiele, eds. 1885. *The Voyage of John Huyghen van Linschoten to the East Indies*. 2 vols. London: Hakluyt Society.

Burnett, D. Graham. 2000. *Masters of All They Surveyed: Exploration, Geography, and a British El Dorado*. Chicago: University of Chicago Press.

Burrows, Simon. 2003. "The Innocence of Jacques-Pierre Brissot." *Historical Journal* 46 : 843–71.

Butterfield, Herbert. 1949. *The Origins of Modern Science 1300–1800*. London: G. Bell.

———. 1957. *The Origins of Modern Science, 1300–1800* Rev. ed. New York: Free Press.

Buttimer, Anne, and Gerard Fahy. 1999. "Imaging Ireland through Geography Texts." In *Text and Image: Social Construction of Regional Knowledges*, ed. Anne Buttimer, Stanley D. Brunn, and Ute Wardenga, special issue of *Beiträge zur Regionalen Geographie* (Leipzig) 49 : 79–191.

Büttner, Manfred. 1978. "Bartolomaus Keckermann, 1572–1609." *Geographers: Biobibliographical Studies* 2 : 73–79.

Caeser, James W. 2000. "Natural Rights and Scientific Racism." In *Thomas Jefferson and the Politics of Nature*, ed. Thomas S. Engeman, 165–89. Notre Dame: University of Notre Dame Press.

Cain, P. J., and A. G. Hopkins. 1993. *British Imperialism: Innovation and Expansion, 1688–1914*. London: Longman.

Callon, Michel. 1986. "Some Elements of a Sociology of Translation: Domestication of the Scallops and the Fishermen of St. Brieux Bay." In *Power, Action and Belief: A New Sociology of Knowledge?* ed. John Law, 196–229. London: Routledge and Kegan Paul.

Camerini, Jane R. 1994. "Evolution, Biogeography, and Maps: An Early History of Wallace's Line." In *Darwin's Laboratory: Evolutionary Theory and Natural History in the Pacific*, ed. Roy MacLeod and Philip F. Rehbock, 70–109. Honolulu: University of Hawai'i Press.

———. 1996. "Wallace in the Field." *Osiris*, 2nd ser., 11 : 44–65.

———. 1997. "Remains of the Day: Early Victorians in the Field." In *Victorian Science in Context*, ed. Bernard Lightman, 354–77. Chicago: University of Chicago Press.

———, ed. 2002. *The Alfred Russel Wallace Reader: A Selection of Writings from the Field*. Baltimore: Johns Hopkins University Press.

Campanella, Tomassso. 1659. *Thomas Campanella: An Italian Friar and Second Machiavel; His Advice to the King of Spain for Attaining the Universal Monarchy of the World*. Trans. Edmund Chilmead. London: Philemon Stephens.

Campbell, John A. 1996. "Modernisation and the Beginnings of Geography in Ireland." In *Geography and Professional Practice*, ed. Vincent Berdoulay and H. van Ginkel, 125–38. Utrecht: Nederlands Geographical Studies.

Campbell, Lorne. 1990. *Renaissance Portraits: European Painting in the Fourteenth, Fifteenth and Sixteenth Centuries*. New Haven: Yale University Press.

Campbell, Tony. 1987. *The Earliest Printed Maps, 1472–1500*. London: British Library.

Candolle, Alphonse de. 1873. *Histoire des sciences et des savants depuis deux siècles*. Geneva: H. Georg.

Cannon, Susan Faye. 1978. "Humboldtian Science." In *Science in Culture: The Early Victorian Period,* ed. Susan Faye Cannon, 73–110. New York: Science History Publications.

Carpenter, Christine, ed. 1996. *Kingsford's Stonor Letters and Papers, 1290–1483.* Cambridge: Cambridge University Press.

Carpenter, Nathanael. 1625. *Geography Delineated Forth in Two Books.* 2 vols. in 1. Oxford: Oxford University Press.

———. 1633. *Achitophel; or, The Picture of a Wicked Politician.* London: J. Okes.

Carruthers, Jane, and Arnold, Marion. 1995. *The Life and Works of Thomas Baines.* Cape Town: Fernwood.

Carter, Paul. 1992. *On Living in a New Country: History, Travelling and Language.* London: Faber and Faber.

Castells, Manuel. 1996. *The Rise of the Network Society.* Oxford: Blackwell.

———. 2001. *The Internet Galaxy: Reflections on the Internet, Business and Society.* Oxford: Oxford University Press.

Cavallo, Guglielmo, and Roger Chartier, eds. 1999. *A History of Reading in the West.* Trans. Lydia G. Cochrane. Cambridge: Polity.

Cave, Terence. 1979. *The Cornucopian Text: Problems of Writing in the French Renaissance.* Oxford: Clarendon.

Censer, Jack R. 1976. *The Prelude to Power: The Parisian Radical Press, 1789–1791.* Baltimore: Johns Hopkins University Press.

Chapman, Allan. 1994. *Dividing the Circle: A History of Critical Angular Measurement in Astronomy, 1500–1850.* London: Wiley.

Chapman, James. 1860. "Notes on South Africa." *Journal of the Royal Geographical Society* 30:17–18.

———. 1971. *Travels in the Interior of South Africa, 1849–1863.* Ed. Edward C. Tabler. 2 vols. Cape Town: A. A. Balkema.

Charles I. [1642] 1999. "His Majesties Answer." In *The Struggle for Sovereignty: Seventeenth-Century English Political Tracts,* ed. Joyce Lee Malcolm, 1:154–78. Indianapolis: Liberty Fund.

Chartier, Roger. 1991. *The Cultural Origins of the French Revolution.* Durham: Duke University Press.

Chew, Matthew K., and Manfred D. Laubichler. 2003. "Natural Enemies—Metaphor or Misconception?" *Science* 301 (4 July): 52–53.

Cipolla, Carlo M. 1973. *The Fontana Economic History of Europe.* 6 vols. London: Fontana.

Clark, William, Jan Golinski, and Simon Schaffer, eds. 1999. *The Sciences in Enlightened Europe.* Chicago: University of Chicago Press.

Clanchy, M. T. 1993. *From Memory to Written Record: England, 1066–1307.* Oxford: Blackwell.

Clapperton, Jane Hume. 1885. *Scientific Meliorism and the Evolution of Happiness.* London: Kegan Paul, Trenchand Co.

Clark, Andy. 2001. *Mindware.* New York: Oxford University Press.

Claxton, Guy. 1997. *Hare Brain, Tortoise Mind: Why Intelligence Increases When You Think Less.* London: Fourth Estate.

———. 1999. *Wise Up: The Challenge of Lifelong Learning.* London: Bloomsbury.

Clendennen, Gary W. 1978. "Charles Livingstone: A Biographical Study with Emphasis on His Accomplishments on the Zambesi Expedition, 1858–1863." PhD diss., University of Edinburgh.

Clifford, D. J. H. 1990. *The Diaries of Lady Anne Clifford.* Stroud: Alan Sutton.

Clow, Archibald. 1952. *The Chemical Revolution: A Contribution to Social Technology.* London: Batchworth.

Coetzee, J. M. 1988. *White Writing: On the Culture of Letters in South Africa.* New Haven: Yale University Press.

Cogswell, Thomas. 1989. "England and the Spanish Match." In *Conflict in Early Stuart England: Studies in Religion and Politics, 1602–1642,* ed. Richard Cust and Ann Hughes, 107–33. London: Longmans.

Cogswell, Thomas, Richard Cust, and Peter Lake, eds. 2002. *Politics, Religion and Popularity in Early Stuart England: Essays in Honour of Conrad Russell.* Cambridge: Cambridge University Press.

Cohen, H. Floris. 1994. *The Scientific Revolution: An Historiographical Inquiry.* Chicago: University of Chicago Press.

———. 1999. "The Scientific Revolution: Has There Been a British View?—a Personal Assessment," *History of Science* 37 : 107–12.

Cohen, I. Bernard. 1980. *The Newtonian Revolution, with Illustrations of the Transformation of Scientific Ideas.* Cambridge: Cambridge University Press.

———. 1982. "The *Principia,* Universal Gravitation, and the 'Newtonian Style,' in Relation to the Newtonian Revolution in Science: Notes on the Occasion of the 250th Anniversary of Newton's Death." In *Contemporary Newtonian Research,* ed. Zev Bechler, 21–108. Dordrecht: Reidel.

———. 1985. *Revolution in Science.* Cambridge: Harvard University Press.

———. 1995. *Science and the Founding Fathers: Science in the Political Thoughts of Thomas Jefferson, Benjamin Franklin, John Adams and James Madison.* New York: Norton.

Cohen, Lester H. 1980. *The Revolutionary Histories: Contemporary Narratives of the American Revolution.* Ithaca: Cornell University Press.

Coleman, William. 2001. "The Strange 'Laissez Faire' of Alfred Russel Wallace: The Connection between Natural Selection and Political Economy Reconsidered." In *Darwinism and Evolutionary Economics,* ed. John Laurent and John Nightingale, 36–48. Cheltenham: Edward Elgar.

Collinson, Patrick. 1982. *The Religion of Protestants: The Church in English Society, 1559–1625.* Oxford: Oxford University Press.

Colp, Ralph Jr. 1978. "Charles Darwin: Slavery and the American Civil War." *Harvard Library Bulletin* 26 : 471–89.

Colwill, Elizabeth. 2000. "Epistolary Passions: Friendship and the Literary Public of Constance de Salm, 1767–1845." *Journal of Women's History* 12 : 39–68.

Connors, Joseph. 1989. "Alliance and Enmity in Roman Baroque Urbanism." *Römisches Jahrbuch der Bibliotheca Hertziana* 25 : 207–94.

Cook, Ian, et al., eds. 2000. *Cultural Turn / Geographical Turn: Perspectives on Cultural Geography.* Harlow: Prentice Hall.

Corbin, Henry. 1993. *A History of Islamic Philosophy.* London: Kegan Paul International.

Cormack, Lesley. 1991. "'Good Fences Make Good Neighbours': Geography as Self-Definition in Early Modern England." *Isis* 82:639–61.

———. 1997. *Charting an Empire: Geography at the English Universities, 1580–1620.* Chicago: University of Chicago Press.

Cosgrove, Denis. 1984: *Social Formation and Symbolic Landscape.* London: Croom Helm.

———. 1994. "Contested Global Visions: One-World, Whole-Earth, and the Apollo Space Photographs," *Annals of the Association of American Geographers* 84:270–94.

———. 2001. *Apollo's Eye: A Cartographic Genealogy of the Earth in the Western Imagination.* Baltimore: Johns Hopkins University Press.

———, ed. 1999. *Mappings.* London: Reaktion.

Cosgrove, Denis, and Luciana L. Martins, 2000: "Millennial Geographics." *Annals of the Association of American Geographers* 90:97–103.

Cosgrove, Denis, and Stephen Daniels, eds. 1988. *The Iconography of Landscape: Essays on the Symbolic Representation, Design and Use of Past Environments.* Cambridge: Cambridge University Press.

Cottingham, John, ed. 1987. *Cambridge Companion to Descartes.* Cambridge: Cambridge University Press.

Coupland, Reginald. 1928. *Kirk on the Zambesi: A Chapter of African History.* Oxford: Clarendon.

Crandell, Gina. 1993. *Nature Pictorialized: "The View" in Landscape History.* Baltimore: Johns Hopkins University Press.

Crary, Jonathan. 1990. *Techniques of the Observer: On Vision and Modernity in the Nineteenth Century.* Cambridge: MIT Press.

Crawford, Elisabeth, Terry Shinn, and Sverker Sörlin, eds. 1993. *Denationalizing Science: The Contexts of International Scientific Practice.* Dordrecht: Kluwer.

Crombie, Alistair. 1994. *Styles of Scientific Thinking in the European Tradition.* 3 vols. London: Duckworth.

———, ed. 1963. *Historical Studies in the Intellectual, Social and Technical Conditions for Scientific Discovery.* London: Routledge and Kegan Paul.

Crosby, Alfred W. 1986. *Ecological Imperialism: The Biological Expansion of Europe, 900–1900.* Cambridge: Cambridge University Press.

Crosland, Maurice. 1967. *The Society of Arcueil: A View of French Science at the Time of Napoleon I.* Cambridge: Harvard University Press.

———. 1969. *Science in France in the Revolutionary Era.* Cambridge: Harvard University Press.

———. 2000. "Styles of Science: National, Regional and Local." In *Encyclopaedia of the Scientific Revolution, from Copernicus to Newton,* ed. Wilbur Applebaum, 622. New York: Garland.

———, ed. 1975. *The Emergence of Science in Western Europe.* Basingstoke: Macmillan.

Cunningham, Andrew and Perry Williams. 1993. "De-Centering the 'Big Picture': The *Origins of Modern Science* and the Modern Origins of Science." *British Journal for the History of Science* 26:407–32.

Cust, Richard. 1986. "News and Politics in Early-Seventeenth Century England." *Past and Present* 112:60–90.

Cust, Richard, and Ann Hughes, eds. 1989. *Conflict in Early Stuart England: Studies in Religion and Politics, 1602–1642*. London: Longmans.

Dacier, Bon-Joseph. 1824. "Notice historique sur la vie et les ouvrages de M. Mentelle." *Histoire et Mémoires de l'Institut Royal de France: Académie des Inscriptions et Belles-Lettres* 7:212–22.

Dainville, François de. 1940. *La géographie des humanistes*. Paris: Beauchesnes.

Dalton, Simon. 2001. "Gender and the Shifting Ground of Revolutionary Politics: The Case of Madame Roland (Jeanne-Marie Roland)." *Canadian Journal of History* 36:259–82.

Daniels, Stephen. 1993. *Fields of Vision: Landscape Imagery and National Identity in England and the United States*. Cambridge: Polity.

———. 1999. *Humphry Repton: Landscape Gardening and the Geography of Georgian England*. New Haven: Yale University Press.

Daniels, Stephen, and Catherine Nash, eds. 2004. *Lifepaths: Geography and Biography*. Special issue of *Journal of Historical Geography* 30.

Darby, H. C. 1962. "The Problem of Geographical Description." *Transactions of the Institute of British Geographers* 30:1–14.

Darnton, Robert. 1968. *Mesmerism and the End of the Enlightenment in France*. Cambridge: Harvard University Press.

———. 1979. *The Business of Enlightenment: A Publishing History of the "Encyclopédie," 1775–1800*. Cambridge: Harvard University Press.

———. 1982. *The Literary Underworld of the Old Regime*. Cambridge: Harvard University Press.

———. 1984. *The Great Cat Massacre and Other Episodes in French Cultural History*. Harmondsworth: Penguin.

———. 1995. *The Corpus of Clandestine Literature in France, 1765–1789*. New York: Norton.

———. 1995. *The Forbidden Best-Sellers of Pre-Revolutionary France*. New York: Norton.

Darnton, Robert, and Daniel Roche, eds. 1989. *Revolution in Print: The Press in France, 1775–1800*. Berkeley and Los Angeles: University of California Press.

Darwin, Charles. 1839. *Journal of Researches into the Geology and Natural History of the Various Countries Visited by H.M.S. "Beagle" under the Command of Captain FitzRoy, R.N., from 1832 to 1836*. London: Henry Colburn.

———. 1859. *On the Origin of Species by Means of Natural Selection; or, The Preservation of Favoured Races in the Struggle for Life*. London: John Murray.

———. 1871. *The Descent of Man, and Selection in Relation to Sex*. 1st ed. 2 vols. London: John Murray.

———. 1874. *The Descent of Man, and Selection in Relation to Sex*. 2nd ed. 1 vol. London: John Murray.

Darwin, Charles, and Alfred Russel Wallace. 1958. *Evolution by Natural Selection*. Ed. Gavin de Beer. Cambridge: At the University Press.

Darwin, Francis, ed. 1887. *The Life and Letters of Charles Darwin, Including an Autobiographical Chapter*. 3 vols. London: John Murray.

Daston, Lorraine, and Michael Otte, eds. 1991. *Style in Science*. Special issue of *Science in Context* 4:221–47.

Daston, Lorraine, and Peter Galison. 1992. "The Image of Objectivity." *Representations* 40:81–128.

Davies, James. 1962. "Towards a Theory of Revolution." *American Sociological Review* 27:5–18.

Dawidoff, Robert. 2000. "Rhetoric of Democracy." In *Thomas Jefferson and the Politics of Nature*, ed. Thomas S. Engeman, 99–122. Notre Dame: University of Notre Dame Press.

Dear, Peter. 1985. "Totius in Verba: Rhetoric and Authority in the Early Royal Society." *Isis* 76:145–61.

———. 1990. "Miracles, Experiments and the Ordinary Course of Nature." *Isis* 81:663–83.

———. 1991. "The Church and the New Philosophy." In *Science, Culture and Popular Belief in Renaissance Europe*, ed. Stephen Pumfrey, Paolo L. Rossi, and Maurice Slawinski, 119–39. Manchester: Manchester University Press.

———. 1995. *Discipline and Experience: The Mathematical Way in the Scientific Revolution.* Chicago: University of Chicago Press.

———. 1998. "The Mathematical Principles of Natural Philosophy: Toward a Heuristic Narrative for the Scientific Revolution." *Configurations* 6:173–94.

———. 2001. *Revolutionizing the Sciences: European Knowledge and Its Ambitions, 1500–1700.* Princeton: Princeton University Press.

———, ed. 1997. *The Scientific Enterprise in Early Modern Europe: Readings from "Isis."* Chicago: University of Chicago Press.

Debus, Allen G., ed. 1972. *Science, Medicine and Society in the Renaissance.* 2 vols. New York: Science History Publications.

"A Declaration of the Parliament of England." [1649] 1999. In *The Struggle for Sovereignty: Seventeenth-Century English Political Tracts*, ed. Joyce Lee Malcolm, 1:372–90. Indianapolis: Liberty Fund.

Dekker, Elly. 1996. *Globes at Greenwich.* Oxford: Oxford University Press and the National Maritime Museum.

Dennett, Daniel. 1995. *Darwin's Dangerous Idea.* New York: Simon and Schuster.

Des Chene, Dennis. 1996. *Physiologia: Natural Philosophy in Late Aristotelian and Cartesian Thought.* Ithaca: Cornell University Press.

Descartes, René. [1644] 1983. *Principles of Philosophy.* Trans. V. R. and R. P. Miller. Dordrecht: D. Reidel.

Dettelbach, Michael. 1996. "Global Physics and Aesthetic Empire: Humboldt's Physical Portrait of the Tropics." In *Visions of Empire: Voyages, Botany and Representations of Nature*, ed. David P. Miller and Peter Hanns Reill, 258–92. Cambridge: Cambridge University Press.

———. 1996. "Humboldtian Science." In *Cultures of Natural History*, ed. Nicholas Jardine, James A. Secord, and Emma C. Spary, 287–304. New York: Cambridge University Press.

Dhombres, N. and J. 1989. *Naissance du pouvoir: Sciences et savants en France, 1793–1824.* Paris: Payot.

Diamond, Jared. 1997. *Guns, Germs and Steel: The Fates of Human Societies.* New York: Norton.

Diffie, Bailey, and George Winius. 1977. *Foundations of the Portuguese Empire, 1415–1580.* Minneapolis: University of Minnesota Press.

Dobbs, Betty Jo Teeter. 1991. *The Janus Faces of Genius: The Role of Alchemy in Newton's Thought.* Cambridge: Cambridge University Press.

Dohrn-van Rossum, Gerhard. 1996. *History of the Hour: Clocks and Modern Temporal Orders.* Chicago: University of Chicago Press.

Donnelly, Michael. 2000. "National Styles of Reasoning." In *Reader's Guide to the History of Science,* ed. Arne Hessenbruch, 500–501. London: Fitzroy Dearborn.

Douglas, Mary. 1991. *Purity and Danger: An Analysis of Pollution and Taboo.* London: Routledge.

Downing, Brian M. 1992. *The Military Revolution and Political Change.* Princeton: Princeton University Press.

Downs, Anthony. 1957. *An Economic Theory of Democracy.* New York: Harper and Row.

Drapeyron, Ludovic. 1887. "L'éducation géographique de trois princes français au XVIIIe. siècle—le duc de Berry et les comtes de Provence et d'Artois (Louis XVI, Louis XVIII, Charles X)." *Revue de Géographie* 11: 241–56.

Drayton, Richard. 2000. *Nature's Government: Science, Imperial Britain, and the "Improvement" of the World.* New Haven: Yale University Press.

Driver, Felix. 1995. "Visualizing Geography: A Journey to the Heart of the Discipline." *Progress in Human Geography* 19: 123–34.

———. 2001. *Geography Militant: Cultures of Exploration and Empire.* Oxford: Blackwell.

Drummond, Gavin. 1708. *A Short Treatise of Geography, General and Special.* Edinburgh: Andrew Symson.

———. 1708. *Bibliotheca Sibbaldiana.* Edinburgh:.

Duhem, Pierre. 1954. *The Aim and Structure of Physical Theory.* Princeton: Princeton University Press.

Dunbar, Gary S. 1960. "Thomas Jefferson, Geographer." *Special Libraries Association, Geography and Map Division, Bulletin* 40 (April): 11–16.

———. 1978. *Elisée Reclus: Historian of Nature.* Hamden, CT: Archon.

Dunmore, John. 1965. *French Explorers in the Pacific.* 2 vols. Oxford: Oxford University Press.

Dwight, Timothy. 1789. *The Duty of Americans.* New Haven: Greens.

———. 1794. *Greenfield Hill: A Poem in Seven Parts.* New York: Childs and Swaine.

———. 1795. *The True Means of Establishing Public Happiness.* New Haven: T. S. Green.

———. [1821–22] 1969. *Travels in New England and New York,* ed. Barbara Miller Solomon. 4 vols. Cambridge: Harvard University Press.

———. 1828. *Sermons.* New Haven. 2 vols.

Dzelzainis, Martin, ed. 1991. *John Milton: Political Writings.* Cambridge: Cambridge University Press.

Earle, Thomas F., and Stephen Parkinson, eds. 1992. *Studies in the Portuguese Discoveries I.* Warminster: Aris.

Edelen, Georges, ed. 1968. *William Harrison's "The Description of England," 1577.* Ithaca: Cornell University Press.

Edney, Matthew H. 1997. *Mapping an Empire: The Geographical Construction of British India, 1765–1843.* Chicago: University of Chicago Press.

Edwards, Elizabeth. 1990. "Photographic 'Types': The Pursuit of Method." *Visual Anthropology* 3: 235–58.

———. 1999. "Photographs as Objects of Memory." In *Material Memories: Design and Evocation,* ed. Marius Kwint, Christopher Breward, and Jeremy Aynsley, 221–36. Oxford: Berg.

———. 2001. *Raw Histories: Photographs, Anthropology and Museums.* Oxford: Berg.

———, ed. 1992. *Anthropology and Photography, 1860–1920.* London: Yale University Press.

Eisenstein, Elizabeth L. 1979. *The Printing Press as an Agent of Change: Communications and Cultural Transformations in Early Modern Europe.* 2 vols. Cambridge: Cambridge University Press.

———. 1983. *The Printing Revolution in Early Modern Europe.* Cambridge: Cambridge University Press.

———. 1986. "On Revolution and the Printed Word." In *Revolution in History,* ed. Roy Porter and Mikuláš Teich, 186–205. Cambridge: Cambridge University Press.

Elias, Norbert. 1976. *The Civilizing Process: State Formation and Civilization.* Oxford: Blackwell.

———. 1992. *Time: An Essay.* Oxford, Blackwell.

Elliott, John H. 1968. *Europe Divided, 1559–1598.* London: Fontana.

———. 1973. "England and Europe: A Common Malady?" In *The Origins of the English Civil War,* ed. Conrad Russell, 246–57. London: Macmillan.

———. 1992. "A Europe of Composite Monarchies." *Past and Present* 137:48–71.

Ellis, Joseph J. 1979. *After the Revolution: Profiles of Early American Culture.* New York: Norton.

———. 1997. *American Sphinx: The Character of Thomas Jefferson.* New York: Knopf.

Elphick, Richard. 1977. *Kraal and Castle: Khoisan and the Founding of White South Africa.* New Haven: University of Yale Press.

Emerson, Roger. 1992. *Professors, Patronage and Politics: The Aberdeen Universities in the Eighteenth Century.* Aberdeen: Aberdeen University Press.

Emerson, Roger, and Paul Wood. 2002. "Science and Enlightenment in Glasgow, 1690–1802." In *Science and Medicine in the Scottish Enlightenment,* ed. Charles W. J. Withers and Paul Wood, 79–142. East Linton: Tuckwell.

Emmison, Frederick G. 1994. *Essex Wills: The Commissary Court, 1560–1574.* Chelmsford: Essex Record Office Publications.

Engels, Frederick. 1941. *Dialectics of Nature.* Ed. Clemens Dutt. London: Lawrence and Wishart.

Etlin, Richard A. 1994. *Symbolic Space: French Enlightenment Architecture and Its Legacy.* Chicago: University of Chicago Press.

Everett, Edward. [1823] 1973. "On the State of the Indians." *North American Review* 258: 10–14.

Fakhry, Majid. 1970. *A History of Islamic Philosophy.* New York: Columbia University Press.

Farini, G. A. 1886. "A Recent Journey in the Kalahari." *Journal of the Royal Geographical Society,* n.s., 8:437–53.

———. 1886. *Through the Kalahari Desert: A Narrative of a Journey with Gun, Camera, and Note-Book to Lake N'gami and Back.* London: Sampson Low, Marston, Searle, and Rivington.

Farrar, Steve. 2003. "Academic Blacklisted over Threat of Invasion." *Times Higher Education Supplement,* 26 September, 1–3.

Febvre, Lucien, and Henri-Jean Martin, eds. 1976. *The Coming of the Book: The Impact of Printing, 1450–1800.* Trans. David Gerard. London: New Left Books.

Feilchenfeldt, Konrad. 1970. *Varnhagen von Ense als Historiker.* Amsterdam: Verlag der Erasmus Buchhandlung.

Feingold, Mordechai. 1984. *The Mathematician's Apprenticeship: Science, Universities and Society in England, 1560–1640.* Cambridge: Cambridge University Press.

Fenton, Edward, ed. 1998. *The Diaries of John Dee.* Charlbury: Day Books.

Ferne, Henry. [1642] 1999. "The Resolving of Conscience." In *The Struggle for Sovereignty: Seventeenth-Century English Political Tracts,* ed. Joyce Lee Malcolm, 1:182–221. Indianapolis: Liberty Fund.

Festinger, Leon. 1967. *The Theory of Cognitive Dissonance.* Stanford: Stanford University Press.

Feyerabend, Paul. 1975. *Against Method: Outline of an Anarchistic Theory of Knowledge.* Atlantic Highlands: Humanities Press.

Fichman, Martin. 1981. *Alfred Russel Wallace.* Boston: Twayne.

———. 2004. *An Elusive Victorian: The Evolution of Alfred Russel Wallace.* Chicago: University of Chicago Press.

Fiering, Norman. 1981. *Moral Philosophy at Seventeenth-Century Harvard: A Discipline in Transition.* Chapel Hill: University of North Carolina Press.

Fincham, Kenneth. 1988. "Prelacy and Politics: Archbishop Abbot's Defence of Protestant Orthodoxy." *Historical Research* 61:36–64.

Fine, Gary Alan, and Lazaros Christoforides. 1991. "Dirty Birds, Filthy Immigrants, and the English Sparrow War: Metaphorical Linkage in Constructing Social Problems." *Symbolic Interaction* 14:375–93.

Finkelstein, David, and Alistair McCleery, eds. 2002. *The Book History Reader.* London: Routledge.

Fischer, Walther. 1918. *Die Persönlichen Beziehungen Richard Monckton Milnes', ersten Barons Houghton, zu Deutschland, unter besonderer Berücksichtigung seiner Freundschaft mit Varnhagen von Ense.* Würzburg: Buchdruckerei Konrad Triltsch.

———, ed. 1922. *Die Briefe Richard Monckton Milnes', ersten Barons Houghton, an Varnhagen von Ense, 1844–1854.* Heidelberg: Carl Winters Universitätsbuchhandlung.

Fissell, Mary, and Roger Cooter. 2003. "Exploring Natural Knowledge: Science and the Popular." In *The Cambridge History of Science,* vol. 4, *Eighteenth-Century Science,* ed. Roy Porter, 129–58. Cambridge: Cambridge University Press.

Fitzmier, John R. 1998. *New England's Moral Legislator: Timothy Dwight, 1752–1817.* Bloomington: Indiana University Press.

Flaherty, Michael G. 1999. *A Watched Pot: How We Experience Time.* New York: New York University Press.

Foran, John, ed. 1997. *Theorizing Revolutions.* London: Routledge.

Forbes, Eric, ed. 1978. *Human Implications of Scientific Advance.* Edinburgh: Edinburgh University Press.

Forman, Paul. 1971. "Weimar Culture, Causality and Quantum Theory, 1918–1927: Adaptation of German Physicists and Mathematicians to a Hostile Environment." *Historical Studies in the Physical Sciences* 3 : 1–115.

Foster, William, ed. 1940. *The Voyages of Sir James Lancaster, 1591–1603*. London: Hakluyt Society.

Foucault, Michel. 1968. *Les mots et les choses*. Paris: Gallimard.

Frasca-Spada, Marina, and Nicholas Jardine, eds. 2000. *Books and the Sciences in History*. Cambridge: Cambridge University Press.

Freeman, Chris. 2002. *As Time Goes By: From the Industrial Revolutions to the Information Revolution*. Oxford: Oxford University Press.

Freeman, Richard B. 1977. *The Works of Charles Darwin: An Annotated Bibliographical Handlist*. London: Dawson.

Freudenthal, Gideon. 1986. *Atom and Individual in the Age of Newton: On the Genesis of the Mechanistic World View*. Dordrecht: D. Reidel.

———. 2000. "A Rational Controversy over Compounding Forces." In *Scientific Controversies: Philosophical and Historical Perspectives*, ed. Peter Machamer, Marcello Pera, and Aristides Baltas, 125–42. Oxford: Oxford University Press.

Furet, François. 1978. *Penser la Révolution française*. Paris: Maspéro.

Gabbey, Alan. 1980. "Force and Inertia in the Seventeenth Century: Descartes and Newton." In *Descartes: Philosophy, Mathematics and Physics*, ed. Stephen Gaukroger, 230–320. Hassocks: Harvester.

Gal, Ofer. 2002. *Meanest Foundations and Nobler Superstructures: Hooke, Newton and the "Compounding of the Celestial Motions of the Planetts."* Dordrecht: Kluwer.

Galassi, Peter. 1981. *Before Photography: Painting and the Invention of Photography*. New York: Museum of Modern Art.

Galison, Peter. 2003. *Einstein's Clocks, Poincaré's Maps: Empires of Time*. London: Sceptre.

Galton, Francis. 1854. "Hints to Travellers." *Journal of the Royal Geographical Society* 24 : 345–58.

———. 1865. "On Stereoscopic Maps, Taken from Models of Mountainous Countries." *Journal of the Royal Geographical Society* 35 : 99–104.

———, ed. 1878. *Hints to Travellers*. London: Royal Geographical Society.

Gamble, Susan. 2002. "An Appealing Case of Spectra: Photographs on Display at the Royal Society, London 1891." *Nuncius* 17 : 635–51.

Garber, Daniel, and Michael Ayers, eds. 1998. *The Cambridge History of Seventeenth-Century Philosophy*. Cambridge: Cambridge University Press.

Gascoigne, John. 1998. *Science in the Service of Empire: Joseph Banks, the British State and the Uses of Science in the Age of Revolution*. Cambridge: Cambridge University Press.

Gatter, Nikolaus. 1996. *"Gift, Geradezu Gift für das Uunwissende Publicum": Der Diaristische Nachlaß von Karl August Varnhagen von Ense und die Polemik gegen Ludmilla Assings Editionen, 1860–1880*. Bielefeld: Aisthesis Verlag.

Gaukroger, Stephen, ed. 1980. *Descartes: Philosophy, Mathematics and Physics*. Hassocks: Harvester.

Gavroglu, Kostas. 2000. "Controversies and the Becoming of Physical Chemistry." In *Scientific Controversies: Philosophical and Historical*, ed. Peter Machamer et al., 177–98. Oxford: Oxford University Press.

Gay, Hannah. 2003. "Clock Synchrony, Time Distribution and Electrical Timekeeping in Britain, 1880–1925." *Past and Present* 181 : 107–40.

Gaziello, Catherine. 1984. *L'éxpédition de Lapérouse, 1785–1788: Réplique française aux voyages de Cook.* Paris: C.T.H.S.

Geison, Gerald. 1995. *The Private Science of Louis Pasteur.* Princeton: Princeton University Press.

George, H. B. 1866. *The Oberland and Its Glaciers: Explored and Illustrated with Ice Axe and Camera.* London: Alfred W. Bennett.

——. 1878. "Photography." In *Hints to Travellers,* ed. Francis Galton, 47–53. London: Royal Geographical Society.

Gerbi, Antonello. 1973. *The Dispute of the New World: The History of a Polemic, 1750–1900.* Rev. ed. Trans. Jeremy Moyle. Pittsburgh: University of Pittsburgh Press.

Gigerenzer, Gerd. 2000. *Adaptive Thinking.* New York: Oxford University Press.

Gigerenzer, Gerd, and Peter M. Todd. 1999. *Simple Heuristics that Make Us Smart.* Oxford: Oxford University Press.

Gigerenzer, Gerd, and Reinhard Selten, eds. 2001. *Bounded Rationality: The Adaptive Toolbox.* Cambridge: MIT Press.

Gilbert, Felix. 1973. "Revolution." In *Dictionary of the History of Ideas,* ed. Philip P. Wiener, 4 : 152–67. New York: Scribner's.

Gilbert, Neil. 1965. *Renaissance Concepts of Method.* New York: Columbia University Press.

Gillespie, Neal C. 1979. *Charles Darwin and the Problem of Creation.* Chicago: University of Chicago Press.

Gillies, John. 1993. *Shakespeare and the Geography of Difference.* Cambridge: Cambridge University Press.

Gillispie, Charles Coulston. [1980] 1981. *Science and Polity in France at the end of the Old Régime.* Princeton: Princeton University Press.

Gilmore, William J. 1989. *Reading Becomes a Necessity of Life: Material and Cultural Life in Rural New England, 1780–1835.* Knoxville: University of Tennessee Press.

Gilson, Étienne. 1930. *Études sur le role de la pensée médiévale dans la formation du système cartésien.* Paris: J. Vrin.

Ginzburg, Carlo. [1976] 1980. *The Cheese and the Worms: The Cosmos of a Sixteenth-Century Miller.* Trans. John and Anne Tedeschi. Harmondsworth: Penguin.

Glacken, Clarence J. 1967. *Traces on the Rhodian Shore: Nature and Culture in Western Thought from Ancient Times to the End of the Eighteenth Century.* Berkeley: University of California Press.

Glander, Philip. 1965. *The Letters of Varnhagen von Ense to Richard Monckton Milnes.* Heidelberg: Carl Winter.

Glennie, Paul D., and Thrift, Nigel J. 1996. "Consumers, Identities, and Consumption Spaces in Early-Modern England." *Environment and Planning A* 25 : 25–45.

——. 2002. "The Spaces of Clock Times." In *The Social in Question: New Bearings in History and the Social Sciences,* ed. Patrick Joyce, 151–74. London: Routledge.

——. 2005. *The Measured Heart: Histories of Clock Times in England.* Oxford: Oxford University Press.

Godlewska, Anne. 1999. *Geography Unbound: French Geographic Science from Cassini to Humboldt.* Chicago: University of Chicago Press.

Goldberg, Peter J. P. 1992. *Women, Work and Life-Cycle in a Medieval Economy.* Oxford: Clarendon.

———. 1995. *Women in England, 1275–1525.* Manchester: Manchester University Press.

Golinski, Jan. 1998. *Making Natural Knowledge: Constructivism and the History of Science.* Cambridge: Cambridge University Press.

Goodman, David. 1992. "The Scientific Revolution in Spain and Portugal." In *The Scientific Revolution in National Context,* ed. Roy Porter and Mikuláš Teich, 158–77. Cambridge: Cambridge University Press.

Gordon, Patrick. 1693. *Geography Anatomiz'd; or, A Compleat Geographical Grammar.* London: Printed by Robert Morden and Thomas Cockerid.

Gossett, Thomas F. 1963. *Race: The History of an Idea in America.* Dallas: Southern Methodist University Press.

Gough, Hugh. 1988. *The Newspaper Press in the French Revolution.* London: Routledge.

Gould, Stephen Jay. 1990. *Wonderful Life.* London: Hutchinson.

———. 1996. *Full House.* New York: Harmony.

Grabiner, Judith V. 2002. "Maclaurin and Newton: The Newtonian Style and the Authority of Mathematics." In *Science and Medicine in the Scottish Enlightenment,* ed. Charles W. J. Withers and Paul Wood, 143–71. East Linton: Tuckwell Press.

Grafton, Anthony. 1993. *Joseph Scaliger: A Study in the History of Classical Scholarship.* Vol. 2, *Historical Chronology.* Oxford: Warburg Institute.

Grafton, Anthony, and Lisa Jardine. 1990. "'Studied for Action': How Gabriel Harvey Read His Livy." *Past and Present* 129:30–78.

Grafton, Anthony, and Nancy Siraisi, eds. 2000. *Natural Particulars: Nature and the Disciplines in Renaissance Europe.* Cambridge: MIT Press.

Graham, G., et al. 1737. "Observations of the Late Total Eclipse of the Sun." *Philosophical Transactions of the Royal Society* 63:175–201.

Grandière, Marcel. 1998. *L'idéal pédagogique en France au XVIIIe siècle.* Studies on Voltaire and the Eighteenth Century. Oxford: Voltaire Foundation.

Grant, Edward. 1996. *The Foundations of Modern Science in the Middle Ages: Their Religious, Institutional and Intellectual Contexts.* Cambridge: Cambridge University Press.

Greely, A. W. 1896. "Jefferson as a Geographer." *National Geographic Magazine* 7:269–71.

Greenberg, John L. 1995. *The Problem of the Earth's Shape from Newton to Clairaut: The Rise of Mathematical Science in Eighteenth-Century Paris and the Fall of "Normal" Science.* Cambridge: Cambridge University Press.

Greenblatt. Stephen. 1980. *Renaissance Self-Fashioning from More to Shakespeare.* Chicago: University of Chicago Press.

Greene, John C. 1984. *American Science in the Age of Jefferson.* Ames: Iowa State University Press.

Greenough, George Bellas. 1841. "Anniversary Meeting Presidential Address." *Journal of the Royal Geographical Society* 11:xxxix–lxxvii.

Gregory, Derek. 1993. *Geographical Imaginations.* Oxford: Blackwell.

Gregory, Stan. 1983. "Quantitative Geography: The British Experience and the Role of the Institute." *Transactions of the Institute of British Geographers* 8:80–89.

Greiling, Werner. 1993. *Varnhagen von Ense—Lebensweg eines Liberalen: Politisches Wirken zwischen Diplomatie und Revolution.* Cologne: Böhlau.

Gribbin, William. 1972. "A Mirror to New England: The *Compendious History* of Jedidiah Morse and Elijah Parish." *New England Quarterly* 45:340–54.

Griewank, Karl. 1955. *Der Neuzeitliche Revolutionsbegriff.* Weimar: H. Böhlaus Nachfolger.

Groening, Gert, and Joachim Wolschke-Bulmahn. 1992. "Some Notes on the Mania for Native Plants in Germany." *Landscape Journal* 11:116–26.

Grove, Richard H. 1995. *Green Imperialism: Colonial Expansion, Tropical Island Edens and the Origins of Environmentalism, 1600–1860.* Cambridge: Cambridge University Press.

Guillaume, James, ed. 1893. *Procès-verbaux du Comité d'Instruction Publique de la Convention Nationale.* Vol. 2. Paris: Ministère de l'Instruction Publique.

Gurr, Ted Robert. 1971. *Why Men Rebel.* Princeton: Princeton University Press.

Hahn, Roger. 1971. *The Anatomy of a Scientific Institution: The Paris Academy of Sciences, 1666–1803.* Berkeley and Los Angeles: University of California Press.

Hale, John R. 1993. *The Civilization of Europe in the Renaissance.* London: Harper Collins.

Hall, A. Rupert. 1954. *The Scientific Revolution, 1500–1800.* London: Longman.

———. 1983. *The Revolution in Science, 1500–1750.* London: Longman.

———. 1993. "Retrospection on the Scientific Revolution." Afterword to *Renaissance and Revolution: Humanists, Scholars, Craftsmen and Natural Philosophers in Early Modern Europe,* ed. J. V. Field and Frank A. L. James, 239–49. Cambridge: Cambridge University Press.

Halley, Edmund. 1715. "Observations of the Late Total Eclipse of the Sun on the Second of April." *Philosophical Transactions of the Royal Society* 29, no. 343: 245–62; no. 345: 314–16.

Hamowy, Ronald. 1979. "Jefferson and the Scottish Enlightenment: A Critique of Gary Wills' *Inventing America.*" *William and Mary Quarterly* 36:502–23.

Hankins, Thomas L. 1965. "Eighteenth-Century Attempts to Solve the *Vis Viva* Controversy." *Isis* 56:281–97.

Hannerz, Ulf. 1986. "Theory in Anthropology: Small is Beautiful." *Comparative Studies in Society and History* 28:362–67.

Hardy, Thomas. 1878. *The Return of the Native.* London: Macmillan.

Harley, J. Brian. 1988. "Maps, Knowledge, and Power." In *The Iconography of Landscape: Essays on the Symbolic Representation, Design and Use of Past Environment,* ed. Denis Cosgrove and Stephen Daniels, 277–312. Cambridge: Cambridge University Press.

———. 1992. "Deconstructing the Map." In *Writing Worlds: Discourse, Text and Metaphor in the Representation of Landscape,* ed. Trevor J. Barnes and James S. Duncan, 231–47. London: Routledge.

Harris, Marvin. 1968. *The Rise of Anthropological Theory: A History of Theories of Culture.* New York: Thomas Y. Crowell.

Harris, Steven J. 1998. "Thinking Locally, Acting Globally." Introduction to *The Scientific*

Revolution as Narrative, ed. Mario Biagioli and Steven J. Harris. Special issue of *Configurations* 6:131–39.

———. 1998. "Long-Distance Corporations, Big Sciences, and the Geography of Knowledge." In *The Scientific Revolution as Narrative*, ed. Mario Biagioli and Steven J. Harris. Special issue of *Configurations* 6:269–303.

Harris, Tim, ed. 1995. *Popular Culture in England, c. 1500–1850*. Basingstoke: Macmillan.

Harrison, Peter. 1998. *The Bible, Protestantism and the Rise of Natural Science*. Cambridge: Cambridge University Press.

Harrison, William. [1578] 1994. *The Description of England*. Ed. Georges Edelen. New York: Dover.

Hart, Henry. 1952. *Sea-Road to the Indies*. London: Hodge.

Hartog, François. 2003. *Régimes d'historicité*. Paris: La Découverte.

Harwood, Jonathan. 1987. "National Styles in Science: Genetics in Germany and the United States between the Wars." *Isis* 78:390–414.

———. 1993. *Styles of Scientific Thought: The German Genetics Community, 1900–1933*. Chicago: University of Chicago Press.

Hatch, Nathan O. 1989. *The Democratization of American Christianity*. New Haven: Yale University Press.

Hatto, Arthur. 1949. "'Revolution': An Enquiry into the Usefulness of an Historical Term." *Mind* 58:495–516.

Haye de Launay, M. de la. [1789]. *Justification du système d'économie politique et financière de Frédéric II, roi de Prusse, pour servir de réfutation à tout ce que le comte de Mirabeau a hazardé à ce sujet dans son ouvrage de la monarchie prussienne*. Paris.

Haym, Rudolf. 1860. "Notizen." *Preussische Jahrbücher* 5:414–16.

Hayward, Abraham. 1860. "Correspondence of Humboldt." *Edinburgh Review* 112:213–36.

Headrick, Daniel R. 2000. *When Information Came of Age: Technologies of Knowledge in the Age of Reason and Revolution, 1700–1850*. Oxford: Oxford University Press.

Heffernan, Michael. 1990. "Rogues, Rascals and Rude Books: Policing the Popular Book Trade in Early Nineteenth-Century France." *Journal of Historical Geography* 16:90–107.

———. 1999. "Historical Geographies of the Future: Three Perspectives from France, 1750–1825." In *Geography and Enlightenment*, ed. David N. Livingstone and Charles W. J. Withers, 125–64. Chicago: University of Chicago Press.

Helgerson, Richard. 1992. *Forms of Nationhood: The Elizabethan Writing of England*. Chicago: University of Chicago Press.

Henry, John. 1986. "England." In *The Scientific Revolution in National Context*, ed. Roy Porter and Mikuláš Teich, 178–210. Cambridge University Press, Cambridge.

———. 1986. "Occult Qualities and the Experimental Philosophy: Active Principles in Pre-Newtonian Matter Theory." *History of Science* 24:335–81.

———. 1997. *The Scientific Revolution and the Origins of Modern Science*. Basingstoke: Macmillan.

———. 2004. "Metaphysics and the Origins of Modern Science: Descartes and the Importance of Laws of Nature." *Early Science and Medicine* 9:73–114.

[Herle, Charles]. [1642] 1999. "A Fuller Answer to a Treatise Written by Doctor Ferne." In

The Struggle for Sovereignty: Seventeenth-Century English Political Tracts, ed. Joyce Lee Malcolm, 1:226–60. Indianapolis: Liberty Fund.

Hermet, Guy, 1996. *Histoire des nations et du nationalisme en Europe*. Paris: Seuil.

Herschel, Sir John F. W. 1861. *Physical Geography: From the Encyclopaedia Britannica*. Edinburgh: Adam and Charles Black.

Hesse, Mary B. 1966. *Models and Analogies in Science*. Notre Dame: University of Notre Dame Press.

———. 1974. *The Structure of Scientific Inference*. London: Macmillan.

———. 1980. *Revolutions and Reconstructions in the Philosophy of Science*. Brighton: Harvester.

Hessenbruch, Arne, ed. 2000. *Reader's Guide to the History of Science*. London: Fitzroy Dearborn.

Hettinger, Ned. 2001. "Exotic Species, Naturalisation, and Biological Nativism." *Environmental Values* 10:193–224.

Heylyn, Peter. 1621. *Microcosmus; or, A Little Description of the Great World: A Treatise Historicall, Geographicall, Politicall, Theologicall*. Oxford: Oxford University Press.

———. 1625. *Microcosmus; or, A Little Description of the Great World: A Treatise Historicall, Geographicall, Politicall, Theologicall*. Rev. ed. Oxford: John Lichfield and William Turner.

———. 1657. *Cosmographie in Four Bookes*. 2nd ed. London: Henry Seile.

Higonnet, Patrice L. R. 1998. *Goodness beyond Virtue: Jacobins during the French Revolution*. Cambridge: Harvard University Press.

Hill, Christopher. 1965. *Intellectual Origins of the English Revolution*. Oxford: Clarendon.

———. 1997. *The Intellectual Origins of the English Revolution Revisited*. 2nd ed. Oxford: Clarendon.

Himmelfarb, Gertrude. 1959. *Darwin and the Darwinian Revolution*. London: Chatto and Windus.

Hindle, Brooke, and Steven Lubar. 1986. *Engines of Change: The American Industrial Revolution, 1790–1860*. Washington: Smithsonian Institution Press.

Hirst, Derek. 1996. "The English Republic and the Meaning of Britain." In *The British Problem, c. 1534–1707: State Formation in the Atlantic Archipelago*, ed. Brendan Bradshaw and John Morrill, 192–219. London: Macmillan.

Hobbes, Thomas. 1843. *English Works*. Ed. William Molesworth. 8 vols. London: John Bohn.

———. 1990. *Behemoth; or, The Long Parliament*. Ed. Ferdinand Tönnies. Chicago: University of Chicago Press.

Hobsbawm, Eric J. 1986. "Revolution." In *Revolution in History*, ed. Roy Porter and Mikuláš Teich, 5–46. Cambridge: Cambridge University Press.

Hodge, M. J. S. 1983. "Darwin and the Laws of the Animate Part of the Terrestrial System, 1835–1837: On the Lyellian Origins of his Zoonomical Explanatory Program." *Studies in the History of Biology* 6:1–106.

———. 1991. *Origins and Species: A Study of the Historical Sources of Darwinism and the Context of Some Other Accounts of Organic Diversity from Plato to Aristotle On*. New York: Garland.

Hodge, M. J. S., and David Kohn. 1985. "The Immediate Origins of Natural Selection." In *The Darwinian Heritage,* ed. David Kohn, 185–206. Princeton: Princeton University Press.

Hoffman, Theodore. 1892. "Exploration in Sikkim: To the North-East of Kanchinjinga." *Proceedings of the Royal Geographical Society* 14:613–18.

Holland, Susan. 1994. "Archbishop Abbot and the Problem of 'Puritanism.'" *Historical Journal* 37:23–43.

Holmes, Clive 1980. "The County Community in Stuart Historiography." *Journal of British Studies* 19:54–73.

Hooke, Robert. 1665. *Micrographia.* London: Printed by J. Martin and J. Allestry.

Hooson, David M., ed. 1994. *Geography and National Identity.* Oxford: Blackwell.

Hooykaas, Reijer J. 1987. "The Rise of Modern Science: When and Why." *British Journal for the History of Science* 20:453–73.

———. 1972. *Religion and the Rise of Modern Science.* Edinburgh: Scottish Academic Press.

Hotson, Howard. 2000. *Johann Heinrich Alsted: Between Renaissance, Reformation and Universal Reform.* Oxford: Clarendon.

Houtman, Cornelis de. 1598. *The Description of a Voyage Made by Certaine Ships of Holland into the East Indies.* Trans. William Philip. London: John Wolfe.

Howell, Wilbur. 1956. *Logic and Rhetoric in England, 1500–1700.* Princeton: Princeton University Press.

Hudson, Patricia. 1989. *Regions and Industries: A Perspective on the Industrial Revolution in Britain.* Cambridge: Cambridge University Press.

Hughes, Ann. 1985. "King, Parliament and the Localities during the English Civil War." *Journal of British Studies* 24:236–63.

———. 1989. "Local History and the Origins of the Civil War." In *Conflict in Early Stuart England: Studies in Religion and Politics, 1602–1642,* ed. Richard Cust and Ann Hughes, 224–53. London: Longmans.

Hume, David. 1748. *Enquiry Concerning Human Understanding.* London.

Humphrey, Chris. 2001. "Time and Urban Culture in Late-Medieval England." In *Time in the Medieval World,* ed. Chris Humphrey and W. M. Ormrod, 105–18. York: York Medieval Press.

Hunt, Lynn. 1984. *Politics, Culture, and Class in the French Revolution.* Berkeley and Los Angeles: University of California Press.

———. 1992. *The Family Romance of the French Revolution.* London: Routledge.

———. 2003. "The World We Have Gained: The Future of the French Revolution." *American Historical Review* 108:1–19.

———, ed. 1989. *The New Cultural History.* Berkeley and Los Angeles: University of California Press.

Hunter, Michael. 1981. *Science and Society in Restoration England.* Cambridge: Cambridge University Press.

———. 2001. *The Occult Laboratory: Magic, Science and Second Sight in Late Seventeenth-Century Scotland.* Woodbridge: Boydell and Brewer.

———, ed. 1989. *Establishing the New Science: The Experience of the Early Royal Society.* Woodbridge: Boydell.

Hunter, Michael, and Annabel Gregory, eds. 1988. *An Astrological Diary of the Seventeenth Century: Samuel Jeake of Rye, 1652–1699.* Oxford: Clarendon.

Huntington, Samuel P. 1968. *Political Order in Changing Societies.* New Haven: Yale University Press.

Hutcheson, Francis. 1755. *A System of Moral Philosophy.* 2 vols. London.

Hutchins, Edwin. 1995. *Cognition in the Wild.* Cambridge: MIT Press.

Iliffe, Rob. 1993. "'Aplattisseur du Monde et de Cassini': Maupertuis, Precision Measurement, and the Shape of the Earth in the 1730s." *History of Science* 31 : 335–75.

———. 2003. "Science and Voyages of Discovery." In *The Cambridge History of Science,* vol. 4, *Eighteenth-Century Science,* ed. Roy Porter, 618–45. Cambridge: Cambridge University Press.

Iltis, Carolyn. 1973. "The Leibnizian-Newtonian Debates: Natural Philosophy and Social Psychology." *British Journal for the History of Science* 6 : 343–77.

Ingold, Tim. 2001. *The Perception of the Environment* London: Routledge.

"Instruction for Intending Travellers." 1905. *Journal of the Royal Geographical Society* 26:viii.

Ivins, William Mills. 1928. "Photography and the Modern Point of View: A Speculation in the History of Taste." *Metropolitan Museum* 1 : 16–24.

———. 1938. *On the Rationalisation of Sight.* New York: Metropolitan Museum of Art.

———. [1953] 1980. *Prints and Visual Communication.* Cambridge: MIT Press.

Jackson, Clare. 2003. *Restoration Scotland, 1660–1690: Royalist Politics, Religion and Ideas.* Woodbridge: Boydell and Brewer.

Jackson, Lewis d'A. 1889. *Aid to Survey-Practice: for Reference in Surveying, Levelling, and Setting-out; and in Route-surveys of Travellers by Land and Sea.* London: Crosby Lockwood.

Jacob, Christian. 1992. *L'empire des cartes.* Paris: Gallimard.

James, Henry, and R. I. Murchison. 1859. *Ordnance Survey: Report of the Committee on the Reduction of the Ordnance Plans by Photography.* London: Ordnance Survey.

James, Lawrence. 1994. *The Rise and Fall of the British Empire.* London: Little, Brown.

James, Preston. 1972. *All Possible Worlds: A History of Geographical Ideas.* Indianalopis: Bobbs-Merrill.

Jay, John. 1987. "Federalist Number II." In *The Federalist; or, The New Constitution,* ed. Max Beloff. Oxford: Blackwell.

Jeal, Tim. 1973. *Livingstone.* London: Heinemann.

Jefferson, Thomas. [1784] 1955. *Notes on the State of Virginia.* Ed. William Peden. Chapel Hill: University of North Carolina Press.

Johns, Adrian. 1998. *The Nature of the Book: Print and Knowledge in the Making.* Chicago: University of Chicago Press.

Johnson, Chalmers. 1983. *Revolutionary Change.* London: Longman.

Jolley, Nicholas. 1987. "The Reception of Descartes' Philosophy." In *Cambridge Companion to Descartes,* ed. John Cottingham, 393–423. Cambridge: Cambridge University Press.

Jones, Eric L. 1981. *The European Miracle: Environments, Economies and Geopolitics in the History of Europe and Asia.* Cambridge: Cambridge University Press.

Jones, Greta. 2002. "Alfred Russel Wallace, Robert Owen and the Theory of Natural Selection." *British Journal for the History of Science* 35 : 73–96.

Joy, Lynn Sumida. 1987. *Gassendi the Atomist: Advocate of History in an Age of Science.* Cambridge: Cambridge University Press.

Jussim, Estelle, and Elizabeth Lindquist-Cock. 1985. *Landscape as Photograph.* New Haven: Yale University Press.

Kaelble, Hartmut. 1986. *Industrialization and Social Inequality in Nineteenth-Century Europe.* New York: St. Martin's.

Kafker, Frank A. 1996. *The Encyclopedists as a Group: A Collective Biography of the Authors of the Encyclopédie.* Studies on Voltaire and the Eighteenth Century. Oxford: Voltaire Foundation.

Kafker, Frank A., and Serena L. Kafker. 1988. *The Encyclopedists as Individuals: A Biographical Dictionary of the Authors of the Encyclopédie.* Studies on Voltaire and the Eighteenth Century. Oxford: Voltaire Foundation.

Kamensky, Jane. 1990. "'In These Contrasted Climes, How Chang'd the Scene': Progress, Declension, and Balance in the Landscapes of Timothy Dwight." *American Quarterly* 43:80–108.

Kaufmann, Thomas DaCosta. 2004. *Toward a Geography of Art.* Chicago: University of Chicago Press.

Kearney, Hugh. 1971. *Science and Change.* London: Weidenfeld and Nicolson.

Keckermann, Bartholomew. 1612. *Systema Geographicum Duobus Libris.* Hanover: Petrus Janichius.

Kennedy, Emmet. 1989. *A Cultural History of the French Revolution.* New Haven: Yale University Press.

Kenney, Padraic. 2002. *A Carnival of Revolution: Central Europe 1989.* Princeton: Princeton University Press.

Kenny, Michael G. 1994. *The Perfect Law of Liberty: Elias Smith and the Providential History of America.* Washington: Smithsonian Institution Press.

Kenny, Neil. 1994. *The Palace of Secrets: Béroalde de Veville and Renaissance Conceptions of Knowledge.* Oxford: Clarendon.

Kent, Dale V., and F. William. 1982. *Neighbours and Neighbourhood in Renaissance Florence.* New York: Academic Press.

Kern, Stephen. 1983. *The Culture of Time and Space, 1880–1913.* Cambridge: Harvard University Press.

Ketcham, Ralph. 1974. *From Colony to Country: The Revolution in American Thought, 1750–1820.* New York: Macmillan.

Keynes, Richard Darwin, ed. 1988. *Charles Darwin's "Beagle" Diary.* Cambridge: Cambridge University Press.

Kidd, Colin. 1999. "Civil Theology and Church Establishment in Revolutionary America." *The Historical Journal* 42:1007–26.

Kimmerl, Michael S. 1990. *Revolution: A Sociological Interpretation.* Oxford: Polity.

Kirk, John. 1865. "Extracts from a Letter from J. Kirk." *Journal of the Royal Geographical Society* 34:290–92.

Kley, Dale van. 1975. *The Jansenists and the Expulsion of the Jesuits from France, 1757–1765.* New Haven: Yale University Press.

Knight, Alan. 1986. *The Mexican Revolution.* 2 vols. Cambridge: Cambridge University Press.

Knorr, Klaus E. 1944. *British Colonial Theories, 1570–1850.* Toronto: University of Toronto Press.

Kock, Victor de. 1953. *By Strength of Heart.* Cape Town: Timmins.

Koerner, Lisbet. 1996. "Purposes of Linnaean Travel: A Preliminary Research Report." In *Visions of Empire: Voyages, Botany, and Representations of Nature,* ed. David Philip Miller and Peter Hanns Reill, 117–52. New York: Cambridge University Press.

———. 1999. *Linnaeus: Nature and Nation.* Cambridge: Harvard University Press.

Kohn, David. 1981. "On the Origin of the Principle of Diversity." *Science* 213:1105–8.

———, ed. 1985. *The Darwinian Heritage.* Princeton: Princeton University Press.

Konvitz, Josef W. 1987. *Cartography in France, 1660–1848: Science, Engineering, and Statecraft.* Chicago: University of Chicago Press.

———. 1990. "The Nation-State, Paris and Cartography in Eighteenth- and Nineteenth-Century France." *Journal of Historical Geography* 16:3–16.

Krasner, James. 1992. *The Entangled Eye: Visual Perception and the Representation of Nature in Post-Darwinian Narrative.* Oxford: Oxford University Press.

Krauss, Rosalind. 1982. "Photography's Discursive Spaces: Landscape/View." *Art Journal* 42:311–20.

Krautheimer. Richard. 1985. *The Rome of Alexander VII.* Princeton: Princeton University Press.

Krise, Raymond, and Bill Squires. 1982. *Fast Tracks: The History of Distance Running.* Lexington, MA: Stephen Greene.

Kristeva, Julia. 1982. *Powers of Horror: An Essay on Abjection.* Trans. Leon Roudiez. New York: Columbia University Press.

Kronenberg, Maria E., ed. 1927. *De Novo Mondo: Antwerp, Jan Van Doesborch; A Facsimile of an Unique Broadsheet Containing an Early Account of the Inhabitants of South America, together with a Short Version of Heinrich Sprenger's Voyage to the Indies.* The Hague: Martinus Nijhoff.

Kropotkin, Pyotr. 1885. "What Geography Ought To Be." *Nineteenth Century* 18:940–56.

———. [1889] 1962. *Memoirs of a Revolutionist.* London: Cresset Library.

Kuhn, Madison. 1966. "Tiffin, Morse, and the Reluctant Pioneer." *Michigan History* 30:111–38.

Kuhn, Thomas S. 1962. *The Structure of Scientific Revolutions.* Chicago: University of Chicago Press.

———. 1970. *The Structure of Scientific Revolutions.* 2nd ed. Chicago: University of Chicago Press.

LaCapra, Dominick. 1989. *Soundings in Critical Theory.* Ithaca: Cornell University Press.

Lakatos, Imre, and Alan Musgrave, eds. 1970. *Criticism and the Growth of Knowledge.* Cambridge: Cambridge University Press.

Landes, David S. 1983. *Revolution in Time: Clocks and the Making of the Modern World.* Cambridge: Harvard University Press.

———. 2003. "Clocks and the Wealth of Nations." *Daedalus* 132, no. 2: 20–26.

Lang, Heinrich. 1860. "Humboldt und unsere Zeit." *Zeitstimmen aus der Reformierten Kirche der Schweiz* 2:230–40, 261–72, 277–86.

Langton, John, and R. J. Morris, eds. 1986. *Atlas of Industrializing Britain, 1780–1914.* London: Methuen.

Larche, P.-J. 1816. "Notice sur Mentelle." *Magazin Encyclopédie* 1:359–71.

Lasky, Melvin J. 1970. "The Birth of a Metaphor: On the Origins of Utopia and Revolution." *Encounter* 48 (February): 30–42; (March): 35–45.

Latham, Robert, and William Matthews, eds. 1970–83. *The Diary of Samuel Pepys: A New and Complete Transcription.* 10 vols. London: Bell.

Latour, Bruno. 1987. *Science in Action: How to Follow Scientists and Engineers through Society.* Cambridge: Harvard University Press.

———. 1988. *The Pasteurization of France.* Trans. Alan Sheridan and John Law. Cambridge: Harvard University Press.

———. 1993. *We Have Never Been Modern.* Hassocks: Harvester Wheatsheaf.

———. 1997. "Trains of Thought: Piaget, Formalism and the Fifth Dimension." *Common Knowledge* 6:170–91.

Laulan, R. 1957. "l'enseignement à l'École Royale Militaire de Paris de l'origine à la réforme du comte de Saint-Germain." *L'Information Historique* 19:152–58.

Laurence, Anne. 1996. *Women in England, 1500–1760.* London: Phoenix.

Laurence, John, and John Nightingale, eds. 2001. *Darwinism and Evolutionary Economics.* Cheltenham: Edward Elgar.

Law, John. 1986. "On the Methods of Long Distance Control: Vessels, Navigation and the Portuguese Route to India." In *Power, Action and Belief: A New Sociology of Knowledge?* ed. John Law, 234–63. London: Routledge and Kegan Paul.

Lawrence, Christopher, and Steven Shapin, eds. 1998. *Science Incarnate: Historical Embodiments of Natural Science.* Chicago: University of Chicago Press.

Lawrence, Susan C. 1988. "Entrepreneurs and Private Enterprise: The Development of Medical Lecturing in London, 1775–1820." *Bulletin of the History of Medicine* 62: 171–92.

Lawson-Peebles, Robert. 1988. *Landscape and Written Expression in Revolutionary America.* Cambridge: Cambridge University Press.

Lefebvre, Georges. 1988. *La grande peur de 1789.* Paris: A Colin.

Lehmann, William C. 1971. *Henry Home, Lord Kames, and the Scottish Enlightenment: A Study in National Character and in the History of Ideas.* The Hague: Martinus Nijhoff.

Lemagny, Jean-Claude, and André Rouillé, eds. 1987. *A History of Photography: Social and Cultural Perspectives.* Trans. Janet Lloyd. Cambridge: Cambridge University Press.

Leris, M. de. 1753. *La géographie rendue aisée; ou, Traité méthodique pour apprendre la géographie.* Paris.

Levi, Giovanni. 2001. "On Micro-History." In *New Perspectives on Historical Writing,* 2nd ed., ed. Peter Burke, 97–119. Cambridge: Cambridge University Press.

Lightman, Bernard, ed. 1997. *Victorian Science in Context.* Chicago: University of Chicago Press.

———. 2000. "The Visual Theology of Victorian Popularizers of Science: From Reverent Eye to Chemical Retina." *Isis* 91:651–80.

Lilley, Samuel. 1973. "Technological Progress and the Industrial Revolution." In *The Fontana Economic History of Europe,* ed. C. M. Cipolla, 3:187–254. London: Fontana.

Lindberg, David C., and Robert S. Westman, eds. 1990. *Reappraisals of the Scientific Revolution.* Cambridge: Cambridge University Press.

Linschoten, Jan Huygen van. 1598. *John Huighen van Linschoten: His Discours of Voyages into ye East and West Indies*. Trans. William Philip. London.

Lipset, Seymour Martin. 1969. *Revolution and Counter-Revolution: Change and Persistence in Social Structure*. London: Heinemann.

"Literature: Letters from Alexander von Humboldt to Varnhagen von Ense." 1860. *Athenaeum*, 17 March, 365–66.

Livingstone, David. 1857. *Missionary Travels and Researches in South Africa*. London: John Murray.

———. 1861. "Extracts from the Despatches of Dr David Livingstone to the Right Honourable Lord Malmesbury." *Journal of the Royal Geographical Society* 31:256–96.

Livingstone, David, and Charles Livingstone. 1865. *Narrative of an Expedition to the Zambesi and Its Tributaries and of the Discovery of the Lakes Shirwa and Nyassa, 1858–1864.* London: John Murray.

Livingstone, David N. 1987. "Human Acclimatization: Perspectives on a Contested Field of Inquiry in Science, Medicine and Geography." *History of Science* 25:359–94.

———. 1988. "Science, Magic and Religion: A Contextual Assessment of Geography in the Sixteenth and Seventeenth Centuries." *History of Science* 26:269–294.

———. 1990. "Geography, Tradition and the Scientific Revolution." *Transactions of the Institute of British Geographers* 15:359–73.

———. 1992. *The Geographical Tradition: Episodes in the History of a Contested Enterprise*. Oxford: Blackwell.

———. 1994. "Science and Religion: Foreword to the Historical Geography of an Encounter." *Journal of Historical Geography* 20:367–83.

———. 1995. "The Spaces of Knowledge: Contributions Towards an Historical Geography of Science." *Environment and Planning D: Society and Space* 13:5–34.

———. 1999. "Geographical Inquiry, Rational Religion and Moral Philosophy: Enlightenment Discourses on the Human Condition." In *Geography and Enlightenment*, ed. David N. Livingstone and Charles W. J. Withers, 93–119. Chicago: University of Chicago Press.

———. 2003. *Putting Science in Its Place: Geographies of Scientific Knowledge*. Chicago: University of Chicago Press.

Lloyd, Geoffrey. 2002. *The Ambitions of Curiosity*. Cambridge: Cambridge University Press.

Lloyd, Trevor O. 1996. *The British Empire, 1558–1995*. Oxford: Oxford University Press.

Loomes, Brian. 1997. *The Clockmakers of Northern England*. Ashbourne: Mayfield.

———. 1998. *Brass Dial Clocks*. Woodbridge: Antiques Collectors Club.

Lozovsky, Natalia. 2000. *"The Earth is Our Book: Geographical Knowledge in the Latin West, c. 400–1000*. Ann Arbor: University of Michigan Press.

Lucas, Colin. 1973. *The Structure of the Terror: The Example of Javogues and the Loire*. Oxford: Oxford University Press.

Luttrell, Barbara. 1990. *Mirabeau*. New York: Harvester-Wheatsheaf.

Lux, David S., and Harold J. Cook. 1998. "Closed Circles or Open Networks? Communicating at a Distance during the Scientific Revolution." *History of Science* 36:179–211.

Lyell, Charles. [1830–32] 1990. *Principles of Geology*. Repr. of 1st ed. 3 vols. Chicago: University of Chicago Press.

Lynch, Michael. 1991. "Science in the Age of Mechanical Reproduction: Moral and Epistemic Relations between Diagram and Photographs." *Biol. Philos.* 6:205–26

M[ay], T[homas]. 1631. *The Mirrour of Mindes; or, Barclay's Icon Animorum Englished.* London: Thomas Walkley.

Machamer, Peter, Marcello Prea, and Aristedes Baltas, eds. 2000. *Scientific Controversies: Philosophical and Historical Perspectives.* New York: Oxford University Press.

MacKenzie, John M., ed. 1996. *David Livingstone and the Victorian Encounter with Africa.* London: National Portrait Gallery.

Mackinder, John Halford. 1900. "A Journey to the Summit of Mount Kenya, British East Africa." *Journal of the Royal Geographical Society* 15:453–86.

———. 1911. "The Teaching of Geography from an Imperial Point of View, and the Use which Could and Should be Made of Visual Instruction." *Geographical Teacher* 6:79–86.

MacLachlan, Alastair. 1996. *The Rise and Fall of Revolutionary England: An Essay on the Fabrication of Seventeenth-Century England.* London: Macmillan.

Maddrell, Avril M. C. 1996. "Empire, Emigration and School Geography: Changing Discourses of Imperial Citizenship, 1880–1925." *Journal of Historical Geography* 22:373–87.

Maienschein, Jane. 1991. "Epistemic Styles in German and American Embryology." *Science in Context* 4:407–27.

Malcolm, Joyce Lee, ed. 1999. *The Struggle for Sovereignty: Seventeenth-Century English Political Tracts.* 2 vols. Indianapolis: Liberty Fund.

Marchant, James. 1916. *Alfred Russel Wallace: Letters and Reminiscences.* 2 vols. London: Cassell.

Marchant, Leslie R. 1987. "Edmunde Mentelle, 1730–1815, and François-Simon Mentelle, 1731–1799." *Geographers Biobibliographical Studies* 11:93–103.

Margadant, Ted. 1992. *Urban Rivalries in the French Revolution.* Princeton: Princeton University Press.

Marggraff, Hermann. 1860. "Die Englische Übersetzung der Briefe Alexander von Humboldt's." *Blätter für Literarische Unterhaltung,* 14 June, 443.

———. 1860. "Stimmen des Auslandes über Alexander von Humboldt's Briefe." *Blätter für Literarische Unterhaltung,* 9 August, 590–93.

———. 1860. "Alexander von Humboldt's Briefe in England und Frankreich." *Blätter für Literarische Unterhaltung,* 27 September, 718–19.

Marshall, Peter, and Ian Walker. 1981. "The First New Nation." In *Introduction to American Studies,* ed. Malcolm Bradbury and Howard Temperley, 45–62. London: Longman.

Martins, Luciana L. 2000. "A Naturalist's Vision of the Tropics: Charles Darwin and the Brazilian Landscape." *Singapore Journal of Tropical Geography* 21:19–33.

Marx, Karl, and Frederick Engels. 1943. *Selected Correspondence, 1846–1895.* Vol. 9. London: Lawrence and Wishart.

Mason, Haydn T., ed. 1988. *The Darnton Debate: Books and Revolution in the Eighteenth Century.* Studies on Voltaire and the Eighteenth Century. Oxford: Voltaire Foundation.

Matless, David. 1998. *Landscape and Englishness.* London: Reaktion.

May, Gita. 1970. *Madame Roland and the Age of Revolution.* New York: Columbia University Press.

Mayhew, Robert J. 1998. "The Character of English Geography, c. 1660–1800: A Textual Approach." *Journal of Historical Geography* 24:385–412.

———. 1998. "Geography in Eighteenth-Century British Education." *Paedagogica Historica* 34:731–69.

———. 1999. "William Guthrie's *Geographical Grammar*, the Scottish Enlightenment and the Politics of British Geography." *Scottish Geographical Journal* 115:19–34.

———. 2000. *Enlightenment Geography: The Political Language of British Geography, c. 1650–1850.* London: Macmillan.

———. 2001. "Geography, Print Culture and the Renaissance: 'The Road Less Travelled By.'" *History of European Ideas* 27:349–69.

———. 2004. "Geography Books and the Character of Georgian Politics." In *Georgian Geographies: Essays on Space, Place and Landscape in the Eighteenth Century,* ed. Miles Ogborn and Charles W. J. Withers, 192–211. Manchester: Manchester University Press, 2004.

Mayr, Ernst. 1994. "The Advance of Science and Scientific Revolutions." *Journal for the History of the Behavioral Sciences* 30:328–47.

McAdoo, H. R. 1965. *The Spirit of Anglicanism: A Survey of Anglican Theological Method in the Seventeenth Century.* London: A. and C. Black.

McCann, Justin. 1976. *The Rule of St. Benedict.* London: Sheed and Ward.

McClintock, Anne. 1995. *Imperial Leather: Race, Gender and Sexuality in the Colonial Contest.* Routledge: London.

McLuhan, Marshall. 1962. *The Gutenberg Galaxy.* London: Routledge.

Medawar, Peter. 1986. *Memoir of a Thinking Radish: An Autobiography.* Oxford: Oxford University Press.

Mentelle, Edme. 1751. *La mort de Polieucte.* Paris.

———. 1757. *Lettre à un seigneur étranger sur les ouvrages périodiques de France, par M. l'abbé D. C. d'H***.* Paris.

———. 1758. *L'amour libérateur.* Amsterdam and Paris: E. Mentelle et Desessarts.

———. [1758] 1783. *Élémens de Géographie, contenant: 1er, les principales divisions des quatre parties du monde; 2ème, une description abrégée de la France, à l'usage des commerçans, avec des cartes.* Paris: By the author.

———. 1761. *Manuel géographique, chronologique et historique, par M. ***, professeur d'histoire et de géographie, dédiée à Mlle. de Fitz-James.* Paris: Dufour.

———. 1766. *Éléments de l'histoire romaine, divisée en deux partie; avec des cartes et un table.* Paris: A. Delalain.

———. [1767] 1769. *Le porte-feuille du R. F. Gillet, ci-devant soi-disant jésuite; ou, Petit dictionnaire, dans laquelle on n'a mis que des choses essentielles, pour servir de supplément aux gros dictionnaires, qui renferment tant d'inutilités; Second édition augmentée, dans laquelle on a ajouté l'entrée triomphante du P.G. aux enfers, suivie de son retour sur la terre.* Madrid.

———. 1772. *Géographie abrégée de la Grèce ancienne, par un professeur d'histoire et de géographie.* Paris: Barbou.

———. 1778–84. *Géographie comparée; ou, Analyse de la géographie ancienne et moderne des peuples de tous les pays et de tous les âges; Accompanée de tableaux analytiques et d'un grand nombre de cartes.* 7 vols. in 8 books, plus atlas. Paris: By the author.

———. 1781. *Cosmographie élémentaire, divisée en parties astronomique et géographique: Ouvrage dans lequel on a tâché de mettre les vérités les plus intéressantes de la physique céleste à la portée de ceux même qui n'ont aucune notion de mathématiques.* Paris: By the author.

———. 1783. *Choix de lectures géographiques et historiques, présentées dans l'ordre qui a paru le plus propre à faciliter l'étude de la géographie de l'Asie, de l'Afrique et de l'Amérique, précédé d'un abrégé de géographie, avec des cartes.* Paris: By the author.

———. 1784. *Géographie; ou, Annonce de quelques ouvrages relatifs à cette science, avec quelques vues sur la manière de l'enseigner.* Paris: P.-G. Simon et N.-H. Nyon.

———. 1784–89. *Dictionnaire de géographie moderne.* 3 vols. Paris: Panckoucke.

———. 1787–93. *Encyclopédie méthodique: Géographie ancienne.* 3 vols. Paris: Panckoucke.

———. 1791. *Lettre d'un auteur citoyen à la commune de Paris, en faveur de la liberté de la presse et de la publication des ouvrages imprimés.* Paris: L. Jorry.

———. 1791. *Méthode courte et facile pour apprendre aisément et retenir sans peine la nouvelle géographie de la France.* Paris: Blachon.

———. 1792. *Tableau élémentaire de la géographie de la République française, à l'usage des écoles du 1er et 2ème âge.* Paris: F. Hocquet.

———. 1792. *Tableau raisonné de la nouvelle division économico-politique, d'après les bases physiques sur lesquelles cette division est établie.* Paris: By the author.

———. 1794. *Déclaration des droits de l'homme et du citoyen, mise en trente strophes, pour être chantée par les hommes libres de tout pays.* Paris: Imprimérie des Sans-Culottes/ Maret.

———. 1795. *Géographie enseignée par une méthode nouvelle; ou, Application de la synthèse à l'étude de la géographie.* Paris: By the author.

———. 1797–98. *Précis de l'histoire des Hébreux depuis Moyse jusqu'à la prise de Jérusalem par les Romains: Ouvrage dans lequel on a tâché de concilier l'exactitude des faits avec les sains lumières de la raison, à l'usage des écoles primaires et centrales de la République française.* Paris: By the author.

———. 1799. *Abrégé de la Géographie universelle de William Guthrie.* Paris: Bernard.

———. 1800–1801. *Cours de cosmographie, de géographie, de chronologie et d'histoire ancienne et moderne, divisé en cent vingt-cinq leçons.* 3 vols. Paris: Bernard.

———. 1801. *Cours d'histoire.* Paris: By the author.

———. 1801. *Précis de l'histoire universelle, pendant les premiers siècles de l'ère vulgaire; ou, Introduction à l'histoire moderne des différents états de l'Europe.* Paris: By the author.

———. 1804. *Abrégé élémentaire de géographie ancienne et moderne.* Paris: Bernard.

———. 1804. *Géographie physique, historique, statistique et topographique de la France en cent huit départements, et de ses colonies.* Paris: Bernard.

———. 1809. *Études convenables aux demoiselles, à l'usage des écoles et des pensions.* Paris: Bossange.

———. 1810. *Exercices chronologiques et historiques.* Paris: Bossange.

———. 1813. *Géographie classique et élémentaire.* 2 vols. Paris: Germain-Mathiot.

Mentelle, Edme, and Conrad Malte Brun. 1803–5. *Géographie mathématique, physique et politique de toutes les partie du monde.* 16 vols., plus atlas. Paris: H. Tardieu.

———. 1816. *Géographie universelle ancienne et moderne, mathématique, statistique, politique et historiquedes cinq parties du monde.* 16 vols., plus atlas. Paris: H. Tardieu.

Mentelle, Edme, and G. Mailhol. [1752] 1773. *Anecdotes orientales.* Paris: P. Vincent.

Menzel, Wolfgang. 1860. "Zeitgeschichte: Briefe von Alexander von Humboldt an Varnhagen von Ense." *Wolfgang Menzels Literaturblatt,* 4 April, 105–8; 7 April, 1860,109–10.

Merchant, Carolyn. 1980. *The Death of Nature: Women, Ecology and the Scientific Revolution.* New York, Harper and Row.

Mertz, J. T. 1896–1914. *A History of European Thought in the Nineteenth Century.* 4 vols. Edinburgh: Blackwood.

Mill, Hugh Robert. 1930. *The Record of the Royal Geographical Society, 1830–1930.* London: Royal Geographical Society.

———. 1951. *An Autobiography.* London: Longmans, Green.

Miller David Philip, and Peter Hanns Reill, eds. 1996. *Visions of Empire: Voyages, Botany, and Representations of Nature.* Cambridge: Cambridge University Press.

Miller, Rory. 1993. *Britain and Latin America in the Nineteenth and Twentieth Centuries.* London: Longman.

Milnes, Richard Monckton. 1860. "Alexander von Humboldt at the Court of Berlin." *Fraser's Magazine* 62:592–98.

Milton, Anthony. 1995. *Catholic and Reformed: The Roman and Protestant Churches in English Protestant Thought, 1600–1640.* Cambridge: Cambridge University Press.

———. 2002. "The Creation of Laudianism: A New Approach." In *Politics, Religion and Popularity in Early Stuart England: Essays in Honour of Conrad Russell,* ed. Thomas Cogswell, Richard Cust, and Peter Lake, 162–84. Cambridge: Cambridge University Press.

Milton, John. 1953–82. "Commonplace Book." In *The Complete Prose Works of John Milton,* ed. Don Wolfe, 1:362–513. New York: Columbia University Press.

———. [1650] 1991. "The Tenure of Kings and Magistrates." In *John Milton: Political Writings,* ed. Martin Dzelzainis, 3–48. Cambridge: Cambridge University Press.

———. [1658] 1991. "A Defence of the English People." In *John Milton: Political Writings,* ed. Martin Dzelzainis, 51–254. Cambridge: Cambridge University Press.

———. [1660] 1999. "The Readie and Easie Way to Establish a Free Commonwealth." In *The Struggle for Sovereignty: Seventeenth-Century English Political Tracts,* ed. Joyce Lee Malcolm, 1:508–25. Indianapolis: Liberty Fund.

Milton, John R. 1981. "The Origin and Development of the Concept of the Laws of Nature." *Archives Européene de Sociologie* 22:173–95.

———. 1998. "Laws of Nature." In *The Cambridge History of Seventeenth-Century Philosophy,* ed. Daniel Garber and M. Ayers, 680–701. Cambridge: Cambridge University Press.

Mirabeau, comte de [Riquetti, H.-G]. 1788. *De la monarchie prussienne sous Frédéric le Grand, avec un appendice contenant des recherches sur la situation actuelle des principales contrées de l'Allemagne.* 7 vols., plus atlas. London.

———. 1789. *Histoire secrète de la cour de Berlin; ou, Correspondence d'un voyageur françois, depuis le mois de juillet 1786 jusqu'au 19 janvier 1787: Ouvrage posthume avec une lettre remise au roi de Prusse regnant, le jour de son avénement au trône.* 2 vols. London: S. Bladon.

Misch, Carl. 1925. *Varnhagen von Ense in Beruf und Politik.* Gotha and Stuttgart: Perthes Verlag.

Mitchell, Timothy, 1989. "The World-as-Exhibition." *Comparative Studies of Society and History* 31:217–36.

Moore, James. 1997. "Wallace's Malthusian Moment: The Common Context Revisited." In *Victorian Science in Context,* ed. Bernard Lightman, 290–311. Chicago: University of Chicago Press.

Moravia, Sergio. 1967. "Philosophie et géographie à la fin du XVIIIe. siècle." *Studies on Voltaire and the Eighteenth Century* 57:937–1011.

Morrill, John. 1987. "The Ecology of Allegiance in the English Civil War." *Journal of British Studies* 26:451–67.

———. 1999. *Revolt in the Provinces: The People of England and the Tragedies of War, 1630–1648.* 2nd ed. London: Longmans.

Morris, Joseph. 1909. "Belfield, East Calder: The Country Mansion of the Lanton Oliphants." *Proceedings of the Society of Antiquaries of Scotland,* 4th ser., 43:324–29.

Morrison-Low, Alison. 2002. "'Feasting My Eyes with the View of Fine Instruments': Scientific Instruments in Enlightenment Scotland, 1680–1820." In *Science and Medicine in the Scottish Enlightenment,* ed. Charles W. J. Withers and Paul Wood, 17–53. East Linton: Tuckwell.

Morse, Jedidiah. 1792. *The American Geography; or, A View of the Present Situation of the United States of America.* 2nd ed. London: John Stockdale.

———. 1795. *Elements of Geography.* Boston.

———. 1822. *A Report to the Secretary of War of the U.S. on Indian Affairs, Comprising a Narrative of a Tour Performed, in the Summer of 1820, under a Commission from the President of the U.S., for the Purpose of Ascertaining, for the Use of the Government, the Actual State of the Indian Tribes, in Our Country.* Newhaven.

———. 1824. *Annals of the American Revolution; or, A record of the Causes and Events which Produced, and Terminated in the Establishment and Independence of the American Republic. . . .* Hartford.

Moss, Ann. 1996. *Printed Commonplace Books and the Structuring of Renaissance Thought.* Oxford: Clarendon.

Moss, Richard J. 1988. "Jedidiah Morse and the Illuminati Affair: A Re-Reading." *Historical Journal of Massachusetts* 16:141–53.

———. 1990. "Republicanism, Liberalism, and Identity: The Case of Jedidiah Morse." *Essex Institute Historical Collections* 126:209–36.

———. 1995. *The Life of Jedidiah Morse: A Station of Peculiar Exposure.* Knoxville: University of Tennessee Press.

Muir, Edward, and Ronald Weissman. 1989. "Social and Symbolic Places in Renaissance Venice and Florence." In *The Power of Place: Bringing Together Geographical and Sociological Imaginations,* ed. John A. Agnew and James S. Duncan, 81–103. Boston: Unwin Hyman.

Mulligan, Lotte. 1984. "'Reason,' 'Right Reason,' and 'Revelation' in Mid-Seventeenth-

Century England." In *Occult and Scientific Mentalities in the Renaissance,* ed. Brian Vickers, 375–401. Cambridge: Cambridge University Press.

Murchison, Roderick. 1858. "Presidential Address." *Journal of the Royal Geographical Society* 28:155.

———. 1859. "Presidential Address," *Journal of the Royal Geographical Society* 29:152.

Musson, Albert E. and Eric Robinson. 1969. *Science and Technology in the Industrial Revolution.* Manchester: Manchester University Press.

Nardi, Bonnie A., and Vicki L. O'Day. 1999. *Information Ecologies: Using Technology with Heart.* Cambridge: MIT Press.

Nash, Gary B. 1965. "The American Clergy and the French Revolution." *William and Mary Quarterly* 22:392–412.

Nasr, Seyyed Hossein, and Oliver Leaman, eds. 1996. *History of Islamic Philosophy.* London: Routledge.

Nathans, Benjamin. 1990. "Habermas's 'Public Sphere' in the Era of the French Revolution." *French Historical Studies* 16:620–44.

Naylor, Simon. 2002. "The Field, the Museum and the Lecture Hall: The Spaces of Natural History in Victorian Cornwall." *Transactions of the Institute of British Geographers* 27: 494–513.

Needham, Joseph. 1963. "Poverties and Triumphs of the Chinese Scientific Tradition." In *Scientific Change: Historical Studies in the Intellectual, Social and Technical Conditions for Scientific Discovery,* ed. Alistair C. Crombie, 117–49. London: Routledge and Kegan Paul.

———. 1979. *The Grand Titration.* London: Allen and Unwin.

Nefftzer, Auguste. 1860. "Correspondence d'Alexandre de Humboldt." *Revue Germanique* 9:656–87.

Nelson, Gareth. 1978. "From Candolle to Croizat: Comments on the History of Biogeography." *Journal of the History of Biology* 11:269–305.

Newitt, Malynn. 1992. "Mixed Race Groups in the Early History of Portuguese Expansion." In *Studies in the Portuguese Discoveries,* 1:35–52. Warminster: Aris.

Newton, Isaac. [1730] 1952. *Opticks; or, a Treatise of the Reflections, Refractions, Inflections and Colours of Light, Based on the Fourth Edition, London, 1730.* New York: Dover.

Nicol, Donald M. 1979. *Church and Society in the Last Centuries of Byzantium.* Cambridge: Cambridge University Press.

Nicolson, Malcolm. 1987. "Alexander von Humboldt, Humboldtian Science and the Origins of the Study of Vegetation." *History of Science* 25:167–94.

———. 1990. "Alexander von Humboldt and the Geography of Vegetation." In *Romanticism and the Sciences,* ed. Nicholas Jardine and Andrew Cunningham, 169–88. Cambridge: Cambridge University Press.

———. 1996. "Humboldtian Plant Geography after Humboldt: The Links to Ecology." *British Journal for the History of Science* 29:289–310.

Nisbett, Richard E. 2003. *The Geography of Thought: How Asians and Westerners Think Differently . . . and Why.* New York: Free Press.

Noll, Mark A. 1989. *Princeton and the Republic, 1768–1822: The Search for a Christian Enlightenment in the Era of Samuel Stanhope Smith.* Princeton: Princeton University Press.

———. 1995. "The Rise and Long Life of the Protestant Enlightenment in America." In

Knowledge and Belief in America: Enlightenment Traditions and Modern Religious Thought, ed. William M. Shea and Peter A. Huff, 88–124. New York: Cambridge University Press.

———. 2002. *America's God: From Jonathan Edwards to Abraham Lincoln.* New York: Oxford University Press.

Nordman, Daniel. 1989. "La pédagogie du territoire, 1793–1814." In *Atlas de la Révolution française,* vol. 4, *Le Territoire: Réalités et représentations,* ed. Daniel Nordman and Marie Vic-Ozouf Marignier, 62–64. Paris.

———, ed. 1994. *L'École Normale de l'an II: Léçons d'Histoire, de Géographie, d'Économie Politique; Édition annoté des cours de Volney, Buache de la Neuville, Mentelle et Vandermonde avec introductions et notes.* Paris: Dunod.

Northeast, Catherine M. 1991. *The Parisian Jesuits and the Enlightenment, 1700–1762.* Studies on Voltaire and the Eighteenth Century. Oxford: Voltaire Foundation, 1991.

Norwich, Oscar. 1983. *Maps of Africa: An Illustrated and Annotated Carto-Bibliography.* Johannesburg: Donker.

Nowotny, Helga. 1994. *Time: The Modern and Postmodern Experience.* Cambridge: Polity.

Numbers, Ronald L., and John Stenhouse, ed. 1999. *Disseminating Darwinism: The Role of Place, Race, Religion, and Gender.* Cambridge: Cambridge University Press.

Nussdorfer, Laurie. 1999. "The Politics of Space in Early Modern Rome." *Memoirs of the American Academy in Rome* 42 : 161–86.

O'Brien, Patrick, ed. 1994. *The Industrial Revolution in Britain.* 2 vols. Oxford: Blackwell.

O'Kane, Rosemary H. T. ed. 2000. *Revolution: Critical Concepts in Political Science.* 4 volumes. London: Routledge.

Ogborn, Miles. 1998. *Spaces of Modernity: London's Geographies 1680–1780.* New York: Guilford.

Ogborn, Miles, and Charles W. J. Withers, eds. 2004. *Georgian Geographies: Essays on Space, Place and Landscape in the Eighteenth Century.* Manchester: Manchester University Press.

Olby, Roger C., Geoffrey N. Cantor, John R. R. Christie, and M. J. S. Hodge, eds. 1996. *Companion to the History of Modern Science.* London: Routledge.

Oldenburg, Henry. 1966–73. *Correspondence.* Ed. A. R. and M. B. Hall. 9 vols. Madison: University of Wisconsin Press.

Ong, Walter. 1954. *Ramus, Method and the Decay of Dialogue: From the Art of Discourse to the Art of Reason.* Cambridge: Harvard University Press.

Ophir, Adir, and Steven Shapin. 1991. "The Place of Knowledge: A Methodological Survey." *Science in Context* 4 : 3–21.

Orme, Nicholas. 1989. *Education and Society in Medieval and Renaissance England.* London: Hambledon.

Osler, Margaret J., ed. 2000. *Rethinking the Scientific Revolution.* Cambridge: Cambridge University Press.

Ospovat, Dov. 1977. "Lyell's Theory of Climate." *Journal of the History of Biology* 10 : 317–39.

———. 1978. "Perfect Adaptation and Teleological Explanation: Approaches to the Problem of the History of Life in the Mid-nineteenth Century." *Studies in History of Biology* 2 : 33–56.

———. 1981. *The Development of Darwin's Theory: Natural History, Natural Theology and Natural Selection, 1838–1859*. Cambridge: Cambridge University Press.

"Our Weekly Gossip." 1860. *Athenaeum*, 10 March, 343–44.

Outram, Dorinda. 1983. "The Ordeal of Vocation: The Paris Academy of Sciences and the Terror, 1793–95." *History of Science* 21:257–73.

———. 1989. *The Body and the French Revolution: Sex, Class, and Political Culture*. New Haven: Yale University Press.

———. 1996. "Life Paths: Autobiography, Science and the French Revolution." In *Telling Lives in Science: Essays on Scientific Biography*, ed. Michael Shortland and Richard Yeo, 85–102. Cambridge: Cambridge University Press.

Overton, Mark. 1996. *Agricultural Revolution in England: The Transformation of the Agrarian Economy, 1500–1850*. Cambridge: Cambridge University Press.

Ozouf, Mona. 1984. "La Révolution française et la perception de l'espace national: Fédérations, fédéralisme, et stéréotypes régionaux." In *L'École de la France: Essais sur la Révolution, l'utopie et l'enseignement*, 27–54. Paris: Gallimard.

———. 1988. *Festivals and the French Revolution*. Cambridge: Harvard University Press.

Pagitt, Ephraim. 1640. *Christianography; or, The Description of the Multitude and Sundry Sorts of Christians in the World, Not Subject to the Pope*. 3rd ed. London: J. Okes.

Paine, Thomas. 1776. *Common Sense*. Philadelphia: J. Humphreys.

———. [1791] 1971. *The Rights of Man*. London: Pelican.

Palmer, Robert R. 1985. *The Improvement of Humanity: Education and the French Revolution*. Princeton: Princeton University Press.

Pang, Alex Soojung-Kim. 1995. "Victorian Observing Practices, Printing Technology, and Representations of the Solar Corona," part 2, "The Age of Photochemical Reproduction." *Journal for the History of Astronomy* 26:63–75.

———. 1997. "Visual Representation and Post-Constructivist History of Science." *Historical Studies in the Physical and Biological Sciences* 28:139–71.

———. 2002. *Empire and the Sun: Victorian Solar Eclipse Expeditions*. Stanford: Stanford University Press.

Papineau, David. 1981. "The Vis Viva Controversy." In *Leibniz: Metaphysics and Philosophy of Science*, ed. Roger S. Woolhouse, 139–56. Oxford: Oxford University Press.

Parish, Elijah. 1812. *A New System of Modern Geography*. 2nd ed. Newburyport, MA: E. Little.

Parker, Geoffrey. 1988. *The Military Revolution: Military Innovation and the Rise of the West, 1500–1800*. Cambridge: Cambridge University Press.

Parker, Henry. [1640] 1999. "The Case of Shipmony Briefly Discoursed." In *The Struggle for Sovereignty: Seventeenth-Century English Political Tracts*, ed. Joyce Lee Malcolm, 1:96–125. Indianapolis: Liberty Fund.

Parker, Kenneth. 1995. "Telling Tales: Early Modern English Voyagers and the Cape of Good Hope." *The Seventeenth Century* 10:121–49.

———. 1996. "Fertile Land, Romantic Spaces, Uncivilized Peoples: English Travel-Writing about the Cape of Good Hope, 1800–50." In *The Expansion of England: Race, Ethnicity and Cultural History*, ed. Bill Schwarz, 198–23. London: Routledge.

Parshall, Karen Hunger. 1982. "Varieties as Incipient Species: Darwin's Numerical Analysis." *Journal of the History of Biology* 15:191–214.

Paterson, James. 1681. *A Geographical Description of Scotland.* Edinburgh.

Paulinyi, Akos. 1986. "Revolution and Technology." In *Revolution in History,* ed. Roy Porter and Mikuláš Teich, 261–89. Cambridge: Cambridge University Press.

Pauly, Philip J. 1996. "The Beauty and Menace of the Japanese Cherry Trees: Conflicting Visions of American Ecological Independence." *Isis* 87:51–73.

———. 2000. *Biologists and the Promise of American Life: From Meriwether Lewis to Alfred Kinsey.* Princeton: Princeton University Press.

Pearson, Michael N. 1987. *The Portuguese in India.* Cambridge: Cambridge University Press.

Pemble, William. 1675. *A Briefe Introduction to Geography; Containing a Description of the Grounds and General Part Thereof.* 5th ed. Oxford: Oxford University Press.

Penn, Simon Andrew Christopher. 1989. "Social and Economic Aspects of Fourteenth-Century Bristol." PhD diss., Birmingham University.

Pennington, Donald, and Keith Thomas, eds. 1978. *Puritans and Revolutionaries: Essays in Seventeenth Century History Presented to Christopher Hill.* Oxford: Clarendon.

"The People's Right Briefly Asserted." [1649] 1999. In *The Struggle for Sovereignty: Seventeenth-Century English Political Tracts,* ed. Joyce Lee Malcolm, 1:362–68. Indianapolis: Liberty Fund.

Pereira, Duarte Pacheco. 1937. *Esmeraldo de Situ Orbis* Trans. G. H. Kimble. London: Hakluyt Society.

Peretti, Jonah H. 1998. "Nativism and Nature: Rethinking Biological Invasion." *Environmental Values* 7:183–92.

Perez, Carlotta. 2003. *Technological Revolutions and Financial Capital.* Cheltenham: Edward Elgar.

Perrin, Carleton E. 1996. "The Chemical Revolution." In *Companion to the History of Modern Science,* ed. Roger C. Olby, Geoffrey N. Cantor, John R. R. Christie, and M. J. S. Hodge, 264–77. London: Routledge.

Perroud, C. 1896. "Jany, le dernier correspondant de Madame Roland." *La Révolution Française* 30:1–36.

———. 1896. "Un dernier mot sur Jany-Mentelle." *La Révolution Française* 30:227–28.

———, ed. 1900–1902. *Lettres de Madame Roland (Jeanne-Marie Roland).* Vol. 1, *1780–1787.* Vol. 2, *1788–1793.* Paris: Imprimérie Nationale.

Petitot, Jean, Francisco J. Varela, Bernard Pachoud, and J Jean-Michel Roy, eds. 1999. *Naturalizing Phenomenology: Issues in Contemporary Phenomenology and Cognitive Science.* Stanford: Stanford University Press.

Philo, Christopher, ed. 1991. *New Words, New Worlds: Reconceptualizing Social and Cultural Geography.* Lampeter: St. David's University College.

Picket, Terry H. 1985. *The Unseasonable Democrat: Karl August Varnhagen von Ense, 1785–1858.* Bonn: Bouvier Verlag.

———. 1986. *Letters of the American Socialist Albert Brisbane to K. A. Varnhagen von Ense.* Heidelberg: Winter.

Pinney, Christopher. 1997. *Camera Indica: The Social Life of Indian Photographs.* London: Reaktion.

Plagnol-Diéval, Marie-Emmanuelle. 1997. *Madame de Genlis et la Théâtre d'Éducation au XVIIIe. siècle.* Studies on Voltaire and the Eighteenth Century. Oxford: Voltaire Foundation.

Pocock, Douglas C. D. 1981. "Sight and Knowledge." *Transactions of the Institute of British Geographers* 6:385–93.

Pole, 1865. "Photography for Travellers and Tourists." *Journal of the Royal Geographical Society* 34:295.

Polišenský, Josef V. 1974. *The Thirty Years War.* London: New English Library.

Popkin, Jeremy D. 1980. *The Right-Wing Press in France, 1792–1800.* Chapel Hill: University of North Carolina Press.

———. 1990. *Revolutionary News: The Press in France, 1789–1799.* Durham: Duke University Press.

Popper, Karl. 1959. *The Logic of Scientific Discovery.* London: Hutchinson.

Porter, Roy. 1986. "The Scientific Revolution: A Spoke in the Wheel?" In *Revolution in History,* ed. Roy Porter and Mikuláš Teich, 290–316. Cambridge: Cambridge University Press.

———, ed. 2003. *The Cambridge History of Science.* Vol. 4, *Eighteenth-Century Science.* Cambridge: Cambridge University Press.

Porter, Roy, and Mikuláš Teich, eds. 1986. *Revolution in History.* Cambridge: Cambridge University Press.

———. 1992. *The Scientific Revolution in National Context.* Cambridge: Cambridge University Press.

Pratt, Mary-Louise. 1992. *Imperial Eyes: Travel Writing and Transculturation.* London: Routledge.

Preuss, Richard, ed. 1892. *Briefe Thomas Carlyle's an Varnhagen von Ense: Aus den Jahren 1837–1857.* Berlin: Paetel.

Price, [William] Lake. [1858] 1973. *A Manual of Photographic Manipulation, Treating of the Practice of the Art; and Its Various Applications to Nature.* 2nd ed. New York: Arno.

"The Publication-of-Letters Nuissance." 1860. *Fraser's Magazine* 61:561–63.

Pumfrey, Stephen, Paolo L. Rossi, and Maurice Slawinski, eds. 1991. *Science, Culture and Popular Belief in Renaissance Europe.* Manchester: Manchester University Press.

Purchas, Samuel. 1905–8. *Hakluytus Posthumus; or, Purchas His Pilgrimes, Contayning a History of the World in Sea Voyages and Lande Travells.* 20 vols. Glasgow: James MacLehose.

Py, Gilbert. 1997. *Rousseau et les éducateurs: Étude sur la fortune des idées pédagogiques de Jean-Jacques Rousseau en France et en Europe au XVIIIe siècle.* Studies on Voltaire and the Eighteenth Century. Oxford: Voltaire Foundation.

Pyenson, Lewis. 2002. "An End to National Science: The Meaning and Extension of Local Knowledge." *History of Science* 40:251–90.

Quint, David. 1993. *Epic and Empire: Politics and Generic Form from Virgil to Milton.* New Jersey: Princeton University Press.

Raby, Peter. 2001. *Alfred Russel Wallace: A Life.* London: Chatto and Windus.

Raj, Kapil. 2000. "Histoire d'un inventaire oublié." *La Recherche* 333 (July/August): 78–83

———. 2005. "Surgeons, Fakirs, Merchants, and Craftspeople: Making L'Empereur's *Jardin* in Early Modern South Asia." In *Colonial Botany: Science, Commerce, Politics,* ed. Claudia Swan and Londa Schiebinger, 252–69. Philadelphia: University of Pennsylvania Press.

Ransford, Oliver, 1978. *David Livingstone: The Dark Interior.* London: John Murray.

Rappaport, Rhoda. 2003. "The Earth Sciences." In *The Cambridge History of Science,* vol. 4, *Eighteenth-Century Science,* ed. Roy Porter, 417–35. Cambridge: Cambridge University Press.

Rashed, Roshdi, ed. 1996. *Encyclopaedia of the History of Arabic Science.* 3 vols. London: Routledge.

Raven-Hart, R. 1967. *Before Van Riebeeck: Callers at South Africa from 1488 to 1652.* Cape Town: Struick.

Ravenstein, E. G. 1893. "Correspondence on 'The Determination of Longitudes by Photography.'" *Geographical Journal* 2:557–58.

Rée, Jonathan. 1999. *I See A Voice: Deafness, Language and the Senses—a Philosophical History.* London: Flamingo.

Reingold, Nathan. 1978. "National Styles in the Sciences: The United States Case." In *Human Implications of Scientific Advance,* ed. Eric G. Forbes, 163–73. Edinburgh: Edinburgh University Press.

———. 1991. "The Peculiarities of the Americans; or, Are There National Styles in the Sciences?" *Science in Context* 4:347–66.

Revel, Jacques. 1991. "Knowledge of the Territory." *Science in Context* 4:133–61.

———. 1992. "La région." In *Histoire de la France,* ed. Jacques Revel, 851–83. Paris: Seuil.

———, ed. 1996. *Jeux d'Échelle: La microanalyse à L'expérience.* Paris: Gallimard.

Reynolds, Robert L. 1961. *Europe Emerges: Transition towards an Industrial World-Wide Society, 600–1750.* Madison: University of Wisconsin Press.

Richard, Hélène. 1986. *Une grande expédition scientifique au temps de la Révolution française: Le voyage de d'Entrecasteaux à la recherche de Lapérouse.* Paris: C.T.H.S.

Richardson, David M., Petr Pyšek, Marcel Rejmánek, Michael G. Barbour, F. Dane Panetta, and Carol J. West. 2000. "Naturalization and Invasion of Alien Plants: Concepts and Definitions." *Diversity and Distributions* 6:93–107.

Richardson, Ralph C. 1998. *The Debate on the English Revolution.* 3rd ed. Manchester: Manchester University Press.

———, ed. 1992. *Town and Country in the English Civil War.* Manchester: Manchester University Press.

Roche, Daniel. 1978. *Le siècle des lumières en Province: Académies et académiciens provincaux, 1680–1789.* Paris: Mouton.

Roloff, Hans-Gert, ed. 1998. *Wissenschaftliche Briefeditionen und ihre Probleme.* Berlin: Weidler Buchverlag.

Rose, Gillian. 1992. "Geography as a Science of Observation: The Landscape, the Gaze and Masculinity." In *Nature and Science: Essays in the History of Geographical Knowledge,* ed. Felix Driver and Gillian Rose, 8–18. Cheltenham: Historical Geography Research Series.

Ross, Ian Simpson. 1972. *Lord Kames and the Scotland of His Day.* Oxford: Clarendon.

Roy, William. 1785. "An Account of the Measurement of a Base Line on Hounslow Heath." *Philosophical Transactions of the Royal Society of London* 75:385–478.

R.P. 1659. *A Geographicall Description of the World.* London: John Streater.

Rubiés, Joan-Pau. 2000. *Travel and Ethnology in the Renaissance: South India through European Eyes, 1250–1650.* Cambridge: Cambridge University Press.

Rudwick, Martin J. S. 1979. "Transposed Concepts from the Human Sciences in the Early Work of Charles Lyell." In *Images of the Earth: Essays in the History of the Environmental Sciences*, ed. L. J. Jordanova and Roy S. Porter, 67–83. Chalfont St. Giles: British Society for the History of Science.

———. 1985. *The Great Devonian Controversy: The Shaping of Scientific Knowledge among Gentlemanly Specialists*. Chicago: University of Chicago Press.

———. 1976. "The Emergence of a Visual Language for Geological Science, 1760–1840." *History of Science* 14:149–95.

———. 1990. Introduction to *Principles of Geology*, by Charles Lyell, 1:[vii–lviii]. Chicago: University of Chicago Press

———. 1992. *Scenes from Deep Time: Early Pictorial Representations of the Prehistoric World*. Chicago: University of Chicago Press.

Rupke, Nicolaas. 1999. "A Geography of Enlightenment: The Critical Reception of Alexander von Humboldt's Mexico Work." In *Geography and Enlightenment*, ed. David N. Livingstone and Charles W. J. Withers, 319–39. Chicago: University of Chicago Press.

———. 2000. "Translation Studies in the History of Science: The Example of *Vestiges*." *British Journal for the History of Science* 33:209–22.

———. In press. *Alexander von Humboldt: A Metabiography*. Bern: Peter Lang.

Ruse, Michael. 1979. *The Darwinian Revolution*. Chicago: University of Chicago Press.

Russell, Bertrand. 1927. *An Outline of Philosophy*. London: Allen and Unwin.

Russell, Conrad. 1990. *The Causes of the English Civil War*. Oxford: Clarendon.

———, ed. 1973. *The Origins of the English Civil War*. London: Macmillan.

Russell, John L. 1974. "Cosmological Teaching in the Seventeenth-Century Scottish Universities." *Journal of the History of Astronomy* 5:122–32 (part 1) and 145–54 (part 2).

Ryan, James R. 1997. *Picturing Empire: Photography and the Visualization of the British Empire*. London: Reaktion; Chicago: University of Chicago Press.

Sabra, A. I. 1996. "Situating Arabic Science: Locality versus Essence." *Isis* 87:654–70.

Sahlins, Marshall. 1988. "Cosmologies of Capitalism: The Trans-Pacific Sector of the 'World System.'" *Proceedings of the British Academy* 74:1–52.

Saint-Hilaire, Y. 1750. *Élémens de géographie; ou, Nouvelle méthode simple et abrégée, pour apprendre en peu de temps et sans peine la géographie*. Lyon.

Salert, Barbara. 1976. *Revolutions and Revolutionaries: Four Theories*. New York: Elsevier.

Salleo, Fernando. 1977. "Mirabeau en Prussie, 1786–1787: Diplomat parallèle ou agent sècret?" *Revue d'Histoire Diplomatique* 3/4:346–56.

Salm, Constance de. 1839. *Notice sur la vie et les ouvrages de Mentelle*. Paris: F. Didot.

Sandys, Edwin. 1629. *Europae Speculum; or, A View or Survey of the State of Religion in the Westerne Parts of the World*. The Hague: M. Sparke.

Schabas, Margaret. 1990. "Ricardo Naturalized: Lyell and Darwin on the Economy of Nature." In *Perspectives on the History of Economic Thought*, vol. 3, *Classicals, Marxians and Neo-Classicals: Selected Papers from the History of Economics Society Conference, 1988*, ed. D. E. Moggridge, 40–49. London: Edward Elgar for the History of Economics Society.

Schaffer, Simon. 1991. "The History and Geography of the Intellectual World: Whewell's Politics of Language." In *William Whewell: A Composite Portrait,* ed. Menachem Fisch and Simon Schaffer, 201–31. Oxford: Clarendon.

Schama, Simon. 1989. *Citizens: A Chronicle of the French Revolution.* London: Viking Penguin.

Schatzki, Theodore R. 2001. *Social Practices.* Cambridge: Cambridge University Press.

Schilder, Günter. 1976. "Organization and Evolution of the Dutch East India Company's Hydrographic Office in the Seventeenth Century." *Imago Mundi* 28:61–78.

———. 1979. "Willem Jansz: Blaeu's Wall Map of the World, on Mercator's Projection, 1606–07, and Its Influence," *Imago Mundi* 31:36–50.

Schlichter, H. 1892. "Celestial Photography as a Handmaid to Geography." *Proceedings of the Royal Geographical Society* 14:714–15.

———. 1893. "The Determination of Geographical Longitudes by Photography." *Geographical Journal* 2:423–29.

Schmidt, Benjamin. 2002. "Inventing Exoticism: The Project of Dutch Geography and the Marketing of the World, circa 1700." In *Merchants and Marvels: Commerce, Science, and Art in Early Modern Europe,* ed. Pamela H. Smith and Paula Findlen, 347–69. New York: Routledge.

Schuster, John A. 1990. "The Scientific Revolution." In *Companion to the History of Modern Science,* ed. Roger C. Olby, Geoffrey N. Cantor, John R. R. Christie, and M. J. S. Hodge, 217–42. London: Routledge.

Schwartz, Joan M. 1996. "The Geography Lesson: Photographs and the Construction of Imaginative Geographies." *Journal of Historical Geography* 22:16–45.

———. 1998. "Agent of Sight, Site of Agency: The Photography in the Geographical Imagination." PhD diss., Queen's University, Kingston, Canada.

Schwartz, Joan M., and James R. Ryan, eds. 2003. *Picturing Place: Photography and the Geographical Imagination.* London: I. B. Taurus.

Schwarz, Bill, ed. 1996. *The Expansion of England: Race, Ethnicity and Cultural History.* London: Routledge.

Scott, Jonathan. 2000. *England's Troubles: Seventeenth-Century English Political Instability in European Context.* Cambridge: Cambridge University Press.

Sears, John F. 1976–77. "Timothy Dwight and the American Landscape: The Composing Eye in Dwight's *Travels in New England and New York.*" *Early American Literature* 9:311–21.

Secord, Anne. 2002. "Botany on a Plate: Pleasure and the Power of Pictures in Promoting Early Nineteenth-Century Scientific Knowledge." *Isis* 93:28–57.

Secord, James A. 1982. "King of Siluria: Roderick Murchison and the Imperial Theme in Nineteenth-Century British Geology." *Victorian Studies* 25:413–42.

———. 2000. *Victorian Sensation: The Extraordinary Publication, Reception, and Secret Authorship of "Vestiges of the Natural History of Creation."* Chicago: University of Chicago Press.

Segovia, Fernando F., and Mary Ann Tolbert, eds. 1995. *Reading from This Place.* Vol. 1, *Social Location and Biblical Interpretation in the United States.* Vol. 2, *Social Location and Biblical Interpretation in Global Perspective.* Augsburg: Fortress Press.

Sellen, Abigail J., and Richard H. R. Harper. 2002. *The Myth of the Paperless Office.* Cambridge: MIT Press.

Shanin, Theodor. 1975. *Peasants and Peasant Societies*. Harmondsworth: Penguin.

Shapin, Steven. 1981. "Of Gods and Kings: Natural Philosophy and Politics in the Leibniz-Clarke Disputes." *Isis* 72:187–215.

———. 1984. "Pump and Circumstance: Robert Boyle's Literary Technology." *Social Studies of Science* 14:481–520.

———. 1988. "Following Scientists Around." *Social Studies of Science* 18:533–50.

———. 1988. "The House of Experiment in Seventeenth-Century England." *Isis* 79:373–404.

———. 1994. *A Social History of Truth: Civility and Science in Seventeenth-Century England*. Chicago: University of Chicago Press.

———. 1996. *The Scientific Revolution*. Chicago: University of Chicago Press.

———. 1998. "Placing the View from Nowhere: Historical and Sociological Problems in the Location of Science." *Transactions of the Institute of British Geographers* 23:5–12.

Shapin, Steven, and Simon Schaffer. 1985. *Leviathan and the Air-Pump: Hobbes, Boyle, and the Experimental Life*. Princeton: Princeton University Press.

Sharif, M. M., ed. 1963. *A History of Muslim Philosophy*. 2 vols. Wiesbaden: Otto Harrassowitz.

Sharpe, Kevin. 2000. *Reading Revolutions: The Politics of Reading in Early Modern England*. New Haven: Yale University Press.

Shea, William R., ed. 1988. *Revolutions in Science: Their Meaning and Relevance*. Canton, MA: Science History Publications.

Shepherd, Christine. 1982. "The Inter-relationship between the Library and Teaching in the Seventeenth and Eighteenth Centuries." In *Edinburgh University Library, 1580–1980*, ed. Jean Guild and Alexander Low, 67–86. Edinburgh: Edinburgh University Library.

———. 1982. "Newtonianism in the Scottish Universities in the Seventeenth Century." In *The Origins and Nature of the Scottish Enlightenment*, ed. Roy H. Campbell and Andrew S. Skinner, 65–85. Edinburgh: John Donald.

Short, John Rennie. 1999. "A New mode of Thinking: Creating a National Geography in the Early Republic." In *Surveying the Record: North American Scientific Exploration to 1930*, ed. Edward C. Carter II, 19–50. Philadelphia: American Philosophical Society.

Sibbald, Sir Robert. 1693. *An Account of the Scotish Atlas; or, The Description of Scotland Ancient and Modern*. Edinburgh: Printed by David Lindsay, James Kniblo, Joshua van Solingen, and John Colmar.

Silver, Morris. 1974. "Political Revolutions and Repression: An Economic Approach." *Public Choice* 14:63–71.

Simon, Eduard. 1860. "La correspondence d'Alexandre de Humboldt avec Varnhagen de Ense, 1827 à 1858." *Revue Contemporaine* 25:128–58.

Sitwell, O. Francis G. 1993. *Four Centuries of Special Geography*. Vancouver: University of British Columbia Press.

Skelton, R. A. 1958. *Explorer's Maps*. London: Routledge.

———. 1964. "The Early Map Printer and His Problems." *Penrose Annual* 57:171–84.

Skocpol, Theda. 1979. *States and Social Revolutions: A Comparative Analysis of France, Russia and China*. Cambridge: Cambridge University Press.

———. 1994. *Social Revolutions in the Modern World*. Cambridge: Cambridge University Press.

Slavin, Morris. 1986. *The Making of an Insurrection: Parisian Sections and the Gironde*. Cambridge: Harvard University Press.

Smith, Andrew B. 1993. "Different Facets of the Crystal: Early European Images of the Khoisan at the Cape, South Africa." *South African Archaeological Society Goodwin Series* 7: 8–20.

Smith, Bernard. 1985. *European Vision and the South Pacific.* 2nd ed. New Haven: Yale University Press.

Smith, Charles H. 1989. "Historical Biogeography: Geography as Evolution, Evolution as Geography." *New Zealand Journal of Zoology* 16: 773–85.

Smith, Crosbie, and Jon Agar, eds. 1998. *Making Space for Science: Territorial Themes in the Shaping of Knowledge.* London: Macmillan.

Smith, Elias. 1811. *The History of Anti-Christ; in Three Books, Written in Scripture Stile, in Chapters and Verses: For the Use of Schools.* Portland.

Smith, Neil. 2003. *American Empire: Roosevelt's Geographer and the Prelude to Globalization.* Berkeley and Los Angeles: University of California Press.

Smith, Pamela H., and Paula Findlen, eds. 2002. *Merchants and Marvels: Commerce, Science, and Art in Early Modern Europe.* London: Routledge.

Smith, Roger. 1996. "The Language of Human Nature." In *Inventing Human Science: Eighteenth-Century Domains,* ed. Christopher Fox, Roy Porter, and Robert Wokler, 88–111. Berkeley and Los Angeles: University of California Press.

Smith, Samuel Stanhope. 1787. *Essay on the Causes of the Variety of Complexion and Figure in the Human Species.* Philadelphia: Robert Aitkin.

Snyder, K. Alan. 1983. "Foundations of Liberty: The Christian Republicanism of Timothy Dwight and Jedidiah Morse." *New England Quarterly* 61: 382–97.

Sobel, Dava. 1995. *Longitude: The True Story of a Lone Genius Who Solved the Greatest Scientific Problem of his Time.* London: Penguin.

Solomon-Godeau, Abigail. 1991. *Photography at the Dock: Essays on Photographic History, Institutions, and Practices.* Minneapolis: University of Minnesota Press.

Sommerville, Johan P. 1986. *Politics and Ideology in England, 1603–1640.* London: Longmans.

Sorrenson, Richard. 1996. "The Ship as a Scientific Instrument in the Eighteenth Century." *Osiris,* 2nd ser., 11: 221–36.

Spary, Emma C. 2000. *Utopia's Garden: French Natural History from Old Regime to Revolution.* Chicago: University of Chicago Press.

Spears, Timothy B. 1989. "Common Observations: Timothy Dwight's *Travels in New England and New York*." *American Studies* 30: 35–52.

Spiller, Michael R. G. 1980. *"Concerning Natural Experimental Philosophie": Meric Casaubon and the Royal Society.* The Hague: Martinus Nijhoff.

Sprague, William Buell. 1874. *The Life of Jedidiah Morse, D.D.* New York: Anson D. F. Randolph.

Sprat, Thomas. 1667. *History of the Royal Society of London.* London: Printed by J. Martyn and J. Allestry.

Stafford, Barbara Maria. 1984. *Voyage into Substance: Art, Science, Nature, and the Illustrated Travel Account, 1760–1840.* Cambridge: MIT Press.

———. 1994. *Artful Science: Enlightenment Entertainment and the Eclipse of Visual Education.* Cambridge: MIT Press.

Stafford, Barbara Maria, and Frances Terpak. 2001. *Devices of Wonder: From the World in a Box to Images on a Screen.* Los Angeles: Getty Research Institute.

Stafford, Robert. 1989. *Scientist of Empire: Sir Roderick Murchison, Scientific Exploration and Victorian Imperialism.* Cambridge: Cambridge University Press.

Stafforde, Robert. 1634. *A Geographicall and Anthologicall Description of all the Empires and Kingdomes, both of Continents and Ilands in this Terrestriall Globe.* London: Simon Waterson.

Stanton, William. 1960. *The Leopard's Spots: Scientific Attitudes Toward Race in America, 1815–59.* Chicago: University of Chicago Press.

Stauffer, Robert C., ed. 1975. *Charles Darwin's "Natural Selection": Being the Second Part of His Big Species Book Written from 1856 to 1858.* Cambridge: Cambridge University Press.

Stauffer, Vernon. 1918. *New England and the Bavarian Illuminati.* New York: Columbia University Press.

Staum, Martin. 1987. "Human Geography in the French Institute: New Discipline or Missed Opportunity?" *Journal of the History of the Behavioural Sciences* 23:332–40.

———. 1996. *Minerva's Message: Stabilizing the French Revolution.* Montreal and Kingston: McGill–Queen's University Press.

Steensgaard, Neils. 1975. *The Asian Trade Revolution of the Seventeenth Century: The East India Companies and the Decline of the Caravan Trade.* Chicago: University of Chicago Press.

Stevenson, Edward Luther. 1921. *Celestial and Terrestrial Globes.* 2 vols. New York: Hispanic Society of America.

Stilgoe, John R. 1986. "Smiling Scenes." In *Views and Visions: American Landscapes before 1830,* Edward J. Nygren, 213–28. Washington, DC: Corcoran Gallery of Art.

Still, Judith. 2000. "Genlis's *Mademoiselle de Clermont:* A Textual and Intertextual Reading." *Australian Journal of French Studies* 37:331–47.

Stoddart, David R. 1986. *On Geography and Its History.* Oxford: Blackwell.

Stone, Laurence. 1966. "Theories of Revolution." *World Politics* 18:159–76.

Suckow, Christian. 1996. "Die Alexander-von-Humboldt-Edition: Ein Projekt der Berlin-Brandenburgischen Akademie der Wissenschaften." *Jahrbuch der historischen Forschung in der Bundesrepublik Deutschland,* Berichtsjahr 1995, 16–21.

Suckow, Christian, and Ingo Schwarz. 1998. "Zur Problematik einer Auswählenden Briefedition: Beispiel: Die Briefe Alexander von Humboldts." In *Wissenschaftliche Briefeditionen und ihre Probleme* ed. Hans-Gert Roloff, 119–22. Berlin: Weidler Buchverlag.

Surface, George T. 1909. "Thomas Jefferson: A Pioneer Student of American Geography," *Bulletin of the American Geographical Society* 41:743–750.

Sutherland, Donald. 1982. *The Chouans: The Social Origins of Popular Counter-Revolution in Upper Brittany, 1770–1796.* Oxford: Clarendon.

Sydow, Carl von. 1948. *Selected Papers on Folklore.* Copenhagen: Rosenkilde and Bagger.

Symson, Andrew. 1823. *A Large Description of Galloway.* Ed. Thomas Maitland. Edinburgh: W. and C. Tait.

Symson, Matthias. 1702. *Geography Compendiz'd; or, The World Survey'd.* Edinburgh: Sold by Mr. Henry Know and John Vallange.

Taillandier, René Gaspard Ernest Saint-René. 1860. "Lettres intimes et entretiens familiers de M. A. de Humboldt." *Revue des Deux Mondes* 28:58–89.

Tallack, Timothy. 1996. *Becoming a Revolutionary: The Deputies of the French Assembly and the Emergence of a Revolutionary Culture.* Princeton: Princeton University Press.

Tapié, Victor-L. 1957. *Baroque et classicisme.* Paris: A. Colin.

Taylor, Eva G. R. 1934. *Late Tudor and Early Stuart Geography, 1583–1650.* London: Methuen.

———, ed. 1963. *A Regiment for the Sea, by William Bourne, and Other Writings on Navigation.* Hakluyt Society, 2nd ser., vol. 121. Cambridge: Hakluyt Society.

Taylor, Stan. 1984. *Social Science and Revolutions.* London: Macmillan.

Terrall, Mary. 2002. *The Man Who Flattened the Earth: Maupertuis and the Sciences in the Enlightenment.* Chicago: University of Chicago Press.

Thomas, Alan. 1978. *The Expanding Eye: Photography and the Nineteenth Century Mind.* London: Croom Helm.

Thomas, Ann. 1997. "The Search for Pattern." In *Beauty of Another Order: Photography in Science,* ed. Ann Thomas, 76–119. New Haven: Yale University Press.

Thompson, E. P. 1963. *The Making of the English Working Class.* London: Gollancz.

———. 1967. "Time, Work-Discipline and Industrial Capitalism," *Past and Present* 38:56–97.

Thomson, John. 1873–74. *Illustrations of China and Its People, a Series of Two Hundred Photographs with Letterpress Description of the Places and People Represented.* 4 vols. London: Sampson Low, Marston, Low, and Searle.

———. 1879. *Through Cyprus with the Camera, in the Autumn of 1878.* London: Sampson Low, Marston, Searle and Rivington.

———. 1885. "Exploration with the Camera," *British Journal of Photography* 32:372–73.

———. 1891. "Photography and Exploration." *Proceedings of the Royal Geographical Society* 13:669–75.

———. 1891. "Photography Applied to Exploration." *Times* (London), 25 August, 5.

———. 1901. "Photography." In *Hints to Travellers: Scientific and General,* ed. John Coles, 52–64. London: Royal Geographical Society.

———. 1921. "Photography." In *Hints to Travellers,* ed. E. A. Reeves, 51–62. London: Royal Geographical Society.

Thomson, John, and Adolphe Smith. 1878. *Street Life in London.* London: Sampson Low, Marston, Searle, and Rivington.

Thomson, Keith Stewart. 1995. *HMS Beagle: The Story of Darwin's Ship.* New York: Norton.

Thrift, Nigel J., and Shaun French. 2002. "The Automatic Production of Space." *Transactions of the Institute of British Geographers,* n.s., 27:309–35.

Throop, William. 2000. "Eradicating the Aliens: Restoration and Exotic Species." In *Environmental Restoration: Ethics, Theory, and Practice,* ed. William Throop, 179–91 . Amherst, NY: Humanity Books.

Tichi, Cecilia. 1979. *New World, New Earth: Environmental Reform in American Literature from the Puritans through Whitman.* New Haven: Yale University Press.

Tilly, Charles. 1978. *From Mobilization to Revolution.* Reading, MA: Addison-Wesley.

Todd, Margot. 1987. *Christian Humanism and the Puritan Social Order.* Cambridge: Cambridge University Press.

Tomlinson, Howard, ed. 1983. *Before the Civil War: Essays on Early Stuart Politics and Government.* London: Macmillan.

Toulmin, Stephen. 1975. "Crucial Experiments: Priestley and Lavoisier." *Journal of the History of Ideas* 18:205–20.

Townsend, Mark. 2003. "Alien Invasion: the Plants Wrecking Rural Britain." *Observer* (London), 2 February, 14.

Toynbee, Arnold. 1884. *Lectures on the Industrial Revolution in England: Popular Addresses, Notes and other Fragments.* London: Rivingtons.

Trenard, Louis 1973. "Manuels scolaires au XVIIIe. siècle et sous la Révolution." *Revue du Nord* 55:99–111.

Tuan, Yi-Fu. 1979. "Sight and Pictures." *Geographical Review* 69:413–22.

Tucker, Jennifer. 1997. "Photography as Witness, Detective, and Imposter: Visual Representation in Victorian Science." In *Victorian Science in Context,* ed. Bernard Lightman, 378–408. Chicago: University of Chicago Press.

Tucker, Robert C. 1969. *The Marxian Revolutionary Idea.* New York: Norton.

Tullock, Gordon. 1974. *The Social Dilemma: The Economics of War and Revolution.* Blacksburg, VA: University Publications.

Turnbull, David. 1993. *Maps are Territories: Science Is an Atlas.* Chicago: University of Chicago Press.

———. 1996. "Cartography and Science in Early Modern Europe: Mapping the Construction of Knowledge Spaces." *Imago Mundi* 46:5–24.

———. 2002. "Travelling Knowledge: Narratives, Assemblage and Encounters." In *Instruments, Travel and Science: Itineraries of Precision from the Seventeenth to the Twentieth Century,* ed. Marie-Nöelle Bourguet, Christian Licoppe, and H. Otto Sibum, 273–94. London: Routledge.

Turner, Gerald L'E. 2000. *London Instrument Makers: The Origins of the London Trade in Precision Instrument Making.* Oxford: Oxford University Press.

Turner, Howard R. 1995. *Science in Medieval Islam.* Austin: University of Texas Press.

Turner, Stephen. 1994. *The Social Theory of Practices: Tradition, Tacit Knowledge, and Presuppositions.* Cambridge: Polity.

"Twenty-three Photographs of Mountain Scenery in Sikkim." 1893. *Proceedings of the Royal Geographical Society* 15:288.

Tyacke, Nicholas. 1978. "Science and Religion at Oxford before the Civil War." In *Puritans and Revolutionaries: Essays in Seventeenth-Century History Presented to Christopher Hill,* ed. Donald Pennington and Keith Thomas, 73–93. Oxford: Clarendon.

———. 1987. *Anti-Calvinists: The Rise of English Arminianism, c. 1590–1640.* Oxford: Clarendon.

Underdown, David. 1979. "The Chalk and the Cheese." *Past and Present* 85:25–48.

———. 1985. *Revel, Riot and Rebellion: Popular Politics and Culture in England, 1603–60.* Oxford: Oxford University Press.

———. 1995. "Regional Cultures?" In *Popular Culture in England, c. 1500–1850,* ed. Tim Harris. Basingstoke: Macmillan

Van Wyk Smith, Malvern. 1986. "'Waters Flowing from Darkness': The Two Ethiopias in the Early European Image of Africa." *Theoria* 68:67–77.

———. 1992. "'The Most Wretched of the Human Race': The Iconography of the Khoisann (Hottentots), 1500–1800." *History and Anthropology* 5:285–330.

Vance, Norman. 1985. *The Sinews of the Spirit: The Ideal of Christian Manliness in Victorian Literature and Religious Thought.* Cambridge: Cambridge University Press.

Vickers, Brian, ed. 1984. *Occult and Scientific Mentalities in the Renaissance.* Cambridge: Cambridge University Press.

Vic-Ozouf Marignier, Marie. 1989. *La formation des départements et la répresentation du territoire français à la fin du XVIIIe. siècle.* Paris: Gallimard.

Vovelle, Michel. 1993. *La découverte de la politique: Géopolitique de la Révolution française.* Paris: La Découverte.

Wace, N. M. 1967. "The Units and Uses of Biogeography." *Australian Geographical Studies* 5 : 15–29.

Waddy, Patricia. 1990. *Seventeenth-Century Roman Palaces.* New York: Architectural History Foundation and MIT Press.

Wallace, Alfred Russel. 1855. "On the Law which has Regulated the Introduction of New Species." *Annals and Magazine of Natural History,* 2nd ser., 16 : 184–96.

———. 1857. "On the Natural History of the Aru Islands." *Annals and Magazine of Natural History,* supplement, 2nd ser., 20 : 473–85.

———. 1858. "On the Tendency of Species to Depart Indefinitely from the Original Type." *Journal of the Proceedings of the Linnean Society: Zoology* 3 : 53–62.

———. 1860. "On the Zoological Geography of the Malay Archipelago." *Journal of the Proceedings of the Linnean Society: Zoology* 4 : 172–84.

———. 1862. "On the Trade of the Eastern Archipelago with New Guinea and Its Islands." *Journal of the Royal Geographical Society* 32 : 127–37.

———. 1864. "On Some Anomalies in Zoological and Botanical Geography." *Natural History Review* 4 : 111–23.

———. 1864. "The Origin of Human Races and the Antiquity of Man Deduced from the Theory of 'Natural Selection.'" *Journal of the Anthropological Society of London* 2 : clviii–clxx.

———. 1869. *The Malay Archipelago: The Land of the Orang-utan and the Bird of Paradise; a Narrative of Travel with Studies of Man and Nature.* 1st ed. 2 vols. London: Macmillan.

———. 1877. *The Malay Archipelago: The Land of the Orang-utan and the Bird of Paradise; a Narrative of Travel with Studies of Man and Nature.* 2nd ed. 1 vol. London: Macmillan.

———. 1905. *My Life: A Record of Events and Opinions.* London: Chapman and Hall.

Wallerstein, Immanuel. 1974. *The Modern World System.* Vol. 1, *Capitalist Agriculture and the Origins of the European World-Economy in the Sixteenth Century.* New York: Academic Press.

Wallis, J. P. R., ed. 1956. *The Zambezi Expedition of David Livingstone, 1858–1863: The Journal Continued with Letters and Dispatches Therefrom.* 2 vols. London: Chatto and Windus.

Warnier, Jean-Pierre. 2001. "A Praxeological Approach to Subjectivation in a Material World." *Journal of Material Culture* 6 : 5–24.

Warntz, William. 1964. *Geography Then and Now.* New York: American Geographical Society Research Series, no. 25. New York: American Geographical Society.

———. 1989. "Newton, the Newtonians and the *Geographia Generalis Varenii.*" *Annals of the Association of American Geographers* 79 : 165–91.

Warwick, Andrew. 1993. "Cambridge Mathematics and Cavendish Physics: Cunningham,

Campbell and Einstein's Relativity." Part 2, "Comparing Traditions in Cambridge Physics." *Studies in History and Philosophy of Science* 24 : 1–25.

Webster, Charles. 1975. *The Great Instauration: Science, Medicine and Reform, 1626–1660.* London: Duckworth.

———, ed. 1981. *Biology, Medicine and Society, 1840–1940.* Cambridge: Cambridge University Press.

Webster, Noah. [1783] 1968. *A Grammatical Institute of the English Language, Part I.* Menston: Scolar Press.

———. [1788] 1965. "On the Education of Youth in America." In *Essays on Education in the Early Republic,* ed. Frederick Rudolph. Cambridge: Harvard University Press.

Weikart, Richard. 1995. "A Recently Discovered Darwin Letter on Social Darwinism." *Isis* 86 : 609–11.

Weindling, Paul. 1981. "Theories of the Cell State in Imperial Germany." In *Biology, Medicine and Society, 1840–1940,* 99–155. Cambridge: Cambridge University Press.

Welch, Cheryl B. 1984. *Liberty and Utility: The French Idéologues and the Transformation of Liberalism.* New York: Columbia University Press.

Welsby, Paul. 1962. *George Abbot: The Unwanted Bishop, 1562–1633.* London: SPCK.

Welshinger, Henri, ed. 1900. *La mission secrète de Mirabeau à Berlin, 1786–1787; d'après les documents originaux des Archives des Affaires Étrangères.* Paris: Plon, Nourrit et Cie.

Wenger, Etienne. 1999. *Communities of Practice.* Cambridge: Cambridge University Press.

Werner, Petra. 2000. *Casanova ohne Frauen? Bemerkungen zu Alexander von Humboldts Korrespondenzpartnerinnen.* Berlin: Alexander-von-Humboldt-Forschungsstelle.

Westfall, Richard S. 1971. *The Construction of Modern Science: Mechanisms and Mechanics.* Cambridge: Cambridge University Press.

———. 1971. *Force in Newton's Physics: The Science of Dynamics in the Seventeenth Century.* London: Macdonald.

———. 1972. "Newton and the Hermetic Tradition." In *Science, Medicine and Society in the Renaissance,* ed. Allen G. Debus, 2 : 183–98. New York: Science History Publications.

———. 1980. *Never at Rest: A Biography of Isaac Newton.* Cambridge: Cambridge University Press.

———. 1984. "Newton and Alchemy." In *Occult and Scientific Mentalities in the Renaissance,* ed. Brian Vickers, 315–35. Cambridge: Cambridge University Press.

———. 1993. "Science and Technology during the Scientific Revolution: An Empirical Approach." In *Renaissance and Revolution: Humanists, Scholars, Craftsmen and Natural Philosophers in Early Modern Europe,* ed. J. V. Field and Frank A. L. James, 63–72. Cambridge: Cambridge University Press.

Whitbread, Helena. 1988. *I Know My Own Heart: The Diaries of Anne Lister, 1791–1840.* London: Virago.

White, Hayden. 1973. *Metahistory: The Historical Imagination in Nineteenth-Century Europe.* Baltimore: Johns Hopkins University Press.

Whitford, Kathryn, and Philip. 1970. "Timothy Dwight's Place in Eighteenth-Century American Science." *Proceedings of the American Philosophical Society* 114 : 60–71.

Whitrow, Gerald J. 1988. *Time in History: Views of Time from Prehistory to the Present Day.* Oxford: Oxford University Press.

Wiener, Philip P., ed. 1973. *Dictionary of the History of Ideas.* 5 vols. New York: Scribner's.

Williams, Raymond. 1976. *Keywords.* London: Fontana.

Wills, Garry. 1978. *Inventing America: Jefferson's Declaration of Independence.* Garden City: Doubleday.

Winchester, Simon. 2001. *The Map that Changed the World: The Tale of William Smith and the Birth of a Modern Science.* London: Viking, 2001.

Winichakul, Thongchai. 1994. *Siam Mapped: A History of the Geo-body of a Nation.* Honolulu: University of Hawai'i Press.

Winslow, Charles-Edward Amory. 1943. *The Conquest of Epidemic Disease: A Chapter in the History of Ideas.* Princeton: Princeton University Press.

Withers, Charles W. J. 1996. "Geography, Science and National Identity in Early Modern Britain: The Case of Scotland and the Work of Sir Robert Sibbald, 1641–1722." *Annals of Science* 53 : 29–73.

———. 1997. "Geography, Royalty and Empire: Scotland and the Making of Great Britain, 1603–1661." *Scottish Geographical Magazine* 113 : 22–32.

———. 1999. "Reporting, Mapping, Trusting: Practices of Geographical Knowledge in the Late Seventeenth Century." *Isis* 90 : 497–521.

———. 2000. "Authorizing Landscape: 'Authority,' Naming and the Ordnance Survey's Mapping of the Scottish Highlands in the Nineteenth Century." *Journal of Historical Geography* 26 : 532–54.

———. 2000. "John Adair, 1660–1718." *Geographers' Biobibliographical Studies* 20 : 1–8.

———. 2000. "Toward a Historical Geography of Enlightenment in Scotland." In *The Scottish Enlightenment: Essays in Reinterpretation,* ed. Paul Wood, 63–97. Rochester: University of Rochester Press.

———. 2001. *Geography, Science and National Identity: Scotland since 1520.* Cambridge: Cambridge University Press.

Withers, Charles W. J., and Robert J. Mayhew. 2002. "Rethinking 'Disciplinary' History: Geography in the British Universities, c. 1580–1887." *Transactions of the Institute of British Geographers* 27 : 11–29.

Withers, Charles W. J., and Paul Wood, eds. 2002. *Science and Medicine in the Scottish Enlightenment.* East Linton: Tuckwell.

Wittman, Reinhard. 1999. "Was There a Reading Revolution at the End of the Eighteenth Century?" In *A History of Reading in the West,* ed. Guglielmo Cavallo and Roger Chartier, trans. Lydia G. Cochrane, 284–312. Cambridge: Polity.

Wolf, Eric. 1969. *Peasant Wars of the Twentieth Century.* New York: Harper and Row.

———. 1975. "Aspects of Group Relations in a Complex Society: Mexico." In *Peasants and Peasant Societies,* ed. Theodor Shanin, 50–66. Harmondsworth: Penguin.

Wolfe, Don, ed. 1953–82. *The Complete Prose Works of John Milton.* 8 vols. New York: Columbia University Press.

Wolff, Larry. 1994. *Inventing Eastern Europe: The Map of Civilization on the Mind of the Enlightenment.* Stanford: Stanford University Press.

Wood, Michael. 1940. *Extracts from the Records of the Burgh of Edinburgh.* HMSO: London.

Wood, Paul B. 1980. "Methodology and Apologetics: Thomas Sprat's *History of the Royal Society.*" *British Journal for the History of Science* 13:1–26.

———. 1992. "The Scientific Revolution in Scotland." In *The Scientific Revolution in National Context,* ed. Roy Porter and Mikuláš Teich, 263–87. Cambridge: Cambridge University Press.

———. 1996. "The Science of Man." In *Cultures of Natural History,* ed. Nicholas Jardine, James A. Secord, and Emma C. Spary, 197–210. Cambridge: Cambridge University Press.

Woods, Mark, and Paul Veatch Moriarty. 2001. "Strangers in a Strange Land: The Problem of Exotic Species." *Environmental Values* 10:163–91.

Woodward, Donald, ed. 1984. *The Farming and Memorandum Books of Henry Best of Elmswell.* Records of Social and Economic History, n.s., 8. London: British Academy.

Wooldridge, Sidney W. 1955 "The Status of Geography and the Role of Field Work." *Geography* 40:73–83.

Woolf, Daniel R. 2000. *Reading History in Early Modern England.* Cambridge: Cambridge University Press.

Woolf, Harry. 1959. *The Transits of Venus: A Study of Eighteenth-Century Science.* Princeton: Princeton University Press.

Woolhouse, Roger S., ed. 1981. *Leibniz: Metaphysics and Philosophy of Science.* Oxford: Oxford University Press.

Wright, Conrad. 1983. "The Controversial Career of Jedidiah Morse." *Harvard Library Bulletin* 31:64–87.

Wright, John Kirtland. 1959. "Some British 'Grandfathers' of American Geography." In *Geographical Essays in Memory of Alan G. Ogilvie,* ed. R. Miller and J. Wreford Watson, 144–65. London: Thomas Nelson.

———. 1966. "What's 'American' about American Geography?" In *Human Nature in Geography: Fourteen Papers, 1925–1965,* 124–39. Cambridge: Harvard University Press.

Wrigley, Anthony. 1988. *Continuity, Chance and Change: The Character of the Industrial Revolution in England.* Cambridge: Cambridge University Press.

Wrigley, Richard. 2002. *The Politics of Appearances: Representations of Dress in Revolutionary France.* Oxford: Berg.

Yaroshevsky, M. G. 1978. "National and International Factors in the Development of Scientific Schools of Thought." In *Human Implications of Scientific Advance,* ed. Eric G. Forbes, 174–81. Edinburgh: Edinburgh University Press.

Yates, JoAnne. 2000. "Business Use of Information Technology during the Industrial Age." In *A Nation Transformed by Information,* ed. Alfred D. Chandler and James W. Cortada, 107–36. Oxford: Oxford University Press.

Young, Robert M. 1985. *Darwin's Metaphor: Nature's Place in Victorian Culture.* Cambridge: Cambridge University Press.

Zandvliet, Kees. 1988. *Mapping for Money: Maps, Plans and Topographic Paintings and their Role in Dutch Overseas Expansion during the Sixteenth and Seventeenth Centuries.* Amsterdam: Batavia Lion International.

Zeller, Suzanne. 1999. "Environment, Culture, and the Reception of Darwin in Canada, 1859–1909." In *Disseminating Darwinism: The Role of Place, Race, Religion, and Gender,* ed. Ronald L. Numbers and John Stenhouse, 91–122. Cambridge: Cambridge University Press.

Zilsel, Edgar. 1942. "The Genesis of the Concept of Physical Law," *Philosophical Review* 51:245–79.

Zuckert, Michael P. 1996. *The Natural Rights Republic: Studies in the Foundation of the American Political Tradition.* Notre Dame: University of Notre Dame Press.

———. 2000. "Founder of the Natural Rights Republic." In *Thomas Jefferson and the Politics of Nature,* ed. Thomas S. Engeman, 11–58. Notre Dame: University of Notre Dame Press.

INDEX

Page numbers in italics refer to figures in the text.

Abbot, George: as geographical author, 248, 250, 252, 262–63; *Brief Description,* 248, 252, 253, 257; on Calvinism, 252, 257–58; on Catholicism, 262–63

Aberdeen, University of. *See* King's College, Aberdeen; Marischal College, Aberdeen

actor-network theory, 35–36. *See also* calculation, center of

Adair, John: geographical and political authority of, 25, 75, 85–86, 87, 90, 99–100; mapping work of, 76, 79, 87, 88, 90

Afghanistan, 114

Africa: in Ptolemy's work, 139; on Blaeu's map, 153; on Speed's map, 153, *154–55;* photographic depictions of, 203, 208–19; trade with, 86, 96

Africa Novo Descriptio (Blaeu), 153

Africae (Speed), 153, *154–55*

Africae Nova Tabula (Hondius), 153

Agar, Jon, 82

agrarian revolution. *See* Agricultural Revolution

Agricultural Revolution, 3, 5, 133, 160

Aid to Survey-Practice (Jackson), 228

air-pump, 24, 32, *33,* 34–35, 54, 94, 98

alchemy, 58

Allen, William, 221; *Picturesque Views on the River Niger,* 221

Almagest (Ptolemy), 42n.31

Almanack and Kalendar containing the Day, Hour and Minute of the Change of the Moon for Ever (Lloyd), 179

Alsted, Johann, 250

America: compared to Europe, 10; genetics in, 45; moral geographies of, 304–29. *See also* American Republic; American Revolution

American Geography (Morse), 241, 309–18, 324

American Republic: idea of, 15; moral geographies of, 304–29

American Revolution, 2, 4, 9, 13, 30, 304, 318–19

American War of Independence. *See* American Revolution

Amsterdam, 144, 151, 180

anatomy, 84

Ancien régime et la révolution, L' (Tocqueville), 40n.4

Anderson, Benedict, 138; *Imagined Communities,* 138

Anglican Church, 24; and Calvinism, 52–54

Angola, 147

Annals of the American Revolution (Morse), 310

Anthropological Society of London, 123–24

anti-Aristotelianism: associations of with Protestantism, 50–51

Aristotle, 39, 49, 50

Aristotelianism: and scientific thought, 13, 24, 46, 49, 51, 77, 93, 98, 250; in England, 13, 24, 46, 49; in France, 24, 49–51, 281; in Scotland, 93, 98. *See also* anti-Aristotelianism

Armageddon, 5

Arminianism: and Anglican Church, 52–54; and Calvinism, 247, 251–61; and Catholicism, 256–58, 262–63; and geography books, 247, 251–57, 258–59; and Protestantism, 243, 244, 356; Peter Heyleyn on, 247, 248, 250–52, 255–57

art: Baroque style in, 356; classicism in, 356; revolution in styles of, 355–58

Art Union, 207

Asia: in Ptolemy's geography, 139; in Wallace's work, 121–23; trade with, 142, 144, 147–48

Assing, Rosa Ludmilla, 344

astrology, 4, 80–81

astronomy: as a "big science," 84; geography and, 39–42, 77, 91, 93, 204; Savilian Chair of in Oxford, 12

Atlantic Ocean, 67

atlases, 138

Athenaeum, 346

Atom and Individual in the Age of Newton (Freundthal), 62–63

Aubrey, John, 176, 179, 190

Australia, 111, 124

Ayr, 183

Bacon, Francis, 52, 309

bacteriology: as a science, 207; and photography, 15, 207, 238n.109

Baghdad, 68

Baines, Thomas, 211, 216, 223–24; *Shifts and Expedients of Camp Life,* 224

Banks, Joseph, 36

Barlow, Roger, 248, 249; *Brief Summe of Geography,* 248

Barlow, Thomas, 51

Barrow, John, 156

Barth, Fredrik, 359

Bataille, George, 148

Batavia (Jakarta), 37

Baudin, Nicolas, 276, 283

Beagle, H.M.S., 109, 110

Behemoth (Hobbes), 243, 251

Beijing, 23

Bennett, Jim, 80

Bentley, Richard, 81

Berlin, 278, 284, 285, 344

Bertius, Petrus, 255–56

biogeography: geography and, 2, 10–11, 13, 25, 107, 108–9; and "Darwinian Revolution," 13; in work of Darwin, 108–20; 125–26; in work of Lyell, 112; in work of Wallace, 13, 26, 117–26; Krotopkin on, 10–11; language of, 108–9; geopolitics of, 25, 107

biography, 297n.8, 336–48

Birmingham (Alabama), 23

Birmingham (England), 23

Blache, Vidal de la, 353, 358

Black, William, 92

Blaeu, Willem: *Africa Nova Descriptio,* 153; Khoisan people on maps of, 134, 138, 151–52, 157; as map maker to Dutch East India Company, 151

Blome, Richard, 95

Boate, Gerald, 245; *Natural History of Ireland,* 245

bookkeeping, 15

books: commonplace, 262; of geography, 13, 15, 75–76, 80, 89–92, 241, 247, 251–57, 258–59; history of, 6, 13; Index of Prohibited, 13

booksellers, 75–76

Boston, 23

botanical gardens, imperialism and, 24

botany: and photography, 204; classification in, 37–39; networks in, 24; tropical, 24

Botero, Giovanni: on Protestantism, 255–56; *Relations of the Most Famous Kingdomes and Common-wealths thorowout the World,* 255; translations of work on, 253

Bourne, Samuel, 219

Bourne, William, 179–80, 196n.50; *Regiment of the Sea,* 179–80

Bowman, Isaiah, 1, 296; *New World,* 296

Boyle, Robert, 24, 51, 54, 90, 94, 308; *General Heads for a Natural History of a Country,* 308

Brahe, Tycho, 91, 94, 95, 180; *Mechanics of*

the New Astronomy, 180; Uraniborg observatory of, 180

Braun, Georg, 152

Brazil, 147

Briefe Description (Abbot), 252, 253, 257–58

Briefe Introduction to Geography (Pemble), 248

Briefe von Alexander von Humboldt von Varnhagen von Ense aus den Jahren 1827 bis 1858 (Ense), 338–40

Brief Summe of Geography (Barlow), 248

Brissot de Warville, Jacques-Pierre, 275, 285; association of with Edme Mentelle, 285

Bristol, 172

British Association for the Advancement of Science, 206

British Journal of Photography, 219

Brittany, 36, 353; mapping in, 36

Broc, Numa, 276; *La Géographie des Philosophes,* 276

Brückner, Martin, 305, 317; on geography and American national identity, 305–6, 317

Buache, Philippe, 278, 281

Buache de la Neuville, Jean Nicholas, 281, 285, 290

Buckle, Henry, 124

Buffon, Georges Louis Leclerc, Comte de, 328, 333n.78

Buffonianism, 317

Buisseret, David, 14

Burckhardt, Jacob, 43–44, 46–47, 64

Burma, 114

Byzantium, science in, 64, 68–69. *See also* Islam

calculation: as cognitive practice, 35, 36; center of, 35, 36, 87, 354

Caledonia's Everlasting Almanack, or, A Prognostication which may serve for ever the Kingdom of Scotland (Symson), 91

Callon, Michel, 35

Calvinism: Abbot on, 252, 257–58; and Arminianism, 247, 251–61; and geography books, 247, 251–57; in relation to Anglican Church, 52–54

Cambridge (England), 80

camera, 199, 201, 208, 210, 211, 217, 219. *See also* photography

Campanella, Tomasso, 258; *De Monarchia Hispania,* 258

Camões, Luis Vaz de: on Khoisan people, 143; *The Lusiads,* 143

Canada, 114

canal building, 8

Candolle, Alphonse de, 110, 355

Cape of Good Hope: as a botanical site, 37; Grey as governor of, 216; Khoisan peoples of, 142 43, 145–52; mapping of, 138–57; British military campaigns in, 114; Portuguese rounding of, 134, 138–39

capitalism: and print culture, 2, 6, 13, 137–38, 151–52, 156–57; as a mode of production, 2, 6

Cardiff, 206

Carpenter, Nathanael: *Geography,* 245, 248, 249, 250, 253–54; politics of his geographical writings, 250–51, 253–54; on Sabbatarianism, 254

Carriera da India, 142

Cartesianism: and magnetism, 57–58; experimentation in, 25, 50; in Newton's work, 55; in Scottish natural philosophy teaching, 86–87, 93–94, 96, 98; in work of Huyghens, 55; in work of Leibniz, 55

Cassini de Thury, César François, 278

Catholic Church, 13, 24, 50, 243

Catholicism: Abbot on, 262–63; and Arminianism, 256–58, 262–63; and geography in the seventeenth century, 256–58, 262–63; associations of with advances in natural philosophy, 51–52; Heyleyn on, 247, 251–52, 253–57

celestial mechanics, 4, 13, 48, 61, 180

Chambers, Robert, 121; *Vestiges of the Natural History of Creation,* 121

Chapman, James, 216, 217; photography in work of, 216, 217

Charles I (king of Great Britain), 243, 256, 265

Charles II (king of Great Britain), 263–64, 265

Chemical Revolution, 3, 8, 30

China: Linnaeus on, 38; photography in, 219–21; politics in, 9, 74n.62, 353–54; science in, 38, 64–66, 355; Scientific Revolution in, 64–66

Chinese revolution, 353–54

chorography: as a form of geographical enquiry, 25, 39, 77, 80–81, 87–89, 94, 99; in Sibbald's work, 87–89, 99

Christianography (Paggitt), 256, 257, 263

Civil War, English. *See* English Civil War

Civilisation of Europe in the Renaissance (Hale), 44

Civitates Orbis Terrarum (Braun and Hogenberg), 152

Clarendon, Earl of, 4; *History of the Rebellion and Civil Wars in England begun in the Year 1641,* 4–5

Clark, James, 88

classicism. *See* art

classification: in botany, 37–39; in zoology, 39

climate: as determinant of scientific development, 23; in Darwin's work, 110–11; in Morse's work, 310–17; in Smith's work, 324–28; in Wallace's work, 123–24

clock makers, 186, *187*

clock time: and embodiment, 163–65; and everyday life, 163–65; in eclipses, 183; in England, 160–94; in Pepys' life, 186; notion of adequation in, 170, 171, 176–77, 196n.45; practices of, 164–65, 168–70, 190–93. *See also* timekeeping

Cluverius, Philip, 96; *Introduction into Geography, both Ancient and Modern,* 96

coal mining, 8

Cohen, H. Floris, 79; *Scientific Revolution, A Historiographical Inquiry,* 79–80

Cohen, I. Bernard, 54–55

Columbus, Christopher, 138, 139

combustion, oxygen theory of, 8

Commercial Revolution, 5

Common Sense (Paine), 9–10

Compendious History of New England (Morse), 313, 317–18

Connecticut, 312, 314; Morse on, 312, 314

Consumer Revolution, 160, 161

Cook, Harold, 88

Copernicanism, 3, 80, 95

Copernican Revolution, 3, 6, 30, 40n.9

Copernicus, Nicolaus, 30, 46, 80, 91, 94; *De Revolutionibus Orbium Caelestium,* 3

Cormack, Lesley, 80–81

correspondence: of Varnhagen von Ense with Humboldt, 336–48; in Darwin's work, 117

Corss, James, 90; *Practical Geometry,* 90

Cosmographie ancienne (Mentelle), 280

Cosmographie (Heyleyn), 252, 253, 256–57, 260, 264

Cosmography and Geography in Two Parts (Blome), 95

cosmology, 94, 95

Cotes, Roger, 81

courts, 82

credibility, 88, 89

Crimea, 115

Crosland, Maurice, 46; *Encyclopedia of the Scientific Revolution,* 46

Cuba, 9, 106, 337

cultural history, 29–30

Cyprus, 219

Czech Republic, 174

Daguerreotype, 204

d'Ailly, Pierre, 138; *Imago Mundi,* 138–39

d'Alembert, Jean Le Rond, 276, 279

d'Alembert, Jean-Baptiste Bougignon, 285

Darnton, Robert: on Grub Street, 274–75; on reading practices, 17n.12; on social differences in the Enlightenment, 274–75

Darwin, Charles: *Beagle* voyage of, 109–10; and biogeography, 2, 45, 108–20, 125–26; and climate, 110–11, 118–20; correspondence of, 117; on division of labor among species, 115, 117–18; on ecology as invasion, 25–26, 106, 109, 117–20; elected to Royal Geographical Society, 113; on evolution by natural selection, 8, 25–26, 109–10, 114–20, 121; on extinction, 109–10, 112, 114–20; on the Galapagos, 111–12; on the geographical distribution of species, 109–10, 111, 114–15; on human variation, 118–20; influence of Lyell on, 110–11, 113, 114, 115, 129n.43, 130n.50; influence of Malthus on, 113–14; on islands, 111, 112–13, 114–15; London lodgings of, 30, *31, 32;* on migration, 111, 114–15; in New Zealand, 111, 129n.42; in Patagonia, 110–111; on the politics of civilization, 119–20; in South America, 110–11, 113,

117; and St. Helena, 111, 143; on transmutation, 110, 111, 112; on the U.S. Civil War, 129n.49; WORKS: *The Descent of Man*, 118, 119, 125; *Journal of Researches*, 111–12, 118, 121; *Natural Selection*, 115; *The Origin of Species*, 115, 117–18, 125

Darwinian Revolution, 2, 6, 13. *See also* Darwin, Charles

Darwinism, 13. *See also* Darwin, Charles

Daston, Lorraine, 44

da Vinci, Leonardo. *See* Leonardo da Vinci

Davis, John: on Khoisan peoples, 146–47, 149, 150; on trade networks, 147, 149

Dear, Peter, 24, 51, 81, 352

Dee, John, 175, 176, 181, 182; timekeeping in work of, 175–76, 181, 182, 197n.57

Defence of the People of England (Milton), 265

de la Blache, Vidal. *See* Blache, Vidal de la

De Houtman, Cornelius. *See* Houtman, Cornelius de

De Monarchia Hispania (Campanella), 258

De Revolutionibus Orbium Caelestium (Copernicus), 3

Descartes, René: on magnetism, 57–58; on matter theory, 50, 51, 56–58, 94; *Principia Philosophiae*, 72n.40

Descent of Man (Darwin), 118, 119, 125

Description and Uses of the Celestial and Terrestrial Globes (Harris), 92

Description of England (Harrison), 182

Description of the County of Angus (Edward), 98–99

Description of the Isles of Orkney (Geddes), 88

Devonian system, 24

Diamond, Jared, 27; *Guns, Germs, and Steel*, 27

Diaz, Bartholomeu, 134, 138–39, 142

Dictionnaire de géographie moderne (Mentelle), 280

Diderot, Denis, 3, 276; and the *Encyclopédie*, 3, 276

display, geography of, 222–25

Donnelly, Michael, 44; *Readers' Guide to the History of Science*, 44

Douglas, Mary, 150–51, 152, 156; *Purity and Danger*, 150–51

Downton, Nicholas, 149–50

Drummond, Gavin, 91, *A Short Treatise of Geography, General and Special*, 91

Durkheim, Emile, 7

Dwight, Timothy: as a Federalist, 3; on New England, 241, 318–24, 328; on pastoral idealism, 241, 318–24; as president of Yale College, 318; *Travels in New England and New York*, 318, 324; *The True Means of Establishing Public Happiness*, 322

Eachard, Laurence, 92; *Gazetteer's, or Newsman's Interpreter*, 92

earth sciences, 8

East India Company, Dutch, 45, 84, 151–52; Blaeu as map maker for, 151

East India Company, English, 145, 148

eclipses, 183, 197n.62

Eclogues (Virgil), 142

École Normale, 240–41, 290; Edme Mentelle in, 240–41, 290

École Royale Militaire, 240, 280, 287; Edme Mentelle in, 240, 280, 287

ecological rationality, idea of in timekeeping, 168, *171*

ecology: associations with invasion in Darwin's work, 25–26, 106–9, 117–20; imperialism and, 120–26; language of, 117–18; politics of, 22–26, 120–26; in Wallace's writings, 108–9, 120–26

Edinburgh: geography in, 25, 75, 86; University of, 25, 75, 86

Edinburgh's True Almanack, or a New Prognostication for the Years 1685–1692 (Paterson), 89

Edward, Robert, 98; *A Description of the County of Angus*, 98

Eisenstein, Elisabeth, 13, 137, 138

Éleméns de Géographie (Mentelle), 279, 299n.27

Elemens de l'histoire romaine (Mentelle), 280

Elements of Geography (Morse), 317

Elias, Norbert, 46–47, 64, 65

embryology, 45

Encheiridion Geographicum (Symson), 90

Encyclopedia of the Scientific Revolution (Crosland), 46

Encyclopédie, 3, 276, 279, 281
Engels, Friedrich, 125, 336
England: Aristotelianism in, 13, 24, 46, 49; clock time in, 160–94; earth sciences in, 8; Enlightenment in, 15; geography in, 2, 243–67; natural philosophy in, 12, 24–25; style of science in, 10, 12, 24–25, 51–55
English Civil War: as a British phenomenon, 243–45; definitions of, 4, 267n.1; Hill on, 30, 353; Hobbes on the revolutionary spaces of, 243–45. *See also* English Revolution
English Revolution: European context of, 243–45; geography and, 3, 9, 15, 243–67; Hill on, 30, 40, 245, 252, 353; local contexts of, 243, 353; national contexts of, 160, 243, 244, 353
Enkhuisen, Jacob Pieterszoon van, 149
Enlightenment: Darnton on social differences in, 274–75; geography and, vii, 10; in England, 15; in Scotland, 97
Ense, Karl August Varnhagen von: correspondence of with Humboldt, 337, 338–40; reception of his work, 337, 340–47; *Briefe von Alexander von Humboldt an Varnhagen von Ense aus den Jahren 1827 bis 1858*, 338–40
Epicurus, 50
eschatology, 5
Essay on the Causes of the Variety of Complexion and Figure in the Human Species (Smith), 324
ethnography, 211, *214*, 215, 324–25
ethnology, 213–14, 215, 235n.52, 324, 325
Europae Speculum (Sandys), 258, 263
evolution, 8, 25–26, 109–26
Exercise chronologique (Mentelle), 295
expeditions, 203. *See also* Zambesi Expedition
experimentation: in Bacon's work, 52; and Cartesianism, 25, 50; cultures of, 51–53; in Galileo's work, 52; geography of, 49; national styles of, 49, 50–56; in Pascal's work, 52; role of theory in, 49–50
exploration: and invasion, 108–9; photography and, 202, 203, 205–15

extinction: Darwin on, 109–10, 112, 114–20; Lyell on, 111, 113; Wallace on, 123–26

factory system, 2, 34
Fakhry, Majid, 67–68
Febvre, Lucien, 352; *La Terre et l'Evolution Humaine*, 352
Feingold, Mordechai, 246
fieldwork, 109–10, 200. *See also* exploration
Finland, 10
Fitzroy, Robert, 111
force, concept of, 48
Foreign Office (British), 222
Forster, Georg, 336
Foucault, Michel, 352
Fox Talbot, William Henry. *See* Talbot, William Henry Fox
France: Aristotelianism in, 24, 49–51, 281; clocks in, 174; geography in eighteenth century, 240, 273–96; styles of natural philosophy in, 12, 24–25
French Revolution: geography and, 3, 4, 9, 15, 28, 160, 273–96, 353; politics of, 15, 36, 160; regional geography of, 36, 160, 239, 277–78, 298n.18, 353
Freudenthal, Gideon, 48; *Atom and Individual in the Age of Newton*, 62–63; on Leibniz, 63

Galapagos, 111, 112, 127
Galileo Galilei, 12, 52, 79, 355
Galton, Francis, 209, 210; photography in work of, 226–27; *Promotion of Scientific Branches of Geography*, 226
Galton, Robert Cameron, 209, *210*
Gama, Vasco da, 142–43
Gassendi, Pierre, 50
Gautier, Théophile, 201
Gavroglu, Kostas, 48
Gazetteer's, or Newsman's Interpreter (Eachard), 92
Geddes, William, 88, 98; *Geographical and Arithmetical Memorials*, 88
General Heads for a Natural History of a Country (Boyle), 308
genetics, 45
Geographia (Ptolemy), 39, 42n.31, 77, 139, *140–41*

Geographia Generalis (Varenius), 12–13
Geographical and Arithmetical Memorials (Geddes), 88
Geographical and Historical Description of the Shire of Tweedale (Pennecuik), 87
Geographical Description of Scotland (Paterson), 89–90
Geographical Description of the Four Parts of the World (Blome), 95
Geographicall Description (Stafford), 248
geographical gazetteers, 15
geographical grammars, 15
Géographie, mathématique, physique et politique de toutes les parties du monde (Mentelle), 294
Géographie ancienne (Mentelle), 280–81
Géographie des Philosophes (Broc), 276
Géeographie physique, historique, statistique et topographique de la France en cent huit départements (Mentelle), 294
Géographie universelle ancienne et moderne (Mentelle), 294
geography: in America, 304–29; and astronomy, 39–42, 77, 84, 91, 93, 204; and biogeography, 2, 10–11, 13, 25, 107, 108–9; and biography, 297n.8; books of, 13, 15, 75–76, 80, 89–92, 241, 247, 251–57, 258–59; and Catholicism in seventeenth-century England, 256–58, 262–63; descriptive, 77, 80, 84, 204; of display, 222–25; in eighteenth-century France, 240, 273–96; and the English Revolution, 240, 243–67; and experimentation, 49, 52; and the French Revolution, 3, 4, 9, 15, 28, 273–96, 353; in Gresham College, 13, 80; and humanist pedagogy, 240, 247; improved by warfare, 1; of the Industrial Revolution, 2, 3, 15, 133, 160, 349; of knowledge, vii, 11–12; Kropotkin on, 10–11; of life, 13, 26, 106–26; mathematical, 77, 80, 84; of meaning, 6; "militant," 108; and national identity, vii, 2; in Newton's work, 81; in Oxford, 80, 240, 245–51; and photography, 14, 135, 199–231; political, 69, 108; and political revolution, 239–42; Quantitative Revolution in, 8; and revolution, 1, 2, 9–16, 23–26; revolution in method in, 14, 240,
248–51; and science, 23–26; and the Scientific Revolution, 75–100; and Technical Revolution, 133–36; theological debates and, 240, 245–47; of thought, 65–66, 74n.63; university teaching of, 25, 42n.30, 75, 86, 92–97, 240
Geography (Carpenter), 245, 250, 253–54
Geography Anatomiz'd; or, a compleat geographical grammar (Gordon), 76, 92
Geography Compendiz'd; or, the World Survey'd (Symson), 90, 96, 99
Geography Made Easy (Morse), 317
Geography of Thought (Nisbett), 65–66
geology, 39, 204
geomagnetism, 39
geopolitics: biogeography and, 25, 107; Wallace's work and, 120–26
Georgia, 313
Germany, 14, 15, 45, 174
Glasgow, University of, 90, 95, 98
globes: celestial, 95; Copernican, 95–96; Ptolemaic, 95–96; terrestrial, 95
Glorious Revolution, 3, 30
Gobi Desert, 67
Godlewska, Anne: on Edme Mentelle, 273, 275–76, 290, 302n.68; on eighteenth-century French geography, 240, 273, 275–76, 290, 356
Golinski, Jan, 82
Gordon, Patrick, 76, 92; *Geography Anatomiz'd; or, a compleat geographical grammar,* 76, 92
Gould, Stephen Jay, 64
Grafton, Anthony, 261
gravity, 54
Grammatical Institute (Webster), 306
Gramsci, Antonio, 7
Grand Titration (Needham), 65
Grant, James Augustus, 208, *209,* 211, 234n.33
Great Devonian Controversy (Rudwick), 30, *31*
Great Trigonometrical Survey, 228
Greece, 247
green imperialism, 108. *See also* biogeography
Greenough, George, 204
Green Revolution, 3

Gregory, David, 75, 76, 85, 93, 96
Gregory, James, 76, 93, 96
Gresham College, 13, 80
Grey, Sir George, 216
Grub Street, 274–75
Guns, Germs, and Steel (Diamond), 27
Guthrie, William, 89

Hakluyt, Richard, 144
Hakluytus Posthumus; or, Purchas His Pilgrimes (Purchas), 147
Hale, John, 44, 64, 65 *Civilisation of Europe in the Renaissance*, 44
Halley, Edmund, 81, 183, *184*
Hamburg, 23
Harris, John, 92; *The Description and Uses of the Celestial and Terrestrial Globes*, 92
Harris, Stephen, 83–84, 87, 103n.53
Harrison, William, 18, 248; *Description of England*, 182
Harwood, Jonathan, 45
Hegel, Georg Wilhelm Friedrich, 4
Henry, John, 24–25, 79–80, 355
Henry VIII (king of England), 52
hermeticism, 6
Herschel, Sir John, 204
Heyleyn, Peter: as an Arminian, 247, 248, 250–51, 255–57; *Cosmographie*, 252, 253, 256–57, 260, 264; *Microcosmus*, 248, 249, 250, 252, 253–54, 255–56, 258–59, 264; on Arminianism, 248, 249, 250, 252–53; on Catholicism, 247, 251–52, 253–57
Higden, Ranulph, 247; *Polychronicon*, 247
Hill, Christopher, 30, 245, 252; on the English Revolution, 30, 40, 245, 252, 353
Hints to Travellers (Royal Geographical Society), 225, 229; and photography, 225–29
History of the American Revolution (Ramsay), 4
History of the Rebellion and Civil Wars in England begun in the Year 1641 (Clarendon), 4–5
History of the Royal Society (Sprat), 53
Hobbes, Thomas: *Behemoth*, 243, 251; idea of revolutionary space in writings of, 243–45, 246–47, 351; on the geographies of the English Civil War, 235, 243–45, 246–47

Hobsbawm, Eric J., 28
Hogenberg, Frans, 152
Hondius, Jodocus, 138, 153, 157; *Africa Nova Tabula*, 153
Hooke, Robert, 54, 58
Hooker, Joseph Dalton, 123
Hooykaas, Reijer, 80
Horological Revolution, 134, 161, 170, 177, 178, 192, 193
Hottentots. *See* Khoisan peoples
Houtman, Cornelis de, 144, 145–46
humanist pedagogy, geography and, 240, 247
Humboldt, Alexander von: assesses Daguerreotypes, 204; biographies of, 338–40; correspondence of with Varnhagen von Ense, 241, 338–40, 344; as a critic, 241; and plant geography, 348n.1; reception of his work, 241, 242, 336–37, 340–47; as a Revolutionary, 241, 336–48; as a scientist, 110, 273; in writings of Kossok, 336–37; in writings of Lang, 347–48; in writings of Milnes, 345
Hutton, James, 8
Huxley, Thomas Henry, 122
Huyghens, Christian, 55

Iberian peninsula, 12. *See also* Spain
Icon Animorum (May), 249
Illustration of China and its People (Thomson), 220, 221
Iltis, Carolyn, 62, 63
Imagined Communities (Anderson), 138
Imago Mundi (d'Ailly), 138–39
imperialism: botanical classification and, 37; ecology and, 13, 120–26; politics of, 120–26
India: British military campaigns in, 114; Mutiny in, 114; photography in, 219; trading posts in, 38, 67, 247
Indian Ocean: and botanical networks, 37; as a trading space, 146; maps of 38; in Ptolemy's geography, 151
Indonesian Archipelago, 144, 147
industrial production, systems of, 5, 15, 17n.10
Industrial Revolution: geography of, 2, 3, 15, 133, 160, 349; technical bases to, 2, 3, 5, 15, 34, 133

information, technologies of, 5–6
Information Revolution, 3, 5–6, 133–34, 137
innovation, technologies of, 5
International Geographical Congress, 221
Internet, 5
Introduction into Geography, both Ancient and Modern (Cluverius), 96
invasion: ecology and, 107, 112–13; Darwin on, 25–26, 106, 112–13; politics of, 25–26, 106–7, 120–26
Iran, 9
Iraq, 109
Ireland, 112, 243
Islam: civilization in, 66–67; science in, 6, 64, 66–69; Scientific Revolution in, 64, 66–67
Italy, 174

Jackson, Lowis D'A., 228; *Aid to Survey-Practice*, 228
Jardin du Roi, 24, 37
Jardine, Lisa, 261
Jefferson, Thomas: as geographer, 306, 328; *Notes on the State of Virginia*, 306, 307–9
Jesuits, 84, 256, 279
John Huighen van Linschoten: His Discours into ye East and West Indies (Linschoten), 144
Johns, Adrian, 14
Jourdain, John, 153
Journal of Researches (Darwin), 111–12, 118, 121
Jurin, James, 81
Jussieu, Antoine de, 37–38

Kaffirs, 114. *See also* Khoisan peoples.
Kalahari desert, 218
Kames, Lord, 325; *Sketches of the History of Man*, 325
Kant, Immanuel, 40n.9
Kashmir, 219
Keckerman, Bartholomew, 250
Ker, John, 97
Kew Gardens. *See* Royal Botanic Gardens (Kew)
Khoisan peoples: on Blaeu's maps, 134, 151–52; in Camões's writings, 143; as cannibals, 152; in Davis's writings, 147,

149, 150; depictions of in sixteenth-century travel narratives, 142–43, 145–46, 147, 149–50; in Downton's writings, 149–50; on Speed's maps, *154–55;* in Sprenger's writings, 142–43; trading capacity of, 143, 145–48; understood as Hottentots, 156, 159n.40
King's College, Aberdeen, 92, 97
Kirk, John, 211–212, *213,* 215, 223–24, 225–26, 234n.40
knowledge: geography of, vii, 11–12, 14; travelling nature of, 9–10, 12, 14
Koerner, Lisbet, 36, 38
Kossok, Manfred, 336–37
Kropotkin, Peter: on biogeography, 10–11; on geography, 10–11
Kuhn, Thomas: on the Scientific Revolution, 27, 28–29; *The Structure of Scientific Revolutions*, 6, 24, 27, 28, 352

laboratories, 12
La géographie enseigné par une méthode nouvelle; ou, Application de la synthèse à l'étude de la géographie (Mentelle), 293
Lagrange, Joseph-Louis, 285
Lakatos, Imre, 48
Lancaster, James, 148
landscape painting, 200, 202
Lang, Heinrich, 347–48
Lapérouse, Jean-François de Galaup, Comte de, 35, 276
Laplace, Pierre-Simon, 285
Lapland, 38, 39, 281
La Terre et l'Evolution Humaine (Febvre), 352
Latour, Bruno: idea of "obligatory passage point" in work of, 35, 36; on nature of science, 35, 36, 82, 84, 354; *Science in Action,* 35, 36
Laud, William, 252, 256
Lavoisier, Antione-Laurent de, 8, 30, 45, 285
Law, John, 35, 84
lecture halls, 25
Lefebvre, George, 353
Leibniz, Gottfried Wilhelm: as a Cartesian, 55; discussed by Freundthal, 63; discussed by Shapin, 62
lending libraries, 14, 25
Leonardo da Vinci, 179

L'Esprit des Lois (Montesquieu), 3

Letters, "Republic" of, 12

Lettre d'un auteur citoyen à la Commune de Paris en faveur de la liberté de la presse (Mentelle), 286

Leviathan and the Air Pump (Shapin and Schaffer), 32, *33*

life, geography of: Darwin on, 13, 26, 106–26; Wallace on, 26, 106–26

Linnaeus, Carl: botanical classifications of, 38; on China, 38; Lapland work of, 281, natural history work of, 37, 38

Linnean Society of London, 115

Linschoten, John Huyghen van, 144; *Itinerario*, 144

Lisbon, 138, 142

Livingstone, Charles: and Zambesi Expedition, 211–13, 216, 222–23; photography in work of, 210–14, 216, 223–24, 225

Livingstone, David (explorer-missionary): *Missionary Travels and Researches in South Africa*, 224; on photography, 211–13, 217, 223, 224; on Zambesi Expedition, 203, 212–15, 223

Livingstone, David (geographer), 82, 199

Lloyd, Humfrey: *Almanack and Kalendar containing the Day, Hour and Minute of the Change of the Moon For Ever*

localism: and scale in geography, 11, 14; in science studies, 11, 14, 34, 82–83, 85

Locke, John, 309

logic, teaching of, 93–94

London: Anthropological Society of, 123–24; Darwin's lodgings in, 30, *31*, 32; Linnean Society of, 115; location of Royal Geographical Society in, 203; Photographic Exhibition in, 218; Royal Society of, 51, 53–54, 55, 57, 88, 94, 183, 197n.62, 203, 211; science in, 31–32, 173, 200, 205, 219; Stereoscopic and Photographic Company of, 210

longitude, 15

Louis XVI (king of France), 281

Louis, XVIII (king of France), 240

Lusiads (Camões), 143

Lux, David, 88

Lyell, Charles: biogeography in work of, 112; on extinction of species, 111, 113; influence of on Darwin, 110–11, 113, 114,

115, 129n.43, 130n.50; influence of on Wallace, 121–22; *Principles of Geology*, 110, 121, 129n.43, 130n.50

Mackinder, Halford, 227, 246

Maclaurin, Colin, 97

macrocosm, 4

Madison, James, 327–28

magnetism: and Cartesianism, 57–58; in work of Petty, 57

Malacca, 142

Malay Archipelago, 121, 123, 142

Malay Archipelago (Wallace), 124–25

Malaysia, 121, 123, 144, 147

Malte Brun, Conrad, 294–96

Malthus, Thomas: influence of on Darwin, 112, 113–14; influence of on Wallace, 121

Malthusianism, 10, 125

Manuel géographique, chronologique, et historique (Mentelle), 280

mapmaking, 76, 77, 78, 80. *See also* mapping

mappae-mundi, 138, 152

mapping: in Brittany, 36; of the Cape of Good Hope, 138–57; as a geographical practice, 1, 14, 30, 77, 78; of the Indian Ocean, 36; and photography, 229–30; and print culture, 137–57; and revolution, 30–36; of scientific networks, *31*, *32*; in the work of Adair, 76, 79, 87, 88, 90; in the work of Wallace, 121

Marischal College, Aberdeen, 98

Markham, Clements, 226

Martin, Martin, 88

Marx, Karl, 4, 125, 336

Massachusetts, 312, 316

mathematical geography, 77, 80, 84

mathematics: Lucasian Chair of, 25, 84; and natural philosophy, 39; Savilian Chair of, 75; teaching of, 25, 75

matter theory: Descartes on, 50, 56–58; natural philosophy and, 56–57

Matthew, Patrick, 45

May, Thomas, 249; *Icon Animorum*, 249

Mayhew, Robert, 353

McLuhan, Marshall, 137

meaning, geography of, 6

mechanical philosophy, 13, 61, 180

Mechanics of the New Astronomy (Brahe), 180

Medawar, Peter, 107
medicine, 204
Mediterranean Sea, 67
Mentelle, Edme: anticlerical views of, 280, 284; on antislavery, 293; association of with Brissot de Warville, 285; association of with Comte d'Artois, 281, 286; association of with Comte de Mirabeau, 284–85, 286, 301n.44; association of with Malte Brun, 294–96; association of with Roland salon, 286–88; in Berlin, 284–85; as a Brissotin, 286–87; and the Conseil d'Administration, 292; in the École Normale, 240–41, 290n.27; in the École Royale Militaire, 240, 280, 287; on geographical method, 282–83; geographical writings of, 240–87; globes made by, 281–84, 300n.37; Godlewska on, 273, 275–76; as "Jany," 289, 302n.59; as playwright, 278–79; on political geography, 282–83; public esteem for, 240, 275; as Republican geographer, 240, 287, 289–90; as Royalist tutor, 282; as satirist, 278–79; WORKS: Cosmographie élémentaire, 280; Dictionnaire de géographie moderne, 280; Éleméns de Géographie, 279, 299n.27; Élémens de l'histoire romaine, 280; Exercise chronologique, 295; Géographie ancienne, 280–81; Géographie, mathématique, physique et politique de toutes les parties du monde, 294; Géographie physique, historique, statistique et topographique de la France en cent huit departments, 294; Géographie universelle ancienne et moderne, 294; La géographie enseigné par une méthode nouvelle: ou, Application de la synthèse à l'étude de la géographie, 293; Lettre d'un auteur citoyen à la Commune de Paris en faveur de la liberté de la presse, 286; Manuel géographique, chronologique, et historique, 280; Méthode courte et facile pour apprendre aisément et retiner sans peine la nouvelle géographique de la France, 287; Précis de l'histoire Universelle, 293; Tableau élémentaire de géographie de la République française, 287; Tableau raisonné de la nouvelle division économico-politique, 287

mercantilism, 37
Mersenne, Marin, 50
metaphysics, teaching of, 93–94. See also natural philosophy
meteorology, 15, 207
method: and humanist pedagogy, 240, 247; in seventeenth-century English geography, 240, 248–51; in writings of Petrus Ramus, 250–51, 300n.34
Méthode courte et facile pour apprendre aisément et retiner sans peine la nouvelle géographie de la France (Mentelle), 287
Mexican Revolution, 354
Mexico, 337
Michelbourne, Edward, 149, 150
microcosm, 4
Microcosmus (Heyleyn), 248, 249, 250, 252, 253–54, 255–56; 258–59, 264
Middle Ages: idea of revolution in, 3; science in, 64–65
"militant" geography, 108
Military Revolution, 3, 14
Mill, H. R., 231; Record of the Royal Geographical Society, 231
Milnes, Richard Moncton (Baron Houghton): as liberal sympathizer, 345; views of toward Humboldt, 345
Milton, John, 265; Defence of the People of England, 265; geographical thought of, 264
Mirabeau, Comte de, 284–86; association of with Edme Mentelle, 284–85, 286, 301n.44
Missionary Travels and Researches in South Africa (Livingstone), 224
Monge, Gaspard, 285
Montesquieu, Charles-Louis de Secondat, Baron de, 3, 325, 327; L'Esprit des Lois, 3
Montrose, 183
Moore, James, 25–26
moral geographies: in Morse's work, 309–18; of post-revolutionary America, 304–29
Morse, Jedidiah: and American patriotism, 331n.34; on anti-Buffonianism, 315; and climate, 310–17; on Connecticut, 312, 314; on European geographies of America, 315; as father of American geography, 241; on Georgia, 313; on idea of

Morse, Jedidiah (*continued*)
    moral topographies, 304, 309–18; on Jef-
        fersonian Republicanism, 315–16; on
        Massachusetts, 312, 316; on native Amer-
        icans, 314–15; on natural theology, 314–
        15; on New England, 310–17; on North
        Carolina, 312; reception of his work,
        305; on South Carolina, 312; and textual
        organization, 266, 310–12; on Virginia,
        312; WORKS: *American Geography,* 240,
        309–18; *Annals of the American Revolu-
        tion,* 310; *Compendious History of New
        England,* 313, 317–19; *Elements of Geog-
        raphy,* 317; *Geography Made Easy,* 317;
        *Report to the Secretary of War of the U.S.
        on Indian Affairs,* 313–14
Moxon, James, 88
Moxon, Joseph: as Hydrographer Royal, 88,
        90; *Tutor to Astronomie and Geographie*
        (1659), 95; *Tutor to Astronomy and Geog-
        raphy* (1665), 95
Mundy, Peter, 147
Murchison, Sir Roderick, 203, 204, 222,
        223
Murray, John, 224, 225
museums, 12, 200, 205

Napier, John, 90
nation, idea of, 2, 12
national identity: and geography, vii, 2; in
        the Rennaissance, 43–44; 64
natural arithmetic, 81
natural history, 24, 36–39, 81, 84, 204, 211
Natural History of Ireland (Boate), 245
Natural History Review, 122
natural philosophy: and Cartesianism, in
        Scotland, 86–87, 93–94, 96, 98; concepts
        of, 3–4, 12, 25, 39, 77; in England, 12,
        24–25, 51–55; in France, 12, 24–25; and
        geography, in Scottish universities, 92–
        97; mathematics and, 39; and matter
        theory, 56–57; mechanistic, 56–57, 60
nature, laws of 3–4, 60
nautical charts, 2
navigation, 12, 15, 25, 77, 78, 96, 161
Needham, Joseph, 65, 66, 355; *Grand Titra-
        tion,* 65
Neolithic Revolution, 3
Netherlands, 3

networks: in botany, 24, 37; of correspon-
        dence, 336–48; scientific, *31,* 32
New England: in Dwight's work, 241, 318–
        24; in Morse's work, 241, 310–17
New Guinea, 121
New System of Modern Geography (Parish),
        310
Newton, Sir Isaac: biographies of, 26, 46,
        58; discussed by Shapin, 62; geography
        in work of, 12–13, 81; on "essential
        properties," 58, 62–63; on gravity, 54;
        and the occult, 54, 58; *Opticks,* 59; on
        planetary motion, 54, 58, 308; *Principia
        Mathematica,* 58, 94; revisions of Vare-
        nius's *Geographia Generalis,* 12–13, 81;
        styles of thought of, 58–60
Newtonianism: in England, 39; and the Sci-
        entific Revolution, 2, 6, 13; in Scotland,
        76, 77, 86, 96, 98
Newtonian Revolution, 2, 6
New Zealand, 111, 112, 114, 129n.42
Nisbett, Richard E., 65; *Geography of
        Thought,* 65
North Carolina, 312
Northeast Passage, 144
North Walsham, 170
Northwest Passage, 144
Notes on the State of Virginia (Jefferson),
        241, 306–9

obligatory passage points. See Latour, Bruno
Observations on the American Revolution
        (Price), 4
occult, 57–58
old regime, 28, 40n.4
Ophir, Adir, 82
Opticks (Newton), 59
Optics, 12
Ordnance Survey, 16
Origin of Species (Darwin), 115, 117–18,
        122, 125; read by Wallace, 122
Ortelius, Abraham, 138
Osler, Margaret, 76, 82, 83
Otte, Michael, 44
Owen, Richard, 222, 223
Oxford: books of geography in, 80, 96,
        240; English Civil War and, 251–57; ge-
        ography in, 80, 240, 247–67; Savilian
        Chair of Astronomy in, 12; Savilian

Chair of Mathematics in, 75; theological debates in, 240, 247–67; University of, 240, 247–67
oxygen, 8
Ozouf, Mona, 277

Paggit, Ephraim, 256, 257, 263; *Christianography*, 256, 257, 263
Paine, Thomas: and French Revolution, 4; *Common Sense*, 9–10; *Rights of Man*, 4
Papineau, David, 62, 63
Paracelsus, 51
Paradigms, 6, 29
Paris: French Revolution in, 37, 250; International Exhibition in, 216; Société de Géographie of, 296
Parish, Elijah, 310; *New System of Modern Geography*, 311
Parliaments, Union of (1707), 79, 86
Pascal, Blaise, 50; experimentalism in work of, 52
pastoral idealism, 241, 318–24
Patagonia, 110–11
Paterson, James, 89; *Edinburgh's True Almanack, or a New Prognostication for the Years 1685–1692*, 89; *Geographical Description of Scotland*, 89–90; *Scots Arithmetician, or Arithmetic in All its Parts*, 89
Paulinyi, Akos, 5
Pemble, William, 248, 250, 252; *Briefe Introduction to Geography*, 248
Pennecuik, Alexander, 87, 98; *Geographical and Historical Description of the Shire of Tweedale*, 87, 99
Pepys, Samuel, 89
Persia. *See* Iraq
Peru, 39
Petty, William, 57, 180; magnetism in work of, 57
Philip, William, 144
Philosophical Transactions, 94, 197n62
phlogiston, 8
Photography: of Africa, 203, 208–19; as an art-science, 200, 201–3; in astronomy, 238n.109; in bacteriology, 15, 207, 238n.109; in botany, 204; *British Journal of*, 219; in China, 219–21; and ethnography, 211–12, *214*, 215, 235n.52; and exploration, 200, 205–15; in fieldwork,

200; and geography, 14, 135, 199–231; and *Hints to Travellers*, 225–29; in India, 219; and landscape depiction, 14, 200, 202; and mapping, 229–30; in meteorology, 15, 207, 238n.109; as a revolutionary technology, 14, 135, 199–200, 202–3; and science, 14–15, 200–203; in spiritualism, 207, 238n.109; truth value of, 135, 222–25, 229–31; in the work of Charles Livingstone, 210–14, 216, 223–24, 225; in the work of Chapman, 216, 217; in the work of David Livingstone (explorer-missionary), 211–13, 217, 223, 224; in the work of Galton, 226–27; in the work of the Royal Geographical Society, 200, 203, 205–6, 208–9, 211, 218, 225–26, 230–31; in the work of Thomson, 205–6, 210, 218, 220–22, 224; on the Zambesi Expedition, 210–14, 216, 223–24, 225
physics, 45–46
physiology, national schools in, 47–48
Picturesque Views on the River Niger (Allen), 221
place, idea of, 37–39
Poland, 174, 181, 264
political arithmetic, 81
political geography, 69, 108
political revolution: in China, 9, 353–54; and geography, 239–42; idea of, 3. *See also* French Revolution; Mexican Revolution; revolution; Russian Revolution
politics: of ecology, 22–26, 119, 120–26; and geography in seventeenth-century England, 245–57
Polychronicon (Higden), 247
Porter, Roy: on science in China, 66; on the Scientific Revolution, 9, 46, 78
Portugal, 12
Practical Geometry (Corss), 90
Précis de l'histoire universelle (Mentelle), 293
Presbyterianism, 243
Price, Richard, 4; *Observations on the American Revolution*, 4
Price, William Lake, 202
Price Revolution, 5
Priestley, Joseph, 45
Principia Mathematica (Newton), 58, 94
Principia Philosophiae (Descartes), 72n.40

Principles of Geology (Lyell), 110, 121, 129n.43, 130n.50
print culture: and capitalism, 2, 6, 13, 137–38, 151–52, 156–57; and mapping, 137–57; geography in, 137; geography of, 13. *See also* mapping
printing press, 1, 137–38
Print Revolution, 2, 13–14, 133–34
Promotion of Scientific Branches of Geography (Galton), 226
Protestantism: associations of with anti-Aristotelianism, 50–51; discussed by Giovanni Botero, 255–56. *See also* Arminianism; Presbyterianism
protoindustrialization, 5
Ptolemy, Claudius: on Africa, 139; *Almagest*, 42n.31; on Asia, 139; astronomical system of, 91, 94; *Geography*, 39, 42n.31, 77, *140–41*, 151; on Indian Ocean, 139, 151
public sphere, 6
Punjab, 114
Purchas, Samuel, 147, 248, 265; *Hakluytus Posthumus*, 147, 248
Puritanism, 6, 161
Puritan Revolution, 30, 40, 245, 252, 253. *See also* English Revolution
Purity and Danger (Douglas), 150–51
Purmerendt, Cornelisz Claesz van, 147

Quantitative Revolution, 8
Quarterly Journal of Forestry, 108

Raj, Kapil, 37
Ramsay, David, 4; *History of the American Revolution*, 4
Ramus, Petrus: on geographical method, 250–51, 300n.34; intellectual reforms of, 250–51
Ratzel, Friedrich, 358
Raynal, Guillaume Thomas François (Abbé), 308
Readers' Guide to the History of Science (Donnelly), 44
reading: Darnton on, 17n.12; geographies of, 14; practices of, 2, 6, 14, 166–67; "Revolution" in, 2, 3, 6
rebellion, 7
Reclus, Elisée, 10

reconnaissance, voyages of, 6
Record of the Royal Geographical Society (Mill), 231
Reformation, 6, 46, 256
Regiment of the Sea (Bourne), 179–80
Reid, Thomas, 97
Relations of the Most Famous Kingdomes and Common-wealths thorowout the World (Botero), 255–56
religion. *See* Anglican Church; Arminianism; Calvinism; Catholicism; Presbyterianism; Protestantism
Renaissance: geography and the, vii; idea of revolution in the, 3; national identity in the, 43–44, 64
Report to the Secretary of War of the U. S. on Indian Affairs (Morse), 313
Restoration, 51; and geography, 265–66, 267
revolution: as an analytical category, 1, 2–6; in art, 355–58; in clock time, 160–94; definitions of, 2–8; economic explanations of, 7; in geographical method, 15; geographies of, 2, 11; geography and, 1, 2, 9–16, 23–26, 351–60, 355–58; in history, 27–30, 160; idea of in the Middle Ages, 3; language of, 4; in perception of Humboldt, 336–48; mapping and, 30–36; political explanations of, 7–8; psychological explanations of, 7; in seventeenth-century English geography, 15, 243–67; sociological conceptions of, 7; theories of, 7–8. *See also revolutions by name*
revolutionary space, idea of, 243–45, 246–47
Revolutions, Year of, 16, 337, 348
Reynolds, Robert, 67, 74n.62
Riebeeck, Jan van, 138, 156
Rights of Man (Paine), 4, 9–10
Rio de Janeiro, 111
rivoluzione. See revolution, idea of in the Middle Ages
Roland de la Platière, Jeanne-Marie, 286–87, 288
Roland de la Platière, Manon Jeanne Phlipon, 286–87, 288–89, 296
Rome, 247, 261
Rousseau, Jean-Jacques, 279, 292
Royal Botanical Gardens (Kew), 36, 212, 222
Royal Geographical Society: Darwin elected

to, 113; and *Hints to Travellers*, 225–29; and photography, 200, 203, 205–6, 208–9, 211, 218, 225–26, 230–31; and promotion of geography, 200, 203, 211, 222, 358; residential locations of in London, 203; Scientific Purposes Committee of, 226–27

Royal Society of London, 51, 53–54, 55, 57, 88, 94, 183, 197n.62, 203, 211

Rudwick, Martin, 30, *31*; *Great Devonian Controversy*, 30, *31*, 32, 34

Russell, Bertrand, 44

Russia, 14

Russian Revolution, 14, 30, 160, 353

Sabra, A. I., 67–68

St. Andrews, University of, 93, 95

St. Helena, 111, 143

St. Petersburg, 10

Samarkand, 68

Sandys, Edwin, 258, 263; *Europae Speculum*, 258, 263

scale, idea of, 11–12, 32–33

Scaliger, Joseph, 253

Schaffer, Simon, 32; *Leviathan and the Air-Pump* (with Shapin), *33*

Scholasticism, 4, 9, 46

Schwartz, Joan, 203

science: Arabic forms of, 64, 67–69; in Byzantium, 64, 68–69; in China, 38, 64, 66, 355; geography of, 23–26; history of, 47, 82, 352–53; international character of, 121, 43–45, 67–69, 70n.3; in Islamic world, 6, 64, 66–69; national styles of, 12, 43–69; and photography, 14–15, 200, 201–3; social constructivism in, 11–12; sociology of, 161; "spatial" turn in the study of, 11–12; styles of in England, 12, 24–25, 51–55; styles of in France, 12, 24–25

Science in Action (Latour), 35

science studies, localist "turn" in, 34, 82–83, 85

scientific networks, maps of, *31*, 32

Scientific Revolution: in China, 64–66; competing ideas about, 76–77, 160; as a geographical phenomenon, vii, 1, 11, 23–26, 75–100; geography and the, 1, 11, 75–100; geography of the, vii, 1, 77–86; as

a historical phenomenon, 3–4, 6, 11, 76, 83–85; in Islam, 6, 64, 66–69; Kuhn on, 27, 28–29; national styles in, 46–74; Newtonianism in, 2, 6, 13; Porter on, 9, 46–47, 78; in Scotland, 25, 86; Shapin on, 6, 11, 82, 89, 101n.5; Teich on, 78

Scientific Revolution, A Historiographical Inquiry (Cohen), 79–80

Scientific Revolution in National Context (Porter and Teich), 46

Scientific Revolution (Shapin), 6

Scotland: Aristotelianism in, 93, 98; Cartesianism in, 86–87, 93–94; Church of, 87; earth sciences in, 8; Enlightenment in, 97; geography in, 75–105; Newtonianism in, 76, 86; Parliament of, 88; Scientific Revolution in, 25, 86

Scots Arithmetician, or Arithmetic in All its Parts (Paterson), 89

Shapin, Steven: on Leibniz, 62, 63; *Leviathan and the Air Pump*, 32–33, *33*, 34; on the localist "turn" in science, 82–83; on Newton, 62, 63; on the Scientific Revolution, 6, 11, 82, 89, 101n.5; *Scientific Revolution*, 6, 11

Shifts and Expedients of Camp Life (Baines), 224

Short Treatise of Geography, General and Special (Drummond), 91

Sibbald, Robert: chorographical work of, 87–89, 90, 99, 103n.53; as Geographer Royal, 25, 75, 79, 86

Siberia, 11

Sicily, 65

Sinclair, George, 98

Sketches of the History of Man (Kames), 325

Skocpol, Theda, 353–54

Smith, Crosbie, 82

Smith, Samuel Stanhope: and climate, 324–28; and ethnological theory, 324–25; *Essay on the Causes of the Variety of Complexion and Figure in the Human Species*, 324; on human variation, 325–26; on human virtue, 326; on moral philosophy, 327–28; as president of Princeton, 324

Smith, William, 8

social history, 29–30

sociology of scientific knowledge, 45, 70n.6

South Africa, 216
South America, 108, 110, 113, 117
South Carolina, 312
Soviet bloc, 9
space, ideas of, 30–39, 81
Spain, 38, 65, 243, 261, 355
spatial history, idea of, 36–39
Speed, John: *Africae*, 153, *154–55*, 157, 268; *Theatre of the Empire of Great Britain*, 268n.11
Speke, John Hanning, 208, 211
Spencer, Herbert, 124
spiritualism, photography and, 207
Sprat, Thomas, 53; *History of the Royal Society*, 53
Sprenger, Balthazar, 142–43; on Khoisan peoples, 142–43
Stafford, Robert, 248, 250; *Geographicall Description*, 248
Stanhope Smith, Samuel. *See* Smith, Samuel Stanhope
Stanley, Henry Morton, 227
Staum, Martin, 276, 292–93
stereoscope, 208–9
Stockholm, 38
Structure of Scientific Revolutions (Kuhn), 6, 24, 27, 28, 352
Sweden, 14, 37
Symson, Andrew, 91–92, 95, 103n.64
Symson, Matthias, 90–91, 96, 99; *Caledonia's Everlasting Almanack*, 91; *Encheiridion Geographicum*, 90; *Geography Compendiz'd; or, the World Survey'd*, 90, 96, 99

Tableau élémentaire de géographie de la République française (Mentelle), 287
Tableau raisonné de la nouvelle division économico-politique (Mentelle), 287
Tahiti, 112–13
Talbot, William Henry Fox, 203, 234n.32
taxonomy, 23, 38
Technical Revolution, 1; and geography, 133–36
technology: innovations in, 5; production of, 2, 5, 15
Teich, Mikuláš, 9; on science in China, 66; on the Scientific Revolution, 77

Theatre of the Empire of Great Britain (Speed), 268n.11
Thompson, E. P., 34
Thomson, John: *Illustrations of China and Its People*, 220, *221*; photographic studios of, 227, 237n.94; as Instructor in Photography, 227; photographic work of, 205–6, 210, 218, 220–22, 224
Thucydides, 247
Tierra del Fuego, 111
time: early modern conceptions of, 15, 134; history of, 358; measurement of, 358–59; revolution in, 134, 358–59; standardization of, 358–59. *See also* timekeeping
timekeeping: communities of practice of, 162, 163–65, 188–91; devices for, 133, 161, 162; ecological rationality in, 168, *171*; foraging behavior in, 168, 176; idea of adequation in, 170, *171*, 176–77, 196n.45; in life of Dee, 175–76, 181, 182, 197n.57; in life of Pepys, 186; revolution in practices of, 170–77
tobacco, 38
trade, 36, 38, 143, 144–45
Trading Revolution, 5
transmutation, idea of in Darwin's work, 110, 111–12
Travels in New England and New York (Dwight), 318, 324
True Means of Establishing Public Happiness (Dwight), 322
Tucker, Jennifer, 207
Turnbull, David, 11–12, 82
Tutor to Astronomie and Geographie (Moxon), 95
Tutor to Astronomy and Geography (Moxon), 95
Tyacke, Nicholas, 246, 266

Ulpius, Euphrosynus, 144
Underdown, David, 244, 353, 358
United Provinces, 260
Uraniborg, 180
Urban Revolution, 3
Uruguay, 113

Vallange, John, 96
van Purmerendt, Cornelisz Claesz, 147

Varenius, Bernhard, 13, 81; *Geographia Generalis,* 13, 81
Vaugondy, Didier Robert de, 276
Venice, 264, 265
Vereenigde Oost-Indische Compagnie (VOC). See East India Company, Dutch
Versailles, 35, 284, 300n.37
Vestiges of the Natural History of Creation (Chambers), 121
Victoria Falls, 216, *217*
Vietnam, 106
Virgil, 142; *Eclogues,* 142
Virginia, 38, 307–9
vis viva controversy, 61–62
Voltaire, François Marie Arouet de, 12

Wales, 121
Wallace, Alfred Russel: addresses Anthropological Society of London, 123, 124; biogeography in writings of, 13, 26, 117–26; climate in works of, 123–24; on evolution, 25; on extinction, 45; on geography of life, 13, 26; geopolitics in his work, 25–26, 120–26; influence of Lyell on, 121–22; influence of Malthus on, 121; on human variation, 123–26, 131n.74; as land surveyor, 26, 120–22; on Malay Archipelago, 121, 123; *Malay Archipelago,* 124–25; on New Guinea, 121; reads *Origin of Species,* 122; role of maps in work of, 121, 122; writing style of, 121, 122. *See also* Wallace's Line
Wallace, James, 88, 98; *Description of the Isles of Orkney,* 88

Wallace's Line, 26, 121
Wallerstein, Immanuel, 354
warfare: improves geography, 1; in ecology, 17–19; 25–26; 106–9, 120–26
War of the Worlds (Wells), 106–7
Warwick, Andrew, 45–46
Washington, George, 308, 313, 315, 328
watches. *See* clocks
Weber, Max, 7
Webster, Charles, 245, 246
Webster, Noah, 305; *Grammatical Institute,* 306
Wells, Charles, 45
Wells, H. G., 107; *War of the Worlds,* 106–7
Westfall, R. S., 58, 80
Wilkins, Thomas, 54–55
Wiltshire, 244, 353
Wishart, John, 94–95
Wolf, Eric, 354, 359
Wolfe, John, 144
Wood, Paul, 86, 103n.53

Yale College, 318
Yaroshevsky, M. G., 47–48, 63
York, 172

Zambesi Expedition: Charles Livingstone on, 211–13, 216, 222–23; David Livingstone (explorer-missionary) on, 203, 212–15, 223; photography and the, 210–14, 216, 223–24, 225
Zambesi River, 216, 217, 226
Zanzibar, 208, *209,* 234n.33
zoology, classification in, 39